T0400028

Hybrid Energy Systems

Sustainable Energy Strategies

by Yatish T. Shah

Hybrid Energy Systems: Strategy for Industrial Decarbonization

Hybrid Power: Generation, Storage, and Grids

Modular Systems for Energy Usage Management

Modular Systems for Energy and Fuel Recovery and Conversion

Thermal Energy: Sources, Recovery, and Applications

Chemical Energy from Natural and Synthetic Gas

Other related books by Yatish T. Shah

Energy and fuel systems integration

Water for energy and fuel production

Biofuels and bioenergy: Processes and Technologies

For more information on this series, please visit: https://www.routledge.com/Sustainable-Energy-Strategies/book-series/CRCSES

Series Preface

While fossil fuels (coal, oil, and gas) were the dominant sources of energy during the last century, since the beginning of the twenty-first century an exclusive dependence on fossil fuels is believed to be a nonsustainable strategy due to (a) their environmental impacts, (b) their nonrenewable nature, and (c) their dependence on the local politics of the major providers. The world has also recognized that there are in fact ten sources of energy: coal, oil, gas, biomass, waste, nuclear, solar, geothermal, wind, and water. These can generate our required chemical/biological, mechanical, electrical, and thermal energy needs. A new paradigm has been to explore greater roles of renewable and nuclear energy in the energy mix to make energy supply more sustainable and environmental friendly. The adopted strategy has been to replace fossil energy by renewable and nuclear energy as rapidly as possible. While fossil energy still remains dominant in the energy mix, by itself, it cannot be a sustainable source of energy for the long future.

Along with exploring all ten sources of energy, sustainable energy strategies must consider five parameters: (a) availability of raw materials and accessibility of product market, (b) safety and environmental protection associated with the energy system, (c) technical viability of the energy system on the commercial scale, (d) affordable economics, and (e) market potential of a given energy option in the changing global environment. There are numerous examples substantiating the importance of each of these parameters for energy sustainability. For example, biomass or waste may not be easily available for a large-scale power system making a very large-scale biomass/waste power system (like a coal or natural gas power plant) unsustainable. Similarly, an electrical grid to transfer power to a remote area or onshore needs from a remote offshore operation may not be possible. Concerns of safety and environmental protection (due to emissions of carbon dioxide) limit the use of nuclear and coal-driven power plants. Many energy systems can be successful at laboratory or pilot scales, but may not be workable at commercial scales. Hydrogen production using a thermochemical cycle is one example. Many energy systems are as yet economically prohibitive. The devices to generate electricity from heat such as thermoelectric and thermophotovoltaic systems are still very expensive for commercial use. Large-scale solar and wind energy systems require huge upfront capital investments which may not be possible in some parts of the world. Finally, energy systems cannot be viable without market potential for the product. Gasoline production systems were not viable until the internal combustion engine for the automobile was invented. Power generation from wind or solar energy requires guaranteed markets for electricity. Thus, these five parameters collectively form a framework for sustainable energy strategies.

It should also be noted that the sustainability of a given energy system can change with time. For example, coal-fueled power plants became unsustainable due to their impact on the environment. These power plants are now being replaced by gas-driven power plants. New technology and new market forces can also change sustainability of the energy system. For example, successful commercial developments of fuel cells and electric cars can make use of the internal combustion engines redundant in the vehicle industry. While an energy system can become unsustainable due to changes in parameters, outlined above, over time, it can regain sustainability by adopting strategies to address the changes in these five parameters. New energy systems must consider long-term sustainability with changing world dynamics and possibilities of new energy options.

Sustainable energy strategies must also consider the location of the energy system. On one hand, fossil and nuclear energies are high-density energies and they are best suited for centralized operations in an urban area, while on the other hand, renewable energies are of low density and they are well-suited for distributed operations in rural and remote areas. Solar energy may be less affordable in locations far away from the equator. Offshore wind energy may not be sustainable if the distance from shore is too great for energy transport. Sustainable strategies for one country may be quite different from another depending on their resource (raw material) availability and local market potential. The current transformation from fossil energy to green energy is often prohibited by required infrastructure and the total cost of transformation. Local politics and social acceptance also play an important role. Nuclear energy is more acceptable in France than in any other country.

Sustainable energy strategies can also be size dependent. Biomass and waste can serve local communities well at a smaller scale. As mentioned before, the large-scale plants can be unsustainable because of limitations on raw materials. New energy devices that operate well at micro- and nanoscales may not be possible on a large scale. In recent years, nanotechnology has significantly affected the energy industry. New developments in nanotechnology should also be a part of sustainable energy strategies. While larger nuclear plants are considered to be the most cost effective for power generation in an urban environment, smaller modular nuclear reactors can be the more sustainable choice for distributed cogeneration processes. Recent advances in thermoelectric generators due to advances in nanomaterials are an example of a size-dependent sustainable energy strategy. A modular approach for energy systems is more sustainable at smaller scale than for a very large scale. Generally, a modular approach is not considered as a sustainable strategy for a very large, centralized energy system.

Finally, choosing a sustainable energy system is a game of options. New options are created by either improving the existing system or creating an innovative option through new ideas and their commercial development. For example, a coal-driven power plant can be made more sustainable by using very cost-effective carbon capture technologies. Since sustainability is time, location, and size-dependent, sustainable strategies should follow local needs and markets. In short, sustainable energy strategies must consider all ten sources and a framework of five stated parameters under which they can be made workable for local conditions. A revolution in technology (like nuclear fusion) can, however, have global and local impacts on sustainable energy strategies.

The CRC Press Series on Sustainable Energy Strategies will focus on novel ideas that will promote different energy sources sustainable for long term within the framework of the five parameters outlined above. Strategies can include both improvement in existing technologies and the development of new technologies.

Series Editor,
Yatish T. Shah

Hybrid Energy Systems
Strategy for Industrial Decarbonization

Yatish T. Shah

CRC Press
Taylor & Francis Group
Boca Raton London New York

CRC Press is an imprint of the
Taylor & Francis Group, an **informa** business

First edition published 2021
by CRC Press
6000 Broken Sound Parkway NW, Suite 300, Boca Raton, FL 33487-2742

and by CRC Press
2 Park Square, Milton Park, Abingdon, Oxon, OX14 4RN

© 2021 Taylor & Francis Group, LLC

CRC Press is an imprint of Taylor & Francis Group, LLC

Library of Congress Cataloging-in-Publication Data
Names: Shah, Yatish T., author.
Title: Hybrid energy systems : strategy for industrial decarbonization / by
Yatish T. Shah.
Description: First edition. | Boca Raton, FL : CRC Press, 2021. | Series:
Sustainable energy strategies | Includes bibliographical references and
index. | Summary: "This book demonstrates how hybrid energy and processes can decarbonize energy industry needs for power and heating and cooling. It describes the role of hybrid energy and processes in nine major industry sectors and discusses how hybrid energy can offer sustainable solutions in each. Sectors include coal, oil and gas, nuclear, building, vehicle, manufacturing and industrial processes, computing and portable electronic, district heating and cooling and water. Written for advanced students, researchers, and industry professionals involved in energy-related processes and plants, this book offers latest research and practical strategies for application of the innovative field of hybrid energy"—Provided by publisher.
Identifiers: LCCN 2020049403 (print) | LCCN 2020049404 (ebook) |
ISBN 9780367747572 (hardback) | ISBN 9781003159421 (ebook)
Subjects: LCSH: Renewable energy sources. | Hybrid power systems. |
Renewable resource integration.
Classification: LCC TJ808 .S53 2021 (print) | LCC TJ808 (ebook) |
DDC 621.31/21—dc23
LC record available at https://lccn.loc.gov/2020049403
LC ebook record available at https://lccn.loc.gov/2020049404

ISBN: 9780367747572 (hbk)
ISBN 9780367747640 (pbk)
ISBN: 9781003159421 (ebk)

Typeset in Times
by codeMantra

The book is dedicated to my sons, James, Jonathan and Keith

Contents

Preface...xix
Author xxi

Chapter 1   Hybrid Energy Systems—Strategy for Decarbonization.....................1

    1.1   Introduction 1
    1.2   Hybrid Energy Systems Defined 4
    1.3   Examples of Hybrid Energy Systems 8
          1.3.1   Hybrid Solar-Wind Renewable Systems...................... 8
          1.3.2   Combined Heat and Power Hybrid Energy System..... 12
    1.4   Outline of the Book 16
    References 18

Chapter 2   Hybrid Energy Systems for Building Industry.................................19

    2.1   Introduction 19
          2.1.1   Concept of Zero-Energy Buildings 20
          2.1.2   Grid Connection 21
          2.1.3   Fuel Switching.. 22
          2.1.4   Renewable Energy Credits 22
          2.1.5   Energy Supply Options and Priorities....................... 22
    2.2   Customer Automation and Energy Management Systems 26
          2.2.1   Dynamic Pricing and Demand Response 28
          2.2.2   Process for Renewable Energy Building
                    Connection to the Electrical Grid 30
    2.3   Role of Hybrid Energy Systems in Net Zero-Energy
          Buildings...32
    2.4   Solar Thermal with Storage... 34
          2.4.1   Solar-Boosted Heat Pump 35
          2.4.2   Building Integrated Solar Thermal Technologies
                    and Their Applications... 36
    2.5   Solar Electric PV with Storage...37
    2.6   Hybrid PV/Solar Thermal Concept 42
    2.7   Building-Integrated Options (BiPVT/a) 43
          2.7.1   Works on Window Systems.. 46
                 2.7.1.1   Building-Integrated Window Systems
                          (BiPVT/w)..47
          2.7.2   Heat-Pump Integration (PVT/Heat Pump)................ 48
          2.7.3   PVT-Integrated Heat Pipe 49
          2.7.4   PVT Trigeneration.. 49
          2.7.5   Commercial Aspects 50
    2.8   Solar PVT with Geothermal Heat Pump................................ 50

2.9   PV/Wind/Storage Hybrid Energy System 54
       2.9.1   Pros and Cons of Hybrid PV-Wind Energy
               Systems...57
       2.9.2   Theoretical Case Studies for PV-Wind Hybrid
               Energy System..57
2.10   Other Issues and Innovations for Hybrid Energy for
       Buildings..59
       2.10.1 Hybrid Electric Building Design.............................59
       2.10.2 Hybrid Energy Modules for Improving Building
               Efficiency in the Future Electric Grid...................... 60
       2.10.3 Economics of Renewable Hybrid System for
               Residential Purpose...61
References 62

Chapter 3   HESs for Carbon-Free District Heating and Cooling 71

3.1   Introduction 71
       3.1.1   Drivers for DHC..76
3.2   Small Hybrid Fossil-Renewable Heating and
       Cooling Grids 76
3.3   DH by Biomass Based HES 78
       3.3.1   DH with CHP 79
       3.3.2   Some Examples of Hybrid Biomass-Based DH in
               Europe 81
       3.3.3   Hybrid-Solar-Biomass DH 82
3.4   Hybrid Geothermal DH... 84
       3.4.1   Hybrid Modular Geothermal Heat Pump for
               District Heating.. 86
       3.4.2   Solar Thermal Recharge and Sewer Heat Recovery 87
       3.4.3   The Multisource Hybrid Concept............................... 87
3.5   Decarbonizing District Heating with Hybrid Solar
       Thermal Energy.. 88
3.6   District Heating with Hybrid Wind Energy 93
3.7   District Heating with Small Modular Nuclear Reactors by
       Hybrid Process of Cogeneration.. 93
       3.7.1   Global Assessment of Modular Nuclear
               Heat-Based District Heating............................... 96
3.8   District Heating by Hybrid Industrial Waste Heat 101
3.9   Optimization Models for Hybrid District
       Heating Systems 107
3.10   Role of TES in District Heating 109
       3.10.1 Energy Central 113
3.11   Hybrid DE in US 114
       3.11.1 Examples of Use of DE in US................................115
References 120

Chapter 4   Hybrid Energy Systems for Vehicle Industry.....................129

          4.1   Introduction 129
          4.2   Hybrid Energy in Maritime Industry 129
                4.2.1   Boats, Yachts, and Ferries...129
                4.2.2   Role of Renewable Sources in Shipping Industry 134
                4.2.3   Hybrid Ships and Roles of Renewable Sources
                        and Energy Storage 137
                4.2.4   Energy Storage and Usage in Ships 140
                4.2.5   GE Naval Vessel Electrification 140
                4.2.6   Hybrid Energy for Large Ships 141
                4.2.7   Use of Hybrid Microgrids for Ships.........................145
                4.2.8   Future Marine Power Systems 147
                        4.2.8.1   Maritime Microgrids 149
                4.2.9   Hybrid Power Module 152
                4.2.10 Hybrid Fuel Cell-Based Ships 153
          4.3   Hybrid Energy for Air Vehicles..154
                4.3.1   Hybrid Aircraft...154
                4.3.2   Unmanned Aerial Vehicles 156
                4.3.3   Manned Solar Aircraft 157
                4.3.4   Solar Electric, Hybrid, and Hydrogen Aircraft........158
                4.3.5   Integrated EMS for Hybrid Electric Aircraft............162
          4.4   Hybrid Trains and Railways 167
                4.4.1   Solar-Powered Train System 168
                4.4.2   Hybrid Electric Railway...170
                4.4.3   Energy Storage Technology for Hybrid
                        Electric Railway 174
          References 175

Chapter 5   Hybrid Energy Systems for Coal Industry 183

          5.1   Introduction 183
          5.2   Coal-Based Hybrid Power Plants 184
                5.2.1   Cocombustion of Coal and Biomass 184
                5.2.2   Cofiring Coal-Natural Gas 184
                        5.2.2.1   Options for Natural Gas Addition 186
                5.2.3   Coal-Solar Hybrid for Power and Fuels 187
                        5.2.3.1   Advantages of Coal-Solar
                                  Hybridization 188
                        5.2.3.2   Disadvantages of Coal-Solar
                                  Hybridization 190
                5.2.4   Role of Wind Energy..193
                5.2.5   Carbon Capture from Biomass and Cofired Plants....... 194
                5.2.6   Conversion of Carbon Dioxide to Power by Fuel
                        Cell Technology or to Diesel Fuel.............................195
                5.2.7   Combined Cycle to Improve Efficiency 195

        5.2.8   Conversion of Waste Heat to Power or
                 Additional Industrial Use 195
    5.3   Coal-Biomass Cogasification...196
    5.4   Hybrid Power by IGCC Plants...198
        5.4.1   Commercial Cogasification IGCC Plants.................201
            5.4.1.1   ELCOGAS IGCC Plant, Puertollano,
                       Spain 201
            5.4.1.2   The Willem Alexander IGCC Plant,
                       Buggenum, The Netherlands 202
            5.4.1.3   Polk IGCC Plant, Florida, USA............... 203
        5.4.2   Other Cogasification Projects and Proposals 204
    5.5   Liquid Synthetic Fuels by Cogasification............................ 204
    5.6   Hybrid Energy Systems for Coal to Chemicals.................... 206
        5.6.1   Nuclear-Coal Integration System 207
        5.6.2   Wind/Solar-Coal Integration System 209
        5.6.3   Biomass-Coal Integration System........................... 209
        5.6.4   Carbon Tax Impact on the Economic
                 Competitiveness of Hybrid Energy System.............210
        5.6.5   Carbon-Neutral Cycle via CO_2 Capture and
                 Conversion System 210
    5.7   Novel Hybrid Processes Combining Coal/Biomass to
           Chemicals and Hydrogen Production.....................................211
        5.7.1   NREL Hybrid Concepts (USA) (Gasification/
                 Cogasification + Electrolysis) 212
        5.7.2   CRL Energy, New Zealand (Coal/Biomass
                 Cogasification + Electrolysis) 214
        5.7.3   Other Hybrid Projects for Chemicals and
                 Hydrogen 216
    References 217

Chapter 6   Hybrid Energy Systems for Nuclear Industry 223

    6.1   Introduction 223
    6.2   Diversity of Hybrid Energy Systems 224
    6.3   Nuclear-Renewable Hybrid Energy Systems........................ 226
    6.4   Nature of Interactions in Components of N-RES
           Hybrid Energy Systems 229
        6.4.1   Tightly Coupled N-R HES for Power and Heat 230
        6.4.2   Thermal Interconnections of Components of
                 N-RES Hybrid Energy Systems 230
        6.4.3   Electricity Interconnections of Components of
                 N-RES Hybrid Energy Systems 236
        6.4.4   Chemical Interconnections of Components of
                 N-RES Hybrid Energy Systems 239
        6.4.5   Hydrogen Interconnections of Components of
                 N-RES Hybrid Energy Systems 241

6.4.6   Mechanical Interconnections of Components of
          N-RES Hybrid Energy Systems 242
6.4.7   Information Interconnections of Components of
          N-RES Hybrid Energy Systems 243
6.4.8   System-Level Considerations for N-R HES
          Development.. 243
6.5   Industrial Applications of N-R HES 246
6.6   Tools Required for Successful N-R HES Applications 250
6.6.1   Dynamic Modeling Tools for N-R HES Impact
          Assessment, Design Optimization, and Nuclear
          Reactor Design Studies 250
6.6.2   Thermal Hydraulics and Electricity
          Interconnections 251
6.6.3   Power Generation and Storage Systems 252
6.6.4   Control, Safety, Security, and Licensing 253
6.7   Case Studies... 253
6.7.1   Case Studies 1 and 2: West Texas Synthetic
          Gasoline and Arizona Desalination Plant 253
6.7.2   Case Study 3: N-R HES for Hydrogen
          Production 257
References 261

Chapter 7   Hybrid Energy Systems for Manufacturing Industry 271

7.1   Introduction 271
7.2   Methods for Improving Energy Efficiency by HESs 276
7.2.1   Process Heating Systems (Including Steam for
          Unit Operations) 276
7.2.2   Motor-Driven Systems.. 280
7.2.3   Process Intensification.. 280
7.3   Hybrid Energy Systems Which Include Waste Heat
       Recovery and Conversion 280
7.3.1   CHP Systems... 283
7.3.2   Cogeneration Using Nuclear Heat 285
7.3.3   Options for Waste Heat to Power 286
          7.3.3.1   Thermodynamic Cycles............................ 286
          7.3.3.2   Thermoelectric Power.............................. 287
          7.3.3.3   Thermophotovoltaic Devices.................... 291
          7.3.3.4   Thermionic Devices................................. 292
          7.3.3.5   Piezoelectric Devices.............................. 293
          7.3.3.6   Heat Pumps for Process Heat 293
7.4   Role of Biomass Systems for Industrial Processes................ 295
7.4.1   Biomass-Based Hybrid Systems............................... 295
7.5   Role of Geothermal Energy.. 299
7.5.1   Vapor Recompression.. 300
7.5.2   Geothermal Heat for Chemical Industry................. 300

7.6   Role of Hybrid Solar Thermal Energy 301
        7.6.1   Solar Cooling.. 303
7.7   Potential of Renewable Energy Technologies for
        Industrial Electricity Use... 304
7.8   Realizable Economic Potential of Renewable Energy
        Integration 305
        7.8.1   Priority Areas of Action.. 307
7.9   Reduction in GHG Emission by Clean Energy Alternatives 308
7.10   Closing Perspectives...312
References 315

Chapter 8   Hybrid Energy Systems for O&G Industries....................................323

8.1   Introduction 323
8.2   Drivers for Hybrid Renewable Energy Systems for Oil
        and Gas Industry 325
        8.2.1   Depletion of High-Quality Oil Reserves....................325
        8.2.2   Environmental Concerns in the O&G Industry 326
        8.2.3   Falling Renewable Energy Costs................................327
8.3   Challenges to Renewable Integration 328
        8.3.1   Variability of Generation...328
        8.3.2   System Reliability 329
        8.3.3   Operational Considerations......................................329
        8.3.4   Government Policies 330
8.4   Hybrid Systems..332
        8.4.1   Evaluation and Successful Case Studies 333
8.5   Hybrid Power Systems for Offshore Units 334
8.6   Upstream: Renewable Integration in Oil and Gas
        Production 335
        8.6.1   Electrification of Drilling and Primary Recovery 336
        8.6.2   Use of Hybrid Renewable-Energy-Powered
                 Secondary Recovery..337
        8.6.3   Concentrating Solar and Geothermal Heat for
                 Tertiary Recovery (EOR) 338
        8.6.4   Examples of Successful Case Studies 339
                 8.6.4.1   Photovoltaic Hybrid Systems....................339
                 8.6.4.2   Wind Power Systems 340
                 8.6.4.3   Use of Geothermal Energy 341
                 8.6.4.4   Solar Thermal Systems............................342
        8.6.5   Closing Perspectives...342
8.7   Midstream: Integration of Hybrid Renewable Energy
        Systems in Oil and Gas Transportation..................................343
        8.7.1   Compressor Electrification, Heat Recovery, and
                 Use of Turbo Expanders 344
8.8   Downstream: Integration of Hybrid Renewable Energy
        Systems in Oil Refining..345

8.8.1   Cogeneration (Heat and Power) and Use of
        Hybrid Renewable Energy Systems 347
8.8.2   Hydrogen Production.. 348
8.8.3   Other Efforts to Hybridized Oil and Gas with
        Renewable Energy...350
8.9   Perspectives on Use of Renewable Energies for Oil..............352
8.10   Natural Gas-Renewable Sources Hybrid Systems.................353
        8.10.1 Temporal Framework 354
        8.10.2 Collaborative Market Redesign...............................354
        8.10.3 Perspectives on Low- and Zero-Emission Hybrid
                Generation 357
        8.10.4 Role of Storage in Hybrid Generation..................... 360
        8.10.5 Low-Carbon Renewable Fuel Storage and
                Transmission...361
        8.10.6 Renewable Fuel Injection in the Grid.......................362
References 367

Chapter 9   Hybrid Energy Systems for Computing and Electronic Industries.... 379

9.1   Introduction 379
9.2   The Case of Hybrid Approach for Data Centers 381
9.3   Hybrid Processes to Improve Energy Efficiency of
       Data Centers 384
9.4   Role of Hybrid Renewable Energy for Data Centers.............386
        9.4.1   Technology Capabilities 388
        9.4.2   Implementation Challenges.....................................388
        9.4.3   Benefits...389
        9.4.4   Hybrid Renewable Energy Green-Works
                Framework for Data Centers 390
                9.4.4.1   Hybrid Renewable Energy Systems...........391
                9.4.4.2   Energy Balance Challenge.......................392
                9.4.4.3   The Green Works Framework 393
9.5   Hybrid Storage Devices for Data Centers 396
9.6   Forms of Hybrid Energy in Data Centers............................. 400
        9.6.1   Heat Integration... 400
        9.6.2   Demand Response 401
        9.6.3   Innovative Use of Backup Power 402
9.7   Hybrid Energy Harvesting for Portable Electronics............. 402
        9.7.1   Hybrid, Multisource Energy Harvesters 403
                9.7.1.1   Magnetic and Kinetic Energy.................. 403
                9.7.1.2   Kinetic and Solar Energy 405
                9.7.1.3   Wind and Thermal Energy 405
                9.7.1.4   Solar, Kinetic, and Radio Frequency
                          Energy.. 406
                9.7.1.5   HCs for the Harvesting of Solar and
                          Mechanical Energy................................... 407

9.7.1.6   HCs for the Harvesting of Biomechanical
          and Biochemical Energy 407
9.7.1.7   HCs for the Harvesting of Solar and
          Thermal Energy 408
9.7.1.8   Hybrid Energy via Microscale Waste
          Heat Applications 408
9.7.1.9   Harvester-Sensor Integrations 410
9.7.2   Self-Powered Hybrid Micro-/Nanosystems 410
9.8   Energy Harvesters Integrated with Energy Storage
       AND/OR End Users 411
9.8.1   Harvester-Storage Integrations 411
9.8.2   Hybrid Nanogenerators 411
9.8.3   Wearable Devices of ESSs and Nanogenerators 415
9.8.4   CMOS Technology-Based Harvesters and
          Systems 415
9.9   Hybrid Energy Storage for Low-Power Embedded
       Systems Applications 416
9.9.1   Integrating Faradaic and Capacitive Storage
          Mechanisms 419
9.10   Power Electronics for Renewable Energy Systems 420
9.10.1   DC-to-DC Converters 422
9.10.2   Inverters 422
References 422

Chapter 10 Hybrid Energy Systems for Water Industry 439

10.1   Introduction 439
10.2   Desalination 441
10.2.1   Desalination Process Alternatives 442
10.2.2   Efficiency Improvement through Hybrid
          Processes 443
10.2.3   Role of Renewable Energy in Desalination 447
10.3   Hybrid Solar Energy for Desalination 448
10.3.1   Solar-Thermal Systems 448
          10.3.1.1 Direct Solar Thermal Desalination 449
          10.3.1.2 Indirect Solar Thermal Desalination 449
          10.3.1.3 Perspectives on Pilot and Commercial
                    Scale Operations 452
10.3.2   Solar PV-Based Application 453
10.3.3   Hybrid Solar Thermal-Solar PV 455
10.4   Hybrid Wind Energy for Desalination 455
10.5   Hybrid Geothermal Energy for Desalination 462
10.6   Hybrid Wave Energy 464
10.6.1   Barge-Wave and Below Water Energy Conversion 465
10.7   Desalination by Cogeneration by Small Modular
       Nuclear Reactors 468

10.8   Future Prospects of Desalination by Hybrid Renewable
        Energy Sources and Cogeneration Processes........................474
10.9   Hybrid Energy Systems for Wastewater Treatment...............476
10.10   Future Role of MFC for Wastewater Treatment....................481
References 483

Chapter 11 Hybrid Energy Systems for Hydrogen Production...........................493

        11.1   Introduction 493
        11.2   Role of Biomass for Hydrogen Production 495
        11.3   Biomass-Based Hybrid Systems... 497
                11.3.1   Coal-Biomass 498
                11.3.2   Wastewater Treatment—Biomass 499
                11.3.3   Concentrated Solar—Biomass 500
                11.3.4   Nuclear—Biomass.. 503
                11.3.5   Fuel Cell-Biomass 504
                11.3.6   Electrolysis—Biomass 505
                11.3.7   Wind—Biomass 506
                11.3.8   Industrial Waste Heat-Biomass Hybridization......... 508
        11.4   Hybrid Biomass Systems Recommended by NREL 508
        11.5   Indirect Gasifier Hybrid System... 509
                11.5.1   Peaking Modifications..510
        11.6   Direct Gasifier Hybrid System 511
        11.7   Hybrid Energy Systems for Hydrogen Production by
                Electrolysis 512
                11.7.1   Hybrid Renewable (Wind, Solar) Electrolysis...........514
                11.7.2   Commercial-Scale Hybrid Nuclear Heat-
                        Based HTE 516
        11.8   Economic Aspects of Green Hydrogen Production 518
        References 519

Index..527

Preface

There are significant pressures to reduce emission of carbon in the environment due to its harmful effect on ozone layer resulting in global warming. Two sources of carbon that are emitted by natural and man-made processes are carbon dioxide and methane. Although methane is significantly more harmful than carbon dioxide, man-made processes are more responsible for carbon dioxide emission. Methane emitted in the atmosphere eventually gets converted to carbon dioxide. Methane is a very reactive substance and can be easily converted if captured. Carbon dioxide, on the other hand, is much more stable and more difficult to convert by chemical and biological processes. The reduction of carbon dioxide by energy industry has therefore become a more urgent necessity.

The reduction of carbon dioxide emission or what we call decarbonization of energy industry can be achieved by a number of strategies. First strategy is the energy conservation which can be achieved by careful and frugal use of energy. This depends on human behavior which is not always easy to control. The second strategy is to improve the efficiency of energy conversion processes and usages. This can be achieved by adopting energy-efficient processes and reducing waste heat. This book will show that this is often achieved by hybrid energy processes like cogeneration and mixing of various unit operations, etc. Since we currently use fossil fuels in many processes, an efficiency increase essentially means less consumption of fossil fuels and thereby less emission of carbon dioxide.

The third strategy is to replace fossil fuel use for raw materials for chemicals and materials by biofuels and thereby reducing carbon emission. Efficiency can also be improved by innovative process changes such as additive manufacturing, hybrid desalination, hybrid wastewater treatment, hybrid unit operations, etc.

Fourth strategy is to replace the use of fossil fuels by renewable fuels such as solar, wind, biomass (waste), geothermal, and water. Renewable fuels are more sustainable because they are bountiful, free, and they do not generate carbon dioxide. While renewable fuels are very desirable for decarbonization of energy industry, they also pose numerous challenges such as intermittency of solar and wind energy and lack of availability of geothermal and hydro energy in some places. Renewable energy sources are also low-density sources and they are more distributed and difficult to harness in large capacities. The capital costs for capturing renewable sources are also high, although operating costs are low because they are free. The book shows that the best way to harness renewable sources is through hybrid energy systems. For power generation, storage, and transport, this was demonstrated in my previous book. Fifth and final strategy is to use hydrogen-based technology like fuel cell which does not emit carbon and they can be used for both static and mobile power generation. The use of fuel cell, however, will also require hybrid operation. While nuclear energy is also clean (less carbon emitting than fossil fuels), it can also be hybridized with renewable sources.

This book is in essence a second part of the previous book and it shows how hybrid energy systems can decarbonize industrial needs for power and heating and cooling.

While the previous book was focused on hybrid power generation, storage, and grids, the present book examines the role of hybrid energy systems for ten different industries: coal, oil and gas, nuclear, building, vehicle, manufacturing and industrial processes, computing and portable electronic, district heating and cooling, water industry, which includes wastewater treatment and desalination, and hydrogen production. The book shows that hybrid energy systems (like cogeneration) can improve efficiency and this concept is widely used globally and across all industries. The book also shows that hybrid energy system has been demonstrated as a very effective strategy for insertion of renewable fuels in the energy industry and thereby decarbonizing the industry. With the help of numerous examples, the book demonstrates that hybrid energy systems are effective strategies for industry decarbonization.

Besides a brief introductory chapter, the book is divided into ten chapters covering ten industries outlined above. While the case made for the use of hybrid energy systems is clear and convincing, more research is aggressively done to improve our understanding and commercial applications of this concept in a variety of practical situations. The book should be very useful for graduate students and researchers in industry and government energy laboratories. The book should also be useful to people involved in energy processes and plants. The book is a good reference book for all people interested in energy.

MATLAB® is a registered trademark of The MathWorks, Inc. For product information,

    please contact:
    The MathWorks, Inc.
    3 Apple Hill Drive
    Natick, MA 01760-2098 USA
    Tel: 508-647-7000
    Fax: 508-647-7001
    E-mail: info@mathworks.com
    Web: www.mathworks.com

Author

Yatish T. Shah received his B.Sc. in chemical engineering from the University of Michigan, Ann Arbor, USA, and MS and Sc.D in chemical engineering from the Massachusetts Institute of Technology, Cambridge, USA. He has more than 40 years of academic and industrial experience in energy-related areas. He was chairman of the Department of Chemical and Petroleum Engineering at the University of Pittsburgh, Pennsylvania, USA; dean of the College of Engineering at the University of Tulsa, Oklahoma, USA, and Drexel University, Philadelphia, Pennsylvania, USA; chief research officer at Clemson University, South Carolina, USA; and provost at Missouri University of Science and Technology, Rolla, USA, the University of Central Missouri, Warrensburg, USA, and Norfolk State University, Virginia, USA. He was also a visiting scholar at University of Cambridge, UK, and a visiting professor at the University of California, Berkley, USA, and Institut fur Technische Chemie I der Universitat Erlangen, Nurnberg, Germany. Dr. Shah has previously written eleven books related to energy, eight of which are under "Sustainable Energy Strategies" book series by Taylor and Francis of which he is the editor. He has also published more than 250 refereed reviews, book chapters, and research technical publications in the areas of energy, environment, and reaction engineering. He is an active consultant to numerous industries and government organizations in the energy areas.

1 Hybrid Energy Systems—Strategy for Decarbonization

1.1 INTRODUCTION

The industrial growth and the increasing desire of growing population to improve quality of life have led to the increasing demand on energy. Energy is extremely important for any economy to generate wealth and it is the key component for GDP growth. It is estimated to grow from 549 quadrillion British thermal units (Btu) in 2012 to 815 quadrillion Btu in 2040, 48% increase in 28 years. The energy demand could have more than doubled without efficiency gain and suitable energy mix. The non-Organization for Economic Cooperation and Development (non-OECD) countries are the major contributor in this drastic energy demand. In these countries, energy demand will rise by 71% from 2012 to 2040 in contrast with only 18% in developed countries in same time span. An average GDP growth of 4.2% per year is estimated between 2012 and 2014 in non-OECD countries as compared to 2.0% per year in OECD countries as estimated by IEO2016. In terms of energy consumption by sectors, industry is leading followed by residential and transport. This trend persists from 1971 to 2014, and overall industrial consumption doubled during this period. Power plant sectors consume 35% of global energy, and it is estimated to grow due to urbanization in developing countries. Global electricity demand is expected to increase over 65% from 2014 to 2040, 2.5 times faster than overall energy demand [1,2].

The competition among countries for the industrial development has severe impact on the environment in terms of CO_2 emission. Every drop of fuel pollutes environment and intensity depends on process efficiency. Global CO_2 emission measured as 40 gigaton (Gt) per year in 2016 is almost double as compared to 1980 emission level. Energy sector is the major contributor, 68%, in CO_2 emission and power generation sector sharing 42% in energy sector emission followed by transport 23% and industry 19%. A systematic diversification of the global energy mix and technology improvement driven by economics and climate policies almost flattens the CO_2 emission rate in 2014, only 0.8% increase, as compared to 1.7% in 2013 and 3.5% in 2000. During 2012–2014, the moderate increase in CO_2 emission, 0.8%–1.7%, is remarkable when global economic growth rate was 3% as compared to 4% emission annually with GDP growth rate of 4.5% in last decade. In other words, partial decoupling of economic growth and CO_2 emission has been observed in 2012–2014 due to shift in energy production and consumption, power generation, technological improvements, and policy implementation. In 2015, the milestone year, 170 countries signed an agreement at the 21st Conference of the Parties (COP21) in Paris for climate action. The Paris Agreement is the first international climate agreement extending mitigation obligations to all developed and developing countries representing over 90% of energy-related CO_2 emissions and approximately 7 billion people. The agreement aims to achieve CO_2 emission peak as soon as possible to cap the increase in the global average temperature to below 2°C. In addition, it also aims to pursue extra efforts to limit the temperature increase to 1.5°C. Business as usual scenario, as most of the countries followed, can lead to over 5°C temperature increase [1,2].

Going forward, along with reduction in carbon emission, affordable cost, commercial viability of technology, and equal access of energy by both urban and rural communities are important drivers for the energy industry. There has been a recognition by the energy industry in recent years that there are ten sources of energy in the world: coal, oil, gas, biomass, waste, nuclear, solar, wind, geothermal, and water. As pointed out in my previous book [3] that none of these sources satisfies completely all the requirements of energy production. Their availability also varies significantly across the world. In order to capture renewable source, particularly non-dispatchable sources like solar and wind, energy storage has become very important. In the past hydropower storage captured about 95% of storage market although at smaller scale, batteries have captured the most attention. Just like multiplicity of energy sources, there are also multiple types of storage devices; like batteries (both non-flow and flow), capacitors (including super capacitors), mechanical storage devices like compressed air and flywheel, magnetic storage device like SMES, thermal storage like, molten salt, ice bricks, phase change materials, thermochemical heat, etc., bulk gravitational storage devices which include technologies such as pumped hydro and gravel in railcars and hydrogen storage. As shown in my previous book [3] that none of these storage devices individually meets all the requirements of the storage needs [3].

In the past, electricity was mainly supplied by macro level utility grids with power generated from centralized large-scale fossil fuel (coal or natural gas) or nuclear energy power plants. These large-scale power plants follow the economy of scale and did not require additional storage devices due to their spinning and nonspinning reserves. While they served well-urban communities, they did not serve rural and isolated communities where utility grid was not accessible. In recent years, the desired access of renewable sources changed this paradigm. Unlike large-scale fossil fuel and nuclear-based power plants, renewable sources are of low density and more distributed. In order to capture these distributed generation sources, a new grid structure, microgrids, is developed which can be connected to utility grid or can operate in islanded mode at medium-to-low voltages more near customers. Furthermore, in order to serve rural and isolated communities, off-grid structures of minigrids, nanogrids, and stand-alone systems are developed which are not connected to the utility grid. These different grid structures are also outlined in my previous book [3]. Thus, just like sources of generation and nature of storage, grid structure has also become hybrid in order to capture distributed renewable energy sources and serve the rural communities.

In recent years, energy industry has been going through major changes. It needs to address issues such as (a) carbon dioxide emission reduction, (b) more balanced use of all energy sources, (c) affordability for all customers, (d) accessible to both urban and rural or isolated communities, (e) more efficient in generation, storage, and distribution, and (f) more balanced between distributed and centralized mode of operation. Rapid development of new nanotechnologies makes industry to be more flexible and adaptable. These desired changes also force industry to operate in more modular form.

The pressure on energy industry to reduce carbon dioxide emission has resulted in the adoption of multiple strategies which all lead to lesser and more efficient use of fossil fuels and more insertion of renewable sources in the overall energy use mix. Low efficiency of large-scale power generation processes by coal, gas, and nuclear energy has forced to use hybrid energy systems like cogeneration or combined heat and power (CHP) to improve their efficiency. Cogeneration and CHP processes are, however, more easily implementable at smaller scale. This has also led to the development of small modular nuclear reactors.

Energy industry has recognized that the best way to respond to the required changes is to make generation, storage, and transport processes more heterogeneous. The insertion of renewable sources in hybrid grid structure is an evolving process and will require power and heating and cooling requirements to be satisfied by multiple generation sources. The evolving storage requirements will also have to be heterogeneous in order to satisfy the needs of a variety of applications such as electric car, portable electronics, hybrid grid structure, etc. The development of stable and workable off-grid structure will require significant use of multiple renewable sources which include one or more energy storage devices. Even if in long term renewable sources completely replace fossil fuels, their use will require backup devices like storage or diesel fuel. More hybrid energy systems will be required to make many systems more efficient. Thus, the energy system will become hybrid in one or other form. It appears that hybrid energy systems are the important parts of the future energy industry.

In recent years, significant efforts are made in the development of fuel cells. They are being made more affordable and durable at both large and small scales and for both static and mobile applications. Fuel cell can be either power generating or storage device. While this carbon-free device has a strong future, it also has low power density and some other limitations and will generally operate best in a hybrid form which includes another source of generation or storage. The successful development and use of fuel cells will also help decarbonize energy industry.

Going forward, energy industry has adopted following strategies to serve its needs in an environmentally acceptable way:

a. Improve efficiency of all energy conversion processes. In particular, for conversion of thermal energy to electrical energy by converting waste heat to other power, fuel, or heating/cooling needs (CHP or cogeneration approach) by unique hybrid energy systems; in this regard, small modular nuclear reactors will find unique place in the future overall energy mix. Waste heat is also very prominent in industrial processes and in mobile industry.

b. Reduce the use of fossil fuel by combining its usage with other forms of low carbon nuclear and renewable technologies for power and heating and cooling needs.

c. Replace fossil fuels by biofuels for the production of raw materials for various chemicals, materials, fuels, and fuel additives.

    d. Replace fossil fuels with multiple sources of renewable energy particularly in rural and isolated off-grid communities.

    e. Insert renewable energy sources in utility power plants and utility grid.

    f. Make more efforts to harness distribute energy sources using microgrid platform.

    g. Make use of hybrid energy storage systems to improve efficiency of energy use in various applications.

    h. Convert CO_2 to chemicals, materials or fuels by additional hybrid chemical or biological processes. This subject is discussed in great details in an excellent national academy of engineering report by national academy of science, engineering, and medicine [4]. This is also the theme of my next book on treatment strategies for carbon emissions.

    i. Make more use of hydrogen (carbon-free) base technologies.

In my previous book on hybrid power-generation, storage, and grids, the importance of hybrid energy, hybrid storage systems, and hybrid grids for power industry was delineated in detail. The book showed that for growing power industry the hybrid power is the future and it offers many positive values toward decarbonization of power industry. The book pointed out that no single source of energy satisfies all the criteria for sustainable energy. The book also pointed out that no energy storage device alone provides all the required characteristics of energy storage, such as energy density, power density, cycle life, etc. Present utility grids are not effective for harnessing distributed energy sources and they are not accessible in the rural and remote areas. Thus sources, storage devices, and grid transport all need to be hybrid and multifaceted to serve the future needs of electric power. The criteria for sustainable energy including (a) reliability and flexibility, (b) affordability, (c) accessibility, (d) high efficiency, and (e) durability and long-term sustainability are all best satisfied by the hybrid power.

    In the present book, we extend the applications of the concept of hybrid energy systems to ten major industries: coal, oil and gas, nuclear, building, vehicle, manufacturing industry, computing and portable electronic industry, district heating and cooling industry, water industry and hydrogen production, and illustrate that the concept of hybrid energy systems also helps decarbonize these industries. The book illustrates various methods used to apply hybrid energy systems in these industries. For each industry, the use of power and heating and cooling needs is considered.

1.2 HYBRID ENERGY SYSTEMS DEFINED

Hybrid energy systems can be defined in a number of different ways. Hybrid energy system as defined here is an umbrella of systems which include multiple sources of energy and multiple storage devices and systems with hybrid energy processes. These systems can be connected to utility grid, microgrid, or they can be off-grid (like mini grid, nanogrid, or stand-alone systems). The pros and cons of hybrid energy systems and related issues and challenges depend on further details on the contents and the methods adopted for their use.

Multiple sources of generation in a hybrid energy system can be nonrenewable-nonrenewable (like coal and gas), nonrenewable-renewable (like coal-solar), renewable-renewable (like solar-wind), nuclear-nonrenewable (like nuclear-gas), or nuclear-renewable (like nuclear-solar). Each of these systems has its own pros and cons and challenges and issues. For renewable sources, further breakdown in nondispatchable (like solar and wind) and dispatchable (like biomass, geothermal, hydro, etc.) is required. This differentiation also requires different strategies for implementation, particularly due to intermittent nature of solar and wind energy. All of these combinations are unique and situation dependent and are aimed toward reduction of carbon emission with different degrees of success. Multiple sources also include considerations of process-generated sources like waste heat or waste product. The present book considers all of these options.

Storage is an example of another method of energy generation when it is required. As pointed out in my previous book [3], there are multiple devices for storage, each with its own pluses and minuses. Many energy applications require multiple storage devices for sustainable and high-quality energy (particularly power) use. The present book includes multiple storage devices as part of hybrid energy systems. Finally, the details of implementations and challenges faced on the use of hybrid energy systems depend on whether these systems are connected to utility grid, microgrid, or they are part of off-grid operations (like minigrid, nanogrid, or stand-alone systems). Hybrid energy systems include both power and heat (cool).

Decarbonization of energy industry essentially means less and more efficient use of fossil fuels (particularly coal, oil, and gas in that order). This requires two-prong strategies: (a) improve efficiency of all energy conversion processes and reduce the unnecessary need for the energy consumption. This can be largely achieved by conservation measures and building more hybrid energy systems like cogeneration and carry out better integration and hybridization of raw materials and process steps so as to reduce material and energy need which are dependent on fossil fuels and (b) replace the use fossil fuel by renewable or nuclear energy sources for all required process energy. This can be done by using hybrid energy systems involving multiple sources of generation with or without storage. In essence the task for energy industry is to move from right to left in Figure 1.1 to the extent possible. It should be pointed out that decarbonization is also possible if produced CO_2 is captured and sequestered or converted to other useful materials. However, as pointed out by national academy report [5], this is a difficult task and significant more R&D is needed. As mentioned earlier, this subject is also examined in details in my next book on treatment strategies for carbon emissions.

General Electric (GE) defined hybrid power as: "Hybrid power plants usually combine multiple sources of power generation and/or energy storage, and a control system to accentuate the positive aspects and overcome the shortcomings of a specific generation type, in order to provide power that is more affordable, reliable and sustainable. Each application is unique, and the hybrid solution that works best for a specific situation will depend on numerous factors including: existing generation assets, transmission and distribution infrastructure, market structure, storage availability and fuel prices and availability" [3]. Similar concept in a broader term

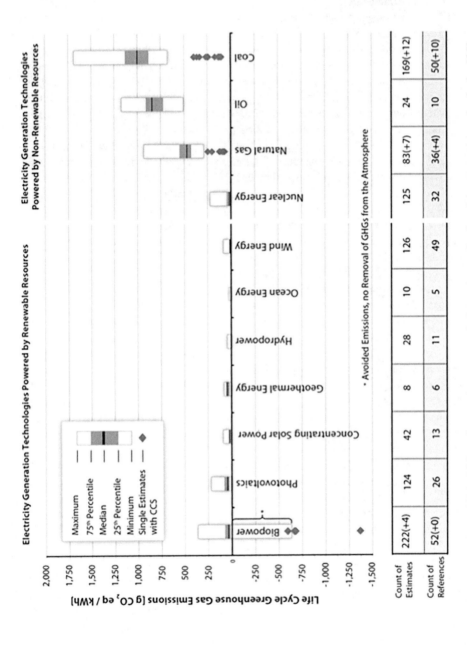

FIGURE 1.1 Comparison of as-published life cycle greenhouse gas emission estimates for electricity generation technologies [6,7].

of hybrid energy systems (which include both power and heat) is adapted in the present book.

There are numerous examples of hybrid energy systems that are worth noting. GE definition of hybrid power includes multiple sources of power generation (like wind and solar) with energy storage with and without grid. The district heating with multiple renewable sources of heat with and without storage is another example of hybrid energy system where power may not be involved. A zero-energy building with solar energy to generate power and heat and geothermal energy for HVAC system with or without storage is also another example of hybrid energy system. Here both power and heating and cooling are parts of hybrid energy system. Cogeneration (combined heat and power) is another example of hybrid energy system to improve energy efficiency. In this case process generates second source of energy (i.e., waste heat). Exxon-Mobil and Fuel Cell energy partnership to generate power by burning coal or natural gas and use fuel cell to generate more power from waste CO_2 is another form of hybrid energy system where power is generated in two different ways within a process. Using multiple sources (including renewable and nuclear) to generate heat with and without thermal storage for industrial processes (such as glass making operation) with conversion of waste heat to generate electricity by the process of thermoelectricity is another form of hybrid energy system where industrial heating results in the waste heat to generate power. This is a reverse cogeneration process once again used to improve energy efficiency. Hybrid vehicles can also generate power by internal combustion (IC) engine, fuel cell, or battery along with power generated via thermoelectricity of waste exhaust heat.Electric vehicle (EV) with energy storage (one or more) is also hybrid operation. In all the cases mentioned above there can be different level of integration of components of hybrid energy systems. Wastewater treatment with combined physical and chemical process is a hybrid energy system. Desalination can be carried out with multiple options of hybrid energy systems all designed to be more efficient and less energy consuming.

The book makes the case that hybrid energy system is an important strategy for decarbonization of various industries. Hybrid energy systems are (a) more reliable and flexible, (b) more efficient, (c) more affordable, (d) more environment friendly, and (e) ultimately more durable and sustainable over long term compared to the single component energy sources or process. Hybrid energy systems are essential for the deeper penetration of renewable and nondispatchable energy (particularly solar and wind which are intermittent by their nature) sources in overall energy mix in order to reduce carbon emission to the environment. Hybrid energy systems are also important for better use of nuclear heat and suitable power generation by a combination of nuclear and renewable sources. The advantages of hybrid storage and hybrid grid transport for power industries are described in detail in my previous book [3].

The important defining components of hybrid energy systems are thus:

1. sources of power generations
2. nature of storage device used
3. nature of connections of hybrid energy systems to the grid

4. nature of use: power, heat (or cool) or both
5. end applications: building, vehicle, manufacturing industry, etc.

Previous book articulated more details on these components with focus on power. The present book goes one step further and articulates strategy of hybrid energy systems for decarbonization of broader industrial applications which include both power and heat (cool). Because hybrid energy systems are so varied, the book articulates how these systems can be effectively used in each industry. Unlike previous book, this book puts less emphasis on the grid structure used and more emphasis on how different industries use hybrid energy systems to decarbonize the particular industry. Both power and heat (cool) are considered.

There are also various ways the components of hybrid energy systems can interact with each other. Besides components outlined above, it is the nature of interactions that sets different hybrid energy systems apart from each other. As shown in detail in Chapter 6, the nature of interactions will dictate the details of the hybrid energy implementations.

1.3   EXAMPLES OF HYBRID ENERGY SYSTEMS

While the book examines numerous hybrid energy systems used in various industries, here we show two important examples of hybrid energy systems that are commercially used in the energy industry.

1.3.1   HYBRID SOLAR-WIND RENEWABLE SYSTEMS

As pointed out in my several publications [3,8,9], in recent years, significant efforts are made to introduce solar-wind sources for power generation in the traditional utility grid. Here nonrenewable and renewable sources are combined. The introduction of the renewable sources in grid poses several challenges [3,8,9]. In order to alleviate some of these challenges, as shown in my previous book [3], platforms of microgrid and energy storage devices are used.

The unpredictable pattern of natural renewable resources makes them individually unsuitable to provide uninterrupted and reliable power supply to its users. The technical difficulties may arise due to uncontrollable weather conditions like wind speed fluctuation, day and night behavior, and summer and winter sun conditions. In other words, renewable energy is usually dependent on the weather conditions for its source of power input. For example, a hydro generator depends on rain to fill dams to supply flowing water, a wind turbine depends on the wind within a specific range of speed to turn the blades and run the turbine, and a solar panel depends on the position of the sun and, preferentially, no clouds to make electricity.

A combination of two or more renewable energy sources is therefore more effective than the single source system in terms of cost, efficiency, and reliability. We can easily reduce the need for fossil fuels by properly choosing a combination of renewable energy sources. The combination of two or more energy sources, working together in order to compensate for each other, is an effective hybrid energy system. The main advantage of a hybrid energy system is the enhancement of reliability and cost-benefit of the system. Due to the fact that some renewable energy sources such as solar radiation and wind are, most of the times, intermittent, they are frequently combined with other power sources such as utility grid, diesel generators, or a storage device. The objective is to ensure continuous supply of power or other energy need.

Nowadays two types of hybrid systems are in operation: one using only renewable energies, which are ideal for applications in isolated systems, and a second type that makes use of the production from diesel and gas generators or a storage device as a backup. Solar-wind hybrid systems have as the main advantage the way they complement each other and the fact that they are exclusively renewable energy resources. The behavior of solar radiation throughout the day follows an approximately constant pattern of production, reaching its peak at noon decreasing until sunset. The wind generator can serve as a complement to the system, in the periods where there is low or nonexistent solar radiation. This characteristic gives this type of system a greater reliability in the matter of continuity of electrical production over time.

Solar-wind-diesel/gas or simply solar-diesel/gas hybrid systems work similarly to the mentioned solar-wind system. This type of system has the advantage of reducing the consumption of fossil fuels. It is even more reliable, because diesel or gas generators work as a backup, thus ensuring the operation of the system even in periods when the remaining energy sources are not available or are not enough to guarantee the energy required.

Around the world, thousands of communities and industrial sites are however not powered by the utility grid. The traditional means of electrification in such locations is diesel generation since it is a more conventional technology and more people are trained in operation and maintenance. Wind turbines and solar panels are better known renewable energy devices used in hybrid power systems. Hybrid systems usually include energy storage (see Figure 1.2), so they can deliver a certain amount of energy on demand. These systems provide a high degree of energy security through

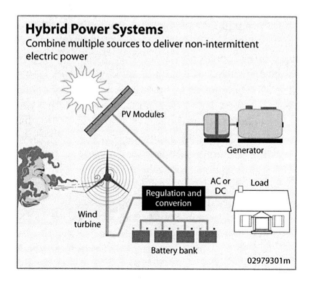

FIGURE 1.2 Solar wind-battery hybrid system [8].

the mix of generation methods and can often incorporate a storage system (battery and/or fuel cell) or to ensure maximum supply reliability and security, a small fueled generator (nonrenewable). Hybrid power systems exhibit higher reliability and lower cost of generation than those that use only one source of energy. Some studies have been conducted concerning the techno-economic assessment of an autonomous hybrid PV/diesel hybrid power system installed in a bungalow complex in Elounda, Crete. In remote areas which are far from the grids, electricity is supplied either by diesel generators or small hydroelectric plants. Under such circumstances, the supply of fuel becomes so expensive that hybrid diesel/photovoltaic generation becomes competitive with diesel-only generation. Many of these studies are reviewed in my previous book [3]. When solar-wind hybrid system is injected in the utility grid, as pointed out in my previous book, additional storage (battery) is required to stabilize grid. Solar-wind hybrid systems can also be effectively used with the microgrids.

The use of solar-wind hybrid power systems in industries is becoming so usual and so advanced that in some cases its users have started requiring design and model stand-alone systems with software such as MATLAB/SIMULINK, to evaluate their performance. One example of such a study was considered to establish a comparison of a hybrid power system based on PV module, wind turbine, and diesel generator with a PV/wind/battery hybrid power system. From the simulation results, it was identified that a hybrid connected system with the battery can perform better in some ways than the diesel connected system. The industry is adapting and there are many cases of implemented hybrid Systems. In Jamaica, for example, it is installed as the world's biggest hybrid system. It was installed by Wind Stream Technologies in an area very close to the coast and takes advantage of winds with an average speed around 96.5 km/h. With this system, an output production of 106 MWh/year and a return of investment in less than 4 years are expected. The plant is designed to save the company approximately $2 million in energy costs over 25 years. The hybrid energy system consists of 50 Solar Mills (patented product with solar panels and vertical axis turbines), making it the largest facility in the world [3].

EFACEC has also developed a Hybrid System Platform (called EFASOLAR3) which has the capability of integrating several energy sources such as diesel, photovoltaic, electrical grid, energy storage, wind, and hydro. This system can be implemented in an existing installation or in a new project, ensuring a stable operation and optimizing the energy produced by the photovoltaic plant. A hybrid system combining floating photovoltaics and hydroelectric power generation has been installed in the Alto Rabagão dam in Portugal (by EDP Group). FPV systems are producing energy on areas that are unused otherwise. The pioneer project has an installed capacity of 220 kW and is expected to produce 332 MWh [3].

Large-scale hybrid systems and solutions combined with heat pumps should be the main bets of the solar thermal industry for years to come. The ISOL Index study conducted annually by the German Soll Agency in 18 of the world's leading solar thermal markets analyzes the responses of more than 340 manufacturers and distributors of the thermal solar present in these countries, in order to evaluate the business climate and expectations for this market. Portugal is one of the markets analyzed. The expectation is to identify an increase in terms of this type of systems, due to the quick return of investments, cost savings, and use of renewable sources replacing fossil fuels. A typical solar/wind hybrid system is illustrated in Figure 1.3 [9]

FIGURE 1.3   Hybrid System on <u>Žirje</u>, Croatia [9].

Although solar-wind hybrid system described here has a bundle of advantages, there are some challenges and problems related to this system that needs to be addressed:

1. Most of hybrid systems require storage devices where batteries are mostly used. These batteries require continuous monitoring and increase the cost, as the battery life is limited to a few years. It is reported that battery lifetime increases to around years before they can be used in hybrid systems.
2. Due to dependence of renewable sources involved in the hybrid system on weather results in the load sharing between the different sources employed for power generation, the optimum power dispatch and the determination of cost per unit generation are not easy. However, optimization of hybrid energy system is essential. This should be obtained with several end objectives.
3. The reliability of power can be ensured by incorporating weather-independent sources like diesel generator or fuel cell.
4. As the power generation from different sources of a hybrid system is comparable, a sudden change in the output power from any of the sources or a sudden change in the load can affect the system stability significantly.
5. Individual sources of the hybrid systems have to be operated at a point that gives the most efficient generation. In fact, this may not occur due to the fact that load sharing is often not linked to the capacity or ratings of the sources. Several factors decide load sharing such as reliability of the source, economy of use, switching required between the sources, availability of fuel, etc. Therefore, it is desired to evaluate the schemes to increase the efficiency to as high level as possible.

Going forward, these challenges will need to be addressed.

1.3.2 COMBINED HEAT AND POWER HYBRID ENERGY SYSTEM

Another widely used commercial hybrid energy system is cogeneration. Cogeneration or CHP is the use of a heat engine or power station to generate electricity and useful heat at the same time (Figure 1.4) Trigeneration or combined cooling, heat, and power (CCHP) (see Figure 1.5) refers to the simultaneous generation of electricity and useful heating and cooling from the combustion of a fuel or a solar heat collector. The terms *cogeneration* and *trigeneration* can also be applied to the power systems simultaneously generating electricity, heat, and industrial chemicals (e.g., syngas). Both cogeneration and trigeneration are classic examples of hybrid processes that improve energy efficiency, reduce the fuel usage, and decarbonize the industry (see Figures 1.6 and 1.7).

Cogeneration is a more efficient use of fuel because otherwise-wasted heat from electricity generation is put to some productive use. CHP plants recover otherwise wasted thermal energy for heating. One example is combined heat and power district heating. Small CHP plants are an example of *decentralized energy*. By-product heat at moderate temperatures (100°C–180°C, 212°F–356°F) can also be used in *absorption refrigerators* for cooling [10].

The supply of high-temperature heat first drives a gas or steam turbine-powered generator. The resulting low-temperature waste heat is then used for water or space heating. At smaller scales (typically below 1 MW), a gas engine or diesel engine may be used. Trigeneration differs from cogeneration in that the waste heat is used

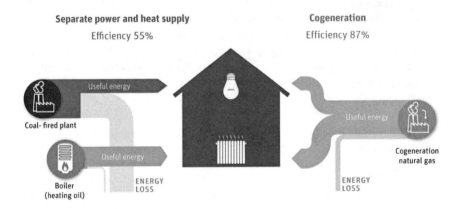

Why cogeneration is more efficient than conventional coal power plants
Comparing the energy efficiency of cogeneration with conventional coal power plant and heating system
Source: ASUE

Separate power and heat supply — Efficiency 55%

Cogeneration — Efficiency 87%

Useful energy — Coal-fired plant

Useful energy — Boiler (heating oil)

ENERGY LOSS

Useful energy — Cogeneration natural gas

ENERGY LOSS

With a coal fired power plant, more than half the energy input is wasted.
Cogeneration reduces the primary energy demand by 36%.

Energy Transition   energytransition.org

FIGURE 1.4   Example of cogeneration [10].

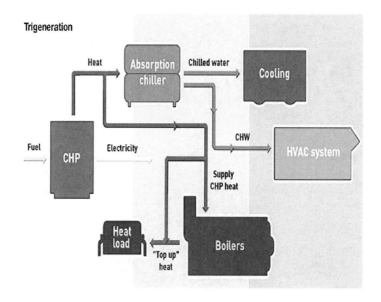

FIGURE 1.5 Trigeneration—combined power, heating, and cooling [10].

FIGURE 1.6 Masendo CHP power station in Denmark. This station burns straw as fuel. The adjacent greenhouses are heated by district heating from the plant [10].

for both heating and cooling, typically in an absorption refrigerator. CCHP systems can attain higher overall efficiencies than cogeneration or traditional power plants. In the United States, the application of trigeneration in buildings is called building cooling, heating, and power. Heating and cooling output may operate concurrently or alternately depending on need and system construction.

Cogeneration was practiced in some of the earliest installations of electrical generation. Before central stations distributed power, industries generating their own power used exhaust steam for process heating. Large office and apartment buildings, hotels, and stores commonly generated their own power and used waste steam for building heat. Due to the high cost of early purchased power, these CHP operations continued for many years after utility electricity became available.

FIGURE 1.7 A cogeneration plant in Metz, France. The 45 MW boiler uses waste wood biomass as an energy source, providing electricity and heat for 30,000 dwellings [10].

Many process industries, such as chemical plants, oil refineries, and pulp and paper mills, require large amounts of process heat for such operations as chemical reactors, distillation columns, steam driers, and other uses. This heat, which is usually used in the form of steam, can be generated at the typically low pressures used in heating, or can be generated at much higher pressure and passed through a turbine first to generate electricity. In the turbine, the steam pressure and temperature are lowered as the internal energy of the steam is converted to work. The low-pressure steam leaving the turbine can then be used to process heat.

Steam turbines at thermal power stations are normally designed to be fed high-pressure steam, which exits the turbine at a condenser operating a few degrees above ambient temperature and at a few millimeters of mercury absolute pressure. (This is called a *condensing* turbine.) For all practical purposes, this steam has negligible useful energy before it is condensed. Steam turbines for cogeneration are designed for *extraction* of some steam at lower pressures after it has passed through a number of turbine stages, with the unextracted steam going on through the turbine to a condenser. In this case, the extracted steam causes a mechanical power loss in the downstream stages of the turbine. Or they are designed, with or without extraction, for final exhaust at *back pressure* (noncondensing). The extracted or exhaust steam is used for process heating. Steam at ordinary process heating conditions still has a considerable amount of enthalpy that could be used for power generation, so cogeneration has an opportunity cost.

A typical power generation turbine in a paper mill may have extraction pressures of 160 psig (1.103 MPa) and 60 psig (0.41 MPa). A typical back pressure may be 60 psig (0.41 MPa). In practice, these pressures are custom designed for each facility. Conversely, simply generating process steam for industrial purposes instead of high enough pressure to generate power at the top end also has an opportunity cost (see: Steam supply and exhaust conditions). The capital and operating cost of high-pressure boilers, turbines, and generators is substantial. This equipment is normally operated continuously, which usually limits self-generated power to large-scale operations.

A combined cycle (in which several thermodynamic cycles produce electricity) may also be used to extract heat using a heating system as condenser of the power plant's bottoming cycle. For example, the RU-25 MHD generator in Moscow heated a boiler for a conventional steam power plant, whose condensate was then used for space heat. A more modern system might use a gas turbine powered by natural gas, whose exhaust powers a steam plant and whose condensate provides heat. Cogeneration plants based on a combined cycle power unit can have thermal efficiencies above 80%. This is also an example of hybrid process.

The viability of CHP (sometimes termed utilization factor), especially in smaller CHP installations, depends on a good baseload of operation, both in terms of an on-site (or near site) electrical demand and heat demand. In practice, an exact match between the heat and electricity needs rarely exists. A CHP plant can either meet the need for heat (*heat-driven operation*) or be run as a power plant with some use of its waste heat, the latter being less advantageous in terms of its utilization factor and thus its overall efficiency. The viability can be greatly increased where opportunities for trigeneration exist. In such cases, the heat from the CHP plant is also used as a primary energy source to deliver cooling by means of an absorption chiller.

CHP is most efficient when heat can be used on-site or very close to it. Overall efficiency is reduced when the heat must be transported over longer distances. This requires heavily insulated pipes, which are expensive and inefficient; whereas electricity can be transmitted along a comparatively simple wire, and over much longer distances for the same energy loss. The requirement that waste heat needs to be used locally, restricts its effectiveness for large-scale power plants because they are generally operated in remote areas. Cogeneration is thus more effective in small-scale, modular, and moveable operation because they can be more easily aligned to where the heat is needed. For this reason, small modular nuclear reactors will have better success with cogeneration and they are now favored over large-scale operation. A car engine becomes a CHP plant in winter when the reject heat is useful for warming the interior of the vehicle. The example illustrates the point that deployment of CHP depends on heat uses in the vicinity of the heat engine.

Thermally enhanced oil recovery (TEOR) plants often produce a substantial amount of excess electricity. After generating electricity, these plants pump left-over steam into heavy oil wells so that the oil will flow more easily, increasing production. TEOR cogeneration plants in Kern County, California, produce so much electricity that it cannot all be used locally and is transmitted to Los Angeles. CHP is one of the most cost-efficient methods of reducing carbon emissions from heating systems in cold climates and is recognized to be the most energy-efficient method of transforming energy from fossil fuels or biomass into electric power. Cogeneration plants are commonly found in district heating systems of cities, central heating systems of larger buildings (e.g., hospitals, hotels, prisons), and are commonly used in the industry in thermal production processes for process water, cooling, steam production, or CO_2 fertilization. Waste heat can also be converted to power by thermoelectric, thermophotovoltaic, or piezoelectric operations. In industrial processes, this leads to reverse cogeneration (combined power and heat). This subject was discussed in detail in my previous book [3]. Many other examples of hybrid processes are outlined in the present book.

While cogeneration improves energy efficiency and thereby reduces fuel use and carbon emission, the excess heat needs to be used locally because heat cannot be transported over long distance. This is why cogeneration for large nuclear reactors is not as effective as for small modular nuclear reactors which can be placed near the point of waste heat use. Hybrid processes for desalination and wastewater treatment also require careful thinking of process hybridization to reduce overall process complexity of the hybridized processes. The implementation of hybrid energy systems requires careful understanding of system level effects of hybridization. Some hybridization may reduce local energy consumption but it can have negative effects on overall system. Each hybridization of process thus needs to be examined both from local and global point of views. Often, there can be multiple options for hybridization that need to be critically assessed. This can be challenging in some complex processes.

1.4 OUTLINE OF THE BOOK

The discussions outlined above show the effectiveness of hybrid energy systems to improve process efficiency and replace use of fossil fuels by renewable sources, both resulting in decarbonization of energy industry. This means that going forward we should be wise and choose generation, storage, and transport mechanisms that best serve our needs of (a) low carbon emission which means lesser use of fossil fuels and (b) affordable and equal access of energy to everybody. We need to be mindful of the fact that there are ten sources of energy and do not need to be overly dependent on fossil fuels. We also need to choose most efficient and less energy consuming processes and materials. We need to be less dependent on fossil fuels for raw materials needed for industrial and commercial processes. The theme of the present book is to demonstrate with the examinations of ten major industries that hybrid energy system is a good strategy to achieve many of these objectives. Most importantly, they can help decarbonize major industries in a sustainable manner. Because of their varied nature, they also allow innovations to make industrial and commercial processes most efficient and less energy consuming and allow gradual and deliberate insertions of renewable energy sources in the industrial processes to better balance the use of fossil, nuclear, and renewable energy sources. Modular hybrid energy systems will also allow flexibility to insert innovations in industrial and commercial processes particularly by nanotechnology to further improve energy efficiency.

Currently, energy industry is skewed toward the use of fossil fuels. Going forward, the best strategy to balance the use of energy sources among fossil, nuclear, and renewables is to devise workable hybrid energy system. This is an evolutionary approach since shifting of energy mix from fossil dominated industry will require reuse or elimination of a vast infrastructure created for the use of fossil fuels and building new infrastructure for generation by renewable sources and small-scale modular nuclear reactors along with the storage and grid transport for renewable sources. Hybrid energy system is the best strategy for transitioning to the carbon emission-free society.

The book demonstrates the effective use of hybrid energy systems concept in ten major industries that are largely responsible for carbon dioxide emission. These are coal, oil, and gas, nuclear, building, vehicle (or transportation), manufacturing and industrial processes, computing and portable electronic industry, district heating and cooling industry, water industry, and hydrogen production industry. The book considers nuclear industry because it can play significant role in hybridization with renewable sources and offering hybrid processes (through cogeneration) in serving energy needs of the other industries with less carbon emission. Ten industries are covered in the ten chapters. Each of these industries is vital and growing as population and societal needs for energy increase. They reflect the present and future nature of our society.

It is important to note that while coal industry will suffer from less use of coal in power industry, it is a vital industry for chemical and material production and use of coal in that arena will continue to grow. Same argument applies to oil industry. Gas industry is a less polluter compared to coal and oil and it is also more versatile for generation and usages. Synthetic gas and hydrogen give gas industry more sustainability. Just like coal and oil, gas industry is also vital for chemicals, fuels, and materials. While nuclear industry produces less carbon dioxide emission, in future, the trend appears to be toward small modular nuclear reactors and their use in providing nuclear heat to the local industry. Building, transportation, and industrial (manufacturing) processes produce a lion share of carbon dioxide emission and they will go through radical changes with less use of fossil fuels. Computing (particularly large-scale data gathering) and portable electronics are the fastest growing and large-scale energy consuming industries that will need to rely on hybrid energy systems for energy source and energy harvesting. Currently energy required for the data storage is largely supplied by fossil fuels. Water industry will take special meaning in the 21st century as water shortage is perceived as a potential problem. Desalination of water is high energy consuming industry and at present heavily depends on fossil energy. District heating and cooling, which is widely used in Europe and North America, is capturing more attention throughout the world. Hybrid energy systems are well suited for this industry. Finally, the production of hydrogen is currently carried out using fossil fuels. Going forward this industry needs to be decarbonized with the use of biomass, electrolysis, and renewable and nuclear sources of energy. With the help of applications of various types of hybrid energy systems to these ten industries, this book makes a point that hybrid energy system is a very workable and practical strategy for the decarbonization of the industrial world.

REFERENCES

1. Energy Analysis, Data and Reports | Department of Energy. (2016). Department of Energy, Washington, DC. www.energy.gov/eere/energy-analysis-data-, yearly reports.
2. EIA reports on "Today in Energy". (2016). U.S. Energy Information Administration, Washington, DC, yearly reports.
3. Shah, Y.T. (2021). *Hybrid Power-Generation, Storage and Grids.* CRC Press, New York.
4. National Academies of Sciences, Engineering, and Medicine. (2019). *Gaseous Carbon Waste Streams Utilization: Status and Research Needs.* The National Academies Press, Washington, DC, Doi: 10.17226/25232.
5. National Academies of Sciences, Engineering, and Medicine. (2019). *Gaseous Carbon Waste Streams Utilization: Status and Research Needs.* The National Academies Press, Washington, DC, Doi: 10.17226/25232.
6. Lee, A., Zinaman, O. and Logan, J. Opportunities for synergy between natural gas and renewable energy in the electric power and transportation sectors *National Renewable Energy Laboratory* Technical Report NREL/TP-6A50-56324 December 2012 Contract No. DE-AC36-08GO28308, Golden, CO.
7. Fulton, M. and Melquist, N. (2010). *Natural Gas and Renewables: A Secure Low Carbon Future Energy Plan for the United States.* DB Climate Change Advisors, New York; Fulton, M. and Melquist, N. (2011). *Natural Gas and Renewables: The Coal to Gas and Renewables Switch is on!* DB Climate Change Advisors, New York.
8. Hybrid Wind and Solar Electric Systems. (2015). Department of Energy, a website report, Department of Energy, Washington, DC.
9. Solar Hybrid Power System. (2020). Wikipedia, *The free Encyclopedia*, last visited 23 January 2020.
10. Cogeneration. (2020). Wikipedia, *The Free Encyclopedia*, last visited 6 September 2020.

2 Hybrid Energy Systems for Building Industry

2.1 INTRODUCTION

As shown in my earlier book [1], the building industry has extensively used the modular approach to cut cost, improve efficiency and quality, and reduce the time for construction. Buildings also account for approximately 40% of the worldwide annual energy consumption [2]. The total global energy consumption in 2007 was 495 quadrillion British thermal units (Btu), out of which the buildings sector consumed about 198 quadrillion Btu. According to the Energy Information Agency, worldwide energy consumption is expected to increase 1.4% per year through 2035, implying that buildings will consume 296 quadrillion Btu by 2035 [3].

Fossil fuels meet a majority of world energy needs, and because buildings are a large energy consumer, they are also a major contributor to global carbon emissions and greenhouse gas (GHG) production. It is now largely recognized that addressing energy use in buildings can reduce total fossil fuel consumption and the associated GHG emissions. Benefits such as decreased building operational energy costs have prompted growing interest among policy-makers, the technical community, and the general public in addressing building energy issues and investigating solutions for reducing fossil energy consumption of buildings.

While energy efficiency is being incorporated into new construction, existing buildings account for a majority of the building stock that will be in place in the foreseeable future. In his 2009 presidential address, American Society of Heating, Refrigerating, and Air-Conditioning Engineers (ASHRAE) presidential member Gordon Holness stated that 75%–80% of the buildings that will exist in 2030 already exist today [4]. This statistic suggests that there is an opportunity for reducing the building sector's contribution toward global energy consumption through reduction of energy use in existing buildings. Significant efforts are needed to at least reduce fossil energy consumption in building industry.

Reducing existing building energy consumption (by fossil energy) consists of two synergistic approaches: (a) to reduce the need for energy through implementation of energy efficiency or conservation measures, and (b) to offset the remaining building energy needs through use of renewable energy systems. It is important to note that building energy efficiency measures should be considered first as the cost to invest in efficiency (and conservation) measures is approximately half the cost of installing renewable energy generating capacity equal to what the efficiency measures offset [5]. It is advised that all energy efficiency opportunities are explored and as many are implemented as is feasible before or in conjunction with renewable energy projects for existing buildings. The subject of energy efficiency and conservation was discussed in detail in my previous book [1]. Numerous efforts have been made in

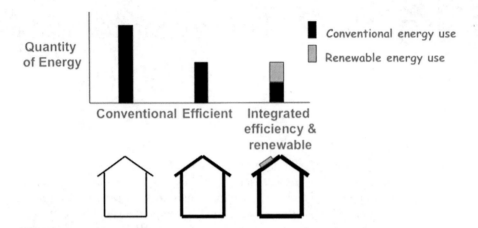

FIGURE 2.1 Demonstration of how combining energy efficiency and renewable energy strategies significantly reduce total building conventional energy use. (National Renewable Energy Laboratory [6].)

the literature [6,7] to examine how renewable energy (largely solar and geothermal) can replace the use of fossil energy to provide the energy needs of the buildings. If possible, wind and water can also be used to provide power for buildings. Often microgrids (or nanogrids) are used to supply the energy needs of buildings or building complexes. These efforts are largely modular in nature. The effectiveness of efficiency measure and use of renewable sources on energy consumption of building is illustrated in Figure 2.1.

2.1.1 Concept of Zero-Energy Buildings

National and local policy is being implemented in both developed and developing countries that require greater amounts of energy to come from renewable energy resources. For example, the 2009/28/EC Renewable Energy Sources (RES) Directive requires that 20% of energy produced within the European Union is from renewable energy systems by 2020 compared to 2010 [8]. Also, the 2002/91/ED Energy Conservation in Buildings Directive requires building energy labeling and sets standards for energy performance, including application of renewable energy resources [9]. As policies such as these are enacted, incentives for installing renewable energy systems are also being developed and regulatory barriers are being removed. The use of renewable energy systems for meeting building energy needs is also becoming a means for demonstrating leadership in environmental sustainability and resource conservation, increasing the reliability of on-site electrical and thermal energy supplies, addressing energy security issues, and other benefits. These actions are encouraging those who are making decisions regarding existing building retrofit projects to seek out ways to use renewable energy systems to meet sustainable building goals. In addition, these actions encourage those who are paying for the energy use associated with these buildings to explore using renewable energy systems as a means to reduce utility costs and, in many cases, the building's carbon footprint.

During the last century, buildings, both residential and commercial, were electrified by the power generated from central sources which largely used fossil fuels and nuclear energy. The heating and cooling needs of the buildings were also provided largely by fossil fuels. This led to the contribution of carbon dioxide emission by buildings to the extent of 25%–30% of the total emission. With the greater emphasis on the reduction of carbon emission to the atmosphere, toward the end of the last century, the use of renewable sources to curtail the use of fossil fuels for buildings was contemplated. This also led to the concept of zero-energy buildings mainly in Europe and North America.

In concept, an net zero energy building (NZEB) is a building with greatly reduced operational energy needs. In such a building, efficiency gains have been made such that the balance of the energy needs can be offset by renewable technologies. Torcellini et al. [10] developed an NZEB definition system to improve the understanding of what zero energy means. They developed four documented definitions—net-zero site energy, net-zero source energy, net-zero energy costs, and net-zero energy emissions. Each NZEB definition, corresponding to a different energy use accounting method, has merits as a zero-energy design goal; however, there is no single best accounting method. The NZEB definition used should align with the owner's goals for the project. NZEB classifications from NZEB:A to NZEB:D are proposed based on the RE type and location with respect to a building by NREL. This classification system recognizes that there are many possible RE supply options, depending on the site constraints and locally available renewable options. Because design goals are so important to achieving high-performance buildings, the way an NZEB goal is defined is crucial to understanding the combination of applicable efficiency measures and RE supply options. This NZEB accounting and classification system is applicable to owners developing design goals, to architects and engineers tracking the modeled performance of the design, and to operators measuring the energy use.

2.1.2 GRID CONNECTION

Before discussing the RE supply options available to NZEBs, we must look at the issue of grid connection. Conceptually, an NZEB produces as much as or more energy than it uses annually and exports excess RE generation to the utility (electricity grid, district hot water system, or other central energy distribution system) to offset the energy used. For NZEBs, a utility connection is allowed for energy balances. A grid-connected NZEB uses traditional energy sources such as electricity and natural gas utilities when on-site generation from RE does not meet the loads. When the on-site generation exceeds the building's loads, excess energy is exported to the utility. By using the utility to account for the energy balance, excess production can offset later energy use. We assume that excess on-site generation can always be sent to the grid to be fully used. However, in high market penetration scenarios, the grid may not always need this energy. In this scenario (and depending on the electricity utility), on-site energy storage would become necessary to maintain the zero-energy status of the building. Off-grid NZEBs are also possible under this classification system; however, they typically require additional on-site generation capabilities combined with significant energy storage technologies. Backup energy sources for off-grid NZEBs would also have to be supplied with RE fuels under this classification system.

2.1.3 FUEL SWITCHING

The NZEB definitions and classifications enable renewable electricity generation to offset various fossil fuel energy uses. For example, natural gas energy use can be offset with excess photovoltaic (PV) or wind energy exported to the grid; the offset level is determined by the energy use accounting method. A site energy accounting allows for a 1-to-1 offset between fuels; source energy can place an additional offset (approximately 3-to-1) for renewable electricity exported to the grid. These source-to-site ratios are well documented and vary depending on the utility mix [11].

2.1.4 RENEWABLE ENERGY CREDITS

Many RE projects are partially financed through the sales of renewable energy credits (RECs). Although this is an important financial tool, once the RECs are sold and then purchased by someone else, the project cannot claim the benefits of the RE produced on-site for the purposes of NZEB classification. In some utility purchase models, the RECs are not resold; rather, they are retired and used by the utility to meet a renewable portfolio standard obligation. In these examples—where a project does not own or retain its RECS—the project's RE is not available as an RE option in the NZEB context. This is to avoid double-counting the RECs. If the project buys back an equivalent amount of certified RECs, such as Xcel's Windsource [12] or Green-E certified renewables [13], the project's on-site RE is available for credit toward an NZEB position for as long as the project purchases equivalent RECs.

2.1.5 ENERGY SUPPLY OPTIONS AND PRIORITIES

Various supply-side RE generation technologies are available for NZEBs [10]. Typical examples of these technologies include PV, solar hot water connected to a district hot water system, wind, hydroelectricity, and biofuels. Demand-side RE and efficiency measures include strategies that save energy but typically are not commoditized. These cannot be included in the supply-side balance for achieving an NZEB. Typical examples of demand-side RE and energy efficiency strategies include passive solar heating, daylighting, solar ventilation air preheaters, and domestic solar water heaters. Guiding principles for RE in NZEBs were developed to minimize the energy transfers from generation source to end use and provide long-term maintainability in the built environment. These include as follows: (a) minimize the overall environmental impact by encouraging energy-efficient building designs, using emissions-free RE, and reducing transportation, transmission, and conversion losses; (b) will be available over the lifetime of the building; (c) are highly scalable, widely available, and have high replication potential for future NZEBs.

NREL developed a hierarchy of RE sources in the NZEB context to reflect these principles. Table 2.1 shows this ranking in order for preferred application. This hierarchy is weighted toward RE technologies that are available within the building

TABLE 2.1

NZEB RE Supply Option Hierarchy [7]

Option Number	NZEB Supply-Side Options	Examples
0	Reduce site energy use through energy efficiency and demand-side renewable building technologies.	Daylighting; insulation; passive solar heating; high-efficiency heating, ventilation, and air-conditioning equipment; natural ventilation, evaporative cooling; ground-source heat pumps; ocean water cooling
	On-Site Supply Options	
1	Use RE sources available within the building footprint and connected to its electricity or hot/chilled water distribution system.	PV, solar hot water, and wind located on the building
2	Use RE sources available at the building site and connected to its electricity or hot/chilled water distribution system.	PV, solar hot water, low-impact hydro, and wind located on parking lots or adjacent open space, but not physically mounted on the building
	Off-Site Supply Options	
3	Use RE sources available off site to generate energy on site and connected to the building's electricity or hot/chilled water distribution system.	Biomass, wood pellets, ethanol, or biodiesel that can be imported from off site, or collected from waste streams from on-site processes that can be used on site to generate electricity and heat
4	Purchase recently added off-site RE sources, as certified from Green-E (2009) or other equivalent REC programs. Continue to purchase the generation from this new resource to maintain NZEB status.	Utility-based wind, PV, emissions credits, or other "green" purchasing options. All off-site purchases must be certified as recently added RE. A building could also negotiate with its power provider to install dedicated wind turbines or PV panels at a site with good solar or wind resources off site. In this approach, the building might own the hardware and receive credits for the power. The power company or a contractor would maintain the hardware.

footprint and at the site. Rooftop PV and solar water heating are the most applicable supply-side technologies for widespread application of NZEBs. Other supply-side technologies such as parking lot-based wind or PV systems may be available for limited applications. A good NZEB should first encourage energy efficiency and then use RE sources that are available within the building footprint—the definitions and classification system support this. However, some high energy use building types such as hospitals and grocery stores cannot realistically apply efficiency measures to the point where they can offset all energy use with a combination of on-building and on-site renewables. Therefore, this NZEB classification system includes off-site supply options so all possible building types can potentially reach an NZEB position.

This hierarchy was initially discussed by Torcellini et al. [10]. Since then it is expanded to a formal classification system, added detail to the RE source hierarchy, and modified it to accommodate specific site details and energy uses. After examining several NZEBs and how people use the definitions, NREL added a discussion to address RE use for campus and neighborhood environments. Table 2.1 provides a brief overview of each energy supply option and is followed by a discussion of each.

Option 0 states that a building must reduce site energy use through demand-side RE and energy efficiency technologies. This includes Passivhaus construction of the house along with all the measures taken such as LED lights, proper insulation, etc., to reduce energy consumption and improve the efficiency. Option 0 is considered a prerequisite and is an essential and fundamental quality of NZEBs. This subject is discussed in great detail in my previous book [1]. Options 3 and 4 are not in control of building owner and they depend more on the extent to which renewable hybrid sources are used by utility owners to generate electricity and heating and cooling needs. The issue of insertion of renewable sources for power generation in utility grid is discussed in my previous book [14], and it will not be repeated here. This issue is, however, important and as pointed out in my previous book [14] significant progress has been made to insert hybrid renewable sources in utility grid. Two important changes in this regard that are made are to insert energy storage technologies (largely batteries) in the grid and development of hybrid microgrids to harness distributed renewable sources by medium- and low-voltage grid which can be connected to utility grid. These changes are described in great detail in my previous book [14]. These changes are meant for decarbonization of energy generation and transport.

The major emphasis and efforts in recent years toward zero-energy buildings are focused toward implementation of largely option 1 and to some extent option 2 described above. For buildings in rural and remote environments that are not connected to the grid, in the past electricity and heating and cooling needs were largely satisfied by diesel or gas generators and oil or gas heating and cooling. These methods produced significant amount of carbon dioxide emission. In recent years, with the help of off-grid systems such as minigrids, nanogrids, or stand-alone systems, and with the use of available renewable sources like solar, wind, geothermal, hydro, and biofuel along with the use of storage devices (like batteries and thermal storage), efforts are made to decarbonize the energy needs. Most off-grid systems have been hybrid and have sometimes used diesel generator as a backup device. Various options of off-grid hybrid renewable energy systems examined in the literature for this purpose are described in detail in my previous books [14,15].

For grid-connected buildings, before we examine types of hybrid systems investigated globally, we first briefly examine changes made in the utility grid to accommodate the insertion of renewable sources within the building (or in proximity owned by the building owner). As shown in Figure 2.2, this made new grid smarter and different than the conventional grid which allowed energy consumer to be prosumer so that energy can flow two ways where prosumer can send excess energy generated in the

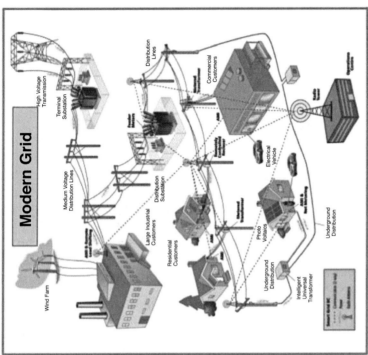

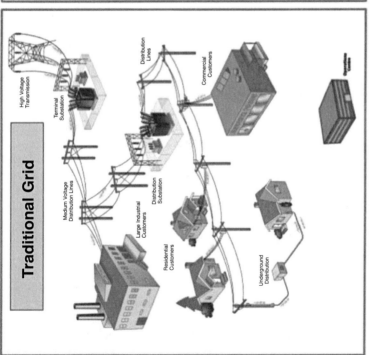

FIGURE 2.2 Changes in grid structure to accommodate prosumer customers [16].

Source: IEEE, Electric Power Grid Modernization Trends, Challenges, and Opportunities, 2017.

building back to the utility owner. For the purpose of this discussion, we use hybrid PV-Wind renewable energy system (which is most used) as an example [16–19].

2.2 CUSTOMER AUTOMATION AND ENERGY MANAGEMENT SYSTEMS

With rising AMI (advanced metering infrastructure) deployments, more than a quarter of U.S. customers now have daily access to their digital energy usage information through their energy providers using mobile apps and web interfaces. Customers can use this information to make smarter energy decisions over time, but a real sea change is coming from advanced control technologies that allow customers to automate changes to their energy use in response to price signals or other inputs. The set-it-and-forget-it nature of smart thermostats makes it easier for customers to participate in demand response and dynamic pricing. Smart home devices, such as NEST, Amazon's Echo, and Google's Home, go even further, allowing users to connect and automate a growing number of technologies (such as lights, thermostats, security cameras, and door locks) with a single device.

NEST, for example, links multiple smart home devices on a network and can activate lights when security cameras detect motion, or turn down lights to save energy when the house is empty. NEST also connects to the utility's metering system through Wi-Fi to respond to time-of-use signals and adjust electric-based heating and cooling systems during peak periods. Amazon's Echo uses voice recognition to check the news, play music, search the web, or purchase services through connected businesses. Capabilities can be expanded by downloading "skills" in the Alexa app from other businesses—including some energy utilities. TXU Energy, a Texas-based energy provider, launched two new Alexa skills in November 2017: one that allows customers to see and manage their account and one that allows them to control their TXU. Of about 117 million U.S. homes in 2016, about 17 million had some type of smart home device. By 2020, 40 million smart thermostats are expected in U.S. homes with 50 million smart light bulbs and 12 million smart water-leak detectors. As shown in Figure 2.3, sales of connected home technologies grew almost 1500% from 2012 to 2017, and this explosive growth is slated to continue as competition increases and vendors expand how devices interact with other businesses and services. This trend will also lead to projected growth in energy management system revenue by fourfold between 2015 and 2025 as depicted in Figure 2.4.

Though connected home technologies are increasing in availability and popularity, none are yet fully "plug and play" to easily enable energy savings or load shifting. Integrating devices such as Amazon Echo and Google Home into home energy systems can be cumbersome and often requires the purchase of additional appliances to fully realize potential cost savings. Nevertheless, the technology maturity (e.g., user interface, controls) is beginning to come together with declining costs to enable future widespread use.

Business owners are also beginning to adopt building energy management systems that allow more precise control and automatic settings to drive down energy use and costs. Since 2011, the market for building energy management systems has grown from $737 million to more than $1 billion and as shown in Figure 2.5, it is

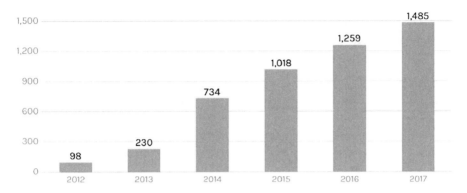

FIGURE 2.3   Connected home technology sales, millions [16].

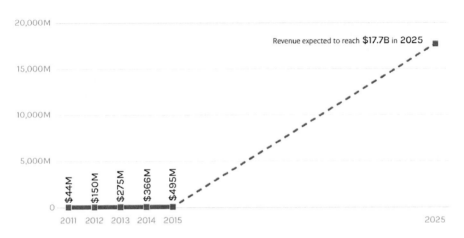

FIGURE 2.4   Growth in home energy management system revenue [16]. (Advanced Energy Now, *2017 Market Report*, prepared by Navigant Research, 2017.)

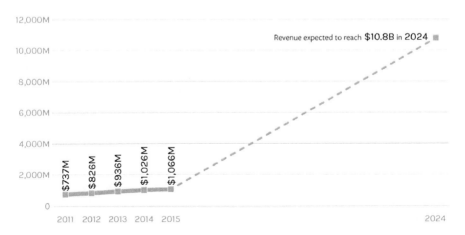

FIGURE 2.5   Growth in building energy management system revenue [16]. (Advanced Energy Now, *2017 Market Report*, prepared by Navigant Research, 2017.)

expected to grow faster—more than ten-fold—in the next decade. These systems employ control technologies to manage appliances, HVAC, and lighting systems, automatically turning them on and off to optimize efficiency or respond to load conditions or pricing forecasts.

Building and home energy management capabilities must be designed to seamlessly coordinate with grid management systems to be effective. Integration will require utilities to foster true bi-directional communication networks that can send and receive price signals, commands, and other data in standard, interoperable formats. Customer privacy and data protection is a growing consideration, as both smart meters and customer-based technologies allow utilities and third-party service providers to collect and analyze vast amounts of energy usage data.

2.2.1 DYNAMIC PRICING AND DEMAND RESPONSE

The combination of AMI and smart customer devices enables customers to effortlessly change their energy use in response to dynamic rates. Time-based rates, or dynamic pricing, include a variety of options for utilities to charge higher rates during peak hours or critical events, and lower rates during off-peak hours. While the electricity industry has been exploring time-based rate options for decades, smart grid technologies make it possible to use dynamic pricing to incentivize significant shifts in customer load during peak periods. Customers can use smart devices to automatically reduce their energy use during peak hours to save money. Direct load control devices—installed in energy-intensive appliances like air conditioners and water heaters—also allow utilities to temporarily turn appliances off during peak periods, often in return for bill credits. Only 5% of U.S. customers participate in dynamic pricing today, as the enabling technologies are being put into place and utilities test program designs.

Recent studies show significant promise. Twenty six utilities who tested various rate programs with more than 400,000 customers under their SGIG projects found that customers reduced their peak demand by up to 23.5%. Peak demand reductions can help utilities defer capital investments in peaking power plants. The transformation to a more distributed future is not happening consistently across the country, but rather is occurring in a patchwork manner driven by favorable policies [16].

Widespread distributed energy resource (DER) penetration implies that a grid control system will have to handle thousands or millions of endpoints. The fundamental control problem is to manage bulk energy system resources (e.g., power generation resources) and dispatchable DERs in a way that will not compromise grid operating requirements, e.g., observing constraints on system frequency, voltages, and the operating limits of grid components. This issue becomes more pronounced as we increase the number and types of DERs. In addition, the integration of DERs that are not owned by utilities further complicates the problem shifting it from one of a direct control to a combination of control and coordination. Coordination is the process that causes or enables a set of decentralized elements to cooperate to solve a common problem, thus becoming a *distributed system*. As shown in Figure 2.6, the various elements of the system will need to coordinate their activities in a way that does not jeopardize the overall reliability and safety of the grid. Coordination

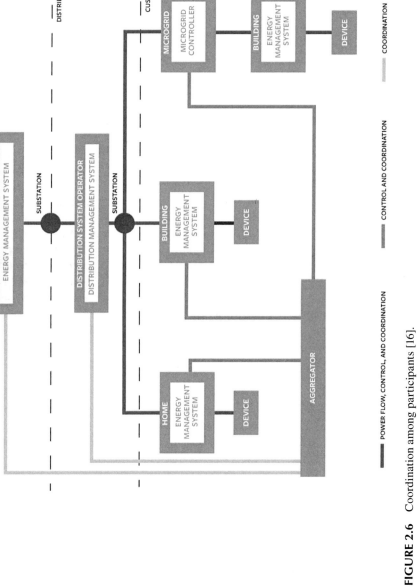

FIGURE 2.6 Coordination among participants [16].

frameworks are required to set the rules governing the interrelationships among the elements (e.g., grid devices and understanding how the various participants and grid devices coordinate will be important for ensuring optimal performance participants) and to enable optimization among them. During normal operations, but also how they would behave during abnormal operations, such as during an unanticipated outage. In this case, a coordination and control strategy will be needed where a microgrid may wish to isolate itself from the rest of the grid or provide ancillary services needed to maintain system operations.

2.2.2 PROCESS FOR RENEWABLE ENERGY BUILDING CONNECTION TO THE ELECTRICAL GRID

Building-sited renewable energy systems generating electricity and connected to the electrical utility grid (known as "grid-tied" systems) are often known as distributed generation (DG) systems. It is important to consider all issues of connecting the DG system to the grid when such systems are part of a building project. Project leads should communicate with the utility on topics such as technical design requirements, which the utility may mandate for interconnection and net metering agreements, including rates the utility will pay for excess renewable energy power generations supplied back to the grid.

In the past, most homes with solar electric systems were not connected to the local utility grid. It made sense to install solar electric systems in areas without easy access to the power grid, where the option of extending a power line from the grid might cost tens of thousands of dollars. In recent years, however, the number of solar-powered homes connected to the local utility grid has increased dramatically. These "grid-connected" buildings, like the one shown in Figure 2.7, have solar electric panels or "modules" that provide some or even most of their power, while still being connected to the local utility. Owners of grid-connected homes can choose to supply a portion of their energy with solar energy, using the utility for power during the night or on cloudy days. Because of the up-front costs of installing a solar electric system, many of these home owners initially install systems that meet about one-quarter to one-half of their energy use.

Net metering allows for the flow of electricity from a grid-connected DG system both to and from the customer typically through a single, bi-directional meter. When a customer's generation exceeds usage, electricity flows back onto the grid. In this way, customer becomes prosumer instead of just consumer. This effectively offsets electricity consumed by the customer at a different time during the same billing cycle or is carried over as a credit on future billing cycles. Net metering policies vary widely. Some net metering programs reimburse customers for excess generation at the wholesale rate, while others reimburse at the retail value. Some policies specify a limit on the capacity of renewable energy systems that can participate in the net metering program.

Interconnection standards specify the technical and procedural process by which a customer connects a renewable energy system to the grid. Such standards include the technical and contractual arrangements by which system owners and utilities must abide. Utilities can be reluctant to allow interconnection of DG systems. The reasons for this are often associated with concerns over ensuring high-quality, reliable power

FIGURE 2.7   Illustration of a home using renewable solar energy [20]. (Photo courtesy of Solar Design Associates, Inc.)

to all customers, load management when considering the intermittency of renewable energy power generation, safety of those maintaining the utility distribution systems, and other similar issues. Even with the increased number of DG systems being added to grid systems, many utilities still have limited experience with these systems. As a result, these utilities address the interconnection questions on a case-by-case basis, which can result in a significant amount of time needed to develop interconnection agreements.

For those pursuing grid-connected renewable energy systems for building projects, they should meet with utility representatives to find out all incentives, work with reputable installer, size the DG system such that there is no back feeding onto the grid, ensure anti-islanding capabilities, undertake engineering studies and negotiate who pays for utility line upgrades, and submit the interconnection proposal early.

When connecting a home energy system to the electric grid, research and consider equipment required as well as your power provider's requirements and agreements. While renewable energy systems are capable of powering houses and small businesses without any connection to the electricity grid, many people prefer the advantages that grid-connection offers. Some of the things you need to know when thinking about connecting your home energy system to the electric grid include (a) equipment required to connect your system to the grid, (b) grid-connection requirements from your power provider, and (c) state and community codes and requirements. Aside from the major small renewable energy system components, additional equipment (called "balance-of-system") such as power conditioning equipment, safety equipment, and meters and instrumentation are required to safely transmit electricity to the desired loads and comply with the power provider's grid-connection requirements [20].

There are also grid connections requirements for renewable energy systems and guidelines (which vary from state to state) for safety and matching of voltage and frequency of the electricity through the grid that needs to be followed. There are also fees and charges and liability insurance requirements. All these need to be handled by home owner with proper dialog with utility provider and appropriate state officials. More details on the process are provided by DOE reports [20–23].

2.3 ROLE OF HYBRID ENERGY SYSTEMS IN NET ZERO-ENERGY BUILDINGS

The 2009/28/EC Renewable Energy Sources (RES) Directive requires that 20% of energy produced within the European Union is from renewable energy systems by 2020 compared to 2010 [8]. Also, the 2002/91/ED Energy Conservation in Buildings Directive requires building energy labeling and sets standards for energy performance, including application of renewable energy resources [9]. As policies such as these are enacted, incentives for installing renewable energy systems are also being developed and regulatory barriers are being removed. The use of renewable energy systems for meeting building energy needs is also becoming a means for demonstrating leadership in environmental sustainability and resource conservation, increasing the reliability of on-site electrical and thermal energy supplies, addressing energy security issues, and other benefits. These actions are encouraging those who are making decisions regarding existing building retrofit projects to seek out ways to use renewable energy systems to meet sustainable building goals. In addition, these actions encourage those who are paying for the energy use associated with these buildings to explore using renewable energy systems as a means to reduce utility costs and, in many cases, the building's carbon footprint.

Primary energy use in buildings accounts for almost 40% of the total energy consumption in the EU40. In U.S. also, building accounts for 30%–40% of total energy consumption. In residential buildings, approximately 80% of the energy used is required for space heating and cooling and sanitary hot water. A significant number and variety of energy supply technologies can be integrated into the built environment. Many of these technologies can be combined in highly efficient hybrid heating (and/or cooling) systems. Hybrid systems, and in particular systems using two or more renewable energy sources, have a huge potential to reduce CO_2 emissions in the building sector through a wide range of applications, depending on the technology chosen, its overall efficiency, and the avoided environmental impact of the relevant fossil fuel alternative. Until recently, the most common hybrid application was the combination of a fossil fuel burner (mainly gas or oil) and solar thermal collectors. Small-scale systems using a combination of two renewable energy sources have gained market share in recent years. The main examples of hybrid renewable energy systems are as follows:

1. Electrically driven heat pumps and solar PV.
2. Electrically driven heat pumps and solar thermal.
3. Thermally driven heat pumps in combination with solar thermal.

One major challenge of hybrid systems is to maximize the combined efficiency of the energy sources employed and at the same time to minimize the operating cost and the environmental impact. This can only be achieved if the system as a whole is considered and not just its various components in isolation. The trade-off between system performance and cost (both related to complexity) is a key thing to understand about hybrid system technology. Improving the relative performance of the individual components is necessary to achieve highly efficient hybrid systems; however, it is not sufficient. To successfully implement scientific research and technological development in small-scale hybrid systems, it is highly important to establish close links with the building and construction industry. In the near future, the built environment needs to be designed, built, and renovated with a clear vision of integrating multiple RES and energy efficiency measures [3,24]. The residential and commercial buildings in U.S. and Europe can be divided into three categories: (a) single-family houses, (b) multifamily houses, and (c) high-rise buildings, which are defined as buildings that are higher than eight stories. The statistics shows that hybrid systems are replaced at a rate which is faster than the rate housing stock is renewed or buildings are refurbished. The systems in about 5% of buildings are replaced each year. A similar phenomenon is observed in nonresidential buildings with the replacement rate depending on the type of building.

Non-residential buildings are more heterogeneous group, used for a great variety of functions, each with different energy demands per unit and each built according to different standards. One peculiar characteristic of this typology of buildings is the higher cooling loads in comparison to housing, which is due to specific appliances (such as computers) and on average higher comfort requirements. The energy performance of each building defines the technical options that can be used to achieve comfortable temperature levels [1]. In order to make hybrid systems more attractive, more efforts should be focused on cost, thermal efficiency, simplification of process implementation, and making process more user-friendly. More efforts should also be put on process prefabrication and integration, automation and control, development of new standards and procedures, all these to make integration of building components in hybrid systems more easy and making hybrid systems 100% renewables [1].

Renewable energy resources commonly used for building applications include solar, wind, geothermal, and biomass. The use of biomass is somewhat indirect in the sense that it is used in the form of biofuels and it is not as prevalent as other sources and we will not evaluate here. Here we consider the following five hybrid sources of energy that are evaluated in the literature and often implemented in practice. These hybrid sources give good picture of the role of hybrid energy in grid-connected building industry. It should, however, be noted that the use of hybrid energy is not always restricted to these options.

1. Solar thermal, including solar hot water (domestic water heating and space heating), and solar ventilation air preheating with thermal storage
2. Solar electric, or PV, systems with battery storage
3. Solar PV/solar thermal hybrid with or without storage or solar heat pump
4. Solar PV/solar thermal hybrid with geothermal heat pump
5. Solar/wind with or without battery storage

2.4   SOLAR THERMAL WITH STORAGE

There are numerous ways hybrid solar thermal energy with storage is used to strive toward zero-energy or zero-carbon buildings. One example is UK's 2008 effort to create zero-carbon emission school building with hybrid solar heating and Passivhaus construction. This type of construction does not allow any waste of energy and requires certain targets to qualify for certification. The standard requires primary energy demand targets for space heating, domestic hot water, lighting, fans, and pumps as well as all the projected appliances along with temperature requirements [1,25]. The concept was first introduced to minimize energy consumption and then use local renewable sources to build zero-carbon school buildings. The Montgomery School was awarded a Quality-Approved Passivhaus certificate in February 2012, essentially for "the comfort and quality of the internal environment with extremely low energy consumption." It was the first Passivhaus school in the United Kingdom. With proper considerations of all aspects of Passivhaus design and the use of solar thermal energy, the school was successful in achieving its mission. More details on the design of this school are outlined in my previous book [1]. Similar results were obtained by Paya-Marin et al. [26] in life cycle analysis of hybrid active and passive solar heating of another school building.

Ceranic et al. [27] pursued a case study which investigated potential of its novel "thermal capacity on demand" energy performance approach. It combined a modular thermal storage solution capable of balancing heating demand and supply for a low-rise, low-mass superstructure with renewable technologies and the level of backup power/services needed [1,28–32]. The study was applied to a smart POD which provided an alternative to traditional classroom planning. In this study, low-temperature diurnal sensible heat storage was used [1,28,29], with a loosely packed rock bed as a medium. Faster response rate, lower temperature, lower energy losses, and lower risk of boiling/freezing and leakage make this option an economical alternative to seasonal storage [1,30]. Furthermore, the medium can often be sourced by recycling the existing waste on the site, giving it an added environmental benefit. In the compact site conditions, the size and thermal performance of diurnal and seasonal store envelope can often be restricted by the available storage space. Hence, to charge the stores to a required temperature level, heat is often added by the heat pumps [1,27–32]. Diurnal stores can provide a significant "load shifting capability" and reduced energy losses, but the required storage volumes are large for the small-to-medium size building typologies and will only be fully resolved by improving the effectiveness and reducing the costs of latent or thermochemical heat storage systems [1,27–32]. Based on the model calculations, the study showed that this storage mechanism can be effective in maintaining the temperature of smart POD at comfortable level all year around. The proposed POD comprises teaching area, storage facilities, entrance lobby area for coat storage and large area to maximize solar gains. The POD was modular, transportable, reusable, autonomous, and sustainable. More details on this study are given in my previous book [1].

Figure 2.8 illustrates a North Carolina home where passive solar design and solar thermal system supply domestic hot water and a secondary radiant floor heating system. Modular units allow one to meet these requirements simply by adding or

FIGURE 2.8 This North Carolina home gets most of its space heating from the passive solar design, but the solar thermal system supplies both domestic hot water and a secondary radiant floor heating system. (Photo courtesy of Jim Schmid Photography, NREL [1].)

subtracting individual units, rather than replacing the entire system. Many business buildings—whether an office space or simply a commercial space—host multiple residents, who may have vastly different heating and air-conditioning needs. Modular units allow office managers to meet these needs simply by routing the output for different units to different locations. More details on this hybrid solar system are given in my previous book [1].

2.4.1 SOLAR-BOOSTED HEAT PUMP

Hybrid solar thermal systems can also use solar-boosted air-source heat pumps for domestic hot water production. The most common configuration for air-source heat pump includes an air-source evaporator and an immersed-coil or wrap-around condenser. An alternative approach, to use a solar thermal collector to boost the evaporator temperature (and energy input) during cold ambient periods. The heat pump is able to cool the solar absorber, reducing heat losses and increasing its efficiency. This has the advantage of providing substantial heat gains even under marginal solar conditions. It also allows for efficient operation over a larger range of seasons and weather conditions, and for more hours of the day. In addition, the solar energy input to heat pipe's (HP) evaporator increases coefficient of performance (COP) and seasonal performance. Various system configurations have been proposed in the past; however, few have gained significant market share [33,34].

The use of a heat pump to drive the heat transfer from a solar collector to the thermal storage allows the operating temperature in the collector to reduce to near or below ambient temperatures. Combining heat pumps with conventional solar systems also has the potential to produce high energy output from low-cost unglazed solar absorber panels. Results indicate that a solar-assisted heat pump (SAHP) domestic hot water system could outperform a conventional solar hot water system even with only half the normally required solar collector area.

An important issue relates to the use of a SB-HPWH in climates with significant seasonal temperature variations. Unglazed solar thermal panels have limited solar-boosting capability at low temperatures. The use of high-performance solar panels (with glazed and insulated absorbers), however, reduces the unit's "non-solar, air-source" capacity making them undesirable in many climates. Consequently, new approaches to these tradition configurations are being developed based on new system configurations and components. These include dual- and tri-mode solar collectors that act as solar or air-source evaporators and may include PV/thermal (PVT) absorbers. New variable speed, high-efficiency DC compressors may also offer significant advantages for a fully integrated solar/HP hybrid heat pump water heater. Careful integration of these components may produce units with unparalleled performance [33,34].

2.4.2 BUILDING INTEGRATED SOLAR THERMAL TECHNOLOGIES AND THEIR APPLICATIONS

Solar energy has enormous potential to meet the majority of present world energy demand by effective integration with local building components. One of the most promising technologies is building integrated solar thermal (BIST) technology. The study by Zhang et al. [1,35] presents a review of the available literature covering various types of BIST technologies and their applications in terms of structural design and architectural integration.

The review covers detailed description of BIST systems using air, hydraulic (water/heat pipe/refrigerant), and phase-changing materials as the working medium. The fundamental structure of BIST and the various specific structures of available BIST in the literature are described. Design criteria and practical operation conditions of BIST systems are illustrated. The state of pilot projects is also fully depicted. Current barriers and future development opportunities are therefore concluded. Based on the thorough review, it is clear that BIST is very promising with considerable energy-saving prospective and building integration feasibility. This review facilitates the development of solar-driven service for buildings and helps the corresponding saving in fossil fuel consumption and the reduction in carbon emission. Solar energy is one of the most important renewable sources locally available for use in building heating, cooling, hot water supply, and power production. Truly BIST systems can be a potential solution toward the enhanced energy efficiency and reduced operational cost in contemporary built environment.

BIST is defined as the "multifunctional energy facade" that differs from conventional solar panels in that it offers a wide range of solutions in architectural design features (i.e., color, texture, and shape), exceptional applicability and safety in construction, as well as additional energy production. It has flexible functions of buildings' heating/cooling, hot water supply, power generation, and simultaneous improvement of the insulation and overall appearance of buildings. This façade-based BIST technology would boost the energy efficiency of buildings and turn the envelope into an independent energy plant, creating the possibility of solar thermal deployment in high-rise buildings.

The system normally comprises a group of modular BIST collectors that receive solar irradiation and convert it into heat energy, whereas the heating/cooling circuits could be further based on the integration of a heat pump cycle, a package of absorption chiller, a modular thermal storage, and a system controller. In case of some unsatisfied weather conditions, a backup/auxiliary heating system (e.g., boiler) is also integrated to guarantee the normal operation of system. In the typical BIST system, the overall energy source is derived from solar heat, which is completely absorbed by modular BIST collectors. This part of heat is then transferred into the circulated working medium and transported to the preliminary heat storage unit, within which heat transfer between the heat pump refrigerant and the circulating working medium occurs. This interaction decreases the temperature of the circulating medium, which enables the circulating medium absorbing heat in the facades for next circumstance.

Meanwhile in the heat pump cycle (compressor–condenser expansion valve–evaporator), the liquid refrigerant vaporizes in the heat exchanger, which, driven by the compressor, subsequently converts into higher temperature and pressure, supersaturated vapor, and further releases heat energy into the tank water via the coil exchanger (condenser of the heat pump cycle), increasing the temperature of the tank water. Further, the heat transfer process within the coil exchanger results in the condensation of the supersaturated vapor, which gets downgraded into lower temperature and pressure liquid refrigerant after passing through the expansion valve. This refrigerant then undergoes the evaporation process within the heat exchanger in the initial heat storage again, thus completing the heat pump operation. When the water temperature in the tank accumulates to a certain level, i.e., 45°C, water can be directly supplied for utilization or under-floor heating system. For the cooling purpose, the system is coupled with an additional appliance of absorption chillers [1,35].

2.5 SOLAR ELECTRIC PV WITH STORAGE

PV arrays convert sunlight to electricity. As shown in Figure 2.9, systems are made up of modules assembled into arrays that can be mounted on or near a building or other structure. A power inverter converts the direct current (DC) generated by the system into grid-quality alternating current electricity [36,37].

Traditional single crystal solar cells are made from silicon, are usually flat-plate, and are generally the most efficient (the solar cell efficiency is an indicator of how well it converts sunlight to DC electricity). Multicrystal solar cells are a similar technology but slightly less efficient. Thin-film solar cells are made from amorphous silicon or nonsilicon materials such as cadmium telluride. Thin-film solar cells use layers of semiconductor materials only a few micrometers thick.

Building-integrated photovoltaic (BiPV) products may be appropriately suited for applications on existing buildings during major renovations. These technologies can double as rooftop shingles (single-ply membrane, standing seam metal roofs, and others.) and tiles, building facades, or the glazing for skylights [38]. Figure 2.10 shows an example of this technology integrated into shingles. In some cases, BiPV

FIGURE 2.9   The Williams Building in downtown Boston, Massachusetts. 372 modules were installed for a total system capacity of 31 kW. (Photo from SunPower, NREL/PIX 08466 [6].)

FIGURE 2.10   Thin-film solar PV shingles [6]. (Photo from United Solar Ovonic, NREL/PIX 13572.)

can add cost and complexity to a project and may not be universally available, but may help enhance acceptance of a project on a visible surface.

Most PV systems installed today are in flat-plate configurations, which are typically made from solar cells combined into modules that hold about 40 cells. A typical American home will use about 10–20 solar panels to power the home. Many solar panels combined together to create one system is called a solar array. For large electric utility or industrial applications, hundreds of solar arrays are interconnected to form a large utility-scale PV system [38]. These systems are generally fixed in a single position but can be mounted on structures that tilt toward the sun on a seasonal basis or on structures that roll East to West over the course of a day [39]. Figure 2.11 shows the components of a typical PV system.

Module Efficiencies	Single Crystal	14%–19%
	Multicrystal	13%–17%
	Thin film	6%–11%

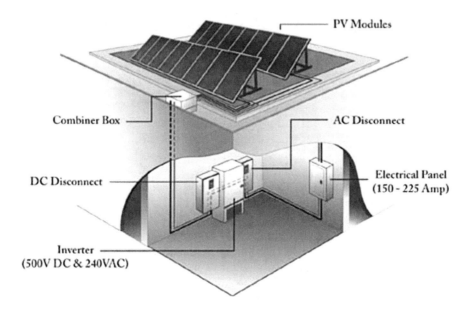

FIGURE 2.11 PV system components [6].

There are typically three scales of solar installations: utility-scale, commercial, and residential.

1. Utility-scale installations are very large arrays located on open lands, and provide power for hundreds or even thousands of homes and businesses.
2. Commercial systems are smaller and may provide power for multiple or single commercial or municipal buildings on campuses, in complexes, neighborhoods, or other special districts. Commercial-scale systems offer potential advantages for locating solar PV. Rather than attempting to find appropriate locations for solar panels on individual structures, a commercial-scale system might be located in a less visible or impactful location, such as above a parking structure or on an open lot. Power can be lost in transmission from these arrays to the end-use location, however; so distances need to be minimized.
3. Residential-scale PV systems produce power for use on a single property. The major challenge with siting solar PV technologies is ensuring appropriate siting for maximum electricity production. An ideal solar installation would be situated in an unshaded, south-facing location with an optimum tilt angle, and would supply electricity to a site where there is a demand for the electricity being produced. Not all sites are suitable for solar technologies, however. The guidelines to determine when solar technologies are appropriate for a site are described by Hayter and Kandt [6].

The work by Bagalini et al. [40] focuses on grid-connected residential PV-battery storage systems, operated with the purpose of maximizing energy self-consumption.

A real system comprising 3 kWp monocrystalline PV modules and 24 kWh advanced lead-acid battery pack (14.4 kWh usable capacity), associated with a grid-connected residential apartment, has been installed at the Green Energy Laboratory of Shanghai Jiao Tong University. The operations of the system have been studied by analyzing experimental data over limited timescales of days, which have also been used to validate a computational model built using the software TRNSYS. The model was used to assess the operations of the system over a full year, giving the possibility to assess its overall energetic performances. Furthermore, the model allowed running additional simulations with different design parameters, such as PV power and battery capacity sizes. An "optimum" configuration of 3.5 kWp of PV and 8 kWh of battery capacity (4.8 kWh usable capacity) has been chosen to carry out experiments to be used in the economic analysis.

With respect to energy performance, it is shown that adding battery energy storage to a domestic PV system associated with an evening-oriented electricity demand would reduce the stress of distributed renewables on the grid by limiting the daily exported power. Additionally, the evening peak demand is also reduced. The reduction in power swings from exporting to importing would lower the difference in power requirements during day and night facilitating balancing operation at higher voltages [32]. From full-year simulations of PV-battery energy storage systems whose size has been optimized according to the load, it can be seen how the amount of self-consumed energy increased from 24% to 79%, the amount of purchased energy decreased by 60%, and the amount of energy sold to the grid decreased by 57% compared to the PV-only scenario.

With respect to economic performance, PV-battery storage systems in the Chinese residential sector are not economically viable in the current context of low electricity tariffs and considerable PV generation incentives which do not take into account storage. However, considering future scenarios of increasing electricity prices and carbon price, decreasing or self-consumption favorable PV generation incentives, and falling technology costs, the economic outlook of PV-battery investments improves. In particular, it is found that doubling the electricity price would make the PV-battery investment profitable. While lowering the export price shows how the PV-only system is less resilient than the PV-battery system. A combined effect of rising electricity prices and falling export prices would reduce the gap in economic performance between those configurations, although it will not be enough to make PV-battery a better investment than PV only. The sudden removal of PV generation incentives in this context would make both investments unworthy. Instead, one of the most effective ways to promote a battery energy storage system in conjunction with PV generation plants is to introduce incentives rewarding only that part of PV-generated electricity that is self-consumed. Another important finding is that the falling battery cost alone is not enough to make PV-battery system preferable over the PV-only case if all other conditions remain the same. This might mean that it is not enough to wait for battery prices to go down if there is the will to push storage significantly. The economic performances of both systems are equally affected in similar measure by increasing carbon price. Finally, three scenarios considering all the above-mentioned effects simultaneously have been tested with the result that for the medium scenario, a combination of doubling electricity tariff, reduction to 25% of current export tariff, PV incentives maintaining the current level but rewarding self-consumption only and technology cost reduction (−66% for battery and −33% for PV and inverter), and increased carbon price (10€/ton CO_2) would result in a good economic return on the PV-battery investments [40] while the PV-only investment has slightly positive NPV. This shows how a combination of naturally evolving conditions, such as rising electricity prices and falling technology costs, and regulator-imposed incentives, such as self-consumption-only PV subsidies, could create conditions for the deployment of more PV but only coupled with battery energy storage.

The changes of energy tariffs and technology costs will lead to different optimum system sizes. The increase in the imported tariff will stimulate the rising of both PV and battery sizes, while the decrease in the exported tariff can only reduce the size of the PV system but has little effect on the battery sizing. The optimum sizes of both the PV and battery increase as the battery cost declines. However, when the battery cost is lower than 73€/kWh, the increase in PV size becomes less obvious. The decrease in PV cost leads to similar trends of the changes of the optimum system sizes as the battery cost declines, despite the fact that the optimum battery size has little variations when the battery cost is lower than 73€/kWh.

The operating strategy of this PV-battery storage system is to maximize self-consumption, hence storing the excess PV power production in the battery, rather than selling it to the grid, in order to use it later when demand cannot be met by solar energy, thus decreasing the amount of energy bought from the grid. Therefore, it is clear in this context that the battery can add a value to a residential PV system, where the demand is hardly matched by PV generation. In the above-described "optimum" configuration, the PV-only system starts from a 21% self-consumption, leaving a large margin for the battery to increase this value, making the PV-battery investment attractive. On the contrary, an office-type load profile, with a daily-only demand, would be less suitable for PV-battery storage application. Therefore, it can be concluded that such system has a better potential for applications in residential buildings.

As far as energy storage is concerned, there are many other operating strategies. In residential PV-battery storage systems, the operation of the battery can be optimized to achieve an economic optimum [41,42], such as lowest electricity bill, when variables such as varying electricity tariff are taken into consideration. Another valuable strategy would be maximizing battery life [43,44] while not compromising too much of the other objectives such as self-consumption. In fact, the battery degradation should be taken into account in more detail in such studies. Despite storage having a great potential in a variety of applications, the authors would like to stress its value in association with renewable energy systems such as PV, either at a residential level or even better at a community level where many PV owners could jointly share the benefits of some energy storage facilities. Overall, self-consumption maximization seems the most natural operating strategy to be followed, and policies, both existing and new, should be pushing toward this direction. Clever operating strategies should be put in place in a smart grid context to optimize the operation of all components while maximizing the overall benefits.

2.6   HYBRID PV/SOLAR THERMAL CONCEPT

Solar cell consisting of its brink photon energy with respect to the particular energy band gap. Below this range electricity conversion is not possible. Photons with higher wavelength range do not generate electron–hole pairs; however, it only dissipates their energy in the form of heat in their cell. Normal PV module converts 4%–17% of the incoming solar radiation into electricity, depending on the type of solar cells in use and the working conditions. This concludes that more than 50% of Solar PV panel which is made from silicon cells exhibit few drawbacks over high solar radiation temperature. As the hour of the day is rising, higher will be the solar radiation. Ultimately, it will tend to increase the surface temperature of the PV cells. Based on research it is observed that, for monocrystalline (C-Si) and polycrystalline (Pc-Si) silicon solar cells, the efficiency decreases by about 0.45% for every degree rise in temperature and for amorphous silicon (a-Si) cells, the effect is less, with a decrease of about 0.25% per degree rise in temperature depending on module design. This undesirable effect can be partially avoided by a proper heat extraction with a fluid (either liquid or air) circulation. Here with a hybrid PVT solar system, the reduction of PV module temperature can be combined with useful fluid heating. Therefore, hybrid PVT systems can simultaneously provide electrical and thermal energy achieving a higher energy conversion rate of the absorbed solar radiation. PVT systems provide a higher energy output than standard PV modules and could be cost effective if the additional cost of the thermal unit is low. Natural or forced air circulation is a simple and low cost method to remove heat from PV modules. However, for air temperature being higher than 20°C, it is not as much effective. Therefore to overcome this effect, the heat may be extracted by circulating low temperature fluid through a heat exchanger that is mounted at the rear surface of PV laminate [45–50].

In the past 3–4 decades, the market of solar thermal and PV electricity generation has been growing rapidly. So were the technological developments in hybrid solar PVT collectors and the associated systems. Generally speaking, a PVT system integrates PV and solar thermal systems for the cogeneration of electrical and thermal power from solar energy. A range of methods are available such as the choices of monocrystalline/polycrystalline/amorphous silicon (c-Si/pc-Si/a-Si) or thin-film solar cells, air/liquid/evaporative collectors, flat-plate/concentrator types, glazed/unglazed designs, natural/forced fluid flow, and stand-alone/building-integrated features. Accordingly, the systems are ranging from PVT air and/or water heating system to hot-water supply through PV-integrated heat pump/pipe or combined heating and cooling and to actively cooled PV concentrator through the use of lens/reflectors. Engineering considerations can be done on the selection of heat removal fluid, the collector type, the balance of system, the thermal to electrical yield ratio, the solar fraction, and so on. These all have determining effects on the system operating mode, working temperature, and energy performance.

Theoretical and experimental studies of PVT were documented as early as in mid-1970s [51–53]. Despite the fact that the technical validity was early concluded, only in recent years that it has gained wide attention. The amount of publications grows rapidly. The following gives an overview of the development of the technology, placing emphasis on the research and development activities in the last decade. Readers may refer to Chow [54] for a better understanding of the early developments.

In the 1990s, the initiative of PVT research was apparently a response to the global environmental deterioration and the growing interest in BiPV designs. Comparing with the separated PV systems, the building integration of PV modules improves the overall performance and durability of the building facade. Nevertheless, building integration may bring the cell temperatures up to 20°C above the normal working temperature [55]. Other than the benefits of cooling, PVT collectors provide aesthetical uniformity than the side-by-side arrays of PV and solar thermal collectors. Alternative cooling schemes of the BiPV systems were examined [56–58]. Hollick [59] assessed the improvement in the system energy efficiency when solar cells were added onto the solar thermal metallic cladding panels on vertical facades.

Concentrator-type (c-PVT) system has demonstrated continued success. Akbarzadeh and Wadowski [60] studied a heat-pipe-based coolant design which is a linear, trough-like system. Luque et al. [61] successfully developed a concentrating array using reflecting optics and one-axis tracking. By that time, facing the conflicting roles of water heating and PV cooling, the design temperature of water that leaves a PVT/w collector is not high. Combining PVT and SAHP technology was then seen as a good alternative. Ito et al. [62] constructed a PVT-SAHP system with pc-Si aluminum roll-bond solar panels.

Generally speaking, in the 20th century, the PVT research works had been mostly focused on improving the cost-performance ratio as compared to the solar thermal and PV systems installed side-by-side. For real-building projects, the PVT/a systems were more readily adopted in Europe and North America, though the higher efficiency of the PVT/w system has been confirmed by that time. Solar houses with PVT/w provision were once sold in Japan in the late 1990s. Unfortunately, such innovative housing was in lack of demand in the commercial market [63]. A summary of the PVT technology in the period, including the marketing potentials, was reported by the Swiss Federal Office [64] and the International Energy Agency (IEA) [65].

2.7   BUILDING-INTEGRATED OPTIONS (BIPVT/A)

In conventional BiPV systems, an air gap is often provided at the rear of the PV arrays for the air cooling of modules by natural convection. The heat recovery from the air stream for a meaningful use constitutes a BiPVT/a system. From a holistic viewpoint, Bazilian and Prasad [66] summarized its potential applications. The multifunctional façade or roof was ideal for PVT integration that produces heat, light, and electricity simultaneously, in addition to the building shelter functionality.

In UK, the Brockstill Environment Centre in Leicester opened in 2001 was equipped with a roof-mounted PVT/a system [67]. To assess the performance of various operational and control modes, a combined simulation approach was adopted with the use of two popular thermal simulation tools: ESP-r and TRNSYS. Monitored actual energy use data of the building show very positive results. Mei et al. [68] studied the dynamic performance of a BiPVT/a collector system constructed in the 1990s at the Mataro Library in Spain. Their TRNSYS model was validated against experimental data from a pc-Si PV facade. The heating and cooling loads for various European buildings with and without such a ventilated facade were then evaluated. The simulation results showed that more winter heating energy can be saved for the use of the preheated ventilation in a building located in Barcelona, but less is for Stuttgart in Germany and Loughborough in UK. The higher latitude locations therefore need a higher percentage of solar air collectors in the combined system. Further, Infield et al. [69] explored different approaches to estimate the thermal performance of BiPVT/a facades, including a design methodology based on an extension of the familiar heat loss and radiation gain factors.

The main difficulty in analyzing BiPVT/a performance lies in the prediction of its thermal behavior. When the temperature profile and the sun shading situation are known, the electrical performance can be readily determined. This is not the case for thermal computation. The estimation of the convective heat-transfer coefficients, for example, is far from direct. The actual processes may involve a mix of forced and natural convection, laminar and turbulent flow, and, simultaneously, the developing flow at the air entrance. The external wind load on the panels further complicates the situation. For a semitransparent facade, thermal energy enters and transmits through the air cavity both directly (for glazing transmission) and indirectly (through convection and radiation exchange). The heat transfer to the ventilating stream is probably most complex, particularly for buoyant flow.

Sandberg and Moshfegh derived analytical expressions for the coolant flow rate, velocity, and temperature rise along the length of the vertical channel behind the PV panels [70]. Their experimental results were well matching the theoretical predictions for constrained flow, but were less accurate for ducts with opened ends. For the latter, Mittelman et al. [71] developed a generalized correlation for the average channel Nusselt number for the combined convective-radiative cooling. Their solution of the governing equations and boundary conditions was computed through CFD analysis. Gan also studied the effect of channel size on the PV performance through CFD analysis [72]. To reduce possible overheating or hot spot formation, the required minimum air gaps were determined. Experimental works on a PVT façade were undertaken by Zogou and Stapountzis [73] for better understanding of the flow and turbulence with natural and forced convection modes. Supported by CFD modeling, the results show that the selection of flow rate and the heat-transfer characteristics of the back sheet are critical.

In Canada, Chen et al. [74,75] introduced a BiPVT/a system to a near net-zero energy solar house in Eastman Quebec. The solar house, built in 2007, featured with ventilated concrete slabs (VCSs). A VCS is a type of forced-air thermoactive building systems in which the concrete slabs exchange thermal energy with the air passage through its internal hollow voids. The BIPVT system is designed to cover one continuous roof surface to enhance aesthetic appeal and water proofing. Outdoor air is drawn by a variable speed fan with supervisory control to achieve the desired supply temperature. On a sunny winter day, the typical air temperature rise was measured 30°C–35°C. The typical thermal efficiency was at least 20% based on the gross roof area. Analysis of the monitored data showed that the VCS was able to accumulate thermal energy during a series of clear sunny days without overheating the slab surface or the living space. Athienitis et al. [76] presents a design concept with transpired collector. This was applied to a full-scale office building demonstration project in Montreal. The experimental prototype was constructed with UTC (open-loop unglazed transpired collector) of which 70% surface area was covered with black-frame PV modules specially designed to enhance solar energy absorption and heat recovery. The system was compared side by side with a UTC of the same area under outdoor sunny conditions with low wind. This project was considered a near-optimal application in an urban location in view of the highly favorable system design. While the thermal efficiency of the UTC system was found higher than the BIPVT/a combined thermal plus electrical efficiency, the equivalent thermal efficiency of the BiPVT/a system (assuming that electricity can be converted to four times as much heat) can be 7%–17% higher.

Pantic et al. [77] compared three different open-loop systems via mathematical models. These include Configuration 1: unglazed BiPVT roof, Configuration 2: unglazed BiPVT roof connected to a glazed solar air collector, and Configuration 3: glazed BiPVT. It was pointed out that air flow in the BIPVT cavity should be selected as a function of desired outlet temperatures and fan energy consumption. Cavity depths, air velocity in the air cavity, and wind speed were found having significant effect on the unglazed BiPVT system energy performance. Development of efficient fan control strategies has been suggested an important step. Configurations 2 and 3 may be utilized to significantly increase thermal efficiency and air outlet temperature. In contrast, Configuration 3 significantly reduces electricity production and may lead to excessive cell temperatures and is thus not recommended unless effective means for heat removal are in place. The unglazed BIPVT system linked to a short vertical solar air collector is suitable for a connection with a rock bed thermal storage.

For warm climate applications, the ventilated BiPV designs are found better than the PVT/a designs with heat recovery. Crawford et al. [78] compared the energy payback time (EPBT) of a conventional c-Si BiPV system in Sydney with two BiPVT/a systems with c-Si and a-Si solar cells, respectively. They found that the EPBT of the above three installations are in the range of 12–16.5 years, 4–9 years, and 6–14 years, respectively. The two BiPVT/a options reduce the EPBT to nearly one-half. Agrawal and Tiwari [79,80] studied a BiPVT/a system on the rooftop of a building, under the cold climatic conditions of India. It is concluded that for a constant mass flow rate of air, the series connected collectors are more suitable for the building fitted with the BIPVT/a system as rooftop. For a constant velocity of air flow, the parallel combination is then the better choice. While the c-Si BiPVT/a systems have higher energy and exergy efficiencies, the a-Si BiPVT systems are the better options from the economic point of view.

Jie et al. [81] studied numerically the energy performance of a ventilated BiPV façade in Hong Kong. It was found that the free airflow gap affects little the electrical performance, but is able to reduce the heat transmission through the PV façade. Yang et al. [82] carried out a similar study based on the weather conditions of three cities in China: Hong Kong, Shanghai, and Beijing. It was found that on typical days the ratio of space cooling load reduction owing to the ventilated PV facade is 33%–52%. Chow et al. [83] investigated the BIPVT/a options of a hotel building in Macau, with the PVT facade associated with a 24-hour air-conditioned room.

The effectiveness of PV cooling by means of natural airflow was investigated with two options: free openings at all sides of the air gap as Case 1 and in Case 2 the enclosed air gap that behaves as a solar chimney for air preheating. These were also compared with the conventional BIPV without ventilation. The ESP-r simulation results showed an insignificant difference in electricity output from the three options. This was caused by a reverse down flow at the air gap at night, owing to the cooling effect of a 24-hour air-conditioned room located behind the PVT facade. It was concluded that both the climate condition and system operating mode affect significantly the PV productivity.

In China, Ji et al. [84] studied theoretically and experimentally the performance of a PV-Trombe wall, which was constructed at an outdoor environmental chamber. This south-facing façade in Hefei was composed of a PV glazing (with pc-Si cells) at the outside and an insulation wall at the inside with top and bottom vent openings. This leaves a natural flow air channel in between for space heating purpose. The results confirmed its dual benefits—improving the room thermal condition (with 5°C–7°C air temperature rise in winter) and generating electricity (with cell efficiency at 10.4% on average).

2.7.1 WORKS ON WINDOW SYSTEMS

In Sweden, a multifunction PVT hybrid solar window was proposed by Fieber et al. [85]. The solar window was composed of thermal absorbers on which PV cells were laminated. The absorbers were building integrated into the inside of a standard window, thus saving frames and glazing and also the construction cost. Reflectors were placed behind the absorbers for reducing the quantity of cells. Via computer simulation, the annual electrical output showed the important role of diffuse radiation, which accounted for about 40% of the total electricity generation. Compared to a flat PV module on vertical wall, this solar window produced about 35% more electrical energy per unit cell area.

Vertical collectors and windows are more energy efficient at high-latitude locations, considering the sun path. Davidsson et al. [86] studied the performance of the above hybrid solar window in Lund, Sweden (55.44°N). Also a full-scale system combining four of these solar windows was constructed in a single family home in Alvkarleo, Sweden (60.57°N). The solar window system was equipped with a PV-driven DC pump. The projected solar altitude is high in summer, and accordingly a large portion of the solar beam falls directly onto the absorber with a minor contribution from the reflector. This is the ideal operating mode of the solar window, with the reflector partly opened and the window delivers heat, electricity, and light altogether. Effects of different control strategies for the position of the rotatable reflector were also studied, so was the performance comparison with roof collector [87].

A ventilated PV glazing consists of a PV outer glazing and a clear inner glazing. The different combinations of vent openings allow different modes of ventilating flow, which can be buoyant/induced or mechanical/driven. The space heating mode belongs to the BiPVT/a category. Besides the popularly used opaque c-Si solar cells on glass, the see-through a-Si solar window can also be used. Chow et al. [88]

analyzed its application in the office environment of Hong Kong. The surface transmissions were found dominated by the inner glass properties. The overall heat transfer however is affected by both the outer and inner glass properties. Experimental comparisons were made between the use of PV glazing and absorptive glazing [89]. The comparative study on single, double, and double-ventilated cases showed that the ventilated PV glazing is able to reduce the direct solar gain and glare effectively. The savings on air-conditioning electricity consumption are 26% for the single-glazing case and 82% for the ventilated double-glazing case. Further, via a validated ESP-r simulation model [90], the natural-ventilated PV technology was found reducing the air-conditioning power consumption by 28%, comparing with the conventional single absorptive glazing system. With daylight control, additional saving in artificial lighting can be enhanced [91].

2.7.1.1 Building-Integrated Window Systems (BiPVT/w)

The research works on BiPVT/w systems have been less popular than the BiPVT/a systems. Ji et al. carried out a numerical study of the annual performance of a BiPVT/w collector system for use in the residential buildings of Hong Kong [92]. Pump energy was neglected. Assuming perfect bonding of PV encapsulation and copper tubing onto the absorber, the annual thermal efficiencies on the west-facing facade were found 47.6% and 43.2% for film cells and c-Si cells, respectively, and the cell efficiencies were 4.3% and 10.3%. The reductions in space heat gain were estimated 53.0% and 59.2%, respectively.

Chow et al. studied a BiPVT/w system applicable to multistorey apartment building in Hong Kong [93]. The TRNSYS system simulation program was used. They also constructed an experimental BiPVT/w system at a rooftop environmental chamber [94]. The energy efficiencies of thermosyphon and pump circulation modes were compared across the subtropical summer and winter periods. The results show a better energy performance of the thermosyphon operation, with thermal efficiency reaches 39% at zero-reduced temperature and the corresponding cell efficiency at 8.6%. The space cooling load is reduced by 50% in peak summer. Ji et al. [95] further carried out an optimization study on this type of installation. The appropriate water flow rate, packing factor, and connecting pipe diameter were determined.

Based on the above-measured data, Chow et al. also developed an explicit dynamic thermal model of the BiPVT/w collector system [96]. Its annual system performance in Hong Kong reconfirmed the better performance of the natural circulation mode. This is because of the elimination of the pumping power and hence better cost saving [97]. The CPBT was 13.8 years, which is comparable to the stand-alone box channel PVT/w collector system. This BiPVT/w application is able to shorten the CPBT to one-third of the plain BiPV application. The corresponding EPBT and greenhouse-gas payback time (GPBT) were found 3.8 and 4.0 years [24]; these are much more favorable than CPBT.

Anderson et al. analyzed the design of a roof-mounted BiPVT/w system [98]. Their BiPVT/w collector prototype was integrated to the standing seam or toughed sheet roof, on which passageways were added to the trough for liquid coolant flow. Their modified Hottel-Whillier model was validated experimentally. The results showed that the key design parameters, like fin efficiency, lamination requirements, and thermal conductivity between the PV module and the supporting structure, affect significantly the electrical and thermal efficiencies. They also suggested that a lower cost material like precoated steel can replace copper or aluminum for thermal absorption since this does not significantly reduce the efficiencies. Another suggestion was to integrate the system "into" (rather than "onto") the roof structure, as the rear air space in the attic can provide a high level of thermal insulation. The effect of nonuniform water flow distribution on electrical conversion performance of BiPVT/w collector of various size was studied by Ghani et al. [99]. The numerical work identified the important role of the array geometry.

Eicker and Dalibard [100] studied the provision of both electrical and cooling energy for buildings. The cooling energy can be used for the direct cooling of activated floors or ceilings. Experimental works with uncovered PVT collector prototypes were carried out to validate a simulation model, which then calculated the night radiative heat exchange with the sky. Large PVT frameless modules were then developed and implemented in a residential zero energy building and tested.

Matuska compared the performance of two types of fin configurations of BiPVT/w collector systems with the BiPV installation using pc-Si cells [101]. Two different European climates and for roof/façade applications were evaluated by computer simulation. Better energy production potential of the BiPVT/w collector systems was confirmed—the results show 15%–25% increase in electricity production in warm climate (Athens) and 8%–15% increase in moderate climate (Prague). The heat production by steady flow forced convection can be up to 10 times higher than the electricity production.

Corbin and Zhai [102] monitored a prototype full-scale BiPVT/w collector installed on the roof of a residential dwelling. Measured performance was used to develop a CFD model which was subsequently used in a parametric study to assess the collector performance under a variety of operating conditions. Water temperature observed during testing reaches 57.4°C at an ambient temperature of 35.3°C. The proposed BiPVT/w collector shows a potential for providing the increased electrical efficiency of up to 5.3% above a naturally ventilated BiPV roof.

2.7.2   Heat-Pump Integration (PVT/Heat Pump)

Conventional air-to-air heat pumps cannot function efficiently in cold winter with extreme low outdoor air temperatures. Bakker et al. [103] introduced a space and tap-water heating system with the use of roof-sized PVT/w array combined with a ground coupled heat pump. The system performance, as applied to one-family Dutch dwelling, was evaluated through TRNSYS simulation. The results showed that the system is able to satisfy all heating demands, and at the same time, to meet nearly all of its electricity consumption, and to keep the long-term average ground temperature constant. The PVT system also requires less roof space and offers architectural uniformity while the required investment is comparable to those of the conventional provisions.

Bai et al. [104] presented a simulation study of using PVT/w collectors as water preheating devices of a SAHP system. The system was for application in sports center for swimming pool heating and also for bathroom services. The energy performances of the same system under different climatic conditions, that included Hong

Kong and three other cities in France, were analyzed and compared. Economic implications were also determined. The results show that although the system performance in Hong Kong is better than the cities in France, the cost payback period is the longest in Hong Kong since there was no government tax reduction.

Extensive research on PVT/heat pump system with variable pump speed has been conducted in China. Experimental investigations were performed on unglazed PVT evaporator system prototype [105,106]. Mathematical models based on the distributed parameters approach were developed and validated [107,108]. The simulation results show that its performance can be better than the conventional SAHP system. With R-134a as the refrigerant, the PV-SAHP system is able to achieve an annual average COP of 5.93 and PV efficiency 12.1% [109].

In the warm seasons, glazed PVT collector may not serve well as PVT evaporator. In cold winter, however, the outdoor temperature can be much lower than the evaporating temperature of the refrigeration cycle. Then the heat loss at the PV evaporator is no longer negligible. The front cover would be able to improve both the photothermic efficiency and the system COP. Pei et al. concluded that for winter operation, the overall PVT exergy efficiency as well as the COP can be improved in the presence of the glass cover [110]. This is beneficial since the space heating demand is higher in winter.

2.7.3 PVT-INTEGRATED HEAT PIPE

These works were basically done in China. Based on the concept of integrating heat pipes and a PVT flat-plate collector into a single unit, Pei et al. [111,112] designed and constructed an experimental rig of heat-pipe PVT (HP-PVT) collector system. The HP-PVT collector can be used in cold regions without freezing, and corrosion can be reduced as well. The evaporator section of the heat pipes is connected to the back of the aluminum absorber plate, and the condenser section is inserted into a water box above the absorber plate. The PV cells are laminated onto the surface of the aluminum plate. Detailed simulation models were developed and validated by the experimental findings. Through these, parametric analyses as well as annual system performance for use in three typical climatic areas in China were predicted. The results showed that for the HP-PVT system without auxiliary heating equipment, in Hong Kong there are 172 days a year that the hot water can be heated to more than 45°C using solar energy. In Lhasa and Beijing, the results are 178 days and 158 days for the same system operation.

In order to solve the nonuniform cooling of solar PV cells and control the operating temperature of solar PV cells conveniently, Wu et al. [113] developed a HP-PVT hybrid system by selecting a wick heat pipe to absorb isothermally the excessive heat from solar cells. The PV modules were in a rectangular arrangement, and below which the wick heat-pipe evaporator section is closely attached. The thermal-electric conversion performance was theoretically investigated.

2.7.4 PVT TRIGENERATION

Calise et al. [114] studied the possible integration of medium-temperature and high-temperature PVT collectors with solar heating and cooling technology, and hence a polygeneration system that produces electricity, space heating and cooling, and domestic hot water. A case study was performed with PVT collectors, single-stage absorption chiller, storage tanks, and auxiliary heaters as the main system components. The system performance was analyzed from both energetic and economic points of view. The economic results show that the system under investigation in Italy can be profitable, provided that an appropriate funding policy is available.

2.7.5 COMMERCIAL ASPECTS

The commercial markets for both solar thermal and PV are growing rapidly. It is expected that the PVT products, once become mature, would experience a similar trend of growth. In future, the market share might be even larger than that for solar thermal collectors. The higher energy output characteristics of the PVT collector suit better the increasing demands on low-energy or even zero-carbon buildings. Nevertheless, although there are plenty reported literatures on the theoretical and experimental findings of PVT collector systems, those reporting on full-scale application and long-term monitoring have been scarce [115]. The number of commercial systems in practical services remains small. The majority involves flat-plate collectors but only with limited service life. The operating experiences are scattered. In the inventory of IEA Solar Heating and Cooling Task 35, over 50 PVT projects have been identified in the past 20 years. Less than twenty of these projects belong to the PVT/w category which is supposed to have better application potential. On the other hand, while most projects were in Europe such as UK and Netherlands, there have been projects realized in Thailand, in which large-scale glazed a-Si PVT/w systems were installed at hospital and government buildings [116]. It is important to have full documentation of the initial testing and commissioning, as well as the long-term monitoring of the real systems performance, including the operating experiences and the problems encountered. Developments in the balance of system are also important—for example, the improvement works in power quality and power factor in PV inverter design [117]. The improvements in power supply stability with power conditioner and better integration of renewable energy sources on to utility grid have been other key research areas [118].

Standard testing procedures for PVT commercial products are so far incomplete. In essence, the performance of PVT commercial products can be tested either outdoor or indoor. The outdoor test needs to be executed in steady conditions of fine weather, which should be around noon hours and preferably with clear sky and no wind. This can be infrequent; say for Northern Europe, it may take six months to acquire the efficiency curve [119]. Indoor test can be quicker and provides repeatable results. To make available an internationally accepted testing standard is one important step for promoting the PVT products.

2.8 SOLAR PVT WITH GEOTHERMAL HEAT PUMP

For building applications for geothermal technologies, geothermal heat pumps are most preferred. Geothermal heat pumps use the constant temperature of the earth as an exchange medium for heat. Although many parts of the world experience seasonal temperature extremes—from scorching heat in the summer to sub-zero cold in the winter—the ground a meter or so below the surface remains at a relatively constant temperature. Geothermal heat pumps are able to heat, cool, and, if so equipped, supply homes and buildings with hot water. There are four types of geothermal heat pump systems. Three of these—horizontal, vertical, and pond/lake—are closed-loop systems. The fourth type of system is open-loop. Which system is best for a particular site depends on the climate, soil conditions, available land, and local installation costs. All of these approaches can be used for residential and commercial building applications [120]. In recent years, numerous efforts are made to hybridize geothermal heat pump with solar PV or solar PVT systems. Some of these studies are briefly described below [48,120,121].

Research out of the SP Technical Research Institute of Sweden has provided proof of concept for a novel hybrid renewable energy system featuring combined hybrid solar PV and geothermal power. The new system advances the hybrid solar PV concept by making use of the output water within a vertical loop ground source heat pump (GSHP) system to which it flows. Having passed through the PV panels, water is heated to around 10°C; it is then directed into the cold side GSHP system and used as heat source; if there is a surplus of heat, this is then directed down into boreholes. Here, the thermal energy of water is absorbed by the surrounding ground as a result of a temperature differential that arises from the ambient temperature of the ground being between 2°C and 3°C. The now-cooled water is then cycled back up the system and re-used in the cooling of PV panels in a closed-loop system.

The system may be used for the purposes of seasonal storage of thermal energy, In Sweden, seasonal temperatures vary greatly, providing options for how the system can be used accordingly. In the summer one can generate solar thermal energy, but it is not required for anything—so we can use boreholes to store this excess energy for use during the winter when it is required. The system stands to be especially useful in Sweden, where geothermal energy is dominated by low temperature, shallow systems featuring GSHPs used for space heating and domestic hot water heating. About 20% of the Swedish buildings use GSHPs, according to the International Geothermal Association. The changes in directing the solar heating first to the GSHP increased the efficiency of the heat pumps due to the increased temperature of the heat source.

Solar panels are effective for powering appliances, air conditioning, and lighting, among other household applications. Heating, on the other hand, requires an additional solar thermal system. When one pairs a solar energy system with a geothermal heat pump, one gains dual advantages. Geothermal systems require a moderate amount of electricity to run. A solar panel installation can generate electricity for the geothermal heat pump at a comparatively low cost. Two systems working together significantly reduce energy expenses. When HVAC system is switched to geothermal heat pump, as much as 80% of energy expense on hot water and heating and cooling is reduced. Mini split systems geothermal heat pumps which use zone approach to heating and cooling is even more efficient and less expensive. Given the respective benefits of solar and geothermal energy, home owners have shown an increasing interest in combined systems [48,120,121]. There are also government tax credits for installation of a geothermal system. Residential buildings are responsible for a significant part of GHG emissions. In Italy, ISPRA [123] revealed for year 2012 an impact on emissions of 18.5% caused by the residential field. The use of geothermal heat pump for space heating and domestic hot water significantly reduces this emission.

Compare to air source heat pump, GSHP technology requires high initial investments but can ensure lower operating costs due to the constant ground temperature. However, the ground energy exploitation can cause the so-called thermal drift: if load profiles are heating-dominated, heat extraction from the ground involves its temperature decrease; on the other hand, if load profiles are cooling-dominated, mean ground temperature increases. Therefore, the heat pump's long-term efficiency decreases. To minimize the thermal drift, the most favorable solution is the combination of solar collectors with GSHPs. The combination of a GSHP with a solar system can be realized in different ways. Eslami-nejad et al. [124] investigated single and multiple borehole configurations and through TRANSYS simulation showed that heat pump efficiency significantly increased in the multiple borehole configurations.

Kjellsson et al. [125] analyzed different alternative systems for domestic hot water (DHW) and space heating, with combinations of solar collectors and GSHPs. By means of TRNSYS simulations, they concluded that the optimal design to decrease the consumption of electricity of the system is when solar heat produces DHW during summertime and recharges the borehole during wintertime. Amin and Hawlader [126] performed analytical and experimental studies on a solar-assisted heat pump located in Singapore, where solar collectors acted as the evaporator of a heat pump. The results outlined that the thermal performance of the system was strongly affected by solar radiation, collector area, storage volume where the condenser was installed, and the speed of the compressor.

Indirect use of solar heat by means of heat pump technology allows the installation of low temperature solar systems, such as unglazed thermal collectors or PVT solar panels. Dott et al. [127] conducted a simulation study to compare nine possible heating systems supplying space heat and DHW by means of both heat pump and solar systems. For the same roof surface, some of the system configurations considered covered flat plate collectors, others the juxtaposition of thermal and PV panels, the last one PVT panels. The highest seasonal energy performance of the heat pump was found in the case with PVT panels, which also generated the highest amount of electricity. Bakker et al. [128] simulated by means of TRNSYS, a solar-assisted GSHP system with 25 m^2 of PVT panels. PVT panels could provide heat to a storage vessel for a direct use or to the evaporator of the GSHP. The authors found that the system was able to cover the total heat demand of DHW and space heating of a one-family dwelling, and 96% of its electricity use. Furthermore, due to the recharge of the ground due to solar heat injection, the mean ground temperature was kept constant, avoiding the decrease of the energy efficiency of the heat pump. Another advantage of this operation was the cooling of the PVT panels, with a consequent increase of the electricity production.

Emmi et al. [47] investigated and compared the energy performances of some multisource systems, obtained the coupling heat pump technology with renewable energy sources. The analysis was carried out considering a single-family house located in Vicenza, Italy. Four system configurations were examined and simulated through TRNSYS tool. Solar energy was provided by PVT panels, whose thermal and electrical performances were calculated by means of a mathematical model developed in Matlab. Two configurations with air, ground, and solar sources were analyzed. One of these did not consider any integration between solar and ground sources, whereas the other one considered the possibility of recharging the ground with solar energy which exceeded the energy demand of the building. The study showed that all the investigated multisource systems reported an increase of energy efficiency between 14% and 26% compared to a standard air to water heat pump system. The seasonal COP of the system where solar and air sources were considered was equal to 3.64, only slightly lower than those obtained for the more complex systems. The simulation results showed that for heating dominated buildings and so limited borehole fields, the configuration involving just solar and air sources was a good solution, despite the variability of those sources.

The study by Nielsen and Borgesen [129] focused on optimization of the thermal part of a geothermal heat pump which is supplemented by a solar energy system that produces both electricity and heat. The geothermal or ground source heat pump (GSHP) consisted of a water/water heat pump connected to a ground loop. The project focused on optimizing the thermal performance of the GSHP for a minimal electrical consumption and additionally has the possibility of covering the rest of the household electrical consumption by means of solar cells. The project was based upon simulation and numerical analysis in Engineering Equation Solver. The solar energy system consists of PVT panels which are hybrid panels with solar cells that are cooled by liquid. By cooling the solar cells, the electric efficiency was increased. The study showed that by dimensioning the energy system correctly, it is possible for a household to be self-sufficient with both heat and electricity. The study showed that geothermal heat pump combined with a hybrid solar energy system secures independency of energy prices for at least 25 years. Also increased electrical efficiency minimizes the need of roof space for the solar panels. Most rooftop residential solar systems today come with a home energy storage system, which further ensures home energy independence. These systems keep everything running (including the geothermal system) in the event of a power outage.

The H L Turner Group Inc. successfully patented its design for the high-efficiency "E-Max Hybrid™" solar-geothermal heating and/or cooling system. The hybrid design can yield geothermal system performances upward of a 7.5 effective COP, about twice that of conventional geothermal applications. Actual net energy cost reductions of up to 75 percent can be accomplished with the E-Max Hybrid™ design, along with total net reduction in CO_2 emissions of nearly 70%. Hybrid solar-geothermal heating and cooling systems are also applicable to commercial, industrial, and institutional projects where the larger economy of scale helps to further decrease the additional cost/sf of the renewable energy features required to balance the total energy use equation. An efficient industrial geothermal heat pump system is also provided by Dandelion Company [130]. Three examples of hybrid high-end HVAC system are described by Taylor [131].

The study by Qu et al. [132] aimed to improve the performance of an office building based on a new operation strategy for an evacuated tube solar collector-U-tube ground heat exchanger system. The main problem with the traditional operation strategy is the mismatch between the chiller capacity and the real cooling demand, which leads to operation of the air-conditioner units under low load rates and causes low system efficiency. In order to solve this problem, the study proposed a new operation strategy, which controlled the unit based on the load distribution of the building. The study investigated the traditional operation strategy of solar thermal collector and geothermal heat pump in detail, adopting the TRNSYS (Transient System Simulation) tool to develop a detailed simulation model that considered the seasonal system performance. The simulation results were validated by experiments. The simulation and experimental results showed good agreement. The study indicated that compared with the traditional strategy, the proposed operation strategy can reduce the ground thermal imbalance in cold regions. The system coefficient of performance can be improved from 3.7 to 3.92, and the total electricity consumption can be decreased by 6.4%.

New advancements make PV solar more effective than solar thermal, which uses a water-based solution that cannot handle freezing conditions without constant maintenance. While solar thermal may work better in some cases, PV solar is the superior choice for the average home. Furthermore, any excess energy goes back to the grid and may qualify for net metering, which creates additional savings for the homeowner *and* utility provider. Modules on PV solar panels do not include moving parts, so these panels typically last more than 30 years and require little to no maintenance. The combination of solar-geothermal allows the use of fewer solar panels. Also hybrid system allows continuous supply of power. Geothermal systems require only one unit of electricity for every five units of heat they produce, because 80% of their heat production comes from the earth through a ground loop system. Transferring that energy from the earth to the home requires a small amount of output, which the PV solar panels collect and convert on their own. All this makes hybrid system very efficient.

2.9   PV/WIND/STORAGE HYBRID ENERGY SYSTEM

As shown in my previous book [14], this is the most widely used hybrid system for both grid connected and off-grid modes. Producing your own clean, renewable energy is one of the most fundamental aspects of becoming energy independent at home. As mentioned earlier, new smart national grid is slowly shifting toward renewable energy systems. In fact, according to a 2012 study by the National Renewable Energy Laboratory, renewable energy sources can provide up to 80% of the electricity needs of the United States by 2050. Fortunately, this can be achieved by using technologies that are already commercially available today. However, the power grid is inherently inefficient. The US Energy Information Administration estimates that electricity transmission and distribution losses average about 5%. These losses occur during the transmission and distribution of electricity from power stations, through power lines to people's homes and offices.

On a national level, we are making progress. In 2017, over 21% of the renewable energy produced in the US came from wind power, while 7% came from solar power. The problem with choosing between wind and solar is that you limit the amount of energy that your home can produce. With solar panels, you obviously will not be able to produce energy during the nighttime hours, and extended cloudy periods will significantly reduce the power you can collect. Similarly, wind turbines can produce energy at all hours of the day, but only when the wind is present. Wind resource is classified according to its potential to produce electricity over an annual basis (Table 2.2). Wind resource maps can determine if an area of interest should be further explored, but wind resource at a micro level can vary significantly. Therefore, it is important to evaluate the specific area of interest before deciding to invest in wind systems.

If the site has a class 3 wind resource, consider small wind turbine (100 kW or less) or large, low wind speed turbine opportunities. If the site has a class 4 or greater wind resource, wind may be a good option and even larger, utility-scale turbines may provide economic options [6].

Most wind turbines are designed for an operating life of up to 20 years. Wind turbines require land area, so on-site wind power generation usually occurs for projects having space for installing the turbines. As shown in Figure 2.12, roof-mounted wind systems are beginning to be used in some building projects. This, however, requires careful considerations of the building's structural integrity, noise, and the added cost.

Hybrid systems combine two (or potentially more) types of renewable energy. The most common hybrid renewable energy system is a combination of rooftop solar panels and a small or medium-sized residential wind turbine. For people looking to go off-grid, hybrid systems allow you to produce energy around the clock. This way, you can decrease the size of the battery system needed to power your home during night. For residential, grid-connected, renewable energy systems, you will be able to decrease your reliance on the energy grid [133,134]. Most grid-connected renewable energy systems give energy to the grid during the day when it is sunny and then pull energy from the grid during the night. In contrast, hybrid systems allow you also to collect wind energy during the nighttime hours. In certain areas, where both wind

TABLE 2.2
Wind Resource Classifications [6]

Wind Power Class	Resource Potential	Wind Speed at 50 m (m/s)
1	Poor	<5.6
2	Marginal	5.6–6.4
3	Fair	6.5–7.0
4	Good	7.0–7.5
5	Excellent	7.5–8.0
6	Outstanding	8.0–8.8
7	Superb	>8.8

FIGURE 2.12   The City of Medford, Massachusetts, USA, owns a Northern Power Systems Northwind 100 wind turbine sited at McGlynn Elementary and Middle School [6]. (Photo from Northern Power Systems, NREL/PIX 16729.)

and sun are abundant, you might even be able to become zero-carbon or net-positive. If you have a net-positive home, it will produce more energy than your household consumes [133,134].

A four-storey office building powered by a combination of thermal and PV solar and wind energy in South Australia has cut its connection to the electricity grid in what its owners claim to be a world first. The $8 million Fluid Solar headquarters in Adelaide, South Australia's northern suburbs contains more than 2 MWh of energy storage capacity, comprising about 90 percent thermal storage with the remaining 10% provided by conventional battery storage. Surplus electricity generated at the site is used as part of Tesla's car-charging network, with provision of 11 electric vehicle bays that will be charged completely by wind and solar power harvested from a 98 kWp array of 378 PV solar panels on the building's roof. The company says that storage of heat is dramatically cheaper than battery storage and because they have the devices that use thermal energy directly for their heating and cooling it means that 60%–70% of the building's energy requirements are met using solar thermal as opposed to solar PV technology that allows them to use the rest of the roof—about 60%—to do a conventional PV. So the system is a hybrid model between a smaller battery pack running the lights, the lift, the fan systems, and so on, and the heavy lifting is done by the solar thermal.

Wind turbines on the roof are also in place to fill the void for about 20 days in winter when long stretches of cloudy weather reduce the effectiveness of solar. The surplus electricity is used to power the car charging station. In a new patented effort, heating rainwater is collected at the site to between 60°C and 90°C and storing it in a 10,000-liter insulated box. The hot water can be used to directly heat the building in winter and indirectly in summer to dry air and run evaporative cooling. The building also has a turbine that turns low pressure steam into electricity. The company is working on a system that can be retrofitted to existing office buildings. South Australia leads the nation in the uptake of wind energy and roof-top solar with renewable sources accounting for more than 40% of the electricity generated in the state.

2.9.1   PROS AND CONS OF HYBRID PV-WIND ENERGY SYSTEMS

The switch to hybrid solar-wind energy system reduces carbon emission and electricity bills for homeowners. The system can also provide US Residential Renewable Energy Tax Credit. This benefit provided a 30% incentive tax credit for wind, solar, and hybrid residential energy systems, with no cap limit, for systems installed by 12/31/19. After that date, the tax credit remains in place but is reduced to 26% for systems installed by the end of 2020 and 22% for those installed before January 1, 2022. With these incentives in place, the savings on your monthly energy bill will quickly offset the cost of installing a renewable energy system. Furthermore, with grid-connected system, one can reduce electricity bill zero or make some profit from utility company. The investment of the capital cost for installation can be a negative for solar-wind hybrid energy. Depending on location, you may not be able to recover this cost rapidly. While rooftop solar panels generally require very little maintenance, wind turbines need regularly scheduled maintenance so they can perform optimally. For off-grid operation, sufficient storage capacity is necessary in order to avoid waste of excess energy generated by the hybrid system.

2.9.2   THEORETICAL CASE STUDIES FOR PV-WIND HYBRID ENERGY SYSTEM

Carefully thought through management strategies are important for hybrid wind-solar system. Stroe et al. [135] evaluated two power and energy management strategies for a hybrid residential PV-wind system with battery energy storage for Denmark. The solar PV total installed capacity in Denmark has exponentially increased in the last years. According to the International Renewable Energy Agency, the cumulative solar PV capacity increased from 17 MW in 2011 to 399 MW in 2012, reaching 790.4 MW by the end of 2016 [136]. The vast majority of the installations (approx. 95%) are represented by residential roof-top PV systems with power levels below 6 kW. This trend is expected to continue as by 2020 the aim is to generate 5% of electricity from residential solar PV systems [137]. Furthermore, Denmark has always been a leader in the wind power production sector. Nowadays, approximately 33% of the installed wind turbines have a rated power below 25 kW and most of the times they are connected to the low-voltage grid [138]. These renewables' grid integration trends combined with the presence of new loads such as heat pumps and electric vehicles are threatening the stable and reliable operation of low-voltage grids causing voltage unbalances, neutral point displacement, voltage flickers, etc. The aforementioned issues can be mitigated using battery energy storage [138,139]. Among the available storage technologies, lithium-ion batteries represent an obvious solution because of their characteristics (e.g., high efficiency, long lifetime, low self-discharge) combined with rapid price decrease [140,141]. Nevertheless, the installations of such systems should be accompanied by the availability of a power and energy management system, which will control and optimize the power flow by providing different services to the grid and/or to the end-user (e.g., power smoothing, peak shaving, self-consumption maximization, etc.) [142].

The aim of the study by Stroe et al. [135] was to investigate different power and energy management strategies for a hybrid residential PV-wind system using a lithium-ion battery energy storage. It is well known that the performance of lithium-ion batteries is very sensitive to the operating conditions (i.e., load current, temperature, state-of-charge, state-of-health). Thus, in order to develop accurate power and energy management strategies, a lithium-ion battery electric model was developed and parameterized based on extensive laboratory tests. Furthermore, the study developed simple and robust performance models for a small wind turbine and solar PV panels. In order to perform realistic study cases, the study has used real-life data for PV generation, wind power generation, and residential household consumption with a one-second resolution. The behavior of the hybrid energy system was evaluated for power smoothing and energy blocks applications for two different scenarios (i.e., a summer day and a winter day). For the power smoothing application, an moving average functionality (MAF) was considered with two averaging time windows of 5 and 15 minutes. In both cases, the desired smoothing effect was achieved, while the battery was subjected to approx. 3.4 and 1.5 full cycles, for a summer and winter day, respectively. Furthermore, the battery was used to maximize the usage of the renewable energy and minimize the electricity bill, by minimizing the energy bought from the utility grid. This was achieved by computing a 15-minutes average power curve for an entire day, by considering the load profile of the house and the produced renewable energy. The difference between the computed curve and the overall house power curve was charged/discharged from the battery. For the considered size of the battery (i.e., 6 kW), the maximum power of the battery was never reached while the Li-ion battery was subjected to 3.75 full cycles during a summer day and 1.75 full cycles during a winter day.

Numerous other theoretical studies for PV-Wind with and without storage for home electricity and heating and cooling need have been reported in the literature [14,143,144]. Nakomcic-Smaragdakis and Dragutinovis [144] analyzed the application of PV-Wind system for electricity and heat supply of a typical household in Serbia, as well as the cost-effectiveness of the proposed system. The influence of feed-in tariff change on the value of the investment was analyzed. Small, grid-connected hybrid system (for energy supply of a standard household), consisting of geothermal heat pump for heating/cooling, solar PV panels, and small wind turbine for power supply was analyzed as a case study. System analysis was conducted with the help of RETScreen software. Results of techno-economics analysis showed that investing in geothermal heat pump and PV panels is cost-effective, while that is not the case with small wind turbine. Bakic et al. [143] examined technical and economic analysis of grid-connected PV/Wind energy stations in the Republic of Serbia under varying climatic conditions. The technical and economic data, of the various grid-connected PV/wind hybrid energy systems for three different locations: Novi Sad, Belgrade and Kopaonik, using the transient simulations software TRNSYS and HOMER were obtained. The results obtained in this paper showed that locations and technical characteristics of the energy systems have an important influence on the amount of delivering electrical power to the grid. The CO_2 emissions reductions, obtained on the basis of delivered electrical power to distribution networks, were also analyzed. Economic analysis was carried out using life cycling cost method. The adoption and implementation of feed-in tariffs have a significant role in enhancing the implementation of technologies that use renewable energy resources. Many other similar studies are described in my previous book [14].

2.10 OTHER ISSUES AND INNOVATIONS FOR HYBRID ENERGY FOR BUILDINGS

2.10.1 HYBRID ELECTRIC BUILDING DESIGN

Today's hybrid electric car uses more than 100 microprocessors to collect data that is feed into a smart operating system. These microprocessors integrate traditional gasoline engines with electric generators and regenerative brakes to achieve superior MPG results. The car's microprocessors also feed information into dashboards, designed to be productivity tools that coach a driver on how their behaviors can reduce costs and emissions. Hybrid electric buildings will mirror this design. Like hybrid cars, they will have extensive sensor systems—collecting big data that smart systems will use to optimize integrated design components—like rooftop solar, onsite batteries, and load-controlling technologies—to achieve occupant comfort, electrical reliability, and lower cost [145].

Hybrid electric buildings are a quantum technology leap. The difference is similar to that between a landline phone and a mobile smartphone. Hybrid electric design is a bridge—linking the values of grid service and onsite electrical systems to deliver guaranteed lower electric bills. By linking utility service and onsite technologies, hybrid electric buildings create owner and tenant value while also offering a value to the utility in the form of stored electricity and/or solar generation accessible during local or system critical-voltage time periods.

The following are the three drivers for the adoption of hybrid electric building designs by the commercial real estate industry:

1. **Escalating electric bills drive adoption**: A hybrid electric building is designed to achieve guaranteed energy bill reductions. Guaranteed results are achieved through a hybrid electric building's ability to assure a measured result that can be financially back-stopped with commercial performance

insurance. An owner of a hybrid electric building will see a lower electric bill either through building operations or through a payment from the insurer. Unlike utility conservation programs, hybrid electric buildings are designed to achieve a guaranteed reduced energy bill.

2. **Increasing utility service interruptions drive adoption**: Disruption of utility service from weather volatility is now a business norm. Hybrid electric buildings are designed to continue operations during a utility's service interruption. The hybrid electric building has onsite renewable generation plus onsite battery systems that can bridge six or more hours of utility service interruptions. They also have predictive operating systems that can optimize building operations to enable continued operations of key activities during a disruption. A hybrid electric building can make the difference between a business making money after a storm or closing down operations.

3. **Millennials drive hybrid electric building designs**: The millennial generation's entry into the workforce is reshaping how commercial workspace is designed. The baby boomer generation's office design legacy is the VP's corner office with a view, interior cubes for work associates and lots of file cabinets. The millennial generation is pushing digital-centric open office designs that increase collaboration between work associates, accelerate learning, and provide a more humane and social experience.

The open office design preferred by the millennial generation also increases productivity-per-square-foot—which enables reduced office sizes, less energy consumption, and fewer emissions. The open office design also enables hybrid electric building innovations by favoring daylight compared to electric light, plus increased deployment of sensor-based smart systems that can conserve electricity by operating electrical equipment in alignment with a building's occupant needs and location.

There are, however, barriers posed by utility and regulatory policies. The utility industry views a hybrid electric building as a microgrid that combines onsite generation and onsite energy storage. Utilities often view a microgrid as a threat to their revenues. One path to align the interests of utilities and building owners is to re-position hybrid electric buildings as a capacity resource that utilities can aggregate like power plants. The aggregation of thousands of micro-grids will require a disruptive transformation in a utility's system controls, the pricing of "behind-the-meter" capacity, and regulatory policy on how utilities should be compensated for their service. These issues of public and utility policy, rather than technology costs or capabilities, are the largest barrier to the adoption of hybrid electric buildings [145].

2.10.2 HYBRID ENERGY MODULES FOR IMPROVING BUILDING EFFICIENCY IN THE FUTURE ELECTRIC GRID

Jahns et al. [146] showed that the Hybrid Energy Module (HEM) is a promising configuration for a DER that combines both thermal and electrical energy storage with a microsource that is capable of producing a significant amount of heat in addition to electricity (e.g., fuel cells, natural gas gen-set). The availability of the energy storage components makes it possible to utilize a larger fraction of the heat and electricity generated by the micro-source during Combined Heating and Power operation in buildings. However, the amount of energy storage required to deliver the desired building efficiency benefits and the resulting installation costs are heavily dependent on the building type, its location, and the time of year. The study employs linear programming to identify the optimum component ratings and dispatch schedules, and calculates the resulting building efficiency benefits and installation costs. Preliminary results from these optimization case studies deliver promising results suggesting that the HEM configuration is capable of providing significant efficiency advantages in apartment and office buildings located in the northern climate zones of the continental US. A dynamic model that is appropriate for analysis of the HEM's fast transient response is developed for the studied HEM configuration using a natural gas gen-set, and experimental results confirm its usefulness.

2.10.3 ECONOMICS OF RENEWABLE HYBRID SYSTEM FOR RESIDENTIAL PURPOSE

Fikru et al. [147] examined economics of renewable hybrid system for residential purpose. Despite advances in small-scale hybrid renewable energy technologies, there are limited economic frameworks that model the different decisions made by a residential hybrid system owner. The study presents a comprehensive review of studies that examine the techno-economic feasibility of small-scale hybrid energy systems, and we find that the most common approach is to compare the annualized life-time costs to the expected energy output and choose the system with the lowest cost per output. While practical, this type of benefit–cost analysis misses out on other production and consumption decisions that are simultaneously made when adopting a hybrid energy system. In this study, authors propose a broader and more robust theoretical framework—based on production and utility theory—to illustrate how the production of renewable energy from multiple sources affects energy efficiency, energy services, and energy consumption choices in the residential sector. Finally, the study discusses how the model can be applied to guide a hybrid-prosumer's decision-making in the US residential sector. Examining hybrid renewable energy systems within a solid economic framework makes the study of hybrid energy more accessible to economists, facilitating interdisciplinary collaborations.

The study finds that a hybrid system increases the percentage of energy consumption derived from onsite energy generation, which in turn reduces the effective price of energy consumption. With a lower price of energy consumption, hybrid-prosumers use more electricity. Furthermore, hybrid-prosumers also increase their consumption of energy services, which is the source of consumer utility. The economic principles derived in this study suggest that the current price of energy storage (battery) is not competitive enough for hybrid-prosumers to pursue it. The study recommends the feasibility of including different types of energy storage systems in a residential HES. The model can also be extended to further capture any utility gains from GHG emission reductions. This study can also be extended by validating some of the model predictions using data from hybrid-prosumers (e.g., data from a survey of the residential sector). The data set should be rich enough to capture observable and measurable variables, such as energy consumption, as well as behavioral parameters, such as the attitude of residents towards energy use.

Authors admit some of the limitations of the model. The model presented in this study is a static model, with no temporal and spatial dimensions. Furthermore, the study has not delved into the impact that policies and regulations could have on hybrid-prosumers. Regulatory standards governing electric rate structures and compensation mechanisms are expected to influence the choices available to hybrid-prosumers. Also, tax incentives and credits are likely critical in the adoption of renewable energy in the residential sector. Studies are needed to understand what types of regulations and incentives, if any, provide the proper motivation for prosumers to make economically optimal decisions.

REFERENCES

1. Shah, Y.T. (2020). *Modular Systems for Energy usage Managemention.* CRC Press, New York.
2. World Business Council for Sustainable Development (WBCSD). (2009). A website report, www.wbcsd.org/.
3. U.S. Energy Information Administration (EIA)—Sector. (2010). Department of Energy, Washington, DC A website report, www.eia.gov/outlooks/aeo/archive.php.
4. Goel, S. (2016, May). ANSI/ASHRAE/IES Standard 90.1–2010 Performance Rating Method Reference Manual, PNNL-25130. Prepared for the U.S. Department of Energy under Contract DE-AC05-76RL01830, Pacific Northwest National Laboratory Richland, Washington 99352, www.pnnl.gov/main/publications/external/technical_reports/PNNL-25130.pdf.
5. International Energy Agency. (2006). A website report, Paris. www.iea.org/newsroom/news/2006/.
6. Hayter, S.J. and Kandt, A. (2011, August). Renewable energy applications for existing buildings. *A Paper Presented at the 48th AiCARR International Conference Baveno-Lago Maggiore,* Italy 22–23 September 2011, NREL/CP-7A40-52172 Contract No. DE-AC36-08GO28308, NREL, Golden, CO.
7. Pless, S. and Torcellini, P. (2010, June). Paul Net Zero energy buildings: A classification system based on renewable energy supply options, Prepared under Task Nos. BEC7.1210, BEC7.1123 Technical report NREL/TP-550-44586, NREL, Golden, CO.
8. European Union. (2009). Directive 2009/28/EC of the European Parliament and of the Council of 23 April 2009 on the promotion of the use of energy from renewable sources and amending and subsequently repealing Directives 2001/77/EC and 2003/30/EC. *The Official Journal of the European Union,* May 2009. http://www.energy.eu/directives/pro-re.pdf.
9. European Union. (2009). Directive 2002/91/EC of the European Parliament and of the Council of 16 December 2002 on the energy performance of buildings. *The Official Journal of the European Union,* April 2003. http://ec.europa.eu/energy/efficiency/buildings/buildings_en.htm. Accessed 25 July 2011.
10. Torcellini, P., Pless, S., Deru, M. and Crawley, D. (2006). Zero energy buildings: A critical look at the definition. *ACEEE Summer Study on Energy Efficiency in Buildings,* August 2006, National Renewable Energy Laboratory, Pacific Grove, CA, Golden, CO, 16 p.
11. Deru, M. and Torcellini, P. (2007). Source energy and emission factors for energy use in buildings (Revised). Technical Report NREL/TP-550-38617. Under Contract No. DE-AC36-99-GO10337. National Renewable Energy Laboratory, Golden, CO. www.nrel.gov/docs/fy07osti/38617.pdf.

12. Windsource for residences, a website report by XCEL Energy, Minnesota (2020). www.xcelenergy.com/Minnesota/Business/RenewableEnergy/Windsource_/Pages/Windsource.aspx. Accessed February 2010.
13. Green, E. (2009). www.green-e.org/. Accessed June 2010.
14. Shah, Y.T. (2021). *Hybrid Power, Generation, Storage and Grids*. CRC Press, New York, NY.
15. Shah, Y.T. (2019). *Modular Systems for Energy and Fuel Recovery and Conversion*. CRC Press, New York.
16. Smart grid system report. (2018, November). A 2018 Report to Congress by U.S. Department of Energy, Washington, DC.
17. Shah, Y.T. (2016). *Energy and Fuel Systems Integration*. CRC Press, New York.
18. Smart grid. (2020). Wikipedia, The free encyclopedia, last visited 14 May 2020.
19. U.S. Department of Energy. Smart Grid Savings and Grid Integration of Renewables in Idaho. http://energy.gov/sites/prod/files/2013/06/f1/IdahoPowerCaseStudy.pdf.
20. Energy Saver. (2015). Grid-connected renewable energy systems. A website report by Department of Energy, Washington, DC,
21. National Energy Technology Laboratory. (2007, August). NETL modern grid initiative—powering our 21st-century economy (PDF). United States Department of Energy Office of Electricity Delivery and Energy Reliability: 17. Retrieved 06 December 2008.
22. Department of Energy and National Energy Technology Laboratory. Environmental impacts of smart grid. DOE/NETL-2010/1428 , DOE, Washington D.C. (2011) http://www.netl.doe.gov/File%20LibraryResearch/Energy%20Analysis/Publications/Envimpact_SmartGrid.pdf.
23. Department of Energy. (2007). The potential benefits of distributed generation and rate-related issues that may impede their expansion. Report Pursuant to Section 1817 of the Energy Policy Act of 2005, Department of Energy, Washington, DC (2007).
24. Chow, T.T. and Ji, J. (2012). Environmental life-cycle analysis of hybrid solar photovoltaic/thermal systems for use in Hong Kong. *International Journal of Photoenergy* 2012 9 p, Article ID 101968, View at: Publisher Site | Google Scholar.
25. Tatchell, A. (2012). Modular zero carbon emission school building with hybrid solar heating and Passivhaus construction, ISSN 2072-7925, CELE Exchange 2012/1 © OECD.
26. Paya-Marin, M., Lim, J. and Sengupta, B. (2013). Life-cycle energy analysis of a modular/off-site building school. *American Journal of Civil Engineering and Architecture*, 1(3): 59–63.
27. Ceranic, B., Beardmore, J. and Cox, A. (2017, October). A novel modular design approach to "Thermal Capacity on Demand" in a rapid deployment building solutions: Case study of smart-POD. *Energy Procedia* 134: 776–786, Doi: 10.1016/j.egypro.2017.09.582.
28. Steijger, L. (2013). Evaluating the feasibility of 'zero carbon' compact dwellings in urban areas. PhD Dissertation, Loughborough University.
29. Eames, P., Loveday, D., Haines, V. and Romanos, P. (2014). The future role of thermal energy storage in the UK energy system: An assessment of the technical feasibility and factors influencing adoption - research report, UK Energy Research Centre.
30. N'Tsoukpoe K.E., Liu H., Le Pierres N. and Luo L. (2009), A review on long-term sorption solar energy storage. *Renewable and Sustainable Energy Reviews* 13: 2385–2396.
31. Hongois S., Kuznik F., Stevens P. and Roux J.J. (2011, July). Development and characterisation of a new $MgSO_4$−zeolite composite for long-term thermal energy storage, *Solar Energy Materials and Solar Cells* 95(7): 1831–1837, ISSN 0927-0248.
32. Ampatzi E., Knight I. and Wiltshire R., (May, 2013). The potential contribution of solar thermal collection and storage systems to meeting the energy requirements of North European Housing. *Solar Energy* 91:402–421. Doi: 10.1016/j.solener.2012.09.008.

33. Martínez-Gracia, A., Del Amo, A., Torné, S., Bayod-Rújula, A., Uche, J. and Usón, S. (2019, July). Solar-assisted heat pump coupled to solar hybrid panels. *International Conference on Renewable Energies and Power Quality (ICREPQ'19) Tenerife (Spain)*, 10–12 April, 2019, Vol. 17, 578, Doi: 10.24084/repqj17.380, RE&PQJ.

34. Harrison, S. (2017). The potential and challenges of solar boosted heat pumps for domestic hot water heating. *12th IEA Heat Pump Conference*, Rotterdam.

35. Zhang, X., Shen, J., Hong, Z., Wang, L. and Yang, T. (2015). A review of building integrated solar thermal (Bist) technologies and their applications. *Journal of Fundamentals of Renewable Energy and Applications* 5: 182, Doi: 10.4172/2090-4541.1000182.

36. Dykes, K., King, J., DiOrio, N. et al., (2020, May). Opportunities for research and development of hybrid power plants national renewable energy laboratory Technical Report NREL/TP-5000-75026, Contract No. DE-AC36-08GO28308.

37. Denholm, P., Eichman, J. and Margolis, R. (2017). *Evaluating the Technical and Economic Performance of PV Plus Storage Power Plants*. National Renewable Energy Laboratory, Golden, CO, NREL/TP-6A20-68737. https://www.nrel.gov/docs/fy17osti/68737.pdf.

38. NREL. (2009). Learning about renewable energy, solar photovoltaic technology basics. National Renewable Energy Laboratory. Golden CO (2011). http://www.nrel.gov/learning/re_photovoltaics.html. Accessed 17 July 2011.

39. Kandt, A., Walker, A. and Hotchkiss, E. (2011). Implementing Solar PV Projects on Historic Buildings and in Historic Districts. NREL. Draft Technical Report. NREL/TP-7A40-51297, May 2011.

40. Bagalini, V., Zhao, B. Y., Wang, R. Z. and Desideri, U. (2019). Solar PV-battery-electric grid-based energy system for residential applications: System configuration and viability. *Research*, 2019: 17 p, Article ID 3838603, Doi: 10.34133/2019/3838603.

41. Ratnam, E.L. and Weller, S.R. (2018). Receding horizon optimization-based approaches to managing supply voltages and power flows in a distribution grid with battery storage co-located with solar PV. *Applied Energy* 210: 1017–1026, View at: Publisher Site | Google Scholar.

42. Vink, K., Ankyu, E. and Koyama, M. (2019). Multiyear microgrid data from a research building in Tsukuba, Japan. *Scientific Data* 6(1), Article 190020, View at: Publisher Site | Google Scholar.

43. Conte, F., Massucco, S., Saviozzi, M. and Silvestro, F. (2018). A stochastic optimization method for planning and real-time control of integrated pv-storage systems: Design and experimental validation. *IEEE Transactions on Sustainable Energy* 9(3): 1188–1197, View at: Publisher Site | Google Scholar.

44. Davies, D.M., Verde, M.G. and Mnyshenko O. et al., (2019). Combined economic and technological evaluation of battery energy storage for grid applications. *Nature Energy* 4(1): 42–50, View at: Publisher Site | Google Scholar.

45. Chow, T. T., Tiwari, G. N. and Menezo C. Hybrid solar: A review on photovoltaic and thermal power integration. Review Article. *Open Access* 2012: 17 p, Article ID 307287, Doi: 10.1155/2012/307287.

46. Aldubyan, M. and Chiasson, A. (2017, December). Thermal study of hybrid photovoltaic-thermal (PVT) solar collectors combined with borehole thermal energy storage systems. *Energy Procedia* 141: 102–108, Doi: 10.1016/j.egypro.2017.11.020.

47. Emmi, G., Tisato, C., Zarrella, A. and De Carli, M. (2016). Multi-source heat pump coupled with a photovoltaic thermal (pvt) hybrid solar collectors technology: A case study in residential application. *International Journal of Energy Production and Management* 1: 382–392, Doi: 10.2495/EQ-V1-N4-382-392.

48. Seyam, S., Dincer, I. and Agelin-Chaab, M. Thermodynamic analysis of a hybrid energy system using geothermal and solar energy sources with thermal storage in a residential building. Research article, Free access, Energy Storage, Wiley, First published: 26 October 2019, Doi: 10.1002/est2.103.

49. Topiwala, J. (2019). A review on solar PV interfaced with thermal collector: The hybrid system. *International Journal of Scientific Research and Reviews* 8(2): 3295–3300, Research article Available online www.ijsrr.org ISSN: 2279-0543.
50. Diwania, S., Agrawal, S., Siddiqui, A. and Singh, S. (2019). Photovoltaic–thermal (PV/T) technology: A comprehensive review on applications and its advancement. *International Journal of Energy and Environmental Engineering*, Doi: 10.1007/s40095-019-00327-y.
51. Wolf, M. (1976). Performance analyses of combined heating and photovoltaic power systems for residences. *Energy Conversion* 16(1–2): 79–90, View at: Google Scholar.
52. Kern, E.C. and Russell, M.C. Combined photovoltaic and thermal hybrid collector systems. *Proceedings of the 13th IEEE Photovoltaic Specialists*, June 1978, 1153–1157, Washington DC, View at: Google Scholar.
53. Florschuetz, L.W. (1979). Extension of the Hottel-Whillier model to the analysis of combined photovoltaic/thermal flat plate collectors. *Solar Energy* 22(4): 361–366, View at: Google Scholar.
54. Chow, T.T. (2010). A review on photovoltaic/thermal hybrid solar technology. *Applied Energy* 87(2): 365–379, View at: Publisher Site | Google Scholar.
55. Davis, M.W., Fanney, A.H. and Dougherty, B.P. (2001). Prediction of building integrated photovoltaic cell temperatures. *Journal of Solar Energy Engineering* 123(3): 200–210, View at: Publisher Site | Google Scholar.
56. Moshfegh, B. and Sandberg, M. (1998). Flow and heat transfer in the air gap behind photovoltaic panels. *Renewable and Sustainable Energy Reviews* 2(3): 287–301, View at: Google Scholar.
57. Clarke, J.A., Hand, J.W., Johnstone, C.M., Kelly, N. and Strachan, P.A. (1996). Photovoltaic-integrated building facades. *Renewable Energy* 8(1–4): 475–479, View at: Google Scholar.
58. Brinkworth, B.J., Marshall, R.H. and Ibarahim, Z. (2000). A validated model of naturally ventilated PV cladding. *Solar Energy* 69(1): 67–81, View at: Google Scholar.
59. Hollick, J.C. (1998). Solar cogeneration panels. *Renewable Energy* 15(1): 195–200, View at: Publisher Site | Google Scholar.
60. Akbarzadeh, A. and Wadowski, T. (1996). Heat pipe-based cooling systems for photovoltaic cells under concentrated solar radiation. *Applied Thermal Engineering* 16(1): 81–87, View at: Publisher Site | Google Scholar.
61. Luque, A., Sala, G., Arboiro, J.C., Bruton, T., Cunningham, D. and Mason, N. (1997). Some results of the EUCLIDES photovoltaic concentrator prototype. *Progress in Photovoltaics: Research and Applications* 5(3): 195–212, View at: Google Scholar.
62. Ito, S., Miura, N. and Wang, K. (1999). Performance of a heat pump using direct expansion solar collectors. *Solar Energy* 65(3): 189–196, View at: Publisher Site | Google Scholar.
63. Ito, S. and Miura, N. (2004, October). Photovoitaic and thermal hybrid systems. *Proceedings of the Asia-Pacific Conference of International Solar Energy Society*, Gwangju, 73–78, View at: Google Scholar.
64. Affolter, P., Ruoss, D., Toggweiler, P. and Haller, B.A. (2000). New generation of hybrid solar collectors. Final Report DIS 56360/16868, Swiss Federal Office for Energy, View at: Google Scholar.
65. IEA. (2002). Photovoltaic/thermal solar energy systems: Status of the technology and roadmap for future development, Task 7 Report, International Energy Agency, PVPS T7–10; Brussels, Belgium
66. Bazilian, M. and Prasad, D. (2002). Modeling of photovoltaiv heat recovery system and its role in design decision support tool for building professionals. *Renewable Energy* 27(1): 57–68.

67. Cartmell, B.P., Shankland, N.J., Fiala, D. and Hanby, V. (2004). A multi-operational ventilated photovoltaic and solar air collector: Application, simulation and initial monitoring feedback. *Solar Energy* 76(1–3): 45–53, View at: Publisher Site | Google Scholar.

68. Mei, L., Infield, D., Eicker, U. and Fux, V. (2003). Thermal modelling of a building with an integrated ventilated PV façade. *Energy and Buildings* 35(6): 605–617, View at: Publisher Site | Google Scholar.

69. Infield, D., Mei, L. and Eicker, U. (2004). Thermal performance estimation for ventilated PV facades. *Solar Energy* 76(1–3): 93–98, View at: Publisher Site | Google Scholar.

70. Sandberg, M. and Moshfegh, B. (2002). Buoyancy-induced air flow in photovoltaic facades effect of geometry of the air gap and location of solar cell modules. *Building and Environment* 37(3): 211–218, View at: Publisher Site | Google Scholar.

71. Mittelman, G., Alshare, A. and Davidson, J.H. (2009). A model and heat transfer correlation for rooftop integrated photovoltaics with a passive air cooling channel. *Solar Energy* 83(8): 1150–1160, View at: Publisher Site | Google Scholar.

72. Gan, G. (2009). Effect of air gap on the performance of building-integrated photovoltaics. *Energy* 34(7): 913–921, View at: Publisher Site | Google Scholar.

73. Zogou, O. and Stapountzis, H. (2012). Flow and heat transfer inside a PV/T collector for building application. *Applied Energy* 91(1): 103–115, View at: Publisher Site | Google Scholar.

74. Chen, Y., Athienitis, A.K. and Galal, K. (2010). Modeling, design and thermal performance of a BIPV/T system thermally coupled with a ventilated concrete slab in a low energy solar house: Part 1, BIPV/T system and house energy concept. *Solar Energy* 84(11): 1892–1907, View at: Publisher Site | Google Scholar.

75. Chen, Y., Galal, K. and Athienitis, A.K. (2010). Modeling, design and thermal performance of a BIPV/T system thermally coupled with a ventilated concrete slab in a low energy solar house: Part 2, ventilated concrete slab *Solar Energy* 84(11): 1908–1919, View at: Publisher Site | Google Scholar.

76. Athienitis, A.K., Bambara, J., O'Neill, B. and Faille, J. (2011). A prototype photovoltaic/thermal system integrated with transpired collector. *Solar Energy* 85(1): 139–153, View at: Publisher Site | Google Scholar.

77. Pantic, S., Candanedo, L. and Athienitis, A.K. (2010). Modeling of energy performance of a house with three configurations of building-integrated photovoltaic/thermal systems. *Energy and Buildings* 42(10): 1779–1789, View at: Publisher Site | Google Scholar.

78. Crawford, R.H., Treloar, G.J., Fuller, R.J. and Bazilian, M. (2006). Life-cycle energy analysis of building integrated photovoltaic systems (BiPVs) with heat recovery unit. *Renewable and Sustainable Energy Reviews* 10(6): 559–575, View at: Publisher Site | Google Scholar.

79. Agrawal B. and Tiwari, G.N. (2010). Optimizing the energy and exergy of building integrated photovoltaic thermal (BIPVT) systems under cold climatic conditions. *Applied Energy* 87(2): 417–426, View at: Publisher Site | Google Scholar.

80. Agrawal B. and Tiwari, G.N. (2010). Life cycle cost assessment of building integrated photovoltaic thermal (BIPVT) systems. *Energy and Buildings* 42(9): 1472–1481, View at: Publisher Site | Google Scholar.

81. Jie, J., Wei, H. and Lam, H.N. (2002). The annual analysis of the power output and heat gain of a PV-wall with different integration mode in Hong Kong. *Solar Energy Materials and Solar Cells* 71(4): 435–448, View at: Publisher Site | Google Scholar.

82. Yang, H., Burnett, J. and Zhu, Z. (2001). Building-integrated photovoltaics: Effect on the cooling load component of building façades. *Building Services Engineering Research and Technology* 22(3): 157–165, View at: Publisher Site | Google Scholar.

83. Chow, T.T., Hand, J.W. and Strachan, P.A. (2003). Building-integrated photovoltaic and thermal applications in a subtropical hotel building. *Applied Thermal Engineering* 23(16): 2035–2049, View at: Publisher Site | Google Scholar.
84. Ji, J., Jiang, B., Yi, H., Chow, T.T., He, W. and Pei, G. (2008). An experimental and mathematical study of efforts of a novel photovoltaic-Trombe wall on a test room. *International Journal of Energy Research* 32(6): 531–542, View at: Publisher Site | Google Scholar.
85. Fieber, A., Gajbert, H., Hakansson, H., Nilsson, J., Rosencrantz, T. and Karlsson, B. (2003). Design, building integration and performance of a hybrid solar wall element. *Proceedings of the ISES Solar World Congress*, Gothenburg, View at: Google Scholar.
86. Davidsson, H. Perers, B. and Karlsson, B. (2010). Performance of a multifunctional PV/T hybrid solar window. *Solar Energy* 84(3): 365–372, View at: Publisher Site | Google Scholar.
87. Davidsson, H. Perers, B. and Karlsson, B. (2012). System analysis of a multifunctional PV/T hybrid solar window. *Solar Energy* 86(3): 903–910, View at: Publisher Site | Google Scholar.
88. Chow, T.T., Fong, K.F., He, W., Lin, Z. and Chan, A.L.S. (2007). Performance evaluation of a PV ventilated window applying to office building of Hong Kong. *Energy and Buildings* 39(6): 643–650, View at: Publisher Site | Google Scholar.
89. Chow, T.T., Pei, G., Chan, L.S., Lin, Z. and Fong, K.F. (2009). A comparative study of PV glazing performance in warm climate. *Indoor and Built Environment* 18(1): 32–40, View at: Publisher Site | Google Scholar
90. Chow, T.T., Qiu, Z. and Li, C. (2009). Potential application of "see-through" solar cells in ventilated glazing in Hong Kong. *Solar Energy Materials and Solar Cells* 93(2): 230–238, View at: Publisher Site | Google Scholar.
91. Chow, T.T., Qiu, Z. and Li, C. (2009). Performance evaluation of PV ventilated glazing. *Proceedings of the Building Simulation 2009, 11th International Building Performance Simulation Association Conference*, July 2009, Glasgow, View at: Google Scholar.
92. Ji, J., Chow, T.T. and He, W. (2003). Dynamic performance of hybrid photovoltaic/thermal collector wall in Hong Kong. *Building and Environment* 38(11): 1327–1334, View at: Publisher Site | Google Scholar.
93. Chow, T.T., Chan, A.L.S., Fong, K.F., Lo, W.C. and Song, C.L. (2005). Energy performance of a solar hybrid collector system in a multistory apartment building. *Proceedings of the Institution of Mechanical Engineers A* 219(1): 1–11, View at: Publisher Site | Google Scholar.
94. Chow, T.T., He, W. and Ji, J. (2007). An experimental study of façade-integrated photovoltaic/water-heating system. *Applied Thermal Engineering* 27(1): 37–45, View at: Publisher Site | Google Scholar.
95. Ji, J., Han, J. Chow, T.T. et al., (2006). Effect of fluid flow and packing factor on energy performance of a wall-mounted hybrid photovoltaic/water-heating collector system. *Energy and Buildings* 38(12): 1380–1387, View at: Publisher Site | Google Scholar.
96. Chow, T.T., He, W., Chan, A.L.S., Fong, K.F., Lin, Z. and Ji, J. (2008). Computer modeling and experimental validation of a building-integrated photovoltaic and water heating system. *Applied Thermal Engineering* 28(11–12): 1356–1364, View at: Publisher Site | Google Scholar.
97. Chow, T.T., Chan, A.L.S., Fong, K.F., Lin, Z., He, W. and Ji, J. (2009). Annual performance of building-integrated photovoltaic/water-heating system for warm climate application. *Applied Energy* 86(5): 689–696, View at: Publisher Site | Google Scholar.
98. Anderson, T.N., Duke, M., Morrison, G.L. and Carson, J.K. (2009). Performance of a building integrated photovoltaic/thermal (BIPVT) solar collector. *Solar Energy* 83(4): 445–455, View at: Publisher Site | Google Scholar.

99. Ghani, F., Duke, M. and Carson, J.K. (2012). Effect of flow distribution on the photovoltaic performance of a building integrated photovoltaic/thermal (BIPV/T) collector. *Solar Energy* 86(5): 1518–1530, View at: Publisher Site | Google Scholar.

100. Eicker, U. and Dalibard, A. (2011). Photovoltaic-thermal collectors for night radiative cooling of buildings. *Solar Energy* 85(7): 1322–1335, View at: Publisher Site | Google Scholar.

101. Matuska, T. (2012). Simulation study of building integrated solar liquid PV-T collectors. *International Journal of Photoenergy* 2012: 8 p, Article ID 686393, View at: Publisher Site | Google Scholar.

102. Corbin, C.D. and Zhai, Z.J. (2010). Experimental and numerical investigation on thermal and electrical performance of a building integrated photovoltaic-thermal collector system. *Energy and Buildings* 42(1): 76–82, View at: Publisher Site | Google Scholar.

103. Bakker, M., Zondag, H.A., Elswijk, M.J., Strootman, K.J. and Jong, M.J.M. (2005). Performance and costs of a roof-sized PV/thermal array combined with a ground coupled heat pump. *Solar Energy* 78(2): 331–339, View at: Publisher Site | Google Scholar.

104. Bai, Y. Chow, T.T. Ménézo, C. and Dupeyrat, P. (2012). Analysis of a hybrid PV/thermal solar-assisted heat pump system for sports center water heating application. *International Journal of Photoenergy* 2012: 13 p, Article ID 265838, View at: Publisher Site | Google Scholar.

105. Ji, J., Liu, K., Chow, T.T., Pei, G., He, W. and He, H. (2008). Performance analysis of a photovoltaic heat pump. *Applied Energy* 85(8): 680–693, View at: Publisher Site | Google Scholar.

106. Ji, J., Pei, G., Chow T.T. et al., (2008). Experimental study of photovoltaic solar assisted heat pump system. *Solar Energy* 82(1): 43–52, View at: Publisher Site | Google Scholar.

107. Ji, J., Liu, K., Chow, T.T., Pei, G. and He, H. (2007). Thermal analysis of PV/T evaporator of a solar-assisted heat pump. *International Journal of Energy Research* 31(5): 525–545, View at: Publisher Site | Google Scholar.

108. Ji, J., He, H., Chow, T.T., Pei, G., He, W. and Liu, K. (2009). Distributed dynamic modeling and experimental study of PV evaporator in a PV/T solar-assisted heat pump. *International Journal of Heat and Mass Transfer* 52(5–6): 1365–1373, View at: Publisher Site | Google Scholar.

109. Chow, T.T., Fong, K.F., Pei, G., Ji, J. and He, M. (2010). Potential use of photovoltaic-integrated solar heat pump system in Hong Kong. *Applied Thermal Engineering* 30(8–9): 1066–1072, View at: Publisher Site | Google Scholar.

110. Pei, G., Ji, J., Chow, T.T., He, H., Liu, K. and Yi, H. (2008). Performance of the photovoltaic solar-assisted heat pump system with and without glass cover in winter: A comparative analysis. *Proceedings of the Institution of Mechanical Engineers A* 222(2): 179–187, View at: Publisher Site | Google Scholar.

111. Pei, G., Fu, H., Zhu, H. and Ji, J. (2012). Performance study and parametric analysis of a novel heat pipe PV/T system. *Energy* 37(1): 384–395, View at: Publisher Site | Google Scholar.

112. Pei, G., Fu, H., Ji, J., Chow, T.T. and Zhang, T. (2012). Annual analysis of heat pipe PV/T systems for domestic hot water and electricity production. *Energy Conversion and Management* 56: 8–21, View at: Publisher Site | Google Scholar.

113. Wu, S.Y., Zhang, Q.L., Xiao, L. and Guo, F.H. (2011). A heat pipe photovoltaic/thermal (PV/T) hybrid system and its performance evaluation. *Energy Build* 43(12): 3558–3567, View at: Publisher Site | Google Scholar.

114. Calise, F., d'Accadia, M.D. and Vanoli, L. (2012). Design and dynamic simulation of a novel solar trigeneration system based on hybrid photovoltaic/thermal collectors (PVT). *Energy Conversion and Management* 60: 214–225, View at: Publisher Site | Google Scholar.

115. Zondag, H., Bystrom, J. and Hansen, J. PV-thermal collectors going commercial. IEA SHC Task 35 paper, 2008, View at: Google Scholar.
116. Hansen, J., Sørensen, H. Byström, J., Collins, M. and Karlsson, B. (2007, September). Market, modelling, testing and demonstration in the framewrk of IEA SHC Task 35 on PV/thermal solar systems. *Proceedings of the 22nd European Photovoltaic Solar Energy Conference and Exhibition*, DE2-–5, Milan, View at: Google Scholar.
117. Shen, C.L. and Su, J.C. (2012). Grid-connection half-bridge PV inverter system for power flow controlling and active power filtering. *International Journal of Photoenergy* 2012, 8 p, Article ID 760791, View at: Publisher Site | Google Scholar.
118. Amorndechaphon, D., Premrudeepreechacharn, S., Higuchi, K. and Roboam, X. (2012). Modified grid-connected CSI for hybrid PV/wind power generation system. *International Journal of Photoenergy* 2012: 12 p, Article ID 381016, View at: Publisher Site | Google Scholar.
119. Elswijk, M.J., Zondag, H.A. and van Helden, W.G.J. (2002). Indoor test facility PVT-panels—a feasibility study. Energy Research Centre of the Netherlands, ECN-I-02–005, View at: Google Scholar.
120. DOE. 2011. Geothermal Heat Pumps. EERE. http://www.eere.energy.gov/basics/renewable_energy/geothermal_heat_pumps.html. Accessed 17 July 2011.
121. Lee, K.S., Kang, E.C. Ghorab, M., Yang, L., Entchev, E. and Lee, E.J. (2017). Smart building heating, cooling and power generation with solar geothermal combined heat pump system, *12th IEA Heat Pump Conference*, Rotterdam.
122. Chiasson, A. and Yavuzturk, C. (2014). Simulation of hybrid solar-geothermal heat pump systems *Proceedings, Thirty-Ninth Workshop on Geothermal Reservoir Engineering*, 24–26 February, 2014, Stanford University, Stanford, CA, SGP-TR-202.
123. De Lauretis, R. Italian Green House Inventory report 1990–2016, ISPRA - Istituto Superiore per la Protezione e la Ricerca Ambientale Via Vitaliano Brancati, 48-00144 Romawww.isprambiente.gov.it ISPRA, Rapporti 283/2018 ISBN 978-88-448-0890-7.
124. Eslami-nejad, P. and Bernier, M. (2012). Simulations of a new double U-tube borehole configuration with solar heat injection and ground freezing. eSim 2012: The Canadian conference on building simulation, Halifax, Nova Scotia.
125. Kjellsson, E., Göran, H. and Bengt, P. (2010). Optimization of systems with the combination of ground-source heat pump and solar collectors in dwellings. *Energy* 35: 2667–2673, Doi: 10.1016/j.energy.2009.04.011.
126. Amin, Z. and Hawlader, M. (2013). A review on solar assisted heat pump systems in Singapore. *Renewable and Sustainable Energy Reviews*. 26: 286–293, Doi: 10.1016/j.rser.2013.05.032.
127. Dott, R., Genkinger, A. and Afjei, T. (2012). System evaluation of combined solar & heat pump systems. *Energy Procedia* 30: 562–570, Doi: 10.1016/j.egypro.2012.11.066.
128. Bakker, M., Zondag, H.A., Elswijk, M.J., Strootman, K.J. and Jong, M.J.M. (2005). Performance and costs of a roof-sized PV/thermal array combined with a ground coupled heat pump. *Solar Energy* 78(2): 331–339.
129. Nielsen, K.W. and Børgesen, J. (2012, June 22). Geothermal heat pump combined with a hybrid solar energy system. *ZENODO, Conference Paper*. Bucharest, Romania.
130. Dandelion Energy. (2020). Wikipedia, The free encyclopedia, last visited 3 August, 2020.
131. Taylor, M. Three examples of hybrid high-end HVAC systems. A website report 17 June 2019.
132. Qu, S., Han, J., Sun, Z. et al., (2019). Study of operational strategies for a hybrid solar-geothermal heat pump system. *Building Simulation* 12: 697–710, Doi: 10.1007/s12273-019-0519-3.
133. Dagdougui, H., Minciardi, R., Ouammi, A., Robba, M. and Sacile, R. Modelling and control of a hybrid renewable energy system to supply demand of a green-building, International Environmental Modelling and Software Society (iEMSs) In: D.A.

Swayne, W. Yang, A.A. Voinov, A. Rizzoli and T. Filatova (Eds.) *Proceedings of 2010 International Congress on Environmental Modelling and Software Modelling for Environment's Sake, Fifth Biennial Meeting*, Ottawa, http://www.iemss.org/iemss2010/index.php?n=Main.

134. Stroe, D.-I., Zaharof, A. and Iov, F. Power and energy management with battery storage for a hybrid residential PV-wind system – a case study for Denmark. *12th International Renewable Energy Storage Conference, IRES 2018, Energy Procedia* 10505(20178): 040604–040707.

135. Stroe, D.-I., Zaharof, A. and Iov, F. (2018, November). Power and energy management with battery storage for a hybrid residential PV-wind system – a case study for Denmark, *Energy Procedia* 155: 464–477.

136. IRENA. Retrieved 5 February 2018, http://resourceirena.irena.org/gateway/countrySearch/?countryCode=DNK.

137. Dansk Solcelle Forening. Retrieved 5 February 2018, http://solcelleforening.dk/.

138. Energistyrelsen. Danish Energy Agency. Retrieved 5 February 2018, https://ens.dk/.

139. Vazquez, S., Lukic, S., Galvan, E., Franquelo, L.G. and Carrasco, J.M. (2010). Energy storage systems for transport and grid applications. *IEEE Transactions on Industrial Electronics* 57(12): 3881–3895.

140. Akhil, A.A., Huff, G., Currier, A.B., Kaun, B.C., Rastler, D.M., Chen, S.B., Cotter, A.L., Bradshaw, D.T., and Gauntlett, W.D. (2014). DOE/EPRI 2013 Electricity Storage Handbook in Collaboration with NRECA. Technical Report SAND2013-5131.

141. Stan, A.-I., Swierczynski, M., Stroe, D.-I., Teodorescu, R. and Andreasen, S.J. (2014). Lithium Ion battery chemistries from renewable energy storage to automotive and back-up power applications. *2014 International Conference on Optimization of Electrical and Electronic Equipment (OPTIM)*, 713–720. Brasov, Romania

142. Zhou, B., Li, W., Chan, K.W., Cao, Y., Kuang, Y., Liu, X. and Wang, X. (2016). Smart home energy management systems: Concept, configurations, and scheduling strategies. *Renewable Sustainable Energy Reviews* 61: 30–40.

143. Bakic, V., Pezo, M. and Stojkovic, S. (2016). Technical and economic analysis of grid-connected PV/wind energy stations in the Republic of Serbia under varying climatic conditions. *FME Transactions* 44: 71–82.

144. Nakomcic-Smaragdakis, B. and Dragutinovic, N. (2015). Hybrid renewable energy system application for electricity and heat supply of a residential building. *Thermal Science* 20: 144–145, Doi: 10.2298/TSCI150505144N.

145. How 'Hybrid Electric Buildings' Will Transform Commercial Real Estate (February 3, 2014). A website report, www.triplepundit.com›story›how-hybrid-electric-bu.

146. Jahns, T.M., Hart, P.J., Lasseter, R.H. and Beihoff, B.C. (2019). Hybrid energy modules for improving building efficiency in the future electric grid. *Energies* 12: 2639.

147. Fikru, M., Gelles, G., Ichim, A. and Smith D. (2019). Notes on the economics of residential hybrid energy system, Energies 12(14): 2639, Doi: 10.3390/en12142639.

3 HESs for Carbon-Free District Heating and Cooling

3.1 INTRODUCTION

The concept of district energy (DE) dates back to ancient Rome, where hot water was used to heat public baths and other buildings. Urban steam systems first became common about 100 years ago (the first North American system was built in 1877 in Lockport, New York), and modern hot water systems have been used extensively in Europe since the 1970s. Today, as modern DE (see Figures 3.1 and 3.2) rapidly gains acceptance, systems are being built in increasing numbers in cities and communities across North America. District heating and cooling (DHC) systems can provide space heating and domestic hot water (DHW) for large office buildings, schools, college campuses, hotels, hospitals, apartment complexes, and other municipal, institutional, and commercial buildings. Systems can also be used to heat neighborhoods and single-family residences [1,2,3–10].

There is much debate about the topic of DHC as an increasingly important means of generating cheap sustainable energy for future generations. DHC or energy is a heat-generating system that is located and distributed centrally, satisfying residential and commercial heating needs such as hot water and space heating. A conceptual diagram of district energy system is illustrated in Figure 3.1. The heat is often generated via fossil fuel plants, but increasingly these are being replaced by "greener" methods, such as geothermal heaters, biomass boilers, and solar arrays. Nuclear power is also a controversial but useful method. Wind energy and industrial waste heat are also considered. The benefits of DE plants is that they have the ability to provide more efficient heating (and cooling) and better control of pollution when compared to local boilers. The system components of a typical district energy system is illustrated in Figure 3.2. Research suggests that DHC, when used with hybrid combined heat and power (CHP) process, is the cheapest way to cut carbon emissions and produces the lowest carbon footprint. These types of power plants are being developed in Denmark and other countries as stores for renewable energies, such as those generated from the wind and sun [1,2,3–9].

There are different types of district heat (DH) systems, and sometimes these are simply heat-only plants. Others are full power-and-heat combined plants, producing electricity as well as heat. These can be very efficient, with the more advanced facilities providing nearly 80% of heat efficiency. Such DH plants are being built in Ukraine, Hungary, Slovakia, Switzerland, Brazil, and Sweden. Some contains energy storage. Most plants are hybrid in nature with different degree of sources

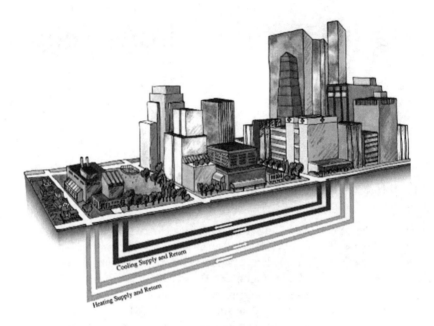

FIGURE 3.1   Conceptual diagram of DE system [10].

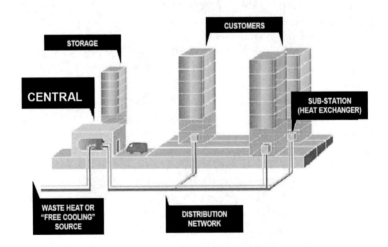

FIGURE 3.2   Schematic of Example DE System Components [10].

and storage integration. The ultimate objective is to totally decarbonize the DHC system. The substations within DHC system distribute the generated heat to the customer network via a series of insulated pipes, comprising of feed and return routes. These pipes are usually laid underground and the heat is usually distributed by water or steam. Steam is the less common method used, but it has the advantage of being available for industrial processing at higher temperatures. The heating network of pipes will then be connected to dwellings by heat substations. A system in Norway is currently operating successfully with this method, losing only 10% of thermal energy through its highly efficient distribution system. Once the heat is within the customer's building, it is usually metered to encourage customers to use energy only when necessary.

DH is currently experiencing a renaissance in the UK. Implemented across Europe during the postwar period, DH remains popular on the continent in places such as Germany, Scandinavia, and much of Eastern Europe. DH in Denmark, for example, currently heats over 60% of homes with that number rising to 95% in Copenhagen. In contrast, the UK, which saw significant growth in DH with the council housing boom in the 1950s–1970s, fell out of love with DH when the North Sea natural gas network was established in the 1980s. The tide is turning, however, and the UK's energy future with regard to DH looks to be falling in line with the rest of Europe's. The last government funded, via DECC, over 140 DH feasibility studies to the tune of over £6m. The Government's Heat Strategy published in 2013 firmly placed DH as the preferable source of heating in urban areas by 2050. Today's figure of 2% of domestic demand in the UK being fed by DH is predicted to rise to a figure of 20% by 2030.

The Canadian District Energy Association surveyed 118 owners and operators of DE systems in Canada in 2008 and subsequently reported the survey results [1,2,3–9]. These results indicate the scope and growth of the technology. The distribution of DE systems in Canada is described by the building floor space served. Ontario, with 43%, has the highest share of DE in Canada. The next highest shares are in Alberta and British Colombia with 12% and 10%, respectively. The survey results also show that 27 million m^3 of floor space is heated by DE in Canada. Ontario has the largest percentage of DE-heated floor space in Canada with 44%; Alberta and Quebec have the next greatest portions of the DE-heated floor space, at 17% each. There have been two major periods of growth for DE in Canada [3–9]. The first was in the 1970s when energy prices drastically increased. Not only was an expansion experienced in DH but also in CHP plants for industrial process applications. The second significant growth occurred in the late 1990s when the Canadian government encouraged the application of DE to foster sustainable energy and community planning.

DH is defined as "a pipe network that allows centralized heat sources to be connected to many heat consumers." A DH network comprises three main components: "one or more energy center(s), the pipe network itself, and connections to heat customers." It allows heat to be used from sources, which would not be normally possible within individual dwellings or buildings. Most DH systems are kept small in order to avoid excessive heat losses by long pipelines. According to IRENA [4,8], in 2014, worldwide DH was largely carried out using fossil fuels with coal and its products (46%), natural gas (43.2%), oil and its products (4.3%) with remaining about 7% by nuclear, biomass, solar, and geothermal. Due to environmental concerns about CO_2 emission, since then, picture has been shifting and more and more renewable energy is introduced in hybrid manner to decarbonize DH. Both hybrid energy and hybrid processes are used for this purpose. The use of thermal storage is also increased. Thus, in recent years, DH is carried out by hybrid energy systems of fossil and renewable energy and hybrid energy process of cogeneration. More efforts are also made to increase contribution of small modular nuclear reactors. Many countries (like Germany) use CHP or cogeneration processes to supply heat for DH. The nature of hybridization can vary significantly depending on how and the extent of renewable energy used in the given situation. In all cases, thermal storage provides backup security. There are several points about use of renewables in DHC that should be noted [3–9]:

1. The use of renewable DHC is driven by several factors. It has environmental (reduced emissions and air pollution) and systemic benefits (positive impacts on the electric grid, DHC infrastructure, and local economy). It can make use of synergies in the urban context (suitable integration into urban environment; reduced space needs) and can increase energy security (reduced fuel imports, diversification of energy mix).

2. The role of DHC is highly diverse across countries, and its use seems to be more influenced by institutional factors and historic developments than climate conditions.

3. In Denmark and Poland, DH satisfies a major share of the total demand for heat (51% and 34%, respectively). Driven by urbanization, China has rapidly increased its use of centralized systems in the country's northern regions. In the other countries assessed, the DH share ranges between 0.4% (Japan) and 8.6% (Germany).

4. District cooling is primarily used to supply space cooling to commercial buildings. Its deployment is largely independent of the climate and systems can be found across a broad range of latitudes. In the United Arab Emirates (UAE), it covers more than 20% of the total space-cooling load, partly in residential buildings. Strong policies are in place to increase this share in the UAE but district cooling has received limited attention in the other countries.

5. Renewable DHC is currently dominated by the use of biomass for DH. The share of renewable energy in DH is highest in Denmark (42%) and Switzerland (40%). In most countries, the bulk of DH is covered by co-generation (CHP) plants. In Germany, almost 90% of DH systems have CHP plants.

Coal plays a leading part in the DH systems in China, Poland, Germany and Denmark, covering between 90% (China) and 24% (Denmark) of heat generation. In other countries, natural gas also accounts for a large share of the mix, such as in the US (73%), Japan, (55%), Germany (45%), and Switzerland (31%). District cooling is mainly supplied by electricity (with the exception of Japan, where natural gas is also used). Policy efforts to promote DHC systems vary widely by country in terms of size and scope. Strong national frameworks are in place in Denmark and Germany, which regulate and provide support for DH systems. In China, specific targets for building up DH are included in the country's five-year plans. The policy focus in Poland is on renovating the existing infrastructure. In other countries, DHC policies tend to be implemented on a subnational level. This is the case in, for example, Switzerland, which subsidizes renewable DH in some cantons. Meanwhile, the Emirate of Abu Dhabi in the UAE sets a goal to raise the share of district cooling in total cooling use to 40%.

DH is widely used in parts of Europe, North America, and Asia. In Europe, it covers 12% of total heat demand [11]. In countries in Scandinavia and Central and Eastern Europe, large heat distribution infrastructures were developed during the second half of the 20th century. In these countries, it remains the principal way to provide energy for space and water heating in urban areas. In countries in Western Europe, such as Germany and Switzerland, DH makes a more modest but nevertheless considerable contribution, and its input into the energy mix has been relatively stable in recent years. In China, district heat is growing due to rapid urbanization and the potential of DE to provide cost-efficient heating services, especially to new urban developments. District cooling is used to provide space cooling to residential and commercial buildings. It can be found in cities with very different climates, such as Stockholm (Sweden) and Manama (Bahrain). This approach is considered to be rather novel in most countries. However, the UAE, for example, has recognized its potential and put in place specific policies to drive expansion [1,2,3–9].

In most countries, fossil fuels are still an important source of energy for total heating while cooling is usually powered by electricity. Coal fulfils a large share of heating energy needs in China, Poland, and Japan. Natural gas is used more in Germany and the US while in Switzerland petroleum products continue to provide a large proportion of heating energy needs. The share of (direct use) of renewable energy for heating ranges from virtually zero in Japan to 14% in Poland (mainly from bioenergy). This shows that expanding DHC systems based on renewables could become an important way to reduce fossil fuel combustion for heating through process of hybridization. DHC uses a wide variety of generation sources across countries. The share of renewable resources in most countries is modest, ranging from close to zero in China and Japan to nearly one-third in Denmark. A few observations from each country are outlined below. More countrywide details are given in excellent reports by IRENA [4,8]. The advantages and disadvantages of DHC are described in great details in my previous book [12] and in IRENA reports [4,8].

Much of the energy efficiency advantages of DE are a result of combining many diverse load profiles, which allows the central energy plant equipment to operate at high load factors with resulting higher levels of efficiency. Aggregation also provides the economies of scale that allows DE systems to employ high-efficiency technologies, such as CHP, and industrial-grade equipment such as condensing economizers that would typically otherwise not be economically or technically feasible for individual buildings. DE using renewable sources is a classic example of hybrid energy and processes to decarbonize heating and cooling industry.

The energy security and resilience benefits of DE infrastructure are often used to support mission-critical operations in places like hospitals, university research centers, military bases, and specialty industries like food processing and pharmaceuticals. With the integration of CHP and microgrid technologies, DE systems can provide high levels of energy reliability providing power, heat, and cooling services without interruption, even during unexpected grid outages due to extreme weather events. The ability for DE systems to use local fuel sources and the flexibility of systems to use multiple types of fuel also contributes to the energy security of communities served by DE. Local operational control ensures that investment decisions are being made close to the point of impact.

3.1.1 DRIVERS FOR DHC

A number of factors drive the expansion of renewable integrated DHC systems. These vary somewhat between regions and might apply rather differently to different renewable technologies. Nevertheless, they can generally be categorized into environmental benefits, systemic benefits, synergies with the urban environment and increased energy security [3–9]. The environment benefits include achievement of clean energy targets, reduction of urban air pollution, fast and cost-effective GHG emission abatement, and reduction in freshwater consumption. Systemic benefits include provision of cross-sectoral benefits, reduction of pressure on the electric grid, leveraging of local resources, provision of economy of scale, leveling of energy demand mix, provision of synergies with other sources of DHC generation and allowance of more cost-effective energy storage. Synergies with the urban environment include removal of construction of decentralized facilities, an optimal integration with the urban environment and in line with the global urbanization trends and reduction of geometric footprint. Increased energy security includes increase in energy diversification and improvement in energy price stability.

The technical properties of DHC systems differ greatly across countries. They have profound implications on both the operation and the economics of systems. Some of them are as follows: (a) Linear heat density describes the ratio of the annual load (in units of energy) and the total length of the network (in meters). (b) Heat losses which are influenced by the quality of the distribution infrastructure and the linear heat density. (c) Storage which is an integral part of modern CHP aligns the production of heat with demand for electricity. (d) Steam which is generally considered an outdated energy carrier for DH (except for industrial applications) has largely been replaced by hot water [13].

3.2 SMALL HYBRID FOSSIL-RENEWABLE HEATING AND COOLING GRIDS

The heating and cooling demand in Europe accounts for around half of EU's final energy consumption. Renewable energy policies often mainly focus on the electricity market, whereas policies for renewable heating and cooling are usually much weaker and less discussed in the overall energy debate. Therefore, it is important to support and promote renewable heating and cooling concepts, the core aim of the CoolHeating project undertaken in Europe [14–26]. The objective of the CoolHeating project, funded by the EU's Horizon 2020 program, is to support the implementation of "small modular renewable heating and cooling grids" for communities in southeastern Europe. This is achieved through knowledge transfer and mutual activities of partners in countries where renewable DH and cooling examples exist (Austria, Denmark, Germany) and in countries which have less development (Croatia, Slovenia, Macedonia, Serbia, Bosnia-Herzegovina). Small modular DH and cooling grids are local concepts to supply households and small and medium industries with renewable heat and/or cooling. In some cases, they may be combined with large-scale DH grids, but the general concept is to have an individual piping grid, which connects a smaller number of consumers. Often, these concepts are implemented for villages or towns. They can be fed by different heat sources, including solar collectors, biomass systems, and surplus heat sources (e.g., heat from industrial processes or from biogas plants that are not yet used, but wasted) [14]. This integrated process makes the overall system hybrid both for energy and processes. Hybrid cogeneration process is of course an integral part of the strategy.

The combination of solar heating and biomass heating is a promising strategy for smaller rural communities due to its contribution to security of supply, price stability, local economic development, local employment, etc. On the one hand, solar heating requires no fuel, and on the other hand, biomass heating can store energy and release it during winter when there is less solar heat available. Thereby, heat storage (buffer tanks for short-term storage and seasonal tanks/basins for long-term storage) needs to be integrated. The main advantages of a biomass/solar heating concept are reduced demand for biomass, reduction of heat storage capacity, and lower maintenance needs for biomass boilers. With increasing shares of fluctuating renewable electricity production (PV, wind), the power-to-heat conversion through heat pumps can furthermore help to balance the power grid. If the planning process is done in a sustainable way, small modular DH/cooling grids have the advantage that, at the beginning, only one part of the system can be realized and additional heat sources and consumers can be added later. This modularity requires good planning and appropriate dimensioning of the equipment (e.g., pipes). It reduces the initial demand for investment and can grow steadily [14,16].

Besides small DH, small district cooling is an important technology with multiple benefits. With increased temperatures due to global warming, the demand for cooling gets higher, especially in southern Europe in which the target countries are located. In contrast to energy demanding conventional air conditioners, district cooling is a good and sustainable alternative, especially for larger building complexes. However, experiences and technologies are much less applied than for DH. The CoolHeating includes both heating and cooling. Especially countries in southern Europe with high solar irradiation need both heating and cooling. The combination of small DH and cooling in the same planning step saves cost and efforts, even if some consumers will demand only either heating or cooling. Thereby, also technical synergies are created (piping, the use of heat pumps). Small modular DH/cooling grids have several benefits. They contribute to increase the local economy due to local value chains of local biomass supply. Local employment as well as security of supply are enhanced. The comfort for the connected household is higher: in the basement of the buildings only the heat exchangers are needed and no fuel storage tank or boiler. Furthermore, no fuel purchase has to be organized. Due to all these benefits, the objective of the CoolHeating project is to support the implementation of small modular renewable heating and cooling grids for communities (municipalities and smaller cities) in southeastern Europe [14–26].

In addition to retrofitting residential buildings in the neighborhood, a biomass DH network is deployed. A two-step implementation process has been developed for the DH network in order to maximize the cost-effectiveness of the investment in the distribution network. This approach emphasizes working toward reducing the distribution length and maximizing the connection rate in a smaller area.

A modular approach is used so that further growth of the network is possible, and distributed generation as well as polygeneration can be more easily integrated. More details on CoolHeating project are described in detail by Rutz et al. [14]. The economics of the hybrid energy system depends on the relative cost of fossil fuels and the cost for insertion of renewable sources. More details on the economics and the benefits of modular approach are described in my previous book and other literature [12,14–26].

3.3   DH BY BIOMASS BASED HES

Prior to the introduction and general implementation of the use of fossil fuels (carbon, oil, and gas), biomass was used for heating buildings. Wood was collected from forests in order to burn it in open fires, heaters, and heating stoves, whereby the heat from the combustion could be used to cook, bake, heat the room, and also to heat water for DHW services and, in certain cases, to send heat to radiators in other parts of the home. The biomass boilers similar to the oil boilers, however, are somewhat larger. Similarly, the volume of fuel is also greater. The biomass boilers produce ash, depending upon the fuel used, of between 0.5% and 2% of the burned fuel. This ash can be used as fertilizer or can be discarded as domestic waste. The initial investment with biomass heat is somewhat higher than fossil fuel heat; however, the savings in fuel costs provide timely amortization of the increased initial investment. Furthermore, public subsidies and tax benefits exist (Personal Income Tax refunds) for the promotion of installation [14].

A DES based on biomass has several advantages for both system customers and the surrounding community [12,14–26] such as (a) low, predictable energy costs, (b) fuel-type flexibility, (c) CHP, (d) better air quality, (e) dollars remain in the local economy, (f) revitalized communities, (g) use of a plentiful and renewable resource, (h) reduced environmental risks, and (i) a meaningful way to address global climate change. For biomass energy to be an effective climate change mitigation strategy, however, the biomass must be harvested in a fashion that sustains the forest resource and increases its vitality and productivity over time. If a forest is clear cut and does not regenerate, there will be no trees to sequester, and carbon and CO_2 levels in the atmosphere will increase. A study by the Oak Ridge National Laboratory found that using part of the forest harvest residue for DH in Vermont has a positive impact on reducing the amount of carbon discharged to the atmosphere [14]. The use of biomass, however, also carries some disadvantages:

1. The heat produced through biomass boilers is somewhat lower than the heat produced by liquid fossil fuel or oil boilers.
2. Biomass has a lower energy density, which means that the storage systems must be larger.
3. The fuel storage systems and ash removal systems may represent higher maintenance costs.
4. The biomass distribution channels are not as developed as the fossil fuel distribution channels: oil and natural gas.

3.3.1   DH WITH CHP

A heat-demand driven CHP unit generates only the amount of heat that is actually needed. If less heat is needed, also less electricity is generated. Ideally, this concept is used, when there is a constant heat demand and 7,500 up to 8,760 full load hours per year. If the heat demand is varying, or if it is decreasing during certain periods, the CHP unit is operated at partial load, according to this definition. This leads to less full load hours (2,000–3,000 h) for DH systems, in which just domestic consumers for space heating are connected [17,18].

A power-demand-driven CHP unit only generates the amount of power that is actually needed or that can be fed into the power grid. Most installed biomass CHP units are designed to generate green electricity according to a guaranteed feed-in tariff. Thus, nearly all power-demand driven systems are either operated at maximum full load hours or at grid-related demand. In some countries, such as in Germany, dedicated incentives were introduced to double the capacity at peak power load (e.g., during the day) and to stop operation at low power load (e.g., during night). Thus power-demand driven CHP units will play an increasingly important role in balancing the power grid. If this power-demand driven concept is applied, the heat supply may not match a potential heat demand, or may be too much. In this case, the surplus heat is often wasted. This led to the situation that units have been installed, which wasted up to 70% of primary energy. After a few years nearly all countries reacted via their legislation and since then the heat utilization of 40%–50% is mandatory for plants that apply for a feed-in tariff. This increases the overall efficiency of the biomass CHP unit to approximately 70%. Hence, the installation of a CHP just makes sense if most of the heat is utilized and if a minimum revenue is gained [1,2,14,16–26].

Historically, biomass technologies for CHP generation were selected according to the thermal and electric capacity of the system. At this time, Organic Rankine Cycle (ORC) systems were selected for small- to medium-scale systems and steam turbines for large-scale systems. Both are thermodynamic processes based on the principle of the Rankine Cycle. The development of highly efficient steam turbines has been driven by large-scale coal or nuclear power plants with an electrical power of a few hundred MW. For smaller scale, ORC processes have been developed, which provide some advantages. The main difference between a steam process and an ORC process is the working media. Water, respectively steam, is replaced by an organic fluid for an ORC process, which has different condensation and evaporation temperatures. With these properties the process can be designed according to the needs of the heat consumer and the heat source. Thus, ORC processes have been optimized for a lower temperature level for the produced thermal energy of 85°C–95°C and a temperature level for the heat source (biomass boiler) of 250°C–350°C. With these parameters, the ORC process is slightly more efficient than a steam cycle as a whole. Another practical reason for choosing this technology is the low effort for operation and maintenance.

Some ORC manufacturers produced standardized ORC modules with complete long-term maintenance contracts. This increased the reliability in such a way that operation with minimum human resources was achieved easily. Another important condition for choosing ORC system was the need for trained operation personnel. In most EU countries, a special education was required for the operation of a steam boiler. Due to lower pressures, temperatures, and different fluid conditions of the organic working media of an ORC process, no such special education was necessary to operate these plants. Finally, the entire ORC plant performed slightly better within the overall economic life cycle assessment compared to different types of steam. Due to this faster market readiness, today ORC plants are widespread. When the market developed rapidly in Europe between 2002 and 2010 due to the provided feed-in tariff for green electricity, some steam turbine manufacturers started to develop small-scale steam turbines, and nowadays, these two technologies, ORC and steam cycle, are performing similar from an economic point of view [23].

Biomass gasification systems are known for more than 100 years, but became mature after 2002 for mid- to large-scale systems and after 2012 for small-scale systems. Based on some demonstration and commercial plants, gasification is nowadays applied to many heating and cooling projects. Gasification is a process that converts the solid biomass into a usable gas consisting mainly of hydrogen, methane, carbon monoxide, and carbon dioxide. The biomass reacts at high temperatures ($>600°C$) with a controlled amount of oxygen ($0<\lambda>1$) to give producer gas. This step is similar to the first step of a combustion process, where gas is converted to gaseous products. On the contrary to a combustion process, this gas is not incinerated in situ. Therefore, up to 80% of the chemical energy of the biomass is contained in the producer gas. This gas is used in a gas engine to generate power and heat in the case of a CHP. If another gasification agent other than air is used, syngas is produced, but for the utilization in a gas engine, it is sufficient to use air as a gasification agent. The mass production led to this price decrease, thus gasification is nowadays also a good option at small scale [12,25,26]. All aforementioned technologies have been assessed within different studies. One report combining the summary of some previous studies updated with latest development shows that the discussed processes follow a common economy of scale, whereas just small differences between the discussed CHP processes occur [12,24]. Use of CHP for DH is rapidly expanding both at small and medium scales. Modular approach has facilitated this momentum.

FVB energy Co. is a Swedish company with branches in USA and Canada specializing in DH. FVB corporation is a world leader in biomass-DH systems. They have installed numerous such systems in many parts of the world [12,14,27,28]. These include 5.4 megawatts-thermal (MWth) biomass combustion boiler wood waste fuel facility at Dalhousie University Agricultural campus, DH for new residential development university, located at Simon Fraser University (SFU) in Burnaby, BC which included 15 MWth biomass module and a 10 MWth natural gas module and supported biomass based DH plant for Prince George DES which provides heating for several landmark buildings in downtown Prince George. FVB also supported Sala-Heby Energi AB to construct a biomass-fueled CHP plant. These and other hybrid biomass-based DH projects supported by FVB energy Co. are described in details in my previous book [12].

3.3.2 SOME EXAMPLES OF HYBRID BIOMASS-BASED DH IN EUROPE

Biomass, including waste wood and plant material, is extensively used for DH as well as CHP production in Denmark, Austria, and Sweden. It has also found application in co-firing with coal for power generation. However, the relatively high cost and limited availability of local biomass, the availability and lower cost of imports, and the inconvenience of handling biomass compared with competing fuels such as oil and gas have prevented use of biomass for DH in Ireland. The greatest potential for DH in Ireland is in large-scale schemes that target apartments and services (in Dublin) as well as small-scale high-occupancy buildings in the rest of the country located where more expensive existing heating sources prevail (i.e., that are not connected to the gas network). Many Central and Northern European countries have DH systems where the community's needs for hot water and space heating are met by metered heat piped from a central source such as a boiler or CHP plant. Such systems are highly suitable for a centralized biomass-fired cogeneration of electricity and heat [1,2,11,14,16,19–26,29,30]. The Spittelau incineration plant to provide DH in Vienna is illustrated in Figure 3.3.

In Torrelago district in Laguna de Duero, Valladolid (ES), Spain, district residents have seen a complete retrofitting including renovation of the buildings' façade and

FIGURE 3.3   The Spittelau incineration plant is one of the several plants that provide DH in Vienna [1].

an upgrade to a DH system that now integrates renewable energy and smart control solutions. Torrelago's DH transition was integrated in the CITyFiED project. Due to the refurbishment in the DH system, the share of energy savings is increased from 40% to 50%, and the system saves 3,392 tCO_2/year compared to the initial gas-fired system, presenting a CO_2 reduction of 94% [11,14,26,29,30].

In Varna, Bulgaria, the heart of the DH is a biomass boiler of 6,500 kW, which works with untreated wood (such as wood chips, bark, and sawdust) from sawmills and forestry workings from the area. In support of this boiler there are two auxiliary boilers of natural gas, each with a thermal power of 7,500 kW, which are used to cover the peaks of demand and as a reserve in case of breakdown or failure of the biomass boiler. The central of Varna also produces power due to a group of ORC turbogenerators. A highly efficient technology generates approximately 5,000 kW of thermal power, 980 kW of electrical power, and a thermal power of about 3,800 kW is retrieved. With the use of the ORC group, the biomass power central can produce 5,000,000 kWh of electricity annually. The DH central is managed by the Department of Public Works of Bressanone, and its network of heat distribution has been linked to the DH network of the adjacent municipality of Bressanone. This way, the DH network serving Varna and Bressanone, reached the length of 120 km, allowing to deploy more than 70 million kWh/year of thermal energy and to serve 1,700 users, including 1,400 in Bressanone and 300 in Varna [11,14,26,29,30]. In England, Euroheat supplies the Energy Cabin; a biomass heating system housed inside a purpose-built box that simply has to be plumbed and wired-in on site—taking days instead of weeks to install. Prefabrication also helps keep costs down—by up to 30%. A Euroheat DH system is currently being installed by "Northdown Wood & Heat" on an estate of over 20 properties; comprising a mansion, let homes, and a church [11,14,17,18,26,29,30].

3.3.3 HYBRID-SOLAR-BIOMASS DH

Biomass may be cheap and carbon-neutral, but a solar upgrade of biomass-fired DH could further improve efficiency and reduce local emissions. For example, solar heat helps avoid having to start up and shut down wood-chip boilers or operate them at partial load. It can even replace backup fossil fuel systems, which provide DH networks with energy in summer. Biomass boiler in Sweden benefits from 10% solar fraction. Larger biomass systems, for instance, between 1 and 100 MWth, have the advantage that their boilers run on wood chips. They take more time to be shut down and restarted, while a partial load results in reduced efficiency, for example, below 20% of nominal power. To prevent partial loading, the systems are usually equipped with a smaller fossil-fuel boiler or buffer storage. Of course, an alternative solution is the integration of a solar thermal field, which can provide a significant boost to boiler efficiency, reducing emissions and costs. One such biomass-solar system in Sweden is the DH network set up in 2010 in Ellös, in the Västra Götaland county. It consists of 4 MWth of biomass boilers, a 1,000 m² solar array, and 200 m³ of buffer storage. The average annual solar fraction is around 10%, and the main benefit of solar integration is the option to switch off the biomass boiler in summer, except for a few rainy or cloudy days in a row. The size of the solar system is planned to be scaled up to 2,000 m² [14].

Solar-assisted biomass DH is quite common in Austria, Moritz Schubert from Austrian-based S.O.L.I.D. said. Sixteen of the 32 SDH plants in operation across the country have been combined with a bio-energy system. Their solar arrays range from 100 to 7,000 m². One factor encouraging these kinds of developments is the national subsidy program for large solar thermal plants above 100 m². Launched in 2010, it has recently been extended to include plants of up to 10,000 m². It provides 40% of the investment cost, and both SMEs and designers of innovative systems can get another 5% on top. Two new SDH systems of between 7,000 and 8,000 m² are planned to be set up this year. Schubert also mentioned a biomass-solar DH system in Mürzzuschlag, a town in the northeast of Austria's Styria region. The plant has solar-based production costs of 35 EUR/MWh, which is slightly below the biomass values of 37–38 EUR/MWh.

Over the last 20 years, Austria's countryside has seen the installation of more than 2,000 biomass DH networks, which makes it quite difficult to find a suitable location for a new grid. Instead, solar collector suppliers are focused on solar upgrades of existing biomass DH. Moreover, some of them have 20–25-year-old boilers, which should be replaced to meet new efficiency requirements. Some also no longer benefit from CHP feed-in tariffs. All these means great potential for integrating solar arrays into existing networks [12,14].

Austrian Institute AEE INTEC indicated that a survey among DH utilities analyzed the main reasons for integrating solar thermal into biomass DH. The survey emphasized that solar could replace backup fossil fuel boilers used exclusively in summer to avoid partial load operation. The DH utilities added that solar thermal could also reduce local emissions, such as dust and nitrogen oxides, created during biomass combustion, which would increase public acceptance of biomass DH. Additionally, Austria had a program to recover up to 35% of the investment in a biomass DH installation. The incentive required that overall network efficiency exceeded 75%—not an easy objective for small networks in areas of low population density, where long pipes are necessary. Solar thermal could help meet the target. In that case, the biomass incentive could serve as indirect support for SDH systems. Hybrid solar-biomass DH systems are also very popular in Southern Europe and China as shown by Ilie and Visa for Romania [31] and Zhang et al. [32] for China.

Ilie and Visa [31] indicated that the energy used in the built-up environment represents at least 40% of the total energy consumed, out of which at least 60% is required for heating, cooling, and DHW. Within the European Union, more than 6,000 communities (i.e., over 9%) use DH systems, the majority of which use the conversion of fossil fuels as a source of energy. The solar-thermal systems that are used on a large (district) scale are becoming more and more efficient from the point of view of their feasibility; however, it is almost impossible to create systems that should satisfy the thermal energy demand throughout the four seasons of the year. The study showed that hybrid solar-biomass (HSB) system is becoming the applicable solution for the majority of the communities that have from this potential, since it can secure independence from the point of view of the use of thermal energy. The study presents the design stages for the implementation of the hybrid solar-biomass systems with a view to identifying the optimal solutions for systems to be integrated into an existing DH system. A case study (Taberei District in Odorheiu Secuiesc City), which provides a detailed description of the feasible technical solutions, is presented.

The method proposed in this study allows the development of the best HSB system required to secure the thermal energy demand for the supply of DHW and DH throughout the year. The proposed steps, the distinct functional situations, and the scenarios proposed for the securing of the thermal energy needed to supply the DHW and DH allow the identification of the optimal operation of the HSB system. The modeling of the functional situations using the TRNSYS software allows the assessment of the thermal energy demand (DH and DHW), the estimation of the solar energy potential, and the simulation of the produced amount of thermal energy (solar-biomass) based on the demand and the potential. The analysis of the feasibility of a HSB system shows that the recovery time is reduced up to 10% due to the use of some common storing/pumping installations.

3.4   HYBRID GEOTHERMAL DH

Geothermal district heating systems (GDHS) include distribution of heat from a central source to individual customers by means of hot pressurized water flowing through distribution pipes. This heat is then used for space heating and DHW preparation but can also be used for low-temperature industry needs. Main advantages of hybrid geothermal energy DHC systems are as follows: (a) Local economy is increased due to local value chains of local geothermal supply; (b) local employment and security of supply are enhanced; (c) comfort of the connected households is increased; (d) security risk due to fuel combustion in households is eliminated and usable space in buildings is increased; (e) environmental pollution is reduced and air quality is improved; and (f) chilled water is provided to the customer by distribution pipes. Hybrid geothermal energy also diversifies energy mix for DHC and provides protection against volatile and rising fuels prices.

Since the share of fluctuating renewable electricity production constantly increases, another way of producing heat could be by utilizing renewable electricity by means of heat pumps and electric boilers. These systems, with further help from heat storage, can also help balancing the power grid. Modularity of these systems enables that only a part of the system can be realized at the beginning with additional heat sources and consumers being added later. Therefore, the initial demand for investment can be reduced and the project can grow steadily [1,12,14,33–36].

With advanced recovery techniques geothermal energy is now accessible in many parts of the world. For example, in Europe, 25% of the population lives in areas directly suitable for geothermal energy for DH (GeoDH, n.d.). Currently, there are around 250 geothermal DH systems (including cogeneration systems) in operation in Europe, with a total installed capacity of about 4,400 MWth and an estimated annual production amounting to some 13,000 GWh/y [1,14,22,33–36]. As of a 2007 study, there were 22 GDHS in the United States. There has been an increase in development of GDHS in the past few years, in particular in France, Germany, and Hungary. With new 200 projects (including upgrading of existing plants), the capacity has grown from 4,500 MWth installed in 2014 to at least 6,500 MWth in 2018.

A key characteristic for geothermal energy is its relatively high investment costs, in particular in areas, where the reservoir is deep underground. Thus, geothermal energy is best feasible in areas with relatively high temperature levels at relatively shallow depths and if it can be supplied as base load capacity to a relatively large DH system. Another key characteristic, in particular regarding the deep reservoirs, is the risk associated with drilling of boreholes of 2–3 km depth. Depending on the accessible temperature level, it may be reasonable to combine geothermal energy with heat pumps in order to increase the temperature levels. These can be either electrical heat pumps or absorption heat pumps, which can be driven by other renewable energies such as by biomass boilers. Hence, the hybrid utilization of geothermal energy sometimes implies considerable additional inputs such as biomass or electricity. This also affects the operation costs, which are relatively low for the geothermal energy itself (pumping costs), but also includes costs for electricity and/or biomass in case of application of heat pumps.

The pumping costs increase with the depth. From experiences in Denmark, it is thus economically more attractive to use heat pumps and extract heat from shallower reservoirs, typically at 1,000–3,000 m depth, where temperatures are 30°C–90°C. This geothermal gradient of 30°C for each 1,000 m depth is a general rule of thumb [35]. When planning geothermal plants, the annual energy production should therefore be relatively large since it must be able to pay back and write off the cost of wells and surface facilities. Based on data from the Danish Energy Agency, a DH system should have an annual sale of at least 400–500 TJ before the hybrid geothermal heating prices are competitive with current price ratio (experiences from Denmark). This may vary from country to country, depending on the geothermal potential. The potential of deep geothermal is significant. There are several support projects organized by CoolHeating project concerning the use of geothermal for DHC, such as the GDHS in Ozaij, Croatia, managed by University of Zegreb, in City of Ozalj (Croatia), by Municipality of Ljutomer (Slovenia), by Municipality of Visoko (Bosnia and Herzegovina), by Municipality of Karposh (Macedonia), and in City of Sabac (Serbia) [1,2,12,14,33–36].

Along with efforts under CoolHeating project, in Stockholm, the first heat pump was installed in 1977 to deliver DH sourced from IBM servers. Today the installed capacity is about 660 MW heat, utilizing treated sewage water, sea water, district cooling, data centers, and grocery stores as hybrid heat sources. Another example is the Drammen Fjernvarme District Heating project in Norway which produces 14 MW from water at just 8°C. Concerns have existed about the use of hydrofluoro-carbons as the working fluid (refrigerant) for large heat pumps. While leakage is not usually measured, it is generally reported to be relatively low, such as 1% (compared to 25% for supermarket cooling systems). A 30-MW heat pump could therefore leak (annually) around 75 kg of R134a or other working fluid. However, recent technical advances allow the use of natural heat pump refrigerants that have very low global warming potential (GWP). CO_2 refrigerant (R744, GWP = 1) or ammonia (R717, GWP = 0) also have the benefit, depending on operating conditions, of resulting in higher heat pump efficiency than conventional refrigerants. An example is a 14 MW(thermal) DH network in Drammen, Norway, which is supplied by seawater-source heat pumps that use R717 refrigerant, and has been operating since 2011.

90°C water is delivered to the district loop (and returns at 65°C). Heat is extracted from seawater (from 60-foot (18 m) depth) that is 8°C–9°C all year round, giving an average coefficient of performance (CoP) of about 3.15. In the process the seawater is chilled to 4°C; however, this resource is not utilized. In a district system where the chilled water could be utilized for air conditioning, the effective COP would be considerably higher. In the future, industrial heat pumps will be further decarbonized by using, on one side, excess renewable electrical energy (otherwise spilled due to meeting of grid demand) from wind, solar, etc., and, on the other side, by making more of renewable heat sources (lake and ocean heat, geothermal, etc.). Furthermore, higher efficiency can be expected through operation on the high-voltage network [1,14,33–36].

3.4.1 HYBRID MODULAR GEOTHERMAL HEAT PUMP FOR DISTRICT HEATING

The DE concept presented here involves heat pumps distributed throughout the district, located in the buildings they serve. GHPs would simply replace conventional furnaces or boilers in buildings and would be installed during construction of each individual building. Individual heat pumps would be sized to meet the intended loads of the building; the hybrid components are only designed to assist the GHX in providing source energy to the heat pumps. Therefore, no supplemental heating is necessary within individual buildings, and emergency back-up heating in buildings would be up to the preference of the individual building owner. The concept of providing low-temperature source water to customer buildings allows for customer flexibility to choose their preferred type of heating system, either ducted forced air or radiant floor heating [37]. Accurate calculation of transient subsurface heat transfer is critically important in sizing ground heat exchangers (GHX) in geothermal heat pump (GHP) systems. Hybrid GHPs have received considerable attention in recent years [38] because they have been shown to significantly improve the economics and energy use of GHP systems. Hybrid GHP systems couple a supplemental heat extraction or rejection subsystem to a conventional GHP system to handle some portion of the building or the ground loads and, as such, permit the use of a smaller, lower-cost GHX. Hybrid GHP systems are more complex in their design than conventional GHP systems due to the transient nature of the supplemental component. Also, recent research on hybrid GHPs identifies more than one method to design a hybrid GHP. For example, Chiasson and Yavuzturk [39,40] describe a method for designing hybrid GHP systems based on annual ground load balancing. Xu [41] and Hackel et al. [42] describe hybrid GHP system design based on lowest life-cycle cost, while Kavanaugh [43] describes a method based on designing the GHX for the nondominant load, and the hybrid component for the balance of the load. Cullin and Spitler [44] describe yet another method based on minimizing first cost of the system, while designing the GHX to supply both the minimum and maximum design entering heat pump fluid temperature over the life cycle of the system. The objective of the literature studies [38–47] is to describe a system simulation approach to examine the feasibility of a multisource hybrid GHP system for an actual proposed DH application in a cold climate, where conventional GHP systems were deemed to be infeasible and impractical.

For consistent delivery temperature of source fluid from which thermal energy can be extracted by heat pumps, a peaking boiler is added to the district system concept. A peaking boiler system serves to offset peak loads on the ground, thus reducing unnecessary GHX size and cost. The optimum size of the boiler depends on the economic trade-off between the avoided GHX cost and the annual operating cost of the boiler. The fuel source for the boiler could be natural gas, biomass, heating oil, or combined fuel. A benefit of a peaking boiler system is that it could be used as a backup in the event of sewage heat recovery interruption (sewer heat exchanger maintenance) or GHX maintenance [12,37].

3.4.2   Solar Thermal Recharge and Sewer Heat Recovery

The role of solar thermal and sewer heat recovery is to offset annually imbalanced loads on the ground by recharging the GHX with thermal energy. Balancing ground loads allows for further reductions in the GHX size and cost. As with a peaking system, the optimum size of the solar collector array and sewer heat recovery system depends on the economic trade-off between the avoided GHX cost and the capital and operating cost of the load balancing systems.

In a DH concept, solar energy would be the "first" energy source added to the district loop, mainly because the most strategic location for solar collector location is on customer roof tops. Solar energy would therefore be added to the district loop immediately downstream of the heat pumps. Similar to the solar recharging concept, sewer heat would be added to the loop at a strategic location as the fluid returns to the GHX. This allows heat to be collected at any time during year and stored underground in the GHX to improve its thermal performance. Useful heat could only be transferred to the district loop when the wastewater temperature exceeds the district loop temperature.

The hybrid district GHP system described here is intended to serve approximately 124,000 m² of mixed residential and commercial floor space in a new subdivision in Whitehorse, Yukon, Canada. Weather conditions at the subject site are sub-arctic. The peak heating load is estimated at 5,840 kW (19.9 million Btu/h), and the annual heating energy load is estimated at 9.4 MWh. The choice among various options described above should be based on parametric analysis with optimization. In making a choice, a life cycle analysis should also be carried out which includes capital cost and other economic data [12,37].

3.4.3   The Multisource Hybrid Concept

Here, the term multisource hybrid is used to describe a hybrid GHP system with multiple heat sources. The basic design concept takes advantage of a modular "plug-and-play" structure such that heat sources or sinks can be added as practical. The concept is centered around a common low-temperature supply pipeline that serves to distribute energy in the form of an aqueous antifreeze solution to the sources and sinks. A low-temperature distribution loop was conceived in this design so that lower-grade heat sources could be rejected to the loop. A lower temperature fluid distribution loop typically requires larger diameter pipe relative to that used in a high-temperature loop, but the added advantage of larger pipe diameter means more fluid volume in the loop and correspondingly more thermal mass (or thermal inertia) of fluid in the pipe, which helps to damp large fluid temperature excursions during peak load times. Amplification of the low-temperature source loop to useful temperatures for space heating is accomplished with water-to-air or water-to-water heat pumps distributed throughout the district in the buildings they serve. The minimum heat pump supply temperature of 0°C was chosen because of the low ground temperatures in cold climates [12,37].

An integral component of the DE system is the GHX, which could consist of one central array or multiple decentralized arrays. The GHX acts to provide a base load heat source for heat pumps, supplemented by a peaking boiler during extreme cold periods. In addition, the GHX acts as a short-term and long-term (i.e., seasonal) storage medium for various waste and other available heat sources, which help to improve the GHX thermal performance during times when heat is needed. The waste and other heat sources considered in this study were limited to solar energy and heat recovered from sanitary sewers. General options exist for other heat sources and sinks, which could conceivably include heat rejection from refrigeration systems (i.e., ice rinks) or any other source deemed practical. This box could also represent another modular GHX as the district system expands and/or additional GHXs are incorporated at decentralized locations.

3.5 DECARBONIZING DISTRICT HEATING WITH HYBRID SOLAR THERMAL ENERGY

DH is a network providing heat, usually in form of hot water. In a DH system, the heat is generated on a larger scale. Therefore, solar thermal, as other technologies, can be scaled up to provide large quantities of hot water. Hence, solar district heating (SDH) plants are a very large-scale application of conventional solar thermal technology. These plants are integrated into local DH networks for both residential and industrial use. During warmer periods they can wholly replace other sources, usually fossil fuels, used for heat supply. Due to developments in large-scale thermal storage, it is now also possible to store heat in summer for winter use. Solar thermal can also meet a share of the heating demand during the winter. These systems are applicable wherever there are DH networks. Usually such networks exist in cold climates, where there is a substantial demand for heat during autumn and winter. These systems consist of solar thermal plants, made up of hundreds of solar thermal collectors. Considering the requirements of such large systems, larger collectors working with bigger loads have been designed specifically for such application. For smaller systems (block heating), normal solar thermal collectors, either flat plate, evacuated tube or even concentrating, can be used. Economic and environmental benefits derived from the acknowledged reliability of this solar thermal application, combined with the technical expertise gained over decades, have contributed to the growing interest in its commercial operation, and currently, there are many plants in operation in Sweden, Denmark, Germany, and Austria.

The solar thermal plants supply heat to a DH network. It can consist of a centralized supply, where a very large collector field delivers heat to a main heating central.

It can also provide, directly or indirectly, a large seasonal heat storage system that can contribute to increasing the input of solar thermal plant to the whole system. The temperature requirements highly depend on the currently used temperature in the grid and the demand requirement. The other possible configuration is a decentralized supply or distributed SDH. In this case, solar collectors are placed at suitable locations (buildings, parking lots, small fields) and connected directly to the DH primary circuit on site. This solution can also be interesting for smaller DH networks or block heating networks. A system is considered as very large when it is over 350 kWth (500 m^2) but SDH systems can reach sizes 100 times bigger, i.e., 35,000 kWth. The benefits of solar thermal systems, in particular for such large systems, cover environmental, political, and economic aspects. Environmental benefits relate to the capacity of reducing harmful emissions. The reduction of CO_2 emissions depends on the quantity of fossil fuels replaced directly or indirectly, when the system replaces the use of carbon-based electricity used for water heating. Depending on the location, a system of 1.4 MWth (2,000 m^2) could generate the equivalent of 1.1 GWhth/year, a saving of around 175 kg of CO_2 [1,36,48–55]. Regarding energy costs, and potential savings, there are three main aspects to consider that have a bigger impact on the comparable costs of the energy produced by a solar thermal system. These are the initial cost of the system, the lifetime of the system, and the system performance. According to the International Energy Agency, for large systems in Europe, the investment costs can go from 350 to 1040 USD/kWth (315–936 EUR/kWth). In terms of energy costs, it can range from 20 to 70 USD/MWhth (18–63 EUR/MWhth) in Southern United States and between 40 and 150 USD/MWhth (36 and135 EUR/MWhth) in Europe [1,36,48–55].

In Europe, the largest progress for the use of solar energy for DH is made in Denmark, Germany, and Austria. In the European Union there are close to 300 systems over 350 kWth in size, feeding into DH. The total capacity installed amounts to 1,100 MW. The combination of solar thermal and DH is a 'very good solution' for reducing CO_2 emissions as well as for reaching the EU goal which calls for an 80% reduction in CO_2 emissions by 2020 compared to 1990 (Heating & Cooling strategy). Solar thermal energy is CO_2-free, solar energy is available everywhere, and the heat generation costs are predictable for the coming 25 years. Furthermore, the technology is fully developed and mature.

For solar thermal systems, required space is a big challenge. In order to keep costs to a minimum, they need to be installed close to the heat consumers which can be expensive in urban environment. Nevertheless, nowadays solar thermal plants can achieve a solar fraction of up to 50% in DH. The generation costs are around 35–60 $/MWh. Use of solar heat for DH has been increasing in Denmark and Germany in recent years. The systems usually include interseasonal thermal energy storage (TES) for a consistent heat output day to day and between summer and winter. Good examples are in Vojensat 50 MW, Dronninglund at 27 MW, and Marstal at 13 MW in Denmark. These systems have been incrementally expanded to supply 10%–40% of their villages' annual space heating needs. The solar-thermal panels are ground-mounted in fields. The heat storage is pit storage, borehole cluster, and the traditional water tank. The scale-up of these systems is possible because of their modular nature [1,12,36,48–55].

Denmark is way ahead in the use of solar energy for DH. By now there are more than 110 systems with about 700 MW of thermal capacity. The reason why Denmark was able to take the leading position lies also with the political and infrastructural circumstances. On the one hand, there are high taxes on fossil fuels like oil and gas. On the other hand, DH is very widely spread in Denmark. The Danish examples show very clearly how renewable energy and CHP can be combined in a "smart" way on a local scale by deploying large heat storage units. They allow CHP plants and power-to-heat systems to operate optimally. As for example, since 2012, solar-assisted DH located in Vojens, Denmark, has met all the expectations of a SDH installation. The pilot scheme with a 17,000 m² (11.9 MWth) large collector field convinced the municipal utility from Southern Denmark to add another 52,500 m² (36.75 MWth) to the field, as well as seasonal storage, which should increase the annual solar share from the 14% measured in 2014 to an expected 45%. Grid temperatures have been set low to match the solar feed-in—they are 75°C–77°C in summer and 37°C–40°C in winter, respectively. Solar heat prices, including seasonal storage, should be around 42 EUR (52 USD)/MWh, compared with 57 EUR (63 USD)/MWh for natural gas supply. Hence, the plant is purely a commercial venture, which has not been supported by direct subsidies.

On the whole in Denmark, more than 1.3 million square meters of solar collectors are connected to DH. In the last 10 years in Denmark, the annual growth in the total collector area has been 42% in average. Similarly, the average increase in the number of DH plants with solar heating has been 29%. As illustrated in Figure 3.4, a central solar heating plant at Marstal, Denmark covers more than half of Marstal's heat

FIGURE 3.4 Central solar heating plant at Marstal, Denmark. It covers more than half of Marstal's heat consumption [1].

consumption. A number of plants have chosen to expand their solar capacity after a few years, and so far, 16 plants have increased their existing solar thermal capacity—some even more than once. As always there are some significant uncertainties in the planned development. The new plan DEVELOPMENT, which was extended to mid-2019 (though with an upper limit of 8,000 MWh), boosted the number of installations again [1,12,36,48–55].

While Denmark is the pioneer country in terms of SDH, Graz, Austria has become a pioneer city for solar thermal energy in DH. The project "BIG SOLAR Graz" (Austria) achieved a major milestone: the land for the construction of a large-scale solar thermal storage with a technical building as well as a relevant part of a future 450,000 m² solar collector field have been secured. Graz is also home to the largest solar thermal plant in central Europe (5.4 MWth) which has been operating for a while now as well as to new flagship projects such as, for example, the "HELIOS project"; a large-scale thermal storage built on a former domestic refuse landfill and fed by three different heat sources: a solar thermal plant, a power-to-heat module, and a CHP plant powered by landfill gas [1,12,36,48–55].

HELIOS plant, a project by Energie Graz, is one best practical example. During the first construction phase, 2,000 m² of flat plate collectors with a 2,500 m³ heat storage were installed on a domestic refuse landfill in the city area of Graz. Heat is also generated by a power-to-heat module with a capacity of 90 kW as well as a 170 kWth CHP plant powered by landfill gas. The feed-in capacity to the DH amounts up to 10 MW. An intelligent storage management system ensures that peak loads in the heating network are diverted, so that heat from renewable sources is priori-tized. The system has been in trial operation since few years. Energie Graz is plan-ning to expand the collector area to 10,000 m². Another practical example is the DH Puchstraße where already today 7,750 m² of solar collectors are feeding into the DH system of Graz. About 5,000 m² of collectors were installed as roof-mounted sys-tems in 2007 and 2,750 m² were added on the ground by the heating plant in 2014. Currently the operator S.O.L.I.D. is adding another 500 m² of solar collectors. In this system, where the collectors are also ground-mounted, collectors from different manufacturers are tested for their application with DH. This expansion is possible because of the modular approach [1,12,36,48–55].

There is also the community Eibiswald, where a 1,250 m² solar collector field and a 105 m³ heat storage have been supplementing the biomass heating plant since 1997. At that time the solar thermal system was able to cover 90% of the local heating network's demand in summer. By adding new customers the yearly heating demand increased to 8 GWh in 2012 and the network grew to 10,000 m in length. This is why an additional 1,200 m² of solar thermal and a 70 m³ buffer storage were installed. The solar fraction is now 12%. The majority of the heating demand is covered by two wood chip boilers with 2.3 and 0.7 MW capacity. In total there are about 35 MWth (ca. 50,000 m² of collector area) of solar thermal feeding into DH systems in the region around Graz and in Styria. The city of Graz is planning to completely decarbonize its heat supply in the medium term and has chosen solar thermal to be one of the main technologies in order to achieve this. In its final expansion stage, the plant 'BIG SOLAR Graz' will provide 20% of the DH demand and will entirely heat about 4,400 buildings.

There is also progress in Germany. There are about 25 large-scale solar thermal plants with integration into DH in operation. More systems with a total capacity of ca. 40 MWth are in the planning and preparation stage. Currently the strongest market segment is made up of energy villages ('Energiedörfer'); five plants are starting in Randegg and Liggeringen (both in Baden-Wuerttemberg), Mengsberg (Hesse), Ellern (Rhineland-Palatinate), and Breklum (Schleswig-Holstein). Utilities in urban areas are also actively involved, as is evident by the biggest network-integrated solar thermal plant in Germany in Senftenberg in Brandenburg and a new pilot plant by the municipal utility in Duesseldorf. The collector area of large-scale systems in Germany is expected to double within the next years. The majority of the currently planned systems are in the segment of urban DH.

The above-mentioned project examples show trends that new generation in DH has begun. DH networks will be the platform for different heat sources: solar thermal, biomass, industrial waste heat, waste incineration, geothermal and heat pumps, and these will reduce the use of fossil fuels by hybrid energy or hybrid processes. In many applications, cogeneration will be used to improve efficiency. There are, however, various options for installing solar thermal collectors. There is a lack of willingness to provide agricultural land for large-scale solar thermal plants. At the "Energy Bunker" in Hamburg-Wilhemsburg, for example, the solar collectors are mounted on an old bunker. They can also be installed on car park buildings, greenhouses, industrial or multifamily buildings, decommissioned landfills, next to wastewater treatment plants, and on noise protection walls. Other options include installations along streets or elevated above agricultural land, a concept already being tested in so-called agro-photovoltaic systems.

There are successful examples of solar-based DH in North America and Middle East. Solar Community, Okotoks, Alberta's the Drake Landing Solar Community project came on stream in June 2007. Since solar radiation is lower there during the winter months, the Drake Landing Solar Community uses a central DH system that stores solar energy in abundance during summer months and distributes the energy to each home for space heating needs during winter months. The Drake Landing Solar Community has achieved a world record 97% annual solar fraction for heating needs, using solar-thermal panels on the garage roofs and thermal storage in a borehole cluster [1,12,36,48–55]. At University in Riyadh, Saudi Arabia, operating since mid-2011, the 25.4 MWth solar district water heating plant provides heat for the Princess Nora Bint Abdul Rahman University, which has a campus for 40,000 students. They use for high-capacity solar thermal systems of over 60 m² and are adapted to suit the Arabian deserts with their heavy sandstorms. Each solar collector is made from special solar glass and equipped with a modified mounting system to withstand unfavorable weather conditions; it has 95% absorption capacity and weighs 170 kg. Throughout its life, the system will save approximately 52 million liters of diesel and reduce the carbon footprint by 125 million kg of CO_2 [1,12,36,48–55].

Above discussion indicates that average investment costs for solar thermal systems can vary greatly from country to country and between different systems. Environmental benefits relate to the capacity of reducing harmful emissions. The reduction of CO_2 emissions depends on the quantity of fossil fuels replaced directly or indirectly, when the system replaces the use of carbon-based electricity used for water heating. Depending on the location, a system of 1.4 MWth (2,000 m^2) could generate an equivalent of 1.1 GWhth/year, a saving of around 175 kg of CO_2 [1,12,36,48–56]. Regarding energy costs and potential savings, there are three main aspects to consider that have a bigger impact on the comparable costs of the energy produced by a solar thermal system. These are the initial cost of the system, the lifetime of the system, and the system performance.

3.6 DISTRICT HEATING WITH HYBRID WIND ENERGY

The concept that curtailed wind power (surplus) contributes to power (which can be converted to heat) need for DH has gained some acceptance [57–59]. Partly, this idea of increasing local renewable energy production in order to convert surplus electricity into thermal energy by hybrid process was presented by Niemi et al. [60]. The analysis they did was based on multicarrier urban energy systems. It was found that for the city of Helsinki, wind power production could be increased by 40%–200% by adding the electricity-to-heat conversion option for using surplus wind energy into the heating network. Long et al. [61] tried to balance wind power variability using electric heat pumps and CHP units. Maegaard [62] presented the Danish example with increased integration of wind energy into the grid through the use of CHP facilities. Hong et al. [63] studied scenarios using EnergyPLAN on wind energy integration in parallel with the heat demand into the existing energy system of Jiangsu province, China. It was revealed that according to the needs of the province, the wind power production could range from 0% to 42% of the total energy demand. The EnergyPLAN model was also used in studies focused mainly in the Danish market aiming at maximizing renewables grid integration [64–66]. Fitzgerald et al. [67] tried to study how power system efficiency can be improved by integrating wind power through an intelligent electric water heating system.

Three countries that have examined linkages between excess wind power for heat pump and electrical heating elements for water in DH schemes are Denmark, Greece, and Scotland. In all cases excess wind power has been sold to grid or neighboring countries like Germany or Norway for Denmark. One possibility that has also been examined is to use CHP for conversion of electricity to heat and back to electricity. Greece study was centered around the city of Kozani and Scotland study was done for island of Orkney. All of these studies were theoretical. In all cases, surplus wind energy was expected. The general conclusion was that while the concept has some merit, the connection between wind energy and DH requires government policy and subsidy to justify capital investment and long-term viability of the general concept [57–59].

3.7 DISTRICT HEATING WITH SMALL MODULAR NUCLEAR REACTORS BY HYBRID PROCESS OF COGENERATION

Nuclear DH is mostly viable in very cold climate, i.e., parts of North America, Europe, and Asia. Vital preconditions for the viability include access to nuclear technology and at least a basic nuclear infrastructure along with the public acceptance.

Besides internationally accepted safety precautions, some additional design features must be adopted for nuclear plants intended for DH to prevent the ingress of radioactive substances into water. For reactors in the SMR size range in cogeneration mode, the possible share of process heat supply would be larger, and heat could even be the predominant product. This would affect the plant optimization and could present more attractive conditions to the potential process heat user. SMRs are more suitable for countries with small or medium electric grids and could be better adapted to cogeneration—generate electricity and process heat than large reactors [68–79].

Historically, nuclear energy sources have been used mainly to produce electricity. The Soviet Union's power generation industry has also followed the path of primarily expanding its nuclear electricity generation capacity. At present, the installed capacity of the 45 nuclear power plant units in the Soviet Union amounts to some 35 gigawatts-electric (GWe). Nuclear power generation is centered, in the main, around the use of water-moderated, water-cooled, pressure-vessel-type reactors (VVERs in Russian means water-water energy reactor) (these series-produced power units have an electrical output of 440 and 1,000 MW), and channel-type, water-cooled, graphite-moderated reactors of the RBMK type (these series-produced power units have an electrical output of 1000 and 1500 MW). The VVER-type reactor units developed by Soviet specialists have also served as a basis for the development of power generation in member countries of the Council for Mutual Economic Assistance (CMEA) [68].

Even when the nuclear power generation industry was at an early stage of development in the USSR, it was clear that focusing solely on the production of electricity would not adequately solve the basic problem which beset the industry, namely, supplanting scarce organic fuel in the country's fuel and energy economy. The only way this problem can be solved is by the extension of nuclear energy into the highly fuel-intensive area of heat production for community and domestic heating and for industrial consumers (1.5 times more fuel is used for this purpose than for the production of electricity). In CMEA countries, which are less well-endowed with fossil fuel resources than the USSR, this problem is even more acute. The plan for developing "nuclear" DH worked out by Soviet specialists to meet the requirements of the country's fuel and energy sector provides for the combination of four different approaches: (a) the use of unregulated steam extraction from the turbines in condensing power plants; (b) the construction of mixed, DH, and condensing nuclear power plants with high back pressure (TK)-type turbines (DH and condensing with regulated steam extraction); (c) the construction of single-purpose nuclear plants which produce only thermal energy for community and domestic heating purposes (DHAPP); and (d) the construction of specialized nuclear plants for industrial heating purposes which, due to new technical features incorporated in the design, could be located in the immediate vicinity of points of consumption and be used to produce heat and electricity, or heat only.

At this point, the first three of these approaches are the most developed technically. The use of unregulated steam extraction from the turbines of nuclear power plants in operation and those under construction holds a special position among these various approaches. In practice, this is the only form of nuclear heating, which has been implemented to date. It was started over 20 years ago when a system was set up to deliver heat from the Beloyarsk plant to supply heat and hot water to the buildings and structures on the plant site itself and to the adjacent living areas. Subsequently, this approach was introduced in other plants as well. The total output of the heating systems in plants already operating is fairly impressive, in excess of 3,000 megawatts-thermal (MWth) at the beginning of 1989. The various turbines in use at present in such nuclear power plants, and those in the process of being manufactured for plants under construction, have varying capacities for the output of heat for DH purposes [68].

The design features of heat delivery systems from nuclear power plants are based on current radiation safety requirements for thermal energy consumers. Thus, the heating water is circulating in a tertiary (in relation to the reactor core) circuit. Pressure in this circuit is kept higher than the maximum possible pressure of the highest unregulated steam extracted, which prevents radioactive products from getting into the heating water should there be a loss of integrity in the heat exchange surface of the boilers. In plants with RBMK reactors, the heating systems have an intermediate coolant circuit between the turbine extraction and the heating water. Pressure in the intermediate circuit is kept higher than the steam pressure but lower than the pressure in the DH circuit. In plants with VVER reactors, the grid water is heated in grid heaters by the steam extracted from the turbine. The highest pressure extraction steam used is lower in pressure than both the pressure in the reactor circuit and the pressure in the grid circuit. To prevent radioactive contamination of the heating water in an accident situation, the heat exchanger is cut off both for the heating steam and for the heated grid water. There is also constant monitoring of radioactivity levels in the heating water. It is important, and an economically attractive idea, that the fullest possible use be made of the existing capacity of DH systems in nuclear power plants. Even allowing for some reduction in electricity output due to the extraction of steam for DH purposes, the dual-purpose (generation of electrical and thermal energy) use of nuclear power plants with VVER-440 power units decreases the volume of inorganic fuel used by 30,000 tons coal equivalent per year for each nuclear unit. Even greater savings are possible when plants with VVER-1000 power units take part in centralized DH. The volume of organic fuel saved by each power unit of this type amounts to 130,000–750,000 tons of coal equivalent per year, depending on the type of turbine installed [12,68].

Despite this important incentive, in the majority of nuclear power plants maximum use is not being made of this capacity to provide heat. The reason for this is that the location of the plant is not always ideal in relation to potential heat consumers. The "guaranteed" consumers (within the plant and plant-living quarters) usually use only a small part of the total potential heat output from a DH unit. The calculated total heat consumption of users on site amounts, as a rule, to approximately 30 MWth for one power unit with an electrical output of 1,000 or 1,500 MWe. Likewise, the total calculated heat consumption of users in the associated plant living quarters does not exceed 260 MWth. In community and domestic heating schemes (heating, hot water, ventilation), nuclear sources are base-loaded and operate jointly with peak sources using organic fuel. As a result, for a 4-unit plant with series-produced VVER-1000 plus K-1000-60/1500-2 turbine, the optimal "l o a d" of the nuclear part of the DH plant due to on-site requirements (within the compound and in living quarters) is not greater than about 230 MW, whereas the total capability is about four times greater. Thus, in modern plants working at design power level, there is a fairly large capacity for heat production, which could be harnessed [12,68,69].

The most sensible course would be to use this available capacity to provide centralized heating for consumers in the adjacent region; i.e., industrial and residential complexes. But there are a number of limitations to such schemes. Technical and economic considerations dictate a specific area of coverage for centralized heating systems for each nuclear power plant. Beyond that area, the total cost of compensating for underproduction of electricity (owing to the fact that the plant is working in a heat-producing regime) and of transporting the heat to the point of consumption exceeds the economies to be made from supplanting organic fuel. The size of this coverage area is determined by climatic conditions in the region (which influences the heat output regime), the cost in relation to the organic fuel supplanted, the amount of heat being delivered by the nuclear power plant, the real conditions involved in the laying of the heat transport line from the plant to the industrial and residential complexes where the heat is to be used, the type of generating equipment installed in the plant, and a few other factors. Another significant consideration is the question of which alternative modes of DH can actually be used in the area to provide a realistic comparison with heating delivered by nuclear power plants. It is not very easy to take into account all of these factors into account when nuclear power plants are to be built since the closeness of a potential heat consumer is not the only, and frequently not the most important, criterion in the authorization process for a construction site. Current radiation safety requirements in the USSR regulate the minimum permissible distance from a nuclear power plant site to major populated areas. Thus, for example, in the case of a plant with an electrical output of 4,000 MWe, this distance varies from around 25 km (if the population density in the area is 100,000–500,000) to 100 km (if the population density is over 2 million). For a large number of reasons, however, nuclear power plant construction sites have to be located far beyond the standard distances mentioned above. Hence, for the majority of nuclear power plants currently operating, there are significant heat output reserves which have not been utilized [68–73].

3.7.1 GLOBAL ASSESSMENT OF MODULAR NUCLEAR HEAT-BASED DISTRICT HEATING

There are several examples of effective modular utilization of nuclear heat both in the Soviet Union and in the rest of the world. The Balakovo nuclear power plant, for instance, provides the town of Balakovo with heating (the town is 12 km from the plant and the heat requirement is over 1,000 MWth), the Rostov plant provides heating for the town of Volgodonsk (13 km from the plant and heat requirement of over 1,000 MWth), the Tatarsk plant supplies heat to the town of Nizhnekamsk (40 km from the plant, heat requirement up to 2,000 MWth), and the Bashkir plant to the town of Neftekamsk. Supplying heat to consumers at some distance from nuclear power plants is also now carried out or is being planned in CMEA countries. In the German Democratic Republic, the Bruno Leuschner nuclear power plant has been providing DH for the town of Greifswald since 1984 (22 km from the plant, heat output—260 MWth). A heating system for the town of Magdeburg, based on the Stendhal plant now under construction, is in the planning stage (distance from the plant—95 km).

In Czechoslovakia, it is planned to equip all plants with VVER-440 reactors (12 power units in all) for DH: the Bohunice plant will supply the town of Trnava with heat, the Dukovany plant the town of Brno, the Mochovice plant the town of Levice. There is a plan to organize heating for the town of Ceske Budejovice from the Temelin plant. The heat transport grids from these plants are very extensive and heat output is fairly high (for example, the heat pipeline from the Dukovany plant to Brno is 40 km long and the calculated heat output is 500 MWth). In Bulgaria, a project for a DH system centered on the Belene nuclear power plant to serve the towns of Pleven (58 km), Svishtov, and Belene, is in the development stage. The total calculated heat output for these towns would be around 700 MWth. It is also planned to use the Kozloduj plant to supply heat for the town of Kozloduj. The technical features intended for inclusion in the heat output systems of these plants are similar in many respects to those already mentioned as being used in the USSR for the delivery of heat from plants with VVER reactors. It should be noted that CMEA countries cooperate closely on the solution of nuclear DH problems. This cooperation is coordinated under a comprehensive program for scientific and technical progress for the period up to the year 2000, which has been adopted by those countries [68–79].

Along with the use of plants now in operation and those under construction for DH purposes, as indicated above, technical decisions have been reached to set up specialized nuclear heating sources—both dual-purpose CHP and single-purpose nuclear district heating atomic power plants (DHAPP). The USSR power generation development program, which has already been adopted, includes plans to construct a number of such plants between now and the year 2000. A CHP design was developed for the European part of the USSR, based on VVER-1000 reactor facilities and new turbine units (TK-450/500-60/3000 turbines) with regulated steam extraction and a high thermal output. A two-unit CHP plant of this type can ensure a heat output of up to 2,100 MW. The plan was to bring at least three plants of this type into service by the year 2000, in Odessa, Minsk, and Khar'kov. It should be pointed out that these plants are not simply DH plants, but mixed heating and condensation units (owing to the fact that the turbines used in them have a large "added" condensing capacity). However, this is inevitable where reactors with a large unit output are involved [12,68–73,78,79].

The review of technical policy with respect to nuclear power generation following the Chernobyl accident brought about a change in attitude to the CHP station designs which had already been developed. As the construction sites for the plants had been chosen in the late 1970s and early 1980s (i.e., when less rigorous requirements were in force than at present), the Odessa and Minsk CHP plants did not satisfy the new approach to safety and their construction was stopped. Design work on the Khar'kov plant was also halted. The design of a single-purpose DHAPP was developed in the USSR simultaneously with the CHP project. This was done because the construction of heating-condensing plants is in some cases limited by a lack of process water, large quantities of which are required to cool the condensers of turbines with large "added" condensing capacity, and because there is no demand in some regions for extra electrical capacity, among other reasons. A water-cooled, water-moderated reactor with a unit output of 500 MWth was specially designed for the DHAPP. This single-purpose type plant, the AST-500, can be located in the immediate vicinity of population centers; i.e., up to 5 km from the city limits. The AST-500 project is being implemented at present in the towns of Gorky and Voronezh. There had been plans to build a number of DHAPPs in the European part of the Soviet Union; however, the negative attitude of the public toward nuclear power after the Chernobyl accident has delayed (and in certain cases cast doubt upon) the implementation of some nuclear power projects which had previously been scheduled for construction [12,68–73,78,79].

There is definite interest in single-purpose nuclear heating plants in CMEA countries. It should be noted, however, that owing to the specific nature of urban construction in these countries, they require facilities with a significantly smaller capacity than those commonly used in the USSR power generation system. In the light of this, the USSR has developed a design for a reactor facility with a unit output of 300 MWth, having technical features similar to those used in the AST-500. In addition, work is being carried out in cooperation with specialized organizations in CMEA countries for building facilities with an even smaller output. In addition to the work on improving the designs for specialized nuclear heating plants, and the development of new types of reactors for them, the search for ways of utilizing more fully the heat output potential of existing nuclear power plants is of great practical significance. Analysis shows that this type of source will dominate the nuclear power infrastructure of the USSR and CMEA countries for a long time into the future [12,68–73,78,79].

Two approaches, in particular, seem highly promising: (a) using the available capacity of the heating facilities of nuclear power plants for long-distance industrial heat supply; and (b) conversion of nuclear power plants which have out-lived their standard service life to a DHAPP system (with the reactor facility operating at reduced power and in a less intensive regime).

The implementation of the first of these approaches will allow the pool of potential customers for thermal energy from nuclear power sources to be extended significantly, and it will introduce nuclear energy into the extremely fuel-intensive area of industrial heat supply. Developmental work done in the USSR and Czechoslovakia has shown that, in the matter of long-distance heat supply to industrial consumers, it makes economic sense to employ a system in which heat is transported in the form of pressurized hot water from the nuclear power plant to the point of consumption. The hot water would then be used to produce steam with the required parameters in the customers' equipment (preferably thermocompressors). The second approach is, at present, at an earlier stage of development, but it is extremely important that an answer be found since it would help solve two problems at once: first, prolonging the operational life of the main equipment in a nuclear energy source, and second, establishing large-scale centralized heat supply sources over a short period of time and at comparatively low cost.

In China, CNNC (China National Nuclear Corporation) launched its Yanlong reactor (referred to as the DHR-400) for DH in November 2017. The move came shortly after the "49-2" pool-type light-water reactor developed by the China Institute of Atomic Energy continuously supplied heat for 168 hours. CNNC indicated that the Yanlong reactor—which an output of 400 MWt—has been developed based upon the company's safe and stable operation of pool-type experimental reactors over the past 50 years. It said the Yanlong is a "safe, economical and green reactor product targeting the demand for heating in northern cities." The reactor can be operated under low temperatures and normal pressures. It can be constructed near urban areas due to the zero risk of a meltdown and lack of emissions. In addition, the reactor is easy to decommission [69]. The use of nuclear energy for DH improves China's energy resource structure. Nuclear energy heating could also reduce emissions, especially as a key technological measure to combat haze during winter in northern China. Thus, it can benefit the environment and people's health in the long run. It can be constructed either inner land or on the coast, making it an especially good fit for northern inland areas, and it has an expected lifespan of around 60 years. In terms of costs, the thermal price is far superior to gas and is comparably economical with coal and CHP [70].

Two types of nuclear heating reactors (NHRs)—one a deep pool type, the other a vessel type—are developed by INET (Institute of Nuclear and New Energy Technology). The vessel-type reactor was selected as the main development direction. Construction of a 5 MWt experimental NHR5 at INET began in 1986 and was completed in 1989. The larger, demonstration-scale NHR200-II was developed from this. A feasibility study on constructing China's first nuclear plant for DH would use the domestically developed NHR200-II low-temperature heating reactor technology. Small modular reactors (SMRs) will be used in the future not just for electrical generation but also for providing heating [70].

There is potential for nuclear DH in multiple regions including Finland and Poland. According to energy for humanity, a number of Finnish cities have received political initiatives to evaluate the feasibility of using SMRs instead of fossil fuels to provide DH. A recent study looked at completely decarbonizing electricity, transport and heating in Helsinki through the use of small, modular advanced reactors. Most of the DH in Finland is supplied by burning coal, natural gas, wood fuels, and peat. It should be noted that while many Finnish cities have progressive climate policies and goals, they have struggled to decarbonize heating and liquid fuels. More than half of the greenhouse emissions of all of Helsinki come from DH, mainly run by fossil fuels. The Society and Energy for Humanity published a report which anticipates that for Helsinki area, future annual energy use in DH at 8 TWh, electricity at 12 TWh and 4 TWh of hydrogen for transportation fuels [68,70].

The initial results seem promising. The scenario modeling assesses the investment in 300 MWth of new DH capacity in the Helsinki Metropolitan area in 2030 either as a CHP plant or as a heat-only boiler. According to Varri [73] and Varri and Syri [74], the SMR with data based on the NuScale reactor fared well against most other options. The preliminary results indicate that a heat-only boiler using the technology would be a fairly profitable while investment in a CHP plant would rely heavily on the form that future electricity markets take. In both cases, the potential investments will still rely on the development of multiple factors, especially the cost of capital. DH can potentially rise to a more prominent role in the future as at least in the EU, and it has been recognized as one of the sources of increased energy efficiency and decarbonization [75]. In the Nordics, DH systems are fairly commonplace already comprising 43% of the total heating market [75]. Even so, the decarbonization of these systems has proven challenging as the efforts so far have been heavily based on increasing the share of biomass.

In Poland, the idea of using the Zarnowec nuclear power plant to provide heating for the Gdansk-Gdynia area is being considered (the plant is 75 km from Gdynia). There is also a plan to provide the town of Poznan with heating from the Warta plant which is scheduled to be built [75–79]. A significant part of the Polish governments plan to diversify the country's energy system and increase its security of supply is the inclusion of nuclear power in the Polish energy mix. The original plans approved in 2009 included the target of two 3 GW plants, the first of which would be online in 2022. While this timetable did not hold, the idea has not been abandoned. The revised timetable and plan include a total of 6 GW of capacity by 2035. Significant progress has been made in reinforcing the necessary institutions and acquiring further knowledge, but further delays seem likely. While new nuclear plants have had issues with their economy elsewhere, it would seem like the strong political backing might partly override the issue. The strong support is also shared by the public as the approval rate near two possible sites has ranged from 60% to 80% [72–79]. Poland also seems highly interested in HTGR technology. While the light water reactors planned could be useful for DH, the country also has a large market for industrial heat with the 13 largest chemical plants already requiring 6.5 GW of heat at the temperature of 400°C–550°C. The initial report on HTGR plans by the Polish Ministry of Energy calls HTGRs "the only practical alternative to replace fossil fuels for industrial heat production" [72–79]. A potential agreement with the Japanese Atomic Energy Agency and multiple Japanese companies to build a 10 MW experimental reactor by 2025 and a proper 160 MW facility by 2030, most likely based on the HTGTR300C mentioned in Section 2.2.3, is expected to be reached at the beginning of 2018 [72–79].

For SMR deployment in Europe, Poland seems like a likely candidate. Beyond the positivity toward nuclear and the need for cleaner energy, the country also has one of the largest DH networks in Europe that covers approximately half of the population. The DH system supplies around two thirds of the overall heat demand of which the share of CHP has ranged between 60% and 65% with the rest powered mostly by often inefficient HOBs. Partly due to this, the Polish government has plans to double the current share of 14% overall electricity production by CHP plants by 2030. The need to reinvest also encompasses the networks themselves as estimated 35% of the Polish DH networks require investments. The system could be developed by not only fixing the old networks but also by linking the smaller systems into existing larger networks. This work is supported by the government and EU-level programs offering funding and loan support for investments in modernization and development of CHP plants and DH networks. Poland also had a certificate system in place for CHP production that is being phased out in 2018. A replacement system has been under discussion with no concrete developments seen yet [72–79].

The SMR technology itself seems suitable for DH. The small scale means that a single module can fit into a fairly small DH network if the base load is large enough.

At the same time, the modular nature of SMRs also means that the plants can be built for a variety of production configurations and they do not necessarily have to be tied down to a single production type. Similarly, the advances in safety seem promising with regard to the emergency planning zone considerations and the plants could hopefully be built fairly close to the networks as long transmission pipelines would bring the costs of the investments up and increase heat losses. The actual economics of the NuScale SMR also show quite a bit of promise, especially for HOB plants. The SMR HOB has the second lowest levelized cost of heat (LCOH) after the MSW CHP in the modeling and the levelized cost is fairly resilient to any changes to the base values based on the sensitivity analysis. Due to the low operating costs of the SMR plants and their nature as base load production, they would also be bringing down the overall cost of heat production and increasing the competitiveness of DH against individual heat pumps and other sources of heating. As the electrification of heating will not only be happening through heat pumps in DH networks, SMRs could be an important tool for energy companies to keep their DH price at a competitive level against heat pump installations for individual buildings [12,68–79]. The potential SMR CHP deployment seems more unlikely but based on the results gained; the same applies to all forms of CHP production outside of MSW CHP. There is some initial promise to the numbers and deployment might be a possibility, but it will be highly dependent on the route that electricity markets take. The levelized cost of electricity (LCOE) and net present value values for the SMR CHP are also clearly more sensitive to any modifications to the base data, especially the cost of capital. The criticality of financing the plants does apply to the HOB plant as well as the overall financial structure of the plants, high initial investments with low production costs, means that most of the LCOH and LCOE is determined by the rate at which the initial investment has to be paid back [12,68–79].

3.8 DISTRICT HEATING BY HYBRID INDUSTRIAL WASTE HEAT

Xia et al. [80] and Fang et al. [81] investigated a method for integrating low-grade industrial waste heat into DH network. This is another example of hybrid cogeneration process. Low-grade industrial waste heat could be a considerable potential energy source for DH, on the condition that the heat from different industrial waste heat sources is integrated properly. This study considers a method for integrating low-grade industrial waste heat into a DH system and focuses on how to improve the outlet temperature of heat-collecting water by optimizing the heat exchange flow for process integration. The pinch analysis concept was considered, and a newly developed thermal theory called entransy analysis was introduced. By using entransy dissipation to describe the energy quality loss during the heat integration, this study analyzed how heat exchange flows influence the final outlet water temperature and attempts to provide an efficient method to optimize the heat exchange flow. Finally, the effectiveness of the proposed methodology was demonstrated by testing it in a project involving the recovery of waste heat from a copper plant for DH [80,81].

Industrial efficiency does not stop at the boundaries of a factory. Applying industrial waste heat in DH systems—including heat from electricity generation—is often mentioned as a promising option for energy savings and CO_2-emission reduction.

Currently, despite a large amount of available waste heat, application of industrial waste heat in the Netherlands is very limited. So, if waste heat is to fulfil an important role in future emissions reductions, there is a major implementation gap to be bridged. The study by Daniels et al. [82,83] presents an analysis of potentials and costs for industrial waste heat utilization in the Netherlands, for low-temperature applications in households, services, and greenhouse horticulture. It starts with identifying the availability of waste heat and the demand for heat. To estimate the technical potential, it evaluates the match between supply and demand: heat should be available at the right temperature level, the right moment, and the right location. As to provide realistic potentials, the study makes a comparison with alternatives, both with regard to alternative application of waste heat on the supply side and with regard to alternative sources of heat on the demand side. It concludes with an assessment of a realistic role of industrial waste heat utilization in DH in the medium to long term, and the roles of major alternatives with regard to their performance in terms of energy savings, emissions reductions, and costs. As the analysis points out, this realistic potential is rather limited: an estimated 10–25 PJ of net energy savings may be realized by the utilization of 25–45 PJ of waste heat in DH.

Large quantities of low-grade waste heat are discharged into the environment, mostly via water evaporation, during industrial processes. Putting this industrial waste heat to productive use can reduce fossil fuel usage as well as CO_2 emissions and water dissipation. The purpose of the study by Daniels et al. [82,83] proposed a holistic approach to the integrated and efficient utilization of low-grade industrial waste heat. Recovering industrial waste heat for use in DH (DH) can increase the efficiency of the industrial sector and the DH system, in a cost-efficient way defined by the index of investment vs. carbon reduction. Furthermore, low-temperature DH network greatly benefits the recovery rate of industrial waste heat. Based on data analysis and in-situ investigations, the study discusses the potential for the implementation of such an approach in northern China, where conventional heat sources for DH are insufficient. The universal design approach to industrial-waste-heat-based DH is proposed. Through a demonstration project, this approach is introduced in detail. This study finds three advantages to this approach: (a) improvement of the thermal energy efficiency of industrial factories; (b) more cost-efficient than the traditional heating mode; and (c) CO_2 and pollutant emission reduction as well as water conservation.

Heating and cooling in the future will utilize energy gained from waste heat which will be distributed at low temperature using DHC networks. It will thus make use of the heat wasted by cooling systems in supermarkets and fruit storage facilities which up to now has simply been released untapped into the atmosphere. South Tyrol's EURAC Institute for Renewable Energy is exploring this new technology in the "FLEXYNETS" project which is financed to the tune of two million euros by the European research program "Horizon 2020." At present DH grids run via high temperatures of around 90°C. To heat individual buildings, the networks have to connect to sizeable thermal plants, such as block thermal plants or waste incinerating plants. The technology which will now be researched by the South Tyrol EURAC Institute for Renewable Energy on the other hand runs at temperatures between 10°C and 20°C. This means that the DH grids can be supplied with energy from sources running at much lower temperatures. Space heating, generated for example from a waste incinerating plant, is intended to be supplemented by heat generated in various everyday processes and which is currently wasted. By using low temperatures when distributing heat, one can reduce the huge heat loss in the underground distribution pipelines, which will make the whole grid much more efficient in the future. According to the experts, the energy consumption for heating and hot water could be reduced by 80%, and for cooling buildings by 40%. Across Europe, this would amount to a reduction of 5 million tons of CO_2 emissions by 2030 [80–101].

Energy intensive industries need to adapt in order to play an important role in the low-carbon economy. Efficient use of energy resources and the minimization of wasted heat will be important. The role of the steel industry in recovering recycled metals means these industries are important for a sustainable economy, providing employment and supporting manufacturing. The aim of the study by Raine [90] was to investigate the potential uses for heat storage in Sheffield, UK for capturing heat which is produced intermittently at a steelworks for both re-use of heat on site at various temperatures and for heat supply to a city-wide heat network. Site visits were followed by calculations using data provided. Heat storage options were investigated to ensure that waste heat could be re-used effectively. The feasibility of using the DH network in the city to carry low grade heat away from the plant was considered. Around 4.7 MW of useful heat could be generated from two steelworks sites in Sheffield and a further 10.9 MW from a site in nearby Rotherham. 22,500 tons of CO_2 could be saved per year by fully exploiting this waste heat resource [80,91].

Sheffield has for decades been a pioneering city in the UK in terms of developing DH to provide heat at lower cost and environmental impact to the city. The city-center network is supplied with up to 60 MW of heat from an Energy from Waste CHP station. The new network offers the prospect of recovering heat from the city's industrial sector. Finney et al. [91,92] identified potential for at least 10 MW of industrial waste heat for the city's expanded DH system. Sheffield and nearby Rotherham have an estimated annual output of 1.07 million tons of steel produced each year from four electric arc furnaces. Approximately 500 kWh of electricity is used in the furnace for each ton of steel produced [17], meaning an average of around 60 MW of electricity consumption and indirect CO_2 emissions of the order 280,000 tons per year from electricity alone. The furnace flue gases from steelworks are very hot, in the range 600°C–1,500°C, and although they are dust-laden, flue gas heat recovery systems in industry are increasingly common, with one example at Port Talbot steelworks [90]. Tenova Group have developed waste heat boilers for electric arc furnace flue gas heat recovery at sites in Germany and South Korea; both examples include the generation of steam. One of the heat recovery systems uses steam to supply heat to a 2.5 MW ORC generator. Higher temperature (higher exergy) energy could be used for power generation or through work processes such as vacuum creation in a steam jet ejector. High-pressure steam accumulators are widely used to balance supply and demand on sites for steam and can be used for storing recovered heat. Regenerator materials mainly made from ceramics are already used on many gas-fired furnaces to recover heat; these materials capture the heat of the exhaust air leaving the furnaces and, when the flow direction periodically changes, use that heat to preheat incoming combustion air. Recuperator systems are an alternative, allowing increased efficiency by exchanging heat from exhaust air with incoming air; this involves a heat exchanger and the heat is not stored.

It is possible to recover heat from cooling water on industry sites, for example, in the water that cools the electric arc furnace walls or the water that cools the hot gas extraction ducts. Residual heat may also be available from steam processes that operate typically at 200°C–250°C. In Graz, Austria heat is recovered from a gas-fired reheat furnace along with high-temperature cooling water at around 90°C from two electric arc furnaces [95–101]. Buffer tanks could be used in this instance to store water at temperatures in the range 70°C–110°C which are suitable for DH. If the water is above its atmospheric pressure boiling point then the tank needs to be pressurized, usually including a cushion of steam or nitrogen at the top of the tank to allow for thermal expansion and contraction processes. Many industry sites have relatively low-temperature cooling systems that are adapted from, or in some cases still use, river water cooling. Circulation of water to cooling towers reduces the need to import cool water. In some cases, cooling circuits with water treatment are necessary to prevent contaminants in the water from escaping. This water is often not hot enough for building heating but heat pumps could be used to draw useful heat from that water. Water tanks can be used to store the cooling water ranging from ambient temperatures up to possibly 70°C if the cooling systems are adjusted to run at higher temperatures, but large capacities are needed for lower temperature stores. Underground TES is another option that uses the thermal capacity of the ground and can be useful for storing large volumes of low-temperature heat for long periods.

Heat pumps can be used to draw heat from the atmosphere, ground, or bodies of water and supply it at a higher temperature suitable for heating. The CoP for the heat pump describes its performance. The emissions factor for electricity in a given country determines the environmental impact of a heat pump using electricity. One recent example in Norway draws heat from the fjords in order to supply a DH system at 90°C [38,95–101]. Sheffield's DH operates at 110°C and delivery of heat from heat pumps at these temperatures is difficult and suitable technology is at an early stage. The Norwegian heat pump uses ammonia as refrigerant which has its critical point at 132°C and pressures of 41 bar are required for the ammonia heat pump condensers to work [38,95–101]. In steelworks, the waste heat sources are primarily flue gases from furnaces (at high temperatures) and the cooling water from the electric furnace (at low temperatures). The electric arc furnace melts scrap metal at very high temperatures and removes impurities. It produces hot and dusty off-gas which must be extracted and filtered before release. Typically, one extraction duct captures gas from close to the furnace, while a canopy duct captures any fugitive emissions from around the furnace. Some furnaces preheat scrap metal in the extraction duct allowing energy to return to the furnace when the scrap is loaded; this reduces electricity consumption but the nature of processes at the sites in Sheffield makes this approach difficult. There are also gas-fired furnaces and cooling water where heat could be recovered [38,95–101].

Industrial sites typically have a range of heating and cooling needs depending upon the stage of the industrial process being carried out. Finding ways to pass heat effectively between parts of the process can save a lot of heating or cooling energy. This method is termed process integration, and the practicality of such steps was considered. The use of DH opens up opportunities for using waste heat streams for a new purpose and this could become more widespread in future, particularly in the UK. The many options for system arrangement, including how components such as heat storage are integrated, create uncertainty over the optimal way to achieve economic and environmental goals. Understanding the potential for using heat pumps is important; they give potential to extract heat from cooling water at the steelworks. If an ammonia heat pump is being used then it may be more effective to only raise the water to 90°C, as high pressures are needed for temperatures above that. The output temperature could then be topped using heat from the flue gases [38,95–101].

High temperature heat is already stored on sites using regenerator materials and in a steam accumulator. The accumulator matches steady production of steam from boilers to the short discharge needs of the vacuum processes. If more heat is to be recovered then the thermal capacities of the ceramic regenerators or the steam accumulator could be used to balance supply and demand of waste heat, although modifications will be needed to make such connections. If flue gas heat is recovered for generating steam then using existing steam storage facilities may be sufficient. Medium to low temperature heat could be stored using hot water tanks which can be linked to DH networks. If the water in the store is circulated in the network then this prevents temperature losses through heat exchangers. However, the water quality needs to be sufficiently high and the temperatures and pressures of operation need to be compatible with the network. In the case of Sheffield's DH, the store would need to operate between 70°C and 110°C. For low-temperature waste heat, underground TES could be used and would give potentially a high heat capacity at low cost. Storing the cooling water is an option, but low temperatures mean a large volume of storage would be needed and this increases expense. If electricity is needed to run a heat pump and upgrade this heat at the time of use then this may be costly if it is needed at peak demand times. Any heat recovery project needs to be evaluated, not just in economic terms but also in terms of its environmental impacts. This can help industry meet environmental objectives [38,95–101].

Modern industry recognizes the need to monitor, control, and minimize emissions of particles and harmful gases from their processes. One example of an emission that is closely monitored is dioxins, and while over 90% of human exposure is from dioxins present in food industry which is regulated to minimize dioxin emissions. High temperatures of over 850°C are required to destroy these particles and there is a chance of reformation between 500°C and 250°C. Many steelworks use quenching of gases with water spray to pass this critical temperature zone [80,84,85,91–100]. With flue gas heat recovery, heat exchanger design should account for this issue. Another important issue for heat recovery is the flue gas level of sulfur oxides, if these gases combine with moisture they create acid and this can corrode any heat exchanger increasing maintenance needs. In some instances, lime can be used to counter acidic gases, and activated carbon can be used to capture dioxins and heavy metals. The amount of water consumption at a steelworks is another issue with environmental implications. Approximately 14–28 m^3 of water is required per ton of steel produced in electric furnaces [95–100]. For some processes, the water is limited to a certain temperature rise in the cooling circuits to prevent corrosion, and high water velocities are needed to prevent particles from settling in cooling systems. The adjustment of cooling systems is a complex issue and will require detailed work by engineers to consider consequences before proceeding; however, use of high-temperature cooling has been achieved in some instances and this would carry benefits in terms of transferring heat to DH as well as reducing water use. Even the current low-temperature cooling could provide water suitable for a heat pump [80].

The amount of heat that can be recovered depends upon how that heat is to be used, for example, if the heat is to be used to generate high-pressure steam then only heat above a certain temperature will be useful. A two-part heat exchanger could be used in order to generate hot water for DH from the partly cooled flue gases allowing for a greater level of heat recovery. These would be aspects to consider in the detailed design of heat recovery equipment. For the contribution of heat to DH, a new heat store would be required. If the heat is just used when demand is available, and discarded when demand is not, then this avoids the need for heat storage; however, it also reduces the environmental benefits. A hot water store could be off-site if connected via DH, although the feed-in temperature and rate needs to be carefully controlled if feed into the heat network is instantaneous. If the heat store is operational as part of the network then it can also provide services to the DH operator [38,80–101]. To effectively use industrial waste heat, the operational principles need to be established early with the DH operator, including what charges and payments apply for supplied heat as well as when and how much heat can be fed-in at various times of day. There will be knock-on effects for other heat sources on the network. For example, a flexible CHP unit can be switched from electricity to heat production and therefore the injection of industrial heat increases the capacity to add electricity production to the grid. If electricity is used for a heat pump then heat injected to DH, this could have a negative effect on emissions associated with delivered heat and the way this is accounted will be important. Altering operational temperatures on the heat network would make the recovery of heat more feasible but the temperatures need to be sufficient to satisfy the needs of all the customers. If the new network connects to the old one then the water needs to be hot enough to be used by an absorption chiller unit in the city center. In the long term, lower temperatures could assist the integration of geothermal and solar energies [38,80–101].

Overall, it is quite probable that the energy spent in running a heat pump is inhibitive to the economics and therefore that only the flue gas heat which is much hotter can be recovered. However, if significant charges are associated with river water use for cooling then saving water by recovering heat may have better economics. The amount of heat that is recovered will depend upon the economics of the project and the sale price which heat can achieve through the network. Recovering heat from high-temperature industrial heat sources can boost the overall system efficiency and give environmental advantages; however, there are barriers to making this feasible. In particular, the up-front cost of constructing DH networks means that connections to industry can be capital-intensive. High-temperature heat pumps are a quickly developing technology, but some designs have practical limits on delivery temperatures which may limit their applications. In the UK, the high carbon intensity of grid electricity reduces the environmental advantages of electricity-driven heat pumps. Adjusting DH networks to run at lower temperatures in future will increase both efficiency and volume of recoverable industrial waste heat. Working to maximize heat recovery will help energy-intensive industry contribute to reducing the economy's carbon intensity [80–101].

3.9 OPTIMIZATION MODELS FOR HYBRID DISTRICT HEATING SYSTEMS

While there is intense global research activity addressing problems underpinning the re-engineering of the electrical power grid, thermal energy grids are likely to play an increasing role in energy systems. The study by Vesaoja et al. [37] described an ongoing effort in hybrid modeling and cosimulation of the physical and control domains of DH networks. The focus was on modeling each domain using semantics and tools natural to each; the study also described the challenges of, and a method for, integration and synchronization of the simulation models in each domain. Here, the dataflow model of computation used by Simulink provides a natural environment for modeling physics-based dynamics of heat energy flows, and the IEC 61499 automation architecture facilitated distributed systems modeling and enabled rapid deployment to field hardware. At the application level, the study showed how this framework enables study of energy flows within a producer/consumer (prosumer) and the analysis of the economic value of integrating distributed solar thermal generation and storage into a prosumer participating in a DH network. Nakomcic-Smaragdakis and Dragutinovic [102] examined hybrid renewable energy system application for electricity and heat supply of a residential building.

The study by Bakken [103] describes a research program that started in 2001 to optimize environmental impact and cost of a small-scale hybrid plant based on candidate resources, transportation technologies, and conversion efficiency, including integration with existing energy distribution systems. Special attention was given to a novel hybrid energy concept fueled by municipal solid waste. The commercial interest for the model was expected to be more pronounced in remote communities and villages, including communities subject to growing prosperity. The study by Bakken [103] describes the hybrid concept for conversion of municipal solid waste in terms of energy supply, as well as the methodology for optimizing such integrated energy systems.

A major challenge when analyzing complex energy systems was to combine the multicriteria objective with the modeling demands of a variety of different energy processes. Adding the complexity and time span of the investment analysis created an optimization problem not easily solved using conventional methods. In this approach, an important goal was to reduce the number of manual assumptions by separate modeling of each energy technology in sufficient detail. It is easy to argue against such an approach because some simplifying assumptions have to be made in any case. It is impossible to account for all physical aspects in one model due to the different properties of the processes involved. The option of combining different optimization methods allowed this approach to be able to obtain these goals without compromising the main physical characteristics of the processes involved [102,103].

Combination of different optimization methods adds new possibilities to the modeling of the energy related problems. It is not necessary to account for everything in one large model, as input from other models can be used in the areas where the "all-in-one models" meet limitations. An example of such a successful hybrid approach is a model that combines the long-term hydropower operation strategy calculated with stochastic dynamic programming with a detailed deterministic subproblem within one week [37,102–105]. It is not possible to account for every aspect of the hydropower system when calculating the long-term strategy. In order to calculate a long-term strategy, one needs to aggregate in time such that start-up cost, time delays, and hydraulic couplings are not properly accounted for. This makes the results less useful if the modeled system does not meet certain assumptions. Typical assumption for aggregated hydropower models are zero start-up cost, intermittent operation allowed for pumps, limited system size, and a nonsequential time description. In many problems, however, these properties can be accounted for when another model is used for implementing the strategy. Adding the results from the long-term strategy as boundary conditions to a deterministic linear mode makes it possible to account for the properties that cannot be included in stochastic optimization. This combination of methods makes it possible to handle details in a proper way despite of the inability of the strategy calculation to handle every hydropower detail.

The approach taken by Bakken [103] can be useful also in the case of energy distribution systems with multiple energy carriers. Detailed process models are created to account for properties that are difficult to combine without simplifying assumptions. A transmission model of either AC or DC power including security constraints is used for the local electricity distribution system. Special models for truck transportation to and from waste plants are used, as well as a DH model which can optimize operation of the DH system taking nonlinear elements into account. An adequate model for hydropower to account for stochastic elements can also be used if there is hydropower in the region. Results from the component models are afterward used in the linear system model to calculate the optimal operation plan for the selected time period (day/ week) for a given topology alternative [102–105]. The optimal operation planning kernel must be integrated with an investment analysis scheme to choose the best possible expansion plan over the planning horizon. So far dynamic programming has been used to find the best expansion plan according to the given alternatives and possible introduction times.

The study thus outlined involved the development of a new methodology for analysis of complex energy distribution systems with multiple energy carriers. The methodology was based on two main levels of modeling. The lower level was used to calculate optimal operation of the system and the upper level handled the investment decisions. In the Graphical User Interface, specific component modules with a standard interface were combined in alternative compositions of the energy system. Each alternative was then generalized to a nodal network with generic energy flow. To enable a multicriteria optimization with a minimum of simplifying assumptions, which might limit the validity of the results, hybrid optimization techniques were implemented, e.g., combinations of stochastic dynamic programming and deterministic short-term optimization. Each energy technology was modeled separately with sufficient detail, supplying the superior linear system model with a simple and unambiguous set of variables such as cost, energy efficiency, and environmental impact. The methodology will enable energy companies to carry out comprehensive analyses of their energy supply systems, and governmental bodies will be able to do comprehensive scenario studies of energy systems with respect to environmental impacts and consequences of different regulating regimes for preserving environmental values [102–105].

3.10   ROLE OF TES IN DISTRICT HEATING

TES is of assistance to energy suppliers in a DE system by allowing [106–126]:

1. the accumulation of thermal energy from off-peak periods for use when demands are high;
2. the storage of excess thermal energy when it is available, for subsequent release to the DE distribution system when thermal demands increase (especially during periods when suppliers are not able to satisfy thermal energy demands with existing facilities). In this way, TES saves thermal energy that would otherwise be wasted;
3. more effective utilization of renewable thermal energy sources like solar than is otherwise possible due to the intermittent nature of the resource supply;
4. Regardless of the source of energy in a DE system, reducing energy losses is usually a main advantage of using TES in DE systems. Incorporating TES in DE systems allows the reduction of thermal losses, resulting in energy savings and increased efficiency for the overall thermal system. Large seasonal TES systems have been built in conjunction with DE technology [112,123–126]. Besides improved efficiency and economics, environmental benefits are another reason for the expansion of TES technology in general and with DE.

Many beneficial applications of TES with DE exist or can be developed. For example, Andersson [113] reports that the application of ATES in one DE system reduced energy use by 90%–95% compared to conventional systems not incorporating TES. DE with hybrid systems can advantageously combine sources of energy such as natural gas, waste heat, wood wastes, and municipal solid waste [106–114]. TES can increase the benefits provided by such systems. Lund et al. [111] point out that "low energy" buildings can be operated with industrial waste heat, waste incineration, power plant waste heat, and geothermal energy in conjunction with a DE network. Incorporating TES can improve designs of such systems. The types of energy sources that can be used with DE and TES systems vary, and systems designs are often tailored to best adapt to the energy source.

Tanaka et al. [112] state the use of TES in conjunction with DE decreases energy consumption compared to a reference system. They also found that seasonal TES is more effective than short-term TES. Andrepont [123] among others [109–111] assessed the economic benefits of using TES technologies in DE systems. He noted that cool TES is applied widely in heating, ventilation, and air-conditioning systems by shifting the cooling load from peak periods during the day to the off-peak periods at night. This time shift significantly reduces operating costs, and this benefit is particularly noteworthy in large-scale DE systems. TES in DE systems also prevents inefficient operation of chillers and auxiliary equipment during low-level operation, enhances system reliability and flexibility, balances electrical and thermal loads in CHP for better economy and lowers accident risks and insurance by enhancing fire protection. The benefits of TES in facilitating the use of renewable energy, especially solar thermal energy for use in heating and cooling buildings, have been pointed out by several investigators [106–113]. Demand is growing for facilities that utilize TES, as they are more efficient and environmentally friendly, exhibiting reductions in (a) fossil fuel consumption, (b) emissions of CO_2, and other pollutants, and (c) chlorofluorocarbon emissions.

Several investigators have reported [109–111,113] that buildings with TES systems in DE applications consume more energy than buildings without TES in many cases, and that all systems are beneficial environmentally. The U.S. Green Building Council did not discourage the use of TES in the first version of the Leadership in Energy and Environmental Design criteria, but also did not deal with the use of TES in DH systems; however, a building with TES is eligible to earn more point for its lower electrical power use. Building a TES requires initial capital for land, construction, insulation, and other items. Determining the appropriate location, designing the proper structure and insulation, and executing the design are important steps in installing a TES, which involve significant costs. The high initial cost is a disadvantage of TES.

The use of TES complements the use of solar energy in the DE system, allowing surplus solar heat in the spring and summer to be stored for subsequent use in the fall. Without TES, this surplus energy would be wasted. TES thereby enhances the benefits of using solar energy in the DE system. It is observed that TES reduces annual fuel use and fuel costs by 30% in a typical DE system, reducing the use of natural gas boilers. The use of TES is also advantageous environmentally, reducing emissions to the atmosphere like CO_2 by 46% in the DE system. The advantages of incorporating TES in the DE system, in terms of enhancing the use of solar thermal energy, suggest TES is likely going to become increasingly important and utilized in industry, power generation, and DE as use of renewable energy expands. Lund et al. [114,126] indicated that renewable energy is the focus of many countries for improving energy security and mitigating climate change. The use of renewable energy in a sustained form for DH is heavily facilitated by the TES. The role of TES for use of renewable energy in DH can be briefly outlined as follows [106–126].

Solar energy can be used for space and water heating. While the advantages of using solar energy are significantly enhanced with the help of TES in the thermal system, the key benefit of TES is in its ability to integrate the solar thermal application, i.e., the energy stored by TES can be made available when thermal energy is in demand and solar availability is uncertain. TES increases the impact of solar collectors by avoiding the loss of solar energy when it exceeds demand. Annual TES is desirable if the excess solar energy is stored for a season or longer, while short-term storage is more appropriate if the solar energy is stored for hours or days. In both cases, however, TES provides a beneficial alternative to the use of conventional fuels as a back-up energy source. Short-term TES applications in conjunction with solar energy in DH systems have been tested in pilot projects and are now used in several countries [112,113–126].

The ground source heat pump is an efficient device and can be used advantageously with DE for HVAC purposes. When a ground source heat pump is the source of energy in DE system, TES can help the energy system store extracted heat from the earth for use when in demand. According to Lund et al. [111], usually, BTES (Borehole Thermal Energy Storage) is an appropriate TES for geothermal energy in conjunctions with ground source heat pumps. Industry can be a supplier of waste heat for a DE system and/or a heat consumer from the DE system. When industry has excess heat or waste heat, it can supply to a DE system, or it becomes a consumer when it requires heat. Holmgren and Gebremedhin [114] point out that cooperation between industry and DE can allow technical and economical factors to be appropriately addressed for both parties in terms of the thermal energy quality and quantity as well as profitability. When the energy for a DE system is supplied by waste heat, the key parameters to be considered in designing DE and TES include (a) heat supply temperature, (b) heat consumption temperature, (c) heat supply time, and (d) heat demand time.

The literature [109–112] also suggested underground thermal storage such as ATES as an appropriate storage for waste heat. This type of system allows stored thermal energy to be made available for consumers and also permits load leveling in the DE system. DE systems can supply waste heat from an existing boiler which has waste heat or from other industrial processes. DE systems using such energy sources are generally more clean, economic, and efficient than DEs using conventional fuels. Holmgren and Gebremedhin [117] also note that the integration of DE with industry reduces not only the cost of heat production but also CO_2 emissions. When a DE system directly operates using fuel, TES is not always needed. Nonetheless, TES provides the possibility of using smaller equipment in designs, by reducing energy need from external energy sources. Electricity is a significant energy source in a DE system. TES helps reduce electricity costs during peak demand periods and allows the system to operate during off-peak periods, e.g., running chillers during the night and storing the cold medium for use in cooling the next day. Short-term cold TES is used in such applications.

According to Holmgren and Gebremedhin [114], waste incineration is a useful technology for recovering the energy content of waste. Alternatively, waste that can be disposed of in landfill sites and accelerated processes (e.g., gasification, pyrolysis) can be used to produce a combustible fuel gas [115]. Holmgren and Gebremedhin [114] believe waste incineration is a preferred option compared to others for supplying heat for DE. Also, TES eliminates energy waste in existing DH network When waste incineration supplies the thermal energy for a DE system, several parameters need to be considered in DE and TES design, including (a) capacity of the waste incineration process, (b) availability of the waste, and (c) time profile of the heat demand.

TES systems are technologies with the potential to enhance the efficiency and the flexibility of the upcoming 4th generation low-temperature district heating (LTDH). Their integration would enable the creation of smarter, more efficient networks, benefiting both the utilities and the end consumers. The study by Espagnet [126] was aimed at developing a comparative assessment of both latent and sensible heat-based TES systems. First, a techno-economic analysis of several TES systems was conducted to evaluate their suitability to be integrated into LTDH. Then, potential scenarios of TES integration were proposed and analyzed in a case study of an active LTDH network. This was complemented with a review of current DH legislation focused on the Swedish case, with the aim of taking into consideration the present situation, and changes that may support some technologies over others.

The results of the analysis showed that sensible heat storage was still preferred to latent heat when coupled with LTDH: the cost per kWh stored was still 15% higher, at least, for latent heat in systems below 5 MWh of storage size; though, they require just half of the volume. However, it is expected that the cost of latent heat storage systems will decline in the future, making them more competitive. From a system perspective, the introduction of TES systems into the network results in an increase in flexibility leading to lower heat production costs by load shifting. It is achieved by running the production units with lower marginal heat production costs for longer periods and with higher efficiency, and thus reducing the operating hours of the other more expensive operating units during peak load conditions. In the case study, savings in the magnitude of 0.5 k EUR/year are achieved through this operational strategy, with an investment cost of 2 k EUR to purchase a water tank. These results may also be extended to the case when heat generation is replaced by renewable, intermittent energy sources; thus increasing profits, reducing fuel consumption, and consequently emissions. This study represents a step forward in the development of a more efficient DH system through the integration of TES, which will play a crucial role in future smart energy system.

The benefits of using TES as a complementary system for solar energy collectors was also examined by Anderpont [123], Reed et al. [117], and Rezaie et al. [119,122,125]. Rezaie et al. [119,122,125] examined this within the context of the Friedrichshafen DH system in Germany, which utilizes renewable energy. The TES in the Friedrichshafen DE system has enhanced the DE's performance. In particular, TES has improved the performance of the solar energy system, reducing annual natural gas consumption and annual fuel cost by 30% and harmful environmental emissions by 46%. DH and/or cooling systems can be augmented through incorporation of TES. Anderpont [123] determined that chilled water TES and low-temperature fluid TES, used in large-scale DE systems, significantly lower installation costs per ton compared with equivalent conventional non-TES chiller plants. In some DE designs, TES is incorporated to store solar energy that would otherwise go to waste during periods when heating is not required. The Friedrichshafen DH system in Germany, for example, uses TES with DE to enhance performance and efficiency [120,121]. The focus of the study by Rezaie et al. [119,122,125] was to demonstrate the role and benefits of TES in DE through a case study, via an energy analysis for Friedrichshafen DE system.

Finally, Romanochenko et al. [127] showed that heat load variations in DH systems lead to increased costs for heat generation and, in most cases, increased greenhouse gas emissions associated with the marginal use of fossil fuels. They investigated the benefits of applying TES in DH systems to decrease heat load variations, comparing storage using a hot water tank and the thermal inertia of buildings (with similar storage capacity). A detailed techno-economic optimization model is applied to the DH system of Göteborg, Sweden. The results show that both the hot water tank and the thermal inertia of buildings benefit the operation of the DH system and have similar dynamics of utilization. However, compared to the thermal inertia of buildings, the hot water tank stores more than twice as much heat over the modeled year, owing to lower energy losses. For the same reason, only the hot water tank is used to store heat for periods longer than a few days. Furthermore, the hot water tank has its full capacity available for charging/discharging at all times, whereas the capacity of the thermal inertia of buildings depends on the heat transfer between the building core *and* its indoor air and internals. Finally, the total system yearly operating cost decreases by 1% when the thermal inertia of buildings and by 2% when the hot water tank is added to the DH system, as compared to the scenario without any storage.

3.10.1 ENERGY CENTRAL

The Energy Central is the result of a cooperation between Grundfos [116] and the local DH company. Few years ago, Grundfos headquarters and the local DH company inaugurated a joint system to store the surplus heat from the Grundfos factories in obsolete groundwater boreholes and use it in the DH network when needed. The new setup was based on three elements:

1. Exploitation of surplus heat from cooling in the factories.
2. Indirect storage of heat in an underground aquifer.
3. Heat pumps supply additional heat, when required.

With the new shared Energy Central, Grundfos productions facilities are cooled by cold water from the DH network. Also, when the cold water needs to be colder, three large cooling compressors with a total cooling capacity of 2.85 MW and thermal power of 3.65 MW handle the job from a central facility. During the heating season, the cold water used to cool the machines at Grundfos becomes hot in the process, and along with the excess heat from the cooling compressors in the energy central, the heat is recycled and sent directly to the DH network. Main features of this waste heat-DE-TES partnership are:

- During summer, when the heating demand is minimal, the surplus heat is stored in underground energy storage—a so-called ATES-stock (Aquifer Thermal Energy Storage) located 80 m underground.

- During the four summer months, a total cooling output of 3,500 MWh is accumulated. And more than 80% of the stored energy during summer will be supplied to the DH network during the heating season.
- The pressure drop across the valves has been completely eliminated and replaced by massive energy savings.
- The ATES system reduces Grundfos' energy consumption for cooling by up to 90%.
- The carbon emission is reduced by some 3,700 tons a year, equivalent to 1.5 tons per household connected to the heating plant.
- The payback time will most likely be shorter than the projected 11.25 years.
- With the new setup, Bjerringbro DH company seizes the opportunity to store energy and becomes less dependent on natural gas.

3.11   HYBRID DE IN US

According to a recent estimate from the U.S. Energy Information Administration, there are more than 660 DE systems operating in the U.S. with installations in every state, and the number of buildings and amount of floor space served by DE is steadily increasing. While there are challenges related to economics, engineering, and education that can pose barriers to potential projects, there are also wide-ranging opportunities for greater use of DE. Research into potential efficiency improvements, system optimization techniques, and integration of advanced technologies has the potential to help accelerate deployment of DE. Cumulative DE industry growth is illustrated in Figure 3.5 [10].

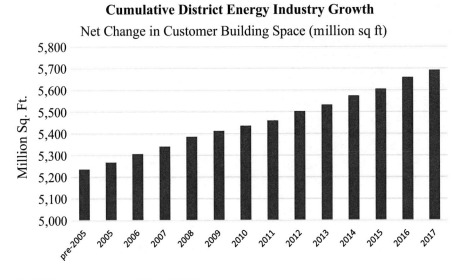

FIGURE 3.5   Growth in DE in US [10].

3.11.1   EXAMPLES OF USE OF DE IN US

Here we illustrate seven examples on the specific use of hybrid DE in US. Many more are illustrated in DOE report [10]. These examples indicate the usefulness of hybrid energy and processes concept that goes in DE and resulting decarbonization. The examples are illustrated with relevant details in in Figures 3.6–3.12 [10].

FIGURE 3.6   Fuel flexibility provides cost savings and resilience to downtown St. Paul. (Photo credit: IDEA) [10].

- **Location**: St. Paul, MN
- **Location**: St. Paul, MN
- **Energy supplied**: hot water, chilled water, electricity

District Energy St. Paul operates the largest wood-fired CHP plant serving a DE system in the Nation. It produces 65 MW of thermal energy and 25 MW of electricity for the local utility. In 2011, the system integrated the Midwest's largest thermal solar installation, which peaks over 1.2 MW and provides hot water and space heating to the Saint Paul River Center and other customers. Fuel flexibility helps avoid market volatility and manage availability constraints. Biomass used in the plant is locally sourced from waste wood including residual from manufacturing, construction waste, tree trimmings, and other sources, providing energy security and supporting the local economy.

FIGURE 3.7 Princeton University integrates solar, CHP, and thermal storage to increase campus resilience. (Photo credit: IDEA) [10].

- **Location**: Princeton, New Jersey
- **Sector**: College/University
- **Energy supplied**: Steam, chilled water, electricity

Princeton University operates an innovative CHP-based microgrid that serves approximately 150 buildings. It includes a 15 MW CHP system, 4.5 MW of on-campus PV, 20,000 tons of chilled water capacity, a large chilled water thermal storage tank, and a sophisticated controls optimization program. Integrating and optimizing these technologies helps save over $2 million/year in operating expenses, reduces emissions, and supports mission-critical.

FIGURE 3.8 Alaska's first geothermal power plant lowers electricity costs. (Photo credit: Chena Power) [10].

- **Location**: Chena, AK
- **Sector**: Municipal
- **Energy supplied**: Steam, hot water, electricity

In July 2006, the State of Alaska began operations at its first geothermal power plant in Chena Hot Springs, AK. Using two 200 kW ORC units, the power plant provides electricity and thermal energy to the surrounding area, including the Chena Hot Springs Resort and its adjacent greenhouse. The system displaced 224,000 gallons of diesel that was previously used to provide power, along with providing 95% reliability and a $0.25/kWh reduction in electricity costs.

FIGURE 3.9   Duel-fuel system powers Tennessee VA Medical Center Campus. (Photo credit: US Department of Veterans Affairs) [10].

- **Location**: Mountain Home, TN
- **Sector**: Healthcare
- **Energy supplied**: Space heating, space cooling, electricity

The Mountain Home VA Healthcare system installed a dual-fuel CHP system at the James H. Quillen Medical Center in 2001 to provide electricity, heating, and cooling to the medical campus. This system uses biogas from the nearby Iris Glen Landfill and can also run on natural gas if the biogas is unavailable. The DE system reduced energy consumption for the campus by 20%, resulting in annual savings over $250,000.

FIGURE 3.10 TECO converts to CHP for enhanced resilience and reduced energy use. (Photo credit: IDEA) [10].

- **Location**: Houston, Texas
- **Sector**: Healthcare
- **Energy supplied**: Steam, chilled water, electricity

Thermal Energy Company (TECO) has been serving the thermal energy needs (heating, cooling, DHW, etc.) of the Texas Medical Center, the largest medical center campus in the world, since 1968. In order to meet the high resiliency requirements of their customers, a 48 MW CHP system was added to the system in 2010. In 2017, TECO maintained continuous operations throughout the severe weather and record flooding from Hurricane Harvey and its aftermath.

FIGURE 3.11   U.S. Navy increases reliability for operations at Portsmouth shipyard. (Photo credit: US Navy) [10].

- **Location**: Kittery, ME
- **Sector**: Military
- **Energy supplied**: Steam, electricity

To provide increased reliability for its operations, the U.S. Army Corps of Engineers and the U.S. Navy installed a microgrid at the Portsmouth Naval Shipyard. The microgrid provides islanding capabilities to the shipyard, eliminating downtime during a disruption on the grid. The project included several upgrades to the shipyard's system, including a new 10.4 MW CHP plant, new boilers, and a battery storage system. These improvements have led to annual energy cost savings of $6.2 million.

FIGURE 3.12   Energy sharing with amazon's DE system. (Photo credit: JORDAN STEAD/ Amazon) [10].

- **Location**: Seattle, Washington
- **Sector**: Corporate Campus
- **Energy supplied**: Hot water

The 60,000 square foot data center at the Westin Building Exchange in downtown Seattle, which houses equipment serving more than 250 telecom and Internet companies, generates enough waste heat from cooling its servers to provide the space heating for Amazon. The central plant for the DE system is located in Amazon's Doppler Tower. As Amazon expands, the company plans to use DE at future campuses. DE was included in the 2017 request for proposal for Amazon's second corporate headquarters (HQ2)

REFERENCES

1. District Heating. (2019). Wikipedia, The free encyclopedia, last visited 17 July 2019.
2. District Heating. (2015). SteepWiki. A website report https://tools.smartsteep.eu/wiki/ District_heating.
3. Matson, C. (2015, July). District-Heating—a real alternative. A website report on Renewable transformation challenge, Elsevier.
4. Renewable Energy in District Heating and Cooling. (2017, March). A Sector Roadmap for Remap, IRENA report, Abu Dhabi.

5. Developing Geothermal District Heating in Europe. (2009). Project coordinator: European Geothermal Energy Council Place du Champ de Mars 2, Brussels E: com@egec.org. WWW.GEODH.EU.
6. Mapping and Analyses of the Current and Future. (2020–2030). Heating/cooling fuel deployment (fossil/renewables), European commission directorate-general for energy directorate C. 2 – New energy technologies, innovation and clean coal, Final report, February 2017 Prepared for: European Commission under contract NENER/C2/2014-641.
7. Dalla Rosa, A., Li, H., Svendsen, S. et al. (2014). IEA DHC Annex X report: Toward 4th generation district heating: Experience and potential of low-temperature district heating.
8. Renewable Energy in District Heating and Cooling Executive Summary. (2017, March). Irena report, Abu Dhabi.
9. District Energy in Cities. (2016). A report by UNEP, Paris, France.
10. Energy Efficiency and Energy Security Benefits of District Energy. (2019, July). DOE report, Washington, DC.
11. Euroheat and Power. (2015). District heating and cooling country by country Survey 2015. Indicator codes: ENER 019, www.euroheat.org/wp-content/uploads/2016/03/2015-Country-by-country-Statistics-Overview.pdf; www.eea.europa.eu/policy-documents/euroheat-and-power-2013-district, 19 June 2019. URL: http://www.euroheat.org/wp-content/uploads/2016/03/2015-Country-by-country-Statistics-Overview.pdf.
12. Shah, Y.T. (2020). *Modular Systems for Energy Usage Management*. CRC Press, New York.
13. Odgaard, O. (2015). China's quest for new district heating reforms. Policy Brief, No. 3, www.thinkchina. ku.dk/documents/2015-12-01ThinkChina_PolicyBrief_district heating_and_CHP_in_China.pdf. Accessed 24 September 2016.
14. Rutz, D., Doczekal, C., Zweiler, R., Hofmeister, M. and Jensen, L.L. (2017). Small modular renewable heating and cooling grids, A Handbook, a report by Cool Heating. eu ISBN: Translations: 978-3-936338-40-9, 2017 by WIP Renewable Energies, Munich. www.coolheating.eu.
15. An Action Plan for Growing District Energy Systems across Canada. (2011, June). A website report by CDEA/ACRT, 75 pages, Ottawa, Canada https://static1.squarespace.com/static/.../t/.../CUIPublication.GrowingDistrictEnergy.pd.
16. Danish Energy Agency, Energinet.dk. (2015). Technology data for energy plants—generation of electricity and district heating, energy storage and energy carrier generation and conversion. May 2012 (certain updates made October 2013, January 2014 and March 2015), ISBN: 978-87-7844-931-3.
17. Ericssona, K. and Wernerb, S. (2009, April). The introduction and expansion of biomass use in Swedish district heating systems. *Energy Institute, Biomass and Bioenergy* 33(4): 659–678, Doi: 10.1016/j.biombioe.2016.08.011.
18. Vallios, I., Tsoutsos, T. and Papadakis, G. (2013, November). Design of biomass district heating systems. *Energy Policy* 62: 236–246, Doi: 10.1016/j.biombioe.2008.10.009.
19. Dimitriou, I. and Rutz, D. (2015). *Sustainable Short Rotation Coppice, A Handbook*. WIP Renewable Energies, Munich, ISBN: 978-3-936338-36-2; www.srcplus.eu.
20. Laurberg, J.L., Rutz, D., Doczekal, C., Gjorgievski, V., Batas-Bjelic, I., Kazagic, A., Ademovic, A., Sunko, R. and Doračic, B. (2016). *Best Practice Examples of Renewable District Heating and Cooling. Report of the Cool Heating Project*, PlanEnergi, Denmark, www.coolheating.eu.
21. Rutz, D. and Janssen, R. (2008). *Biofuel Technology Handbook*. 2nd Version; Biofuel Marketplace Project funded by the European Commission (EIE/05/022). 10 November 2016, WIP Renewable Energies, Munich, 152. www.wip-munich.de/images/stories/6_publica-tions/books/biofuel_technology_handbook_version2_d5.pdf.

22. Rutz, D., Mergner, R. and Janssen, R. (2015). *Sustainable Heat Use of Biogas Plants—A Handbook*, 2nd ed. WIP Renewable Energies, Munich; Handbook elaborated in the framework of the Biogas Heat Project; ISBN: 978-3-936338-35-5 translated in 8 languages; www.wip-munich.de/images/stories/6_publications/books/ Handbook-2ed_2015-02-20-cleanversion.pdf, 10 November 2016.

23. Zweiler, R., Doczekal, C., Paar, K. and Peischl, G. (2008). Endbericht Energetisch und wirtschaftlich optimierte Biomasse-Kraft-Wärmekopplungssysteme auf Basis derzeit verfügbarer Technologien, Energiesysteme der Zukunft, bmvit, FFG-Projektnummer 812771. www.get.ac.at.

24. Zweiler, R. (2013). ToughGas (Entwicklung eines innovativen Wirbelschichtvergasungssystems kleiner Leistung zur Nutzung biogener Reststoffe) (Endbericht No. 834621). Dresden, Germany.

25. Vladimir, N. (2014, April). *Biomass Combined Heat and Power (CHP) for Electricity and District Heating*, Ph.D. Thesis. Norwegian University of Science and Technology (NTNU), Trondheim.

26. District Heating and Cooling, Combined Heat and Power and Renewable Energy Sources, BASREC—Best Practices Survey. (2014, January). COWI, NUORKIVI Consulting, Denmark. C:\Users\CHOE\Desktop\Dokumenter\BASREC\DHC_CHP_RES_survey_BASREC_Countries.docx.

27. FVB Projects I District Energy I FVB Energy Inc. (2019, April 8). www.fvbenergy.com/projects/, a website report.

28. FVB Energy Inc. (2019). District Energy, District Heating. www.fvbenergy.com/.

29. Galatoulas, F., Frere, M. and Ioakimidis, C. An overview of renewable smart district heating and cooling applications with thermal storage in Europe. *Proceedings of the 7th International Conference on Smart Cities and Green ICT Systems (SMARTGREENS 2018)*, 311–319, ISBN: 978-989-758-292-9. 2018 by SCITEPRESS – Science and Technology Publications, Lda. All rights reserved.

30. District Heating System—An Overview I ScienceDirect Topics. (2011). A website report www.sciencedirect.com/topics/engineering/district-heating-system.

31. Ilie, A. and Visa, I. (2019). Hybrid solar-biomass system for district heating. *E3S Web of Conferences* 85: 04006, Doi: 10.1051/e3sconf/20198504006.EENVIRO 2018 www.e3s-conferences.org/articles/e3sconf/pdf/.../e3sconf_enviro2018_04006.pdf.

32. Zhang, C., Sun, J., Ma, J., Xu, F. and Qiu, L. (2019, June). Environmental assessment of a hybrid solar-biomass energy supplying system: A case study. *International Journal of Environmental Research and Public Health* 16(12): 2222. Published online 24 June 2019, Doi: 10.3390/ijerph16122222. PMCID: PMC6617335, PMID: 31238546.

33. Fjernvarme, D. (2016, November 11). Technology. www.geotermi.dk/english/deep-geothermal-energy-in-denmark/technology, a website report.

34. Danish Geothermal District Heating. (2016, November 9). The geothermal concept. www.geotermi.dk/english/deep-geothermal-energy-in-denmark/technology, a website report.

35. GeoDH. (n.d.). Developing geothermal district heating in Europe. www.geodh.eu, https://ec.europa.eu/energy/intelligent/projects/sites/iee-projects/files/projects/documents/geodh_final_publishable_results_oriented_report.pdf (10 November 2016), a website report.

36. Frederiksen, S. and Werner, S. (2013). District heating and cooling. Studentlitteratur, 205.

37. Vesaoja, E., Yang, C.-W., Nikula, H. and Sierla, S. (2014). Hybrid modeling and co-simulation of district heating systems with distributed energy resources. Conference Paper (PDF Available) April 2014 Conference: *Workshop on Modeling and Simulation of Cyber-Physical Energy Systems*, Berlin.

38. Hackel, S. (2008). Development of design guidelines for hybrid ground-coupled heat pump systems. ASHRAE Technical Research Project 1384, American Society of Heating, Refrigerating, and Air Conditioning Engineers (ASHRAE), Atlanta, GA.

39. Chiasson, A.D. and Yavuzturk, C., (2009). A design tool for hybrid geothermal heat pump systems in heating-dominated buildings. *ASHRAE Transactions* 115(2): 60–73.
40. Chiasson, A.D. and Yavuzturk, C. (2009). A design tool for hybrid geothermal heat pump systems in cooling-dominated buildings. *ASHRAE Transactions* 115(2): 74–87.
41. Xu, X. (2007). *Simulation and Optimal Control of Hybrid Ground Source Heat Pump Systems*, Ph.D. Thesis, Oklahoma State University, Stillwater, OK.
42. Hackel, S., Nellis, G. and Klein, S. (2009). Optimization of cooling-dominated hybrid ground-coupled heat pump systems. *ASHRAE Transactions* 115(1): 565–580.
43. Kavanaugh, S.P. (1998). A design method for hybrid ground-source heat pumps. *ASHRAE Transactions* 104(2): 691–698.
44. Cullin, J. and Spitler, J.D. (2010). Comparison of simulation-based design procedures for hybrid ground source heat pump systems. *Proceedings of the 8th International Conference on System Simulation in Buildings 2010*, Liege.
45. Numerical Logics, Inc. (2008). Solar domestic hot water system sizing for Whitehorse, YT and Dawson, YT, Waterloo, ON.
46. SEL. (2000). *TRNSYS, A Transient Systems Simulation Program, Version 15*. Solar Engineering Laboratory (SEL), University of Wisconsin-Madison, Madison, WI.
47. Chiasson, A. (2011, May). A feasibility study of a multi-source hybrid district geothermal heat pump system. *GHC Bulletin*: 9–14. a website report https://oregontechsfcdn.azureedge.net/...source/geoheat.../art3e476ee4362a663989f6f....0.
48. Decarbonising District Heating with Solar Thermal Energy—Solar. (2018, June 7). www.solar-district-heating.eu/solar-district-heating-on-the-roof-of-the-world-4/.
49. Central Solar Heating. (2019). Wikipedia, The free encyclopedia, last visited 26 June 2019.
50. Kempener, R. (2015). Solar heating and cooling for residential applications: Technology brief. IEA-ESTAP and IRENA Technology Brief E21—January 2015. www.irena.org/documentdownloads/publications/irena_etsap_tech_brief_r12_solar_thermal_residential_2015.pdf, 4 August 2016. ISBN: 978-92-95111-60-8.
51. Schrøder, P.A., Elmegaard, B., Christensen, C.H. et al., (2014). Status and recommendations for RD&D on energy storage technologies in a Danish context. www.energinet.dk/SiteCollectionDocuments/Danske%20doku-menter/Forskning%20-%20PSO-projekter/RDD%20Energy%20storage_ex%20app. Pdf, 9 November 2016. A website report.
52. Solair Project. (2009). Increasing the market implementation of solar-air-conditioning systems for small and medium applications in residential and commercial buildings (SOLAIR). Project website www.solair-project.eu/142.0.html. Accessed 4 August 2016.
53. Pauschinger, T. (2016). Solar thermal energy for district heating—ScienceDirect. A website report www.sciencedirect.com/science/article/pii/B9781782423744000057.
54. Trier, D., Skov, C.K., Sørensen, S.S. and Bava, F. Solar district heating trends and possibilities, characteristics of ground-mounted systems for screening of land use requirements and feasibility technical report of IEA SHC Task 52, Subtask B—methodologies, tools and case studies for urban energy concepts prepared by PlanEnergi, Copenhagen, June 2018 for IEA, Paris France. iea-shc.org/.../SDH-Trends-and-Possibilities-IEA-SHC-Task52-PlanEnergi-20180619.
55. Solar District Heating—An Overview I ScienceDirect Topics. (2013). A website report www.sciencedirect.com/topics/engineering/solar-district-heating.
56. Hopkins, A.S., Takahashi, K. and Melissa, D.G. (2018, October). Whited, decarbonization of heating energy use in California buildings technology, markets, impacts, and policy solutions, A report by Synapse, Energy economics Inc. This report was prepared for the Natural Resources Defense Council (NRDC), Decarbonization of Heating Energy Use in California Buildings, A website report www.synapse-energy.com/.../Decarbonization-Heating-CA-Buildings-17-092-1.

57. Xydis, G. (2015). Wind energy integration through district heating. *A Wind Resource Based Approach* 4: 110–127, Doi: 10.3390/resources4010110, ISSN 2079-9276. www. mdpi.com/journal/resources.

58. Wang, J., Zong, Y., You, S. and Træholt, C. (2017). A review of Danish integrated multi-energy system flexibility options for high wind power penetration. *Clean Energy* 1(1): 23–35, Doi: 10.1093/ce/zkx002. Advance Access Publication Date: 24 November 2017. Homepage: https://academic.oup.com/ce.

59. Wind Power and District Heating. (2011). A website report www.pfbach.dk/firma_pfb/forgot-ten_flexibility_of_chp_2011_03_23.pdf.

60. Niemi, R., Mikkola, J. and Lund, P.D. (2012). Urban energy systems with smart multi-carrier energy networks and renewable energy generation. *Renewable Energy* 48: 524–536.

61. Long, H., Xu, R. and He, J. (2011). Incorporating the variability of wind power with electric heat pumps. *Energies* 4: 1748–1762.

62. Maegaard, P. Balancing fluctuating power sources. *Proceedings of 2010 World Non-Grid-Connected Wind Power and Energy Conference (WNWEC)*, 5–7 November 2010, Nanjing, 4–7.

63. Hong, L., Lund, H. and Möller, B. (2012). The importance of flexible power plant operation for Jiangsu's wind integration. *Energy* 41: 499–507.

64. Mathiesen, B.V., Lund, H. and Connolly, D. (2012). Limiting biomass consumption for heating in 100% renewable energy systems. *Energy* 48: 160–168.

65. Lund, H., Möller, B., Mathiesen, B.V. and Dyrelund, A. (2010). The role of district heating in future renewable energy systems. *Energy* 35: 1381–1390.

66. Mathiesen, B.V. and Lund, H. (2009). Comparative analyses of seven technologies to facilitate the integration of fluctuating renewable energy sources. *Renewable Power Generation* 3: 190–204.

67. Fitzgerald, N., Foley, A.M. and McKeogh, E. (2012). Integrating wind power using intelligent electric water heating. *Energy* 48: 135–143.

68. Losev, V.L., Sigal, M.V. and Soldatov, G.E. (1989). Nuclear district heating in CMEA countries. *IAEA Bulletin* 3: 46–49.

69. Margen, P. (1978). The use of nuclear energy for district heating. *Progress in Nuclear Energy* 2(1): 1–28, Doi: 10.1016/0149-1970(78)90010-0.

70. CNNC Completes Design of District Heating Reactor—a website report by World Nuclear News. (2018, September 7). A website report by www.world-nuclear-news. org/.../CNNC-completes-design-of-district-heating-reactor. A model of the Yanlong reactor (Image: CNNC).

71. Csik, B.J. and Kupitz, J. (1997). Nuclear power applications: Supplying heat for homes and industries. *IAEA Bulletin*; ISSN 0020-6067; CODEN IAEBAB; v. 39(2): 21–25, IAEA, Vienna, Austria reference no. 29002477.

72. Tuomisto, H. (2013, April 4–5). Nuclear district heating plans from Loviisa to Helsinki metropolitan area. A paper at Joint NEA/IAEA Expert Workshop on the "Technical and Economic Assessment of Non-Electric Applications of Nuclear Energy" OECD Headquarters, Paris, France. www.oecd-nea.org/ndd/.../3_Tuomisto_Nuclear-District-Heating-Plans.pdf., a website report.

73. Värri, K. (2018). *Market Potential of Small Modular Nuclear Reactors in District Heating*, M.S. Thesis, Aalto University, Finland.

74. Varri, K. and Syri, S. (2019). The possible role of modular nuclear reactors in district heating: Case Helsinki Region, Fortum, Keilalahdentie, 2–4, 02150 Espoo, Finland. *Energies* 12(11): 2195, Doi: 10.3390/en12112195. www.mdpi. com/1996-1073/12/11/2195.

75. Connolly, D., Lund, H., Mathiesen, B. et al., (2014). Heat roadmap Europe: Combining district heating with heat savings to decarbonise the EU energy system. *Energy Policy* 65: 475–489 [Online]. Available at: www.sciencedirect.com/science/article/pii/S0301421513010574.

76. International Energy Agency. (2017). Energy policies of IEA countries—Poland-2016 review.
77. International Energy Agency. Poland: Electricity and Heat for 2015 (cited 9 January 2018) [Online]. Available at: www.iea.org/statistics/statisticssearch/report/?country=P OLAND=&product=electricityandheat. A website report.
78. Sobolewski, J. (2017, September). *HTR Plans in Poland*. Ministry of Energy, Poland.
79. Nikkei Asian Review. (2017). Japan to export safer nuclear reactor to Poland, cited 9 January 2018 [Online]. Available at: https://asia.nikkei.com/Business/Companies/Japan-to-export-safer-nuclear-reactor-to-Poland?page=2.
80. Xia, J., Zhu, K. and Jiang, Y. (2016, April). Method for integrating low-grade industrial waste heat into district heating network. *Building Simulation* 9(2): 153–163. First Online: 3 December 2015, Doi: 10.1007/s12273-015-0262-3.
81. Fang, H., Xia, J. and Jiang, Y. (2015). Key issues and solutions in a district heating system using low-grade industrial waste heat. *Energy* Elsevier, 86: 589–602.
82. Daniëls, B., Wemmers, A. and Wetzels, W. (2011). Dutch industrial waste heat in district heating: Waste of effort? Panel: 3. Matching Policies and Drivers: Policies and Directives to Drive Industrial Efficiency. This is a peer-reviewed paper. Also 2016 Dutch Industrial Waste Heat in District Heating: Waste of Effort?—ECEEE. A website report www.eceee. org/...Industrial.../3-matching-policies-and-drivers-policies-and-dir, a website report
83. Daniëls, B.W., Wemmers, A.K., Tigchelaar, C. and Wetzels, W. (2011, November). Restwarmtebenutting. Potenti.len, besparing, alternatieven, ECN-E-11-058.
84. Papapetrou, M., Kosmadakis, G., Cipollina, A., Commare, U.L. and Micale, G. (2018, 25 June). Industrial waste heat: Estimation of the technically available resource in the EU per industrial sector, temperature level and country. *Applied Thermal Engineering* 138: 207–216, Doi: 10.1016/j.applthermaleng.2018.04.043.
85. Sohani, A.A. *Waste Heat Recovery from SSAB's Steel Plant in Oxelösund Using a Heat Pump*, M.S. Thesis KTH School of Industrial Engineering and Management Energy Technology EGI_2016–082 MSC Division of ETT SE-100 44 Stockholm.
86. Heating and Cooling with Waste Heat from Industry. (2015, July 8). A website report European Academy of Bozen/Bolzano (EURAC). https://phys.org/news/2015-07-cooling-industry.html#jCp.
87. Raine, R., Sharifi, V., Swithenbank, J., Hinchcliffe, V. and Segrott, A. Sustainable steel city: Heat storage and industrial heat recovery for a district heating network. *The 14th International Symposium on District Heating and Cooling*, 7–9 September 2014, Stockholm.
88. Fang, H., Xia, J., Zhu, K., Su, Y. and Jiang, Y. Industrial waste heat utilization for low temperature district heating. *Energy Policy* 62(C): 236–246. www.sciencedirect.com/science/article/pii/S0301421513006113, Doi: 10.1016/j.enpol.2013.06.104.
89. Tong, K., Fang, A., Yu, H., Li, Y., Shi, L., Wang, Y., Wang, S. and Ramaswami, A. (2017, December 11). Estimating the potential for industrial waste heat reutilization in urban district energy systems: Method development and implementation in two Chinese provinces. © 2017. Published by IOP Publishing Ltd. *Environmental Research Letters*, 12(12).125008. Doi: 10.1088/1748-9326/aa8a17.
90. Raine, R.D. (2016, September). *Sheffield's Low Carbon Heat Network and its Energy Storage Potential*, Ph.D. Thesis, University of Sheffield, Sheffield.
91. Finney, K. (2011). *Sheffield University Waste Incineration Centre (SUWIC)*, University of Sheffield, Sheffield Heat Mapping and Feasibility Study of Decentralised Energy, http://research.ncl.ac.uk/pro-tem/components/pdfs/Sheffield_EPSRC_progress_report_3_Sheffield_heat_maps_July2011.pdf. Accessed 4 April 2015.
92. Finney, K.N., Sharifi, V.N., Swithenbank, J., Nolan, A., White, S. and Ogden, S. (2012). Developments to an existing city-wide district energy network—Part I: Identification of potential expansions using heat mapping. *Energy Conversion and Management* 62: 165–175.

93. Doračić, B., Novosel, T., Pukšec, T. and Duić, N. (2018). Evaluation of excess heat utilization in district heating systems by implementing levelized cost of excess heat. *Energies* 11: 575, Doi: 10.3390/en11030575. www.mdpi.com/journal/energies.
94. Koh, S.C.L., Acquaye, A.A., Rana, N., Genovese, A., Barratt, P., Kuylenstierna, J., Gibbs, D. and Cullen, J. (2011). *Supply Chain and Environmental Analysis.* Centre for Low Carbon Futures. www.shef.ac.uk/polopoly_fs/1.153241!/file/SCEnAT-Report.pdf. Accessed 31 July 2014.
95. Remus, R., Monsonet, M.A.A., Roudier, S. and Sancho, L.D. Best available techniques reference document for iron and steel production, European Commission, 2013. *The 14th International Symposium on District Heating and Cooling,* 7–9 September 2014, Stockholm. Sweden, a website report http://eippcb.jrc.ec.europa.eu/reference/BREF/IS_Adopted_03_2012.pdf. Accessed 31 July 14.
96. Dixon, J. and S. Bramfoot. (1985). Design of waste heat boilers for the recovery of energy from arc furnace waste gases. Commission of the European Communities.
97. Jones, J. (1997). Understanding electric arc furnace operations, RI enter for Materials Production. http://infohouse.p2ric.org/ref/10/09047.pdf. Accessed 31 July 2014, a website report.
98. Zuliani, D., Scipolo, V. and Maiolo, J. (2010). Opportunities for increasing productivity, lowering operating costs and reducing greenhouse gas emissions in EAF and BOF steel making. *Proceedings of AISTech,* www.millennium-steel.com/articles/pdf/2010%20India/pp35–42%20MSI10.pdf. Accessed 16 July 2013.
99. World Health Organisation. (2010). Dioxins and their effects on human health, Fact Sheet No. 225. www.who.int/mediacentre/factsheets/fs225/en/. Accessed 31 July 2014.
100. Born, C. and Granderath, R. (2013, February). Benchmark for heat recovery from the offgas duct of electric arc furnaces. *MPT Metallurgical Plant and Technology International* 36(1): 32–35.
101. World Steel Association. (2011). Water management in the steel industry. Article, www.worldsteel.org/media-centre/press-releases/2011/water-management-report.html. Accessed 31 July 2014, a website report.
102. Nakomcic-Smaragdakis, B. and Dragutinovic, N. (2015, January). Hybrid renewable energy system application for electricity and heat supply of a residential building. *Thermal Science* 20(00): 144–145. 237 Reads, Doi: 10.2298/TSCI150505144N.
103. Bakken, B.H., Fossum, M. and Belsnes, M.M. (2001). Small-scale hybrid plant integrated with municipal energy supply system MSc SINTEF energy research, N-7465 Trondheim, Norway, bjom.bakken@energy.sintef.no Ossiach, International Energy symposium. www.iaea.org/inis/collection/NCLCollectionStore/_Public/33/009/33009933.pdf?
104. Holmgren, K. (2006). A system perspective on district heating and waste incineration, Division of Energy Systems Department of Mechanical Engineering Linköpings universitet, Linköping, Sweden 2006, Linköping Studies in Science and Technology Dissertation No. 1053,ISBN: 91-85643-61-0. ISSN 0345-7524 Printed in Sweden by LiU-Tryck, Linköping. https://pdfs.semanticscholar.org/b2e8/167a72eb12a853b1254a7e4a498948e2313a.pdf.
105. Ulloa, P. and Themelis, N.J. Doubling the energy advantage of waste-to-energy: District Heating in the Northeast U.S. © 2007 by ASME. *15th North American Waste to Energy Conference,* 21–23 May 2007, Miami, FL NA EC15-3201 https://pdfs.semanticscholar.org/de71/f5322ef03abcb2cd5d48ae8a2134337ca593.pdf.
106. Kiviluoma, J., Heinen, S., Qazi, H., Madsen, H., Strbac, G., Kang, C., Zhang, N., Patteeuw, D. and Naegler, T. (2017). Harnessing flexibility from hot and cold. *IEEE Power & Energy Magazine* 15(1): 25–33, Doi: 10.1109/MPE.2016.2626618, . ISSN: 1540-7977. Contribution to journal Journal article–Annualreportyear: 2017 Research peer-reviewhenrikmadsen.org/wp.../harmenessing_flexibility_from_heating_and_cooling.pdf.

107. Reed, A.L., Novelli, A.P., Doran, K.L., Ge, S., Lu, N. and McCartney, J.S. (2018). Solar district heating with underground thermal energy storage: Pathways to commercial viability in North America. *Renewable Energy* 126: 1–13.

108. Schuchardt, G.K. (neé BESTRZYNSKI). (2016, December). Integration of decentralized thermal storages within district heating (DH) networks. *Environmental and Climate Technologies* 18: 5–16, Doi: 10.1515/rtuect-2016-0009. www.degruyter.com/view/j/rtuect.

109. Dincer, I. and Rosen, M. (2010). *Thermal Energy Storage: Systems and Applications*, 2nd ed, Doi: 10.1002/9780470970751.

110. Andersson, O. (1997). ATES utilization in Sweden: An overview. *Proceedings of MEGASTOCK'97 7th International Conference on Thermal Energy Storage*, 2, 925–930.

111. Lund, H., Werner, S., Wiltshire, R., Svendsen, S., Thorsen, J.E., Hvelplund, F. and Mathiesen, B.V. (2014). 4th Generation District Heating (4GDH): Integrating smart thermal grids into future sustainable energy systems. *Energy* 68: 1–11.

112. Tanaka, H., Tomita, T. and Okumiya M. (2000). Feasibility study of a district energy system with seasonal water thermal storage. *Solar Energy* 69(6): 535–547.

113. Rezaie, B., Reddy, B.V. and Rosen, M.A. (2017, June). Assessment of the thermal energy storage in Friedrichshafen district energy systems. *Energy Procedia* 116: 91–105, Doi: 10.1016/j.egypro.2017.05.058.

114. Holmgren, K. and Gebremedhin, A. (2004). Modelling a district heating system: Introduction of waste incineration, policy instruments and co-operation with an industry. *Energy Policy* 32: 1807–1817, Doi: 10.1016/S0301-4215(03)00168-X.

115. Sahlin, J., Knutsson, D. and Ekvall, T. (2004). Effects of planned expansion of waste incineration in the Swedish district heating systems. *Resources, Conservation and Recycling* 41: 279–292.

116. Energy Central Bjerring bro District Heating and Grundfos. (2007). A website report www.districtenergy.org/HigherLogic/.../DownloadDocumentFile.ashx?...0.

117. Schmidt, T. and Mangold, D. (2006). Seasonal thermal energy storage in Germany. *Structural Engineering International* 14, Doi: 10.2749/101686604777963739.

118. Schmidt, D. (2009). Low exergy systems for high-performance buildings and communities. *Energy and Buildings* 41: 331–336.

119. Rezaie, B., Reddy, B.V. and Rosen, M.A. (2011). Role of thermal energy storage in district energy systems. In: Rosen, M.A. (ed.) *Energy Storage*. NOVA Publishers, Hauppauge, NY.

120. Dincer, I. and Rosen, M.A. (2011). *Thermal Energy Storage: Systems and Applications*. Wiley, Chichester.

121. Rosen, M.A., Le, M.N. and Dincer, I. (2005). Efficiency analysis of a cogeneration and district energy system. *Applied Thermal Engineering* 25: 147–159.

122. Rezaie, B. and Rosen, M.A. (2012). District heating and cooling: Review of technology and potential enhancements. *Applied Energy* 93: 2–10.

123. Andrepont, J.S. (2006). Developments in thermal energy storage: Large applications, low temps, high efficiency, and capital savings. *Energy Engineering* 103: 7–18.

124. Griffin, T. (2010). LEED® district energy. The new version 2.0 guideline: It's here!. *District Energy* 96: 55.

125. Rezaie, B., Reddy, B.V. and Rosen, M.A. (2015). Exergy analysis of thermal energy storage in a district energy application. *Renewable Energy* 74: 848–854.

126. Rossi Espagnet, A. (2016). Techno-Economic Assessment of Thermal Energy Storage Integration into Low Temperature District Heating Networks, M.S. Thesis KTH School of Industrial Engineering and Management Energy Technology EGI-2016-068 Division of Heat and Power Technology, SE-100 44 Stockholm.

127. Romanchenko, D., Kensby, J., Odenberger, M. and Johnsson, F. Thermal energy storage in district heating: Centralized storage *vs.* storage in thermal inertia of buildings. *Energy Conversion and Management* 162: 26–38, Doi: 10.1016/j.enconman.2018.01.068.

4 Hybrid Energy Systems for Vehicle Industry

4.1 INTRODUCTION

The widespread application of hydrocarbon-based transportation has been raising global issues such as an increase in the demand for nonrenewable petroleum production, high gasoline prices, inefficient power production, and CO_2 emission leading to climate change. Hence, searching for highly efficient, safe, and clean alternative solutions to this method of transport have been among the most emphasized challenges attracting the attention of researchers in both the environment and transportation sectors. A well-knit and coordinated transportation provides mobility to people and goods. The transportation sector mainly consists of road, railway, ships, and aviation, where road transportation consumes 75% of the total energy spent on transportation. The automobile industry plays a significant role in economic growth of the world and hence affects the entire population. Since vehicles mostly run on internal combustion engine (ICE), the transportation industry is accountable for 25%–30% of the total greenhouse gases (GHGs) emission [1,2]. ICE works with the process of fuel combustion resulting in the production of various gases like CO_2, NO_2, NO, and CO [2,3] which cause environmental degradation in the form of greenhouse effect and are responsible for their adverse effect on human health.

The role of hybrid energy (or power) for road vehicles (such as bicycles, automobiles, trucks, buses, etc.) as well as space vehicles has been treated extensively in my previous two books [4,5] and it will not be repeated here. They illustrate that the hybrid power makes very important contribution to the decarbonization of automobiles, buses, bike, and space industries. The acceptance of hybrid and electrical cars is rapidly increasing. In this chapter, we examine the role of hybrid energy for boats and ship industry, aviation industry, and rail industry.

4.2 HYBRID ENERGY IN MARITIME INDUSTRY

The use of hybrid energy in ship industry is widespread from small boats to medium size vessels to global transport and military vessels. It is now expanded to the stage where hybrid microgrids are used for maritime global export and large military vessels. Here we briefly examine all these applications.

4.2.1 BOATS, YACHTS, AND FERRIES

While as shown in Figure 4.1, photovoltaic (PV) is used on sail boats. There are limits to using PV cells for vehicles such as power density, cost, and some design restrictions. These are all illustrated well in my previous books for automobile industry and they also to some degree apply to the maritime industry [6–10].

FIGURE 4.1   PV used for auxiliary power on a yacht [6].

Solar-powered boats (see Figure 4.2) [8] get their energy from the sun. Using electric motors and storage batteries charged by *solar panels* and *PV cells*, solar-powered boats can significantly reduce or eliminate their use of *fossil fuels*. In 2007, five Swiss sailors piloted a solar-powered boat across the Atlantic Ocean. Using solar power only (via solar panels), the "sun21" made the first motorized crossing of the Atlantic

FIGURE 4.2   Solar Sailor of Australia [8].

Ocean in order to promote the great potential of renewable energy (RE) for ocean navigation and to combat climate change.

The "sun21" is a 45.9 ft long specially built solar-powered boat known as a catamaran. On its canopy-like roof are 48 silicon PV cells, which collect energy from sunlight and transmit it to a device in one of the narrow cabins. This device transmits the energy to the 3,600 pounds of storage batteries below the deck. The 11-ton solar boat was powered on the energy needed to light 10,100 W light bulbs. The typical speed was 3.5 knots. Much like a hybrid car, large batteries onboard the solar energy boat store electricity generated by the diesel generators and collected by the solar panels. The electricity then powers the electric motors [5,7,8]. In 2010, the Tûranor Planet Solar, a 30 m long, 15.2 m wide catamaran yacht powered by 470 m^2 of solar panels, was unveiled. It is, so far, the largest solar-powered boat ever built. In 2012, Planet Solar became the first-ever solar EV to circumnavigate the globe.

Hybrid ferries combine multiple sources of power (for example, traditional diesel with electric battery power), resulting in reductions in fossil fuel consumption, carbon emissions, and other pollutants. Three hybrid roll-on/roll-off ferries are in operation on The Clyde, Scotland [10]. The hybrid ferry serving Alcatraz has a PV panel and wind turbine on top that helps power the vessel. Washington State Ferries plans to introduce 22 diesel-electric ferries by 2040, cutting its annual diesel use from 19 to 9.5 million gallons. The New York Hornblower is powered by diesel, hydrogen fuel cells, AGM batteries, and wind and solar energy. Three other hybrid ferries in operations are Stena Jutlandica commissioned in 1996, with battery capacity 1 MW; Berlin and Copenhagen (each) commissioned in 2016, with battery capacity of 1.5 MWh and MS color hybrid, commissioned in 2018 with battery capacity of 4.7 MWh.

A *Danish Ferry* operator has begun converting a diesel-electric ship into a diesel-electric hybrid, the world's largest hybrid passenger ferry (see Figure 4.3). Scandlines commissioned the retrofit of the *Prinsesse Benedikte*, a ship that has 2.7 MWh batteries and can carry 300 vehicles and 900 passengers. The ship carries modular lithium polymer energy storage packs that can recharge in 30 minutes from shore power or generator power and can propel the 16,000 ton ship for about 30 minutes without diesel fuel. Since the *Prinsesse Benedikte* is already a diesel-electric ship and uses a series of diesel generators to power electric motors, the battery packs are most useful in load-leveling [10]. In addition to saving fuel, the conversion helps Scandlines save in maintenance costs.

Eco Marine Power (EMP) company has built *Tonbo Solar-Electric HMP Ferry*. This cutting edge green passenger ferry includes a hybrid marine power (HMP) system that uses a specially designed solar module panel array. The Tonbo is able to use energy collected via the solar modules to charge onboard batteries in order to reduce fuel consumption and is also be able to rapidly re-charge these batteries by using shore power. HMP can also be used to provide power for facilities on a ship such as catering equipment, fans, and lighting. Another innovative solar ferry design by EMP is the *Medaka Eco Ferry*. This smaller solar-electric ferry is being developed to operate as an eco-friendly urban commuter ferry especially in cities where noise and air pollution levels are high.

Globally, 90% of world trade by volume is carried by shipping. The merchant ships globally consume 250 million tons of bunker fuel annually. Just one ship can

FIGURE 4.3 Scandline hybrid ferry [10].

emit 32,988 tons of CO_2 and 959 tons of SO_x. Shipping uses around 140 million tons of oils and generates estimated 4%–5% of the global total of carbon dioxide emitted by human activity. In addition, global shipping produces significant amount of SO_x and NO_x. Shipping is also a major source of pollution around ports and coastal areas. Clearly there is an opportunity to reduce airborne pollution (including particulate matter (PM)) carbon emissions and fossil fuel consumption on a global scale. Solar panel array installed as part of an Aquarius MAS + Solar solution provided by EMP [9] is illustrated in Figure 4.4. EMP has also developed the *Aquarius MRE System*—a revolutionary wind and solar power solution for shipping that will slash fuel consumption plus reduce carbon and noxious gas emissions and reduce operating costs. Depending on the size, type, and operating profile of a ship, the Aquarius MRE System will reduce fuel consumption from 10% to 40%. At the core of the Aquarius MRE system is the EnergySail—a rigid sail designed by EMP which will allow ships to tap into the power of the sun and wind. The EnergySail can be used as a stand-alone unit or be installed as part of an array. This innovative device can also be used by a wide range of vessels including large oil tankers, cargo carriers, passenger ferries, coastguard ships, and even naval vessels.

The EnergySail [11] is a rigid sail and wind-assisted (or sail assisted) propulsion device designed by EMP that allows ships to harness the power even when a ship is at anchor or in port and has been designed to withstand high winds or even sudden micro bursts. The EnergySail can be fitted to a wide variety of vessels from large Capesize bulk ore carriers, RoRo vessels, and cruise ships to naval and coastguard patrol vessels. A variation of the EnergySail that is suitable for Unmanned Surface Vessels and smaller ships such as passenger ferries or fishing vessels is also being developed. The EnergySail is not just a sail device; it is also a flexible platform for ship-RE technologies. This combination of technologies will enable ships

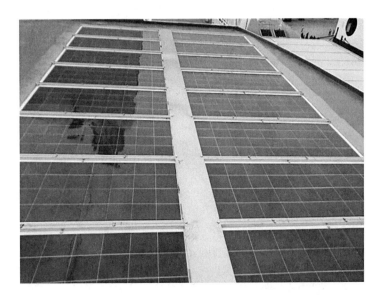

FIGURE 4.4 Solar panel array installed as part of an Aquarius MAS + Solar solution provided by EMP [9].

to lower their airborne emissions including GHG and PM. The flexible design of the EnergySail design also allows for it to be upgraded during the life-cycle of the ship so that newer technologies can be incorporated if required. It has also been designed so that it will require little maintenance and be robust enough to withstand the harsh conditions of an operational life at sea over many years. Although the EnergySail [11] has primarily been designed for EMP's *Aquarius MRE* solution, other applications for its use are also being studied including land based applications. It could also be used as a stand-alone unit on a cable laying vessel, coastal tanker, or oceanographic ship. In addition, a simplified version of *Aquarius MRE* including a modified EnergySail would be suitable on a range of smaller vessels. An alternative version of the EnergySail could also be used horizontally.

An example of how the EnergySail could be incorporated into a modern green ship design is the *Aquarius Eco Ship*. This low emission, sustainable ship design concept includes an EnergySail array. This allows for the propulsive power from multiple sails to be harnessed. The use of these rigid sails at that time proved that ships using sail assisted propulsion could achieve fuel savings of more than 30% depending on the prevailing weather conditions. These sails are not aimed at replacing the ship's main engines however, but rather were a source of supplementary or auxiliary propulsion. The use of sails for this purpose is commonly referred to as sail-assisted or wind-assisted propulsion. Key features of Energy Sail include [11] (a) flexible patented design, (b) use of renewable energies to reduce fossil fuel consumption, (c) automatic positioning via computer control system, (d) can be integrated with other systems and equipment onboard ships, (e) safe, reliable, and robust design, (f) can be lowered and stored, (g) lowers carbon and other pollutant emissions, (h) designed to be used as a stand-alone device or part of a computer controlled sail array, and (i) suitable for wide range of ships and vessels, due to flexible PV (solar) panels or wind power devices. The Aquarius Eco Ship design concept is an example how the EnergySail can be used on a large ship to provide a source of emissions-free supplementary power or sallow maintenance and ongoing costs, recyclable sail-assisted propulsion. The center piece of the Aquarius Eco Ship is *Aquarius MRE* (Marine Renewable Energy)—an innovative and patented fuel saving and emission reduction system that incorporates a variety of elements including solar panels (or PV modules), energy storage modules or hybrid VRLA (valve-regulated lead acid) battery packs, computer control systems, and an advanced rigid sail design [9,12].

4.2.2 ROLE OF RENEWABLE SOURCES IN SHIPPING INDUSTRY

The overall contribution of RE technologies to international shipping is unlikely to achieve a dominant or even major role in the near future. Still, it has strong and increasingly proven capacity to make a modest contribution in many sectors over the short- and medium-term. For selected applications, the role of renewables can be significant, even dominant. Of the various RE options, advanced biofuels have a very high potential to transform energy choices of the shipping sector. However, this potential will depend on a number of factors, including the global availability of sustainable feedstock for the production of biofuels. Hydrogen fuel cells as a power source for shipping also hold great potential but the sustainability of the energy source used to produce the hydrogen, as well as lack of cost-effective and reliable low-pressure storage options for the fuel remain as critical issues to be addressed. Overall, the greatest potential lies in using a combination of RE solutions that maximizes the availability and complementarity of energy resources in hybrid modes. In this sense, achieving the full potential of renewables in the shipping sector will require an integrated systems engineering approach that also addresses the barriers to their deployment.

The barriers to the adoption of RE in the shipping sector are complex. These can be categorized under organizational/structural, behavioral, market, and nonmarket factors. This complexity, in part, reflects the unique and international nature of the shipping industry, with underlying constraints and factors that lie beyond the ability of individual states to effect incentives plus the policy and regulatory framework needed to overcome the barriers. With regard to organizational, structural and behavioral barriers, limited R&D financing, particularly for initial proof-of-concept technologies, is a major factor, together with ship owners' concerns over the risk of hidden and additional costs, as well as opportunity costs of RE solutions. This is particularly true since, historically, there has been a lack of reliable information on costs and potential savings of specific operational measures or RE solutions for this sector. Concerning market barriers, the fundamental problem is that of split incentives between ship owners and hirers, limiting the motivation of owners to invest in clean energy solutions for their shipping stock since the benefits do not always accrue to the investing party and hence savings cannot be fully recouped. Another barrier is the risk adversity of investors in the sector, especially following the collapse of the shipping boom in 2006. Furthermore, the shipping sector is seldom visible to the general public, resulting in less societal pressure on the industry to transition to cleaner energy solutions. Of the nonmarket barriers, the different classes and scales of ships, the markets and trade routes served, and the lack of access to capital are some of the key barriers that must be addressed [13].

Presently, RE options are being considered for the global shipping fleet at all levels and in varying magnitudes, including international and domestic transport of goods, people, and services; fishing; tourism and other maritime pursuits. Renewable options can be used in ships of all sizes to provide primary, hybrid, and/or auxiliary hybrid propulsion, as well as onboard and shore-side energy use. These clean energy solutions are being integrated through retrofits to the existing fleet or incorporated into new shipbuilding and design, with most applications deploying RE as part of an integrated package of efficiency measures which include hybrid processes. The current focus of RE application in shipping is on [13]:

1. *Wind energy*: for example, using *soft-sails*, such as Greenheart's 75 dwt freighter, B9 Shipping's 3,000 dwt bulker and Dykstra/Fair Transport's 7,000 dwt Ecoliner; *fixed-sails*, such as in the UT Wind Challenger and the EffShip's project; *Flettner Rotors*, such as in the Alcyone and Enercon's 12,800 dwt E-Ship 1; *kite sails*, such as in the MS Beluga Skysails; *wind turbines* (no successful prototypes to date).
2. *Solar PVs*: mainly in *hybrid* models with other energy sources on small ships, such as NYK's Auriga Leader and Solar Sailor by OCIUS Technology (formerly Solar Sailor Holdings Ltd.).
3. *Biofuels*: such as the Meri cargo ship which claims to be the first of its size to use 100% bio-oil).

Hydrogen fuel cells have also been used as a clean energy technology for shipping: for example, in the FCS Alsterwasser, a 100-pax fuel-cell-powered passenger vessel based in Hamburg Port (Germany), as well as a number of other small ferries and river boats. In 2012, as part of the Fellow SHIP project, a 330 kW fuel cell was successfully tested onboard the offshore supply vessel, Viking Lady, operating for more than 7,000 hours. This was the first fuel cell unit to operate on a merchant ship, with the electric efficiency estimated to be 44.5% (when internal consumption was taken into account), with no NO_x, SO_x, and PM emissions detectable. In 2012, Germanischer Lloyd set out design concepts for a zero-emissions 1,500 passenger "Scandlines" ferry and a 1,000 TEU (twenty foot equivalent unit) container feeder vessel with a 15-knot service speed and using hydrogen fuel cells. Other RE propulsion systems include WWL's (Wallenius Wilhelmsen Logistics (lines)) proposed ambitious "Orcelle" car carrier that will use a series of underwater flaps, modeled on the tail movements of Irrawaddy dolphins, to create propulsion and generate electricity and hydraulic power for the ship [13].

The enormous variety in global shipping vessel types, usage, and routes means that different applications favor the use of different energy sources and technologies. Although the role and extent of RE technology adoption by the shipping sector varies depending on the scale, function, and operational location of the particular vessel, technology providers contend that research and innovation efforts on the use of RE options, together with efficient designs, are already achieving significant results for immediate and near-term energy savings for a number of selected applications. For example, Enercon reported in 2013 that its prototype E-Ship 1 had achieved 25% savings and OCIUS Technology Ltd reported 5%–100% savings depending on application for the Solar Sailor, which claims a RE solution cost of 10%–15% of the capital cost of the vessel and a return on investment (ROI) of between 2 and 4 years. B9 Shipping and Fair Transport BV predicted additional building and maintenance costs of 10%–15% of total asset costs for a projected 60% savings in fuel, as well as significant reductions in main engine and propeller wear. Fuel savings vary from nearly 100% (fuel switching to renewables) for designs such as *Greenheart* down to only 0.05% main energy and 1% ancillary energy savings of NYK's (Nippon Yusen Kabushiki Kaisha) solar array retrofitted car carrier *Auriga Leader*. The University of Tokyo has predicted that fuel costs could be reduced by as much as one-third with the 60,000 gross tonnage *UT Wind Challenger*. OCIUS Technology contends that by retrofitting opening wing sails to a "motor-sail," without altering the primary propulsion system of a modern tanker or bulker, ship operators can expect 20%–25% fuel savings on cross-equator shipping routes and 30%–40% on same-hemisphere shipping routes, with a payback period of only 2 years, based on 2013 average fuel prices.

In the case of *rotor technology*, the amount of fuel savings decreases as the ship size increases: Savings of 60% for small ships have already been achieved while savings of up to 19% on Very Large Crude Carriers are being modeled. The *Ulysses Project* has focused on ultraslow steaming scenarios to demonstrate that the efficiency of the world's fleet of ships can be increased such that an 80% emissions reduction by 2050 against 1990 baselines can be achieved, with ships of the future travelling at speeds as slow as five knots. In such a scenario, RE technologies could play a dominant role [13].

Roh et al. [14] examined fuel consumption and CO_2 emission reductions of ships powered by a fuel-cell-based hybrid power source. The need for technological development to reduce the impact of air pollution caused by ships has been strongly emphasized by many authorities, including the International Maritime Organization (IMO). This has encouraged research to develop an electric propulsion system using hydrogen fuel with the aim of reducing emissions from ships. This study describes the test bed they constructed to compare their electric propulsion system with existing power sources. The system uses hybrid power and a diesel engine generator with a combined capacity of 180 kW. To utilize scale-down methodology, the linear interpolation method is applied. The proposed hybrid power source consists of a molten carbonate fuel cell (MCFC), a battery, and a diesel generator, the capacities of which are 100, 30, and 50 kW, respectively. The experiments they conducted on the test bed were based on the outcome of an analysis of the electrical power consumed in each operating mode considering different types of merchant ships employed in practice. The output, fuel consumption, and CO_2 emission reduction rates of the hybrid and conventional power sources were compared based on the load scenarios created for each type of ship.

This study analyzed the fuel consumption and CO_2 emission reduction rates when a fuel-cell-based hybrid power source instead of a conventional commercial diesel power source was used in ships. The results showed that under the rated output on a test bed with a load bank of 180 kW, the conventional commercial diesel generator consumed fuel at 43.5 kgoe/h and emitted CO_2 at 148.5 kg/h, whereas the fuel-cell-based hybrid power source consumed fuel at 35.6 kgoe/h and emitted CO_2 at 57.7 kg/h. The hybrid power source reduced fuel consumption by 18% and CO_2 emissions by 61% at part load in the port period. These results indicate that it is possible to reduce CO_2 emissions by up to 61% if a hybrid power source of the same capacity is used to power a ship.

In the study, the actual electric load analysis values of the 5,500 TEU Reefer Container, 13 k TEU Container, 40 k Bulk Carrier, 130 k DWT LNG Carrier, and 300 k DWT Crude Oil Tanker were scaled down according to the operation mode, and the control logic and systems of the test bed developed in this study were operated normally according to the respective load scenarios. Because the output characteristics and control time of the diesel generator, according to the power source of the hybrid system, were reduced, according to the load variation pattern of the ship and the ship's type, the CO_2 emissions of the hybrid system, as compared with the case of the diesel generator, alone operated for each load scenario with an average of 70%~74% less. In order to apply the hybrid system to ships, it is possible to maximize the CO_2 emission reduction effect by setting the capacity of the fuel cell + battery to be able to take charge of the base load of the ship through analysis of the base load of each ship type [14].

4.2.3 HYBRID SHIPS AND ROLES OF RENEWABLE SOURCES AND ENERGY STORAGE

Global trade and the socio-economic activities of island communities depend heavily on shipping, transporting approximately 90% of the tonnage of all traded goods. The global shipping tonnage loaded annually increased from 2.6 billion to 9.2 billion tons between 1970 and 2012, and the demand for shipping is predicted to grow further, owing to the changing configuration of global production, the increasing importance of global supply chains and the expected growth in many economies. As the demand for shipping services continues to grow, research on the use of RE options for the sector—although still relatively immature—is growing fast. Between 2007 and 2012, the world's marine fleet consumed between 250 and 325 million tons of fuel annually, accounting for approximately 2.8% of global annual greenhouse gas emissions (3.1% of CO_2 annual emissions), amid rising bunker fuel prices in a globally volatile fossil fuel market and increasing requirements to significantly reduce emissions from the sector. The International Convention for the Prevention of Pollution from Ships (MARPOL) [15] has stipulated, among other measures, low sulfur emission control areas in the marine environment, as well as mandatory technical and operational measures requiring ships to be more efficient in energy use and to reduce emissions. The MARPOL regulations make the Energy Efficiency Design Index mandatory for new ships and the Ship Energy Efficiency Management Plan for all ships. These economic and environmental constraints, therefore, constitute key drivers for adopting the use of renewables in the shipping sector [16].

Approximately 80% of world trade by volume is carried by sea [17,18]. In 2007, it is estimated that international shipping was responsible for approximately 870 million tons of CO_2 emissions, or 2.7% of global anthropogenic CO_2 emissions [19]. Domestic shipping and fishing activity bring these totals to 1,050 million tons of CO_2, or 3.3% of global anthropogenic CO_2 emissions. Despite the undoubted CO_2 efficiency of shipping in terms of grams of CO_2 emitted per ton-km, it is recognized within the maritime sector that reductions in these totals must be made [19]. Shipping is responsible for a greater percentage share of NO_x (about 37%) and SO_x (about 28%) emissions (AEA Energy & Environment, 2008) and recent legislation is aimed at reducing these emissions through the introduction of emission control areas and requirements on newly built marine diesel engines.

The expected changes in CO_2 emissions from shipping from 2007 to 2050 were modeled for the IMO with reference to the emissions scenarios developed for the UN IPCC. These scenarios are based on global differences in population, economy, land use, and agriculture [19]. The base scenarios indicate annual increases of CO_2 emissions in the range 1.9%–2.7%, with the extreme scenarios predicting changes of 5.2% and −0.8%, respectively. The increase in emissions is related to predicted growth in seaborne transport. If global emissions of CO_2 are to be stabilized at a level consistent with a 2°C rise in global average temperature by 2050, it is clear that the shipping sector must find ways to stabilize, or reduce, its emissions—or, these projected values will account for 12%–18% of all total permissible CO_2 emissions.

CO_2 emissions from world shipping are directly related to the fuel consumption of the fleet. In 2007, approximately 277 million tons of fuel were consumed by international shipping. Three categories of ship account for almost two-thirds of this consumption. The liquid bulk sector accounts for about 65 million tons fuel/year, container vessels for about 55 million tons fuel/year, and the dry bulk sector for about 53 million tons fuel/year [19, p. 42]. Many of the present efforts to reduce CO_2 emissions from global shipping are aimed at the container vessel sector since this contains relatively large vessels travelling at comparatively high speeds, leading to high fuel consumptions. Significant reductions in fuel consumption and hence emissions may be made through introducing slower operational speeds in a practice referred to as "slow steaming" [20,21]. Less attention has been paid to the dry and liquid bulk sectors, where operational speeds are much slower and vessel design optimized over many decades.

Hybrid Power or HMP systems lower fuel consumption, reduce airborne pollution, and are energy efficient. EMP has focused on HMP or "Hybrid" solutions that include RE technologies that are not only kind to the environment but are also cost effective and offer an attractive ROI for vessel owners and operators [22].

For larger vessels and ships, EMP along with a group of partner companies is developing the patented *Aquarius MRE*. Aquarius MRE is an advanced integrated system of rigid sails, solar panels, energy storage modules, and marine batteries that will enable ships to tap into RE by harnessing the power provided by the wind and sun. The array of rigid sails is automatically positioned to best suit the prevailing weather conditions and can be lowered and stored when not in use or in bad weather. The array of rigid sails is based on EMP's patented Energy Sail technology and they can even be used when a ship is at anchor or in harbor. Clearly the Energy Sail is unlike any other sail. EMP's HMP and MRE solutions also incorporate the Aquarius Management and Automation System (Aquarius MAS) (see Figure 4.4).

"The Hour Of Power" has been well received by the marine industry worldwide. This simple concept enables vessels to run in and out of port for an hour on electric with battery power—then carry out open sea work on diesel power. The aim of this innovative hybrid solution is to enhance conventional propulsion systems. Vessels can reduce emissions and improve fuel consumption while extending engine maintenance periods and engine life. "The Hour Of Power" focuses on hybrid solutions linked to viable business cases. For commercial and professional organizations, the concept of running vessels with zero emissions at up to 10 knots for 1 hour will shape decisions that lead to improvements of in-service systems and procurement of next generation vessels. The overall objective is fuel saving and improved efficiency by all means.

For the marine industry to move forward, it needs to utilize expertise from aviation and other sectors to drive this innovation and support relevant safety standards. Automotive manufacturers in Europe, the Far East, and the US have recognized that hybrid technologies such as Plug-in Hybrid Electric Vehicle utilizing lithium-ion batteries will be dominant for the next decade. Reducing emissions from busses and trucks in the world's major cities has been a major driver for lithium-ion battery power storage. The need for self-sufficient land-based grid applications has further extended the capabilities of next generation battery and hybrid technology.

Diesel/electric systems have been used in marine industry for large ships and submarines for many years but these are not true or complete hybrid systems. The diesel/electric vessels use its engines to connect directly to an electric generator. The power in the system is then transferred electrically to the propeller shaft via a motor controller and electric motor. The system may have multiple generators and multiple motors. There is, however, no storage of electricity.

There are currently two main types of hybrid systems. First, a serial hybrid where the engine in the system only powers a generator and is not mechanically connected to the propeller shaft. Second, a parallel hybrid, where the engine is mechanically connected along with an electric machine that can operate as both a propulsion motor and a generator. The reduced electric propulsion generator and battery demands of a parallel system substantially reduce the cost compared to a serial system. Parallel systems are more likely to get initial market acceptance because of perceived greater reliability, as the trusted diesel engine is still connected to the propeller shaft with the electric propulsion adding a redundant system. The parallel hybrid system has been successfully implemented in small- and medium-sized marine vessels, such as survey vessels, petrol vessels, superyacht tenders, unmanned craft and Tugboats, workboats, and OSV (Diving Support Vessel). The addition of energy storage in these vessels will make them truly hybrid. The next generation of cells and batteries (like lithium-ion) is making marine hybrid systems very viable.

Ships with both mast-mounted sails and steam engines were an early form of hybrid vehicle. Another example is the diesel-electric submarine. This runs on batteries when submerged and the batteries can be recharged by the diesel engine when the craft is on the surface. Newer hybrid ship-propulsion schemes include large towing kites manufactured by companies such as SkySails. Towing kites can fly at heights several times higher than the tallest ship masts, capturing stronger and steadier winds. There are many applications of PVs in transport either for motive power or as auxiliary power units, particularly where fuel, maintenance, emissions, or noise requirements preclude ICEs or fuel cells. Due to the limited area available on each vehicle, either speed or range or both are limited when used for motive power.

4.2.4 ENERGY STORAGE AND USAGE IN SHIPS

Energy storage is a major green investment for a ship owner. Returns are maximized when the system is correctly dimensioned for the specific ship, and includes intelligent power control. Rolls-Royce has been delivering ship energy storage systems (ESSs) since 2010; however, the actual energy storage units were previously supplied by an external party. Rolls-Royce now offers SAVe Energy, a cost-competitive, highly efficient, and liquid-cooled battery system with a modular design that enables the product to scale according to energy and power requirements. SAVe Energy comply with international legislations for low and zero-emission propulsion systems. The development work has been partly funded by the Norwegian Research Council of Norway's ENERGIX program. The three ship owning companies Color Line, Norled, and the Norwegian Coastal Administration Shipping Company have been partners in the development, ensuring that the ESS covers a wide variety of marine applications, including ferries, cruise vessels, and multipurpose vessels [23–25]. SAVe Energy can be applied to several areas including peak shaving, spinning reserve, and battery powered vessels. Combined with an LNG (Liquid Natural Gas) or diesel powered engine in a hybrid solution, it increases efficiency and reduce emissions, and can be coupled with most types of propulsion units. In a hybrid set-up, SAVe Energy handles the peak load, while the main power generators will relate to the average load and not reduce the propulsion units thrusting capabilities. SAVe Energy is an Energy Storage Unit system and was recently class approved by DNV GL, confirming that SAVe Energy has been developed in compliance with the newest 2018 ruleset, and is accepted for installation on all vessels classed by DNV GL [23–26].

The electrification of ships is building momentum. Since 2010, Rolls-Royce has delivered battery systems representing about 15 MWh in total. However, due to SAVe Energy, in 2019 alone this number is 10–18 MWh. Battery systems have become a key component of ship power and propulsions systems, and SAVe Energy is being introduced on many of the projects Rolls-Royce is currently working on. This includes the upgrade program for Hurtigruten's cruise ferries, the advanced fishing vessel recently ordered by Prestfjord and the ongoing retrofits of offshore support vessels [23–25].

4.2.5 GE NAVAL VESSEL ELECTRIFICATION

GE (General Electric) Energy Connections signed a contract with French electrical engineering company Cegelec in Brittany, France, to supply a complete static power solution for the French Navy. The solution uses GE's advanced Static Frequency Converters (SFCs) technology to provide high conversion efficiency and a safe and reliable power transfer from the electric power grid to the French naval vessels while in port. In order to keep vessels charged fast and on time, the French Navy requires a reliable and stable supply of electricity. GE's converter technology allows an efficient and secured transition of energy [27,28]. The public grid in France operates on a frequency of 50 Hz and must be converted for use of electrical systems onboard of French naval vessels, which can function on different frequencies according to the type of ship. GE's power solution has very low harmonics levels, ensuring high power quality. This clean electricity helps avoid disturbance on the transmission line, which could create network faults and energy losses. In fact, the conditions of coupling a vessel to the onshore grid are critical and it is necessary to ensure that during these operations the power demand does not lead to the destabilization of the grid upstream. This power demand must be gradual, controlled, and synchronized through "intelligent" automation [27,28].

Modular design allows a smaller footprint and is therefore highly flexible during installation and also allows for future power upgrades. The compact design also means higher power density of the equipment, ensuring the conversion of electricity in a more efficient, secured, and reliable manner. GE's SFC allows several loads, meaning several ships can be charged at the same time. Each converter can hold an overload of 150% for a short period, leaving enough time for the automatic coupling of a supplementary SFC. This means that when a new ship is charged, the immediate increased load will not disturb other ships that are already charging, allowing the smooth operation without load impact. This project has the possibility of seamless future expansion. One can adapt and optimize the Harmonic Filter upstream to the converter frequency according to the grid evolution, allowing easy implementation of extra power capacity if needed in the future, all while keeping the benefit of a simple installation. A special storage device and smart control are also being developed to provide reliable energy supply when microcuts occur, ensuring continuation of the power supply. GE's smart management during events of microcuts is proven to be able to avoid tripping clients' installations several times per year [27,28].

4.2.6 HYBRID ENERGY FOR LARGE SHIPS

The study by Dedes et al. [29] concentrates on the dry bulk sector as one of the major contributors to CO_2 emissions of international shipping and a key sector underpinning global seaborne trade. Between 1986 and 2006, average annual growth in the transport of coal and iron ore was greater than that in the transport of oil and oil products and outstripped global GDP growth [19]. It may be noted that the design of Post-Panamax bulk carriers shares significant similarities to liquid-carrying tankers of a similar size (Aframax tankers: carrying capacity between 80,000 and 120,000 tons, breadth 41–44 m, and ship length 270 m) and the conclusions of this study may thus be of relevance to this sector. Aframax tankers account for approximately one-third of all tankers [30].

Hybrid technology which combines a prime mover and energy storage is successfully used in vehicles in the automotive industry [31] and has been shown to contribute to reduced CO_2 emissions [32,33]. The concept of stored energy onboard ships has been established since the Second World War, where submarines used electric power for propulsion underwater, where the operation of diesel engines was not possible. Typically, the optimization of marine diesel engines is aimed at reducing fuel consumption for a single point of operation, while the selection of components such as turbochargers is made for a wide range of operational conditions, affecting the overall efficiency [16,34]. Marine engines operate in constantly changing conditions at sea due to waves and wind, and the request of charterer's commands to alter voyage speed and destination. Thus the main propulsion engines do not operate at their optimum point and as a result the specific fuel oil consumption (SFOC) is increased [35,36].

The study by Dedes et al. [29] aims to uncouple the propeller demand from the thermodynamic process of the production of rotational speed and torque by the diesel engine. The concept of a re-engineered ship propulsion system that includes batteries as part of an All Electric Ship (AES) concept is considered as the basis of this work. The aim is to produce the required electrical energy at the optimum point for the diesel generators, where the SFOC is a minimum and the power train can be designed to further reduce consumption. An energy storage medium is proposed for installation onboard dry bulk carrier vessels to store the excess of energy and allow load levelling to the mean requirement for propulsive power and auxiliary or "hotel" load. This will result in a decrease in the size of the total installed power as the peak demands can be covered by the storage medium. The potential of such an approach has not yet been investigated for the main propulsion, or the auxiliary system, of large merchant ships. It is expected that the proposed alternative propulsion system will allow a more flexible approach to the propulsion system, permitting further optimization of external emission reduction techniques.

The study uses as its basis actual consumption data for a fleet of dry bulk carrier vessels to calculate the production of CO_2, NO_x, and SO_x emissions and estimate the engine loading for laden and ballast voyages. The correlated data provide the required information to size and select the storage medium and the propulsion engines using an overall daily energy consumption approach. The connection of batteries and operational parameters is considered, along with a comparison of potential storage devices. The energy storage media and diesel generators will not make major changes in the ship weight and longitudinal distribution [29] that would reduce the carrying capacity of the vessel. An initial economic assessment is performed to demonstrate the feasibility of the proposed system.

For the proposed hybrid system, batteries are selected as the energy storage medium since they are a reliable and, depending on type, low cost solution that can store a large amount of energy comparing to other storage media [37]. Their power density (kW/kg) is relatively high and electrical energy efficiency remains high at practical discharge currents. Batteries have been used in the automotive industry in the development of hybrid car applications and have a significant role in the overall efficiency of the power train. The success of the system is dependent on the development of the appropriate battery technologies [31,38]. The weight and operational characteristics are considered to play a significant role in the selection of an appropriate storage system for use in the marine environment.

It is observed that lithium-ion batteries have by far the highest energy density. The additional advantages of lithium-ion cells are the flat characteristic curve of voltage drop during most of the discharge period, the absence of memory effects, and a superior life but there are possible environmental and human health implications [39]. The principal disadvantage of lithium-ion batteries in this application is their large cost which exceeds 600$/kWh. Lead acid batteries appear to be a more economical solution. However, the low material resistance in the marine environment, corrosive failures, and the short life period of 400 complete charges and discharges make them more expensive in the life cycle of the ship. Lead acid batteries suffer from a quick voltage drop and in a long period of storage from self-discharging [40].

Vanadium Redox flow cells and sodium-nickel chloride (Na-NiCl$_2$) batteries are considered viable potential storage types. The key characteristics of Na-NiCl$_2$ batteries are their low weight, high energy and power density, the flat state of charge curve, and a charging efficiency that reaches 92% at favorable charging currents [41,42]. Redox flow cell batteries are an efficient way to store energy with zero self-discharge problems. The membrane that is used in this technology to separate the reactants from each other has to be replaced after a period of time, increasing the maintenance cost of the battery system [43]. Redox flow batteries have the potential to be fast recharging contenders, while refueling can be achieved by rapid pumping of new reactants into the cells, or by recharging it using an electric current.

The required response time in ship applications is larger than in the automotive industry where fast charging and discharging is required due to the drive cycle (e.g., in urban environment where many acceleration/deceleration cycles are present [33]). For a ship typically once port areas have been left the set point of the engine remains within a narrow range for prolonged periods. The maximum depth of discharge is critical in selecting an appropriate technology. It is related to the energy density and the discharge time. The marine application of battery storage requires high energy density and large discharge time with an almost flat curve of voltage drop versus time. Similar to the operation of hybrid electric automotive vehicles, the storage medium (battery) may be operated in two modes [44]. The first mode is called charge depletion mode and is the case where the electrical power required by the vehicle is provided by the stored energy until its depletion. In the case of the proposed hybrid ship, it may be envisaged that the vessel operated inside harbor in this mode, while performing cargo handling operations. Increasingly port authorities are requiring that an external power source is connected to berthed vessels, in order to reduce emissions locally and meet the requirements of Emission Control Areas (ECAs)—a practice referred to as "cold ironing" [45].

When the batteries reach the point of depletion, the system switches to the second, charge sustaining (CS), mode where the vehicle operates as a conventional one. In CS mode, the electricity produced by the generator sets supply both the propulsion unit and recharge the batteries. In an assessment of the "life-cycle" emissions of the vessel, emissions associated with the production, supply, and recycling of the batteries must also be accounted for [46]. However, such issues are beyond the scope of the present study, where the focus is on assessing the potential of hybrid technology to improve the operation of vessels.

The potential fuel savings and emission reduction, through use of an alternative, hybrid system have been investigated in this study. A hybrid propulsion system is a flexible and efficient propulsion system for any type of vessel having a mission profile with extensive high and low load operation. The minimization of engine transient loading and the use of load levelling are shown to result in fuel savings and hence emissions reductions. These may further be reduced through optimization of propulsion and engine components. The overall AES concept proved not to be feasible for laden operation as the engine fluctuation fuel savings do not overcome the electromechanical losses. In ballast voyages, where engine operate away from the optimum point, the AES hybrid concept is feasible. Moreover, the hybrid AES system is more suitable in maneuvring operation or while sailing inside an ECA zone. The possible scenario that combines a two-stroke Diesel direct drive and hybrid propulsion would be an attractive propulsion system due to the high efficiency of a two-stroke engine, in terms of total fuel oil consumption. However, a combination of slow-speed diesel and a hybrid PTO/PTI (power take out/power take in) system with fully integrated auxiliary power generation should also be considered. It was observed that for most of the time the auxiliary electric demand could be covered by one out of three installed generator sets. The application of storage systems allows for an immediate reduction in the installed auxiliary engine capacity and offers a safety net in case of peak loads that in other cases would result in a total ship blackout. Thus a detailed calculation of auxiliary fuel savings will be performed in the future and this will lead to more efficient power handling while the ship is in berth, at sea, or in maneuvring.

Based on its study, authors concluded that the operation of the fleet greatly affects the fuel consumption and hence the emissions. It was observed that for similar designs (Handysize and Handymax), although they were expected to have a similar engine profile and similar consumption, Handymax vessels seem to have negligible savings whereas significant savings were found for Handysize. This is primarily due to chartering commands resulting in variations in voyage speeds. A secondary reason is that the relatively small dimensions of the ship make it suitable for a wide range of loads that greatly affect the engine operation. To extract a more universal conclusion, a larger sample has to be investigated, using detailed operational profiles for the voyages. The installed power requirement is seen to be highest for the Panamax category. On the other hand, bigger ships (e.g., Capesize bulkers) operate their engines closer to the optimum points and there is less speed variation. The installation of a large energy storage medium is thus not necessary. Analysis of the Panamax type, being between the HandyMax and Small Capes (also known as Post-Panamax), shows a further increase in the energy storage demand.

Of the considered battery technologies, the potential of installing Na-NiCl$_2$ and Redox flow cell batteries was examined. It was observed during the initial sizing of the system that installing Na-NiCl$_2$ batteries closely balances the weight saving due to the reduction in carried fuel. For other batteries, a decrease of payload is required to keep the same total displacement. Na-NiCl$_2$ batteries are already tested and certified for applications in the marine environment. However, cooling issues arise and have to be examined in future work that will focus on the practical design of the installation. On the other hand, Redox flow cells do not have constraints in the operational temperature. However, the storage of reactants in different tanks creates potential issues of stability due to the presence of uncontrolled free surfaces and issues of constructional strength due to sloshing effects inside these tanks.

The future application of hybrid systems will require further design considerations for ship electric networks. It is proposed that this will require a revision of safety regulations concerning the propulsion systems and machinery. Moreover, the legislation has to fully understand the purpose and the capability of such hybrid all electric vessels alongside modification to electric power quality regulations. Furthermore, the legislation has to include economic incentives for ship-owners to invest in such systems and reduce the perceived risks to the smooth operation of their fleet inherent in adopting technical innovations.

From an economic point of view, and not accounting for the cost of the protection of the environment, the storage medium with Na-NiCl$_2$ proved to be the most feasible method with a potentially high return on initial investment. Post-Panamax-type fuel savings showed that the rate of ROI for both storage media cases examined is less than 3 years. Other ship types indicate the system is economically feasible over a 25-year period, except Handymax type which needs further investigation, with high rate of return in most cases for the Na-NiCl$_2$ batteries. Meanwhile, while other storage medium is still at high cost as these products reach the market, their cost is likely to drop.

This study has shown that installing hybrid power technology onboard dry bulk ships can save fuel cost up to 1.27 million USD (at the price of 520$/ton) per vessel and per year, assuming that the 60% of the time ship sails in laden and 40% in ballast condition. This value depends as well on the ship's dimensions, the storage medium adopted, and the demand for energy availability. The emission reduction is achieved primarily through reducing the consumption of fuel and further reductions could be achieved by optimization of the combustion process or the operation of other engine components. The combination of a hybrid energy storage and the flexibility that offers in the coupling with the propulsor, along with other possible improvements in hydrodynamics based improvement in ship energy efficiency should allow a step improvement in overall efficiency of ship propulsion systems (estimated between 2% and 10%) although this requires further systematic design studies.

4.2.7 USE OF HYBRID MICROGRIDS FOR SHIPS

The demands for sufficient electrical power in future ship designs varying from warships and naval aircraft carriers to oil tankers and transport vessels are of main importance to current ship designers. From AC generators to the propellers, today's loads predominantly exist within the DC section of the system. With the introduction of advanced communication systems, electromagnetic aircraft launch systems, and electrical weaponry onto naval vessels, high-reliability integrated power systems that can meet varying electrical load demands are highly sought. Power system architectures are dependent upon load types, and the viability of DC distribution, such as medium- or low-voltage DC power systems, in meeting upcoming shipboard electrical demand forecasts is of focus. The paradigm shift of traditional AC system constructions to DC entails both operational benefits and, on the other hand, any accompanying risks. One form of system design does not by any means entitle an industry standard, as vessel functions vary from ship to ship and boat to boat. Ideal system configurations coincide with the shipboard loads and are characterized by high reliability, ease of maintenance, both fuel and payload efficiency, and lower costs and low, or optimally absent, emissions [47].

ABB [48] reported that an Onboard DC Grid can allow for fault current clearance within a time window of a maximum of 40 ms, as compared to 1 second for traditional AC circuits. Furthermore, because the speed of prime movers must be locked at approximately 60 Hz in AC marine systems, the lowest fuel consumption will be achieved operation about 85% of the rated load. Because DC buses are at no frequency, the prime mover speed can be adjusted according to demand, all the while correspondingly adjusting the generator excitation current in the same respects. Given this commodity of optimization speed, the operating window of the prime movers can be brought down to about 50% of the rated load with no increase in fuel consumption [49].

In recent years, there has been an increasing interest in the design and analysis of microgrids in general [50–52], with particular interest on the implementation of hybrid microgrids (*AC and DC microgrids*) within the maritime industry [53]. For instance, the Italian Navy, by the use of a system nicknamed naval package (NP), has successfully corroborated simulation results with experimental data in determining the validity of a medium-voltage DC shipboard IMPS with the use of dual three-phase 2.15 MVA generators. The results showed that the generator produced quality DC outputs with the introduction of faults and specified rectifier arrangements reduced performance degradation and increased system fault tolerance [53].

Several operations are constantly performed on shipboard power systems. Of these, service restoration and reliability improvement may be of most concern. Numerous marine power distribution systems are characterized by radial or weakly meshed topological structure. The unidirectional power flow in the radial distribution systems facilitates the coordination of the protective devices used at the distribution system level. Nevertheless, this radial topological structure makes distribution systems less reliable, compared to transmission systems, which are highly interconnected systems. Despite the unidirectional power flow in the radial distribution systems, the failure of any single component between the load point and the source node may cause service interruptions, which could lead to disconnection of several load points. Distribution systems are constantly equipped with two types of switches: sectionalizing switches and tie switches. The sectionalizing switches are normally closed; however, the tie switches are normally open but can be closed to routing the power and meeting the power demand during abnormal conditions. Distribution system reconfiguration can be used to minimize the duration and frequency of service interruptions and thereby enhance the reliability of the distribution system and the quality of service. By distribution system reconfiguration, one denotes the process of changing the topology of the distribution network by altering the status of sectionalizing and tie switches to achieve certain objectives [51]. Of these objectives, reliability improvement, service restoration, and survivability enhancement are of great concern. Generally speaking, reconfiguration and service restoration of marine power systems are carried out in a similar manner to that used for terrestrial power systems, with few restrictions and constraints.

Wei and Yu [54] presented a method for load sharing in hybrid microgrid systems using control loops. A method for maritime microgrid reconfiguration is presented by Shariatzadeh et al. [55] using genetic algorithms and heuristic techniques. The work presented in Ref. [55] is implemented on a small shipboard power system, and a modified CERTS (Consortium for electric reliability technology solutions) microgrid system including distributed generators has been used. Methods for optimal feeder reconfiguration for terrestrial distribution systems using intelligent and heuristic methods are also proposed in Refs. [50–52]. Ahmad et al. [56] presented a method for optimal sizing of RE resources for next-generation seaports. The RE resources used in Ref. [56] are modeled using HOMER platform. A two-stage technique for isolated microgrid systems is introduced by Hari kumar and Ushakumari [57]. The method proposed in Ref. [57] used graph theory and binary firefly algorithm to perform minimum load curtailment in the isolated microgrid. Methods for service restoration of marine power systems have also been presented in Refs [58–60].

4.2.8　FUTURE MARINE POWER SYSTEMS

Due to the added regulations regarding the increase in efficiency and decrease in pollution of all vessels, it has become evident why most companies desire and are actively searching for better shipboard microgrid architectures. The overall goal of future marine power systems would be to establish and use an almost completely DC marine power system. This is due to the numerous potential benefits it would allow in both efficiency and pollution if properly designed [61–66]. Numerous vessels now deploy diesel engine propulsion; however, they can likely be made more efficient in the future if they were to be switched to electric propulsion under the umbrella of AESs. Electric propulsion may improve fuel consumption as well as the dynamic performance of the ship [67]. Since ships that use electric propulsion consume most of the power generated for propulsion, it is imperative to ensure that the shipboard microgrid of such a vessel is capable of handling the load dynamics, which are connected to electric propulsion as well as the immense amount of power needed to propel a vessel utilizing electric propulsion technologies.

With the use of ESSs, such as battery banks and fuel cells, synchronous generators can be operated at optimal speeds, which allow for improved efficiency and reduced pollution, as a result, allowing a greater reduction in distribution losses as well as increased reliability [54]. The use of ESSs is beneficial as they allow for continuous power flow to the connected loads in the event of a power generator failure.

Transformers, which are AC/AC converters, are used in conventional shipboard AC power systems. Such devices would be replaced with DC/DC power converters for the shipboard DC distribution systems of the future, which would allow the voltage to be stepped down (buck converters, for instance) or stepped up (boost converters, for instance), to the required levels. The voltage levels may also be stepped up and stepped down to achieve certain requirements by the means of buck/boost or cuck converters. On the other hand and for specific AC loads, DC/AC inverters would be required. Benefits of using such converters may include increased power compensation and better frequency regulation.

It is appropriate to mention that the modern and future maritime systems have many similarities in structure; however, they also possess few differences. For instance, as shown in Figure 4.5, one of the major noticeable differences is that in future marine power systems, both the main and the auxiliary or emergency switchboards receive and distribute DC power as opposed to the AC power utilized on the modern marine power system. Another key difference is the use of an ESS, which powers the switchboard in the next-generation marine power systems as can be seen from Figure 4.5. Though only few ships are currently deploying electric propulsion, the ships of the future should solely use electric propulsion due to its higher efficiency. Some advantages of using DC-powered systems instead of the conventionally used AC systems on next-generation shipboard systems are summarized in Table 4.1 [48,49,65,66].

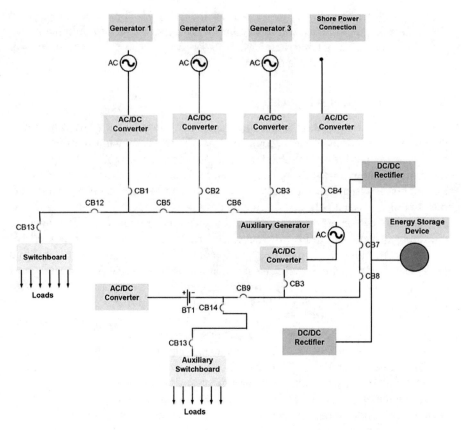

FIGURE 4.5   Single-line diagram of a next-generation marine power system [47].

TABLE 4.1

Advantages of Using DC-Powered Marine Systems [47]

Item	Advantages of Using DC Systems vs. AC Systems
1	Eradication of frequency-related issues and synchronization of sources
2	Proper control of drive systems over a wide range of speeds due to the advancement in power electronic devices
3	Potential reduction in overall size and rating of switchgear
4	Generators operate at or near unity power factor and thereby reactive power compensation may be controlled properly
5	Overall reduced system size in general, allowing for larger cargo space
6	Virtual inertia of power electronic devices may be deployed to enhance the overall system stability
7	Replacement of large distribution transformers with small-sized power electronic converters

4.2.8.1   Maritime Microgrids

In general, DC microgrids characteristically are self-sustaining given the ability to work given a grid-connected or islanded-mode condition. The feature of these self-thriving distribution systems emanates from the connected RE resources and energy storage devices. The general architecture of the emerged maritime microgrids is depicted in Figure 4.6. As shown in this figure, most maritime microgrids consist of conventional generators, several power electronic circuitries, and numerous AC and DC loads such as the propellers and propulsion system as a whole.

In Figure 4.6, the maritime microgrid system consists of two synchronous generators operating in parallel and connected via a sectionalizing switch. The switch is normally open and can be closed during harsh conditions, as in case of breaking,

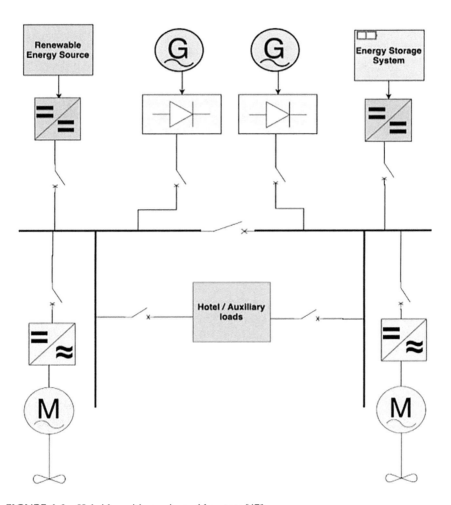

FIGURE 4.6   Hybrid maritime microgrid system [47]

for example. The hybrid maritime microgrid system shown in Figure 4.6 is driven by power electronic circuitry, which is mainly consisting of the following:

- AC/DC full-bridge inverter system.
- ESS EES, and in our case we assumed a battery system.
- DC/DC boost converter(s).
- DC/AC inverter(s): one of the main functions of the DC/AC inversion systems is to feed the propellers.

A lot of research is being conducted to optimize the operation of maritime microgrids, such as enhancing reliability and survivability, improving stability using virtual synchronous generators, and minimizing power curtailment. References [48–60,66] provide literature on some of these techniques. The study of Caravella et al. [47] has reviewed marine power system components which include load model, cogeneration model, static VAR model, and transformer model.

The study concluded that maritime power systems are somewhat different from their typical terrestrial counterparts. In particular, generally speaking, maritime systems consist of generation and distribution, but no transmission, which is very distinct from the territorial electric power systems. The generation and distribution are coupled with distribution feeders in order to transfer the power from the synchronous generators to the loads. Hybrid maritime distribution systems (*hybrid maritime microgrids*) consist also of several power electronic devices such as converters, inverters, rectifiers, and switchboards. Instrumentation and control circuits are used on ships to ensure safe operation of the ship. A typical maritime power system is in fact an isolated power system, which is largely different from the terrestrial power systems. Therefore, survivability of such a system during abnormal circumstances is of great concern for marine power system planners. Enhancing reliability and survivability through feeder reconfiguring of marine distribution systems, however, needs flexible and reliable techniques to accommodate all of the aforementioned characteristics of modern marine power systems. In order to perform studies on any marine power system, a power flow solution is constantly required. Nonlinear power flow and DC power flows are commonly used in the literature to carry out optimization studies on terrestrial and shipboard systems. In recent literature, enhanced linearized power flows [68,69] have also been used as they compensate for most of the drawbacks of the conventional DC power flows. A lot of research needs to be performed in order to secure transition from AC marine power systems to hybrid and then DC-powered marine power systems. This research may include developing methods for enhancing reliability and survivability, improving stability using virtual synchronous generators, and minimizing power curtailment on ships using real-time data and testbeds. Table 4.2 summarizes some challenges and also lists examples of research that need to be carried out in order to accommodate the emerged maritime power systems.

In summary, the study of Caravella et al. [47] has detailed some potential ways for the maritime industry to begin to phase out AC power generation and distribution on new vessels over a short period of time. Over short distances, DC power provides less power loss, no harmonics, and more control over equipment, which are some of

TABLE 4.2
Examples and Challenges of Future Research in Marine Systems [47]

Item	Challenges
1	The penetration of intermittent RE resources and storage devices has increased in nowadays shipboard systems. Consequently, accurate models need to be developed to accommodate such an increase.
2	To ensure reliable operation of next-generation shipboard systems, numerous studies should be carried out in a real-time frame using high-computation facilities and parallel computing.
3	More innovative techniques for optimal load flows, service restoration, and reconfiguration need to be developed. This requires the development of more testbeds, particularly for large-scale realistic marine power systems.
4	Frequency-related issues of power electronic devices are another area of research for future shipboard systems. Effective inverter topologies that minimize stress ratios and switching losses constitute another research path for future maritime systems.
5	To stabilize operation of the hybrid power systems, more work needs to be carried out in the area of power system stability using the concept of virtual inertia. Proper control design and innovative communication protocols would help utilizing the concept of virtual synchronous generators to maintain the stability limits of a given marine electric system.

the major reasons why many industry officials are conducting vast research in this specific area. Therefore, the vessels of the future should consider transitioning into DC power generation and distribution. Nonetheless, during the transition from an AC shipboard power system to a DC shipboard power system, there will be a time during which the vessels will be run by a hybrid shipboard power system, which utilizes a mixture of both AC and DC power. These hybrid systems are known as integrated marine power systems or hybrid maritime microgrids.

Some advantages of hybrid maritime microgrids include higher efficiency, optimum fuel consumption, improved reliability, and more importantly improved survivability. This book chapter provides a quest for future maritime microgrids and integrated marine power systems. Though the power system literature is rich for terrestrial microgrids, it has fallen behind in providing a solid background for the future maritime systems. One of the main objectives of this review study is to pave the way for the researchers by supplying them with recent technologies, visions, and applications for future maritime microgrids. This study has presented a state of the art for maritime microgrids, emphasizing on the design aspects of hybrid maritime microgrids and summarizing the advantages, disadvantages, and the challenges that planners face when integrating RE resources into existing marine power systems. Moreover, this review study has paved the way for mariners and researchers by supplying them with recent technologies, visions, and applications for future maritime integrated power systems and maritime microgrids. This work has also presented and discussed issues associated with the design and control of future maritime microgrids as envisioned by the US Navy near-term development plan of 2025 and the long-term plan of 2035. In addition, this study presented some of the challenges that both current and future IMPS are facing and reviewed some of the remedies that have been recently proposed in the literature to overcome such challenges. This study has also reported on the problem of feeder reconfiguration and service restoration of shipboard power systems and introduces directions on how to enhance the reliability and survivability of maritime power systems using distribution system reconfiguration.

Tang et al. [70,71] examined optimization of hybrid energy systems for maritime vessels. The decarbonization agenda in maritime transport requires that asset owners and operators adopt greener technologies within their existing and new vessels. The primary drivers within this agenda relate to improved environmental metrics, efficient energy performance, and improved asset management. However, the integration of new technologies always presents technical and financial risks. Here, utilizing energy and environmental monitoring from real vessels, the authors propose an energy system optimization architecture, hybrid fusion energy management system (EMS), that optimizes the key performance indicators of energy performance, reduction of diesel engine nitrogen oxide (NO_x), and PM, and prognostic state of health assessment of energy storage technologies. The study presents an integrated approach to the optimization of hybrid vessel performance. Using a data-driven approach to prognostics, the study estimates the remaining useful life and SOH (state of health) of lithium-ion and lead-acid batteries with satisfactory accuracy. The study also deployed MATLAB to generate a third-order polynomial function, which determines the optimal operating curve for diesel engine, in the case where both electricity and fuel as power sources. Using state-of-the-art machine-learning techniques, the authors are able to determine the onboard lithium-ion and lead acid batteries' state of health with accuracy >8% and 4%, respectively. Dependent on the mode of operation, optimization of energy performance indicates fuel saving of between 70% and 80% for the vessel operator. Future research should focus on the integration of more assets into the optimization architecture and increased vessel journey use cases.

4.2.9 HYBRID POWER MODULE

The technology group Wärtsilä introduced a unique hybrid product, the Wärtsilä HY, representing an unprecedented innovation in marine propulsion systems. By leveraging its technical strengths in both engine design and electrical & automation (E&A) systems, Wärtsilä is launching a fully integrated hybrid power module combining engines, an ESS, and power electronics optimized to work together through a newly developed EMS. According to Warysila, this is the marine sector's first hybrid power module of this type produced, thereby establishing a new industry benchmark in marine hybrid propulsion. The Wartsila HY was officially launched at the Nor-Shipping conference and exhibition being held in Oslo, Norway, in 2017 [72].

The Wartsila HY has dedicated versions for each category of vessels. While the first version being made available is designed for tugs and medium sized ferries. Wartsila also sees big potential in other types of vessels as well. This is the first launch of a new product of this type where each individual version is dedicated to a specific market to secure an optimal fit to the requirements of the specific application. There is a notable trend in the marine sector toward hybrid propulsion solutions, which are anticipated to represent a significant percentage of all contracted ships within the coming ten years. The new EMS represents the latest generation integrated control system and has been specifically designed for this application. It creates an outstanding means of interaction with the ship's onboard systems [72].

The Wartsila HY provides a wide range of customer benefits through increased operational efficiency and flexibility, resulting in lower fuel consumption, reduced emissions, and improved vessel performance. When operating in "Green Mode," zero emissions can be achieved. Smokeless operation is also achievable at all load points and in all operating modes, due to a new automation procedure. Furthermore, the reduction in engine operating hours lowers maintenance requirements and extends the intervals between overhauls.

The Wartsila HY ensures that the overall vessel performance is greatly improved compared to operating on conventional machinery solutions or hybrid solutions, while the higher level of redundancy promotes increased safety. Other benefits include instantaneous load acceptance with rapid response to step-load changes, entire system certification, and guaranteed performance. Maritime classification society, Lloyds Register (LR), has issued an Approval in Principle certificate for the Wartsila HY. The certificate is based on technical material and safety analyses, concerning normal operation of the system and a presentation of risk scenarios [72].

4.2.10   HYBRID FUEL CELL-BASED SHIPS

The need for technological development to reduce the impact of air pollution caused by ships has been strongly emphasized by many authorities, including the IMO. This has encouraged research to develop an electric propulsion system using hydrogen fuel with the aim of reducing emissions from ships. Roh et al. [14] examined fuel consumption and CO_2 emission reductions of ship powered by a fuel cell-based hybrid power source. This study analyzed the fuel consumption and CO_2 emission reduction rates when a fuel-cell-based hybrid power source instead of a conventional commercial diesel power source was used in ships. The system built by the study uses hybrid power and a diesel engine generator with a combined capacity of 180 kW. To utilize scale-down methodology, the linear interpolation method was applied. The proposed hybrid power source consisted of a MCFC, a battery, and a diesel generator, the capacities of which are 100, 30, and 50 kW, respectively. The experiments that the study conducted on the test bed were based on the outcome of an analysis of the electrical power consumed in each operating mode considering different types of merchant ships employed in practice. The output, fuel consumption, and CO_2 emission reduction rates of the hybrid and conventional power sources were compared based on the load scenarios created for each type of ship. The results showed that under the rated output on a test bed with a load bank of 180 kW, the conventional commercial diesel generator consumed fuel at 43.5 kgoe/h and emitted CO_2 at 148.5 kg/h, whereas the fuel-cell-based hybrid power source consumed fuel at 35.6 kgoe/h and emitted CO_2 at 57.7 kg/h. The hybrid power source reduced fuel consumption by 18% and CO_2 emissions by 61% at part load in the port period. These results indicate that it is possible to reduce CO_2 emissions by up to 61% if a hybrid power source of the same capacity is used to power a ship [14,73].

In this study, the actual electric load analysis values of the 5,500 TEU Reefer Container, 13 k TEU Container, 40 k Bulk Carrier, 130 k DWT LNG Carrier, and 300 k DWT Crude Oil Tanker were scaled down according to the operation mode, and the control logic and systems of the test bed developed in this study were operated normally according to the respective load scenarios. Because the output characteristics and control time of the diesel generator, according to the power source of the hybrid system, were reduced, according to the load variation pattern of the ship and the ship's type, the CO_2 emissions of the hybrid system, as compared with the case of the diesel generator, alone operated for each load scenario with an average of 70%~74% less. In order to apply the hybrid system to ships, it is possible to maximize the CO_2 emission reduction effect by setting the capacity of the fuel cell + battery to be able to take charge of the base load of the ship through analysis of the base load of each ship type [14,73].

4.3   HYBRID ENERGY FOR AIR VEHICLES

Kuhn et al. [74] presented RE perspectives for aviation. The transformation of the energy supply from fossil to RE is the single most important challenge for the aviation industry's long-term future.

The substitution of fossil kerosene by suitable, sustainable, and scalable alternative energy carriers is a key requirement for which strategies in three categories, i.e., the drop-in fuel, non-drop-in fuel, and fully electric energy path, are presented. For long-term RE perspectives, drop-in fuels from solar thermal reactors, the use of renewable hydrogen for either synthetic paraffinic kerosene or fuel cells, and the material advancements in electric energy storage promise substantial innovation potential. The analysis of the energy options shows that (a) solar fuels are able to overcome known sustainability and/or scalability limitations of bio-to-liquid and other drop-in bio-fuels, that (b) the (non-drop-in) hydrogen fuel perspective has inherent limitations that are either solved by feeding into the solar hydrocarbon fuel process or by serving the electric propulsion paradigm shift via Proton Exchange Membrane (PEM) fuel cells, and that (c) the all-electric regional aircraft is well within the physical regime of feasibility and potentially closer to realization than generally assumed, mainly driven by a few key advancements in material science and nanotechnology.

4.3.1   HYBRID AIRCRAFT

The Boeing Fuel Cell Demonstrator Airplane (see Figure 4.7) has a PEM fuel cell/lithium-ion battery hybrid system to power an electric motor, which is coupled to a conventional propeller. The fuel cell provides all power for the cruise phase of flight. During take-off and climb, the flight segment that requires the most power, the system draws on lightweight lithium-ion batteries. The demonstrator aircraft is a Dimona motor glider, built by Diamond Aircraft Industries of Austria, which also carried out structural modifications to the aircraft. With a wing span of 16.3 m (53 ft), the airplane will be able to cruise at about 100 km/h (62 mph) on power from the fuel cell. Hybrid FanWings have been designed. A FanWing is created by two engines

FIGURE 4.7   In 2008, the Boeing fuel cell demonstrator [75].

with the capability to autorotate and landing like a helicopter. The first passenger aircraft HY4 powered by fuel cell is illustrated in Figure 4.8.

An Australian-based company is working on a project to develop an air crane called the SkyLifter, a "vertical pick-up and delivery aircraft" being capable of lifting up to 150 tons.

A Canadian start-up, Solar Ship Inc., is developing solar powered hybrid airships that can run on solar power alone. The idea is to create a viable platform that can travel anywhere in the world delivering cold medical supplies and other necessitates to locations in Africa and Northern Canada without needing any kind of fuel or infrastructure. The hope is that technology developments in solar cells and the large surface area provided by the hybrid airship are enough to make a practical solar powered aircraft. Some key features of the Solar ship are that it can fly on aerodynamic lift alone without any lifting gas and the solar cells along with the large volume of the envelope allow the hybrid airship to be reconfigured into a mobile shelter that can recharge batteries and other equipment.

The Hunt *Gravity Plane* (not to be confused with the ground-based gravity plane) is a proposed gravity-powered glider by Hunt Aviation in the USA. It also has aero foil wings, improving its lift-drag ratio, and making it more efficient. The Gravity Plane requires a large size in order to obtain a large enough volume-to-weight ratio to support this wing structure, and no example has yet been built. Unlike a powered glider, the Gravity Plane does not consume power during the climbing phase of flight. It does however consume power at the points where it changes its buoyancy between positive and negative values. Hunt claims that this can nevertheless improve the energy efficiency of the craft, similar to the improved energy efficiency of underwater gliders over conventional methods of propulsion. Hunt suggests that the low

FIGURE 4.8   HY4 – the world's first passenger aircraft powered by a hydrogen fuel cell [75].

power consumption should allow the craft to harvest sufficient energy to stay aloft indefinitely. The conventional approach to this requirement is the use of solar panels in a solar-powered aircraft. Hunt has proposed two alternative approaches. One is to use a wind turbine and harvest energy from the airflow generated by the gliding motion, the other is a thermal cycle to extract energy from the differences in air temperature at different altitudes.

4.3.2 UNMANNED AERIAL VEHICLES

There are several unmanned hybrid vehicles in operation [76]:

1. **Pathfinder and Pathfinder-Plus**: This UAV demonstrated that an airplane could stay aloft for an extended period of time fueled purely by solar power.
2. **Helios**: Derived from the Pathfinder-Plus, this solar cell and fuel cell powered UAV set a world record for flight at 96,863 ft (29,524 m).
3. **Qinetiq Zephyr**: The Zephyr, a lightweight unmanned aerial vehicle (UAV) engineered by the United Kingdom defence firm QinetiQ, claimed the endurance record for an UAV. It flew in the skies of Arizona for over two weeks (336 hours). It has also soared to over 70,700 ft (21.5 km).
4. **China's** designed and manufactured UAV successfully reached an altitude of 20,000 meters during a test flight in the country's northwest regions. Named "Caihong" (CH), or "Rainbow" in English, it was developed by a research team from CASC (China Aerospace Science and Technology).

4.3.3 MANNED SOLAR AIRCRAFT

A number of manned solar aircraft have been developed. These are Gossamer Penguin, solar challenger, sunseeker, solar impulse, and solar stratos [77,78,79]. Here we briefly describe the development of Solar Impulse.

Solar Impulse is a Swiss long-range experimental solar-powered aircraft project, and also the name of the project's two operational aircraft. Bertrand Piccard initiated the Solar Impulse project in November 2003 after undertaking a feasibility study in partnership with the École Polytechnique Fédérale de Lausanne [74,77,80]. The aircraft is a single-seated monoplane powered by PV cells; they are capable of taking-off under their own power. The prototype, often referred to as *Solar Impulse 1*, was designed to remain airborne up to 36 hours. A second aircraft, completed in 2014 and named *Solar Impulse 2*, carries more solar cells and more powerful motors, among other improvements. The aircraft's major design constraint is the capacity of the lithium polymer batteries. Over an optimum 24-hour cycle, the motors can deliver a combined average of about 8 hp (6 kW), roughly the power used by the Wright brothers' Flyer, the first successful powered aircraft, in 1903. In addition to the charge stored in its batteries, the aircraft uses the potential energy of height gained during the day to power its night flights [77–79].

The wingspan of *Solar Impulse* 2 is 71.9 m (236 ft), slightly less than that of an Airbus A380. The carbon-fiber Solar Impulse weighs only about 2.3 tons (5,100 lb), little more than an average SUV. It features a nonpressurized cockpit 3.8 m^3 (130 cu ft) in size and advanced avionics, including an autopilot to allow for multiday transcontinental and trans-oceanic flights (see Figure 4.9). Supplemental oxygen

FIGURE 4.9 *Solar Impulse 2* at the Payerne air base in November 2014 [78].

and various other environmental support systems allow the pilot to cruise up to an altitude of 12,000 m (39,000 ft).

4.3.4 Solar Electric, Hybrid, and Hydrogen Aircraft

Solar ships can refer to solar-powered airships or hybrid airships. There is considerable military interest in UAVs; solar power would enable these to stay aloft for months, becoming a much cheaper means of doing some tasks done today by satellites. In September 2007, the first successful flight for 48 hours under constant power of a UAV was reported. This is likely to be the first commercial use for PVs in flight. Many demonstration solar aircraft have been built, some of the best known by AeroVironment [77–79].

The aviation industry has noted a consistent increase in the electrification of aircraft systems, research on electrical propulsion, and investments in electric or hybrid aircraft designs. Projects are also ongoing on liquid hydrogen research for civil aviation purposes. Electric, hybrid, and hydrogen aircraft may help ICAO (International Civil Aviation Organization) meet its major environmental goals on climate change, local air quality, and noise. This article describes the possible environmental benefits that may result from these new technologies and provides an overview of the current status of their development and implementation in aircraft.

Substituting jet fuel with electricity or hydrogen can have a notable impact on the climate change impacts of aviation, as the operation of electric or hydrogen aircraft will not be associated with CO_2 emissions from fuel combustion. However, it is important to note that such CO_2 benefits need to be considered on a life cycle basis and will only occur if the electric energy or hydrogen is obtained from lower carbon sources. For example, as of 2015, 98 airports around the world had installed solar power projects, and this number has continued to grow in the years since. The continued expansion of RE capacity and availability at airports could provide an opportunity for hybrid or electric aircraft to recharge in such a way that CO_2 benefits could be achieved. Similarly, such RE could be used to produce hydrogen with a low CO_2 impact on a life cycle basis. The climate benefits of electric aviation may come not only from its reduced CO_2 emissions but also from the elimination of contrails—the long, thin clouds that form in the wake of jet engines. Although no scientific consensus exist on the radiative forcing effect of contrails, some studies point out that they may have further warming impacts on the global climate.

Beyond electric and hydrogen propulsion, it should be noted that there are various ways to use electricity and hydrogen in aircraft operations. One example is electric taxiing (E-taxi), which could save almost 33 kg of CO_2/minute of use, according to ICAO's Rules of Thumb [80]. Hybrid aircraft can also help to reduce fuel consumption and contrail generation by using electric motors as a supplementary thrust source during the take-off phase, which allows the use of smaller and more efficient jet engines during the cruise phase of flight. Airports around the world have also demonstrated the feasibility of hydrogen for ground support/transport vehicles. For example, initiatives in Heathrow, Berlin, and Los Angeles installed hydrogen fueling stations that produce hydrogen onsite from RE sources, using the electrolysis process.

Full electric aircraft promises significant benefits for local air quality, as the pollutants emitted on the fuel combustion process are avoided. Hybrid-electric aircraft may similarly help improve local air quality impacts of aviation due to its lower fuel burn. However, while looking at air pollution impacts from all types of aircraft including electric ones, brake abrasion, in addition to tire abrasion and road surface erosion, still needs to be considered as these factors are a source of PM emissions. In addition, similarly to the CO_2 emissions, the source of the electricity should be considered when assessing the local air quality impacts of electrification, since different processes of electricity production may still be associated with air pollution. Other factors also need to be considered when looking at overall trends. While becoming more fuel efficient, aircraft tend to increase in size and weight, carrying more passengers and more fuel. This increase in carried fuel could offset the fuel reduction achieved through energy efficiency improvements due to hybrid systems. Therefore, it is clear that hybrid-electric aircraft help reduce air pollutant emissions when looking at the per-passenger figures, but not necessarily when looking at total figures. Moreover, most hybrid-electric aircraft are equipped with batteries for electricity storage and supply. Due to battery energy density and the required power supply, these batteries are currently very heavy, thus can substantially increase the weight of aircraft [80].

A life cycle approach to electric aircraft could be useful to assess the overall impact of electric aircraft on the environment and its sustainability benefits. This approach goes from inception of an aircraft to its end-of-life and helps to avoid environmental and social risks. Batteries used in electric aircraft are currently made of mostly lithium. Air pollutants emitted during processes associated with the production of lithium batteries may affect air quality and health. Moreover, the lifetime of batteries is still short and induces battery waste containing toxic or corrosive materials such as lithium. This hazardous waste could pose threats to health and the environment if improperly disposed. Nevertheless, there are opportunities for improvements in the batteries' life cycles that will reduce possible impacts to the environment and health, as their use increases. Sustainable alternatives to lithium batteries are also being developed [80].

Electric propulsion may also result in lower aircraft noise levels, since electric engines will not have some of the noise sources associated with jet or piston engines, such as combustor and turbine noise. Depending on the design of the aircraft, jet noise may be also reduced substantially due to the lower jet speeds required for aircraft operation. The lower noise levels associated with electric aircraft may facilitate its use in densely populated areas. For example, the low noise of the Pipistrel Alpha Electro is being used to justify its use by flight schools in urban areas, and the Uber Elevate project is aiming at a 15 dB noise reduction when compared with typical helicopter of similar weight. The Pipistrel Taurus G4 taking off from the Sonoma County Airport in California is illustrated in Figure 4.10.

The ICAO Secretariat is currently following the industry developments in electric and hybrid aircraft designs by means of the Electric and Hybrid Aircraft Platform for Innovation (E-HAPI). This website is being maintained with a non-extensive list of projects that have been identified globally, ranging from general aviation or recreational aircraft; business and regional aircraft; large commercial

FIGURE 4.10 The Pipistrel Taurus G4 taking off from the Sonoma County Airport in California [81].

aircraft; and vertical take-off and landing (VTOL) aircraft (also called electric urban air-taxis). Most of them target an entry-in-service date between 2020 and 2030, and some are already commercially available. Four of the projects had their first flights in 2019 (Lilium, City Airbus, Boeing Aurora eVTOL, and Bye Aerospace Sun Flyer 2).

Currently there are no specific ICAO environmental standards in Annex 16 to cover such aircraft types. ICAO is monitoring the developments around these new entrants, and the need for SARPs (Standard and Recommended Practice) and guidance. The general aviation/recreational aircraft group consists of aircraft with MTOW (Maximum take-off weight) from 300 to 1000 kg. These are mostly electric powered aircraft with a seat capacity of two. This category includes aircraft which are already produced and certified, for example, the Pipistrel Alpha Electro. The Pipistrel Alpha Electro is a 2-seat trainer with an endurance of one hour + 30 minute reserve. It is the first certified all-electric aero plane, with about 60 aircraft currently in operation over the world. Energy-cost associated with its operation is around 1 EUR/h, which makes it suitable for use by flight schools. The aircraft under the business and regional aircraft category claims longer flight range close to 1,000 km with increased seat capacity (around ten). A full-scale prototype of the Eviation Alice was displayed at the Le Bourget Air Show in Paris. The Eviation Alice is designed to take 9 passengers + 2 pilots up to 650 miles at a cruise speed of 240 knots. It is powered by three 260 kW (350 hp) electric motors developed by the Siemens eAircraft business, which was recently acquired by Rolls-Royce. At 3,700 kg, the battery accounts for 60% of the aircraft take-off weight. Eviation announced that U.S. regional airline Cape Air is to buy the Eviation Alice, which has a list price of around $4 million each. Eviation expects to receive certification by late 2021, with deliveries predicted for 2022 [80].

Significant progress has also been made on the VTOL category over recent years, with seat capacities from one to five, MTOWs between 450 and 2,200 kg, and projected flight ranges from 16 to 300 km. These aircraft projects are only electric powered and aim to enter into service in the period of 2020–2025. The large commercial aircraft category includes Airbus and Boeing initiatives focused on hybrid-electric, single-aisle aircraft with seat capacities of 100–135 and targeted entry into service after 2030. The E-Fan X is an Airbus project, in partnership with Siemens and Rolls-Royce, which is developing a flight demonstrator testing a 2 MW hybrid-electric propulsion system. The project's aim is to replace one of four gas turbines on a British Aerospace RJ100 with a 2 MW electric motor. Flight testing is expected to start in 2020. With the E-Fan X, Airbus intends to investigate the thermal effects, electric thrust management, altitude and dynamic effects on electric systems, and electromagnetic compatibility issues, as well as facilitate the establishment of certification requirements for electrically powered aircraft. The Lilium Jet is a tilt-jet aircraft with 36 electric motors mounted on its flaps. It will be capable of traveling up to 300 km in 60 minutes, carrying 4 passengers + 1 pilot. The ducted design of the electric motors is expected to provide noise benefits when compared with traditional helicopter designs. The Lilium Jet completed its maiden flight in May 2019 and is expected to be fully operational in various cities around the world by 2025 [80].

Hydrogen-powered aircraft were successfully flown in the past. The Tupolev 155 (Tu-155) was tested in the late 1980s powered by cryogenic hydrogen and liquefied natural gas. This aircraft had a number of fundamental differences from the original version (Tu-154), such as a cryogenic fuel tank along with the fuel supply system and an experimental turbofan engine which operated together with the kerosene engines. The cryogenic complex on the plane was operated using several innovative systems, such as a helium control system for the power plant and a nitrogen system to replace the air in the compartments with the risk of cryogenic fuel leakage. To allow that, nitrogen and helium tanks were installed in the cargo compartment and cabin area. In the late 1990s, the initiative to create Tu-156 as a serial aircraft was proposed but the project has not been completed. However, the Tu-155 flight tests confirmed the possibility of safe operation of the aircraft powered by cryogenic fuel. To date, several factors still hinder a possible use of hydrogen in commercial flights, such as onboard storage, safety concerns, the high cost of producing the fuel, and the need for dedicated infrastructure at airports. Research projects are ongoing to demonstrate the feasibility of hydrogen propulsion and to overcome these challenges, in support of longer term environmental objectives for civil aviation [80].

One of these projects is the $ENABLEH_2$ (ENABLing CryogEnic Hydrogen-Based CO_2-free Air Transport), a recently launched project funded by the European Union and led by Cranfield University. This project aims to revitalize enthusiasm for liquid hydrogen (LH_2) research for civil aviation, demonstrate its feasibility, and the need for more R&D into advanced airframes, propulsion systems, and air transport operations as part of an LH_2 future. The project will include experimental and numerical work for two key enabling technologies: H_2 micromix combustion (for ultra-low NO_x emissions) and fuel system heat management (to exploit the heat sink potential of

LH_2 to facilitate advanced turboelectric propulsion technologies). These technologies will be evaluated and analyzed for competing aircraft scenarios; for advanced short-to-medium range aircraft and for long-range aircraft, both featuring distributed turbo-electric propulsion systems. The study will include mission energy efficiency and life cycle CO_2 and economic viability studies of the technologies under various fuel price and emissions taxation scenarios. $ENABLEH_2$ will also deliver a comprehensive safety audit characterizing and mitigating hazards in order to support integration and acceptance of LH_2. The project will provide a roadmap to develop the key enabling technologies and the integrated aircraft and propulsion systems to TRL 6 in the 2030–2035 timeframe. Table 4.3 summarizes ICAO reported progress on hybrid electric aircraft [80].

Finally, Kuhn et al. [74] presented a RE perspective for aviation. They pointed out that the transformation of the energy supply from fossil to RE is the single most important challenge for the aviation industry's long-term future.

The substitution of fossil kerosene by suitable, sustainable, and scalable alternative energy carriers is a key requirement for which strategies in three categories, i.e., the drop-in fuel, nondrop-in fuel, and fully electric energy path, are presented. The analysis of the energy options shows that [4] solar fuels are able to overcome known sustainability and/or scalability limitations of bio-to-liquid and other drop-in bio-fuels, that [1] the (nondrop-in) hydrogen fuel perspective has inherent limitations that are either solved by feeding into the solar hydrocarbon fuel process or by serving the electric propulsion paradigm shift via PEM fuel cells, and that [3] the all-electric regional aircraft is well within the physical regime of feasibility and potentially closer to realization than generally assumed, mainly driven by a few key advancements in material science and nanotechnology.

4.3.5 INTEGRATED EMS FOR HYBRID ELECTRIC AIRCRAFT

In the past decades, the trend toward "More Electric Aircraft" has materialized in new airliners such as the Boeing 787. This trend is powered by the high reliability and low maintenance requirements of modern mechatronic systems [82–84] and, so far, has targeted only nonpropulsive systems such as the ice protection system. One possible future evolution for this trend is electric propulsion: the hybrid process that allows the use of electricity available onboard to power the propulsion system. This opens new possibilities for aircraft design, such as distributed electric propulsion or radically new aircraft configurations [85,86]. This trend may be accelerated in the near future by several factors: higher oil prices, carbon taxes, changes in politics, etc.

Several research groups around the world are doing design exercises on hybrid aircraft with different configurations [87–90]. Unfortunately, the results of these exercises indicate only marginal improvements in fuel efficiency when hybrid electric designs are compared against conventional propulsion designs. Besides, these marginal improvements are conditionally dependent on the future evolution of battery energy density. To improve these results, it has been proposed to integrate onboard energy management in the aircraft design methodology [91].

In the context of hybrid vehicles, onboard energy management is defined as the management of the different energy sources available onboard during a mission

TABLE 4.3

ICAO Electric and Hybrid Aircraft Platform for Innovation (E-HAPI) [80]

Project	Type	Category	MTOW (KG)	Pax	Target Entry in Service	Cruise Altitude (FT)	Cruise Speed (kt)	Payload (KG)	Range (KM)	Engine Power (kW)
Airbus/Siemens/Rolls Royce E-Fan X	Hybridelectric	Large commercial aircraft	NA	100	2030	NA	NA	6,650	NA	2,000
Zunum Aero ZA10	Hybridelectric	business aircraft	5216.3	12	2020	max. 25,000	295	1134	1,127	1,000 + 500
Magnus Aircraft/ Siemens eFusion	Hybrid dieselelectric	General aviation/ recreational aircraft	600.1	2	NA	NA	100–130	NA	1,100	60
Boeing Sugar Volt	Hybridelectric	Large commercial aircraft	NA	135	2030–2050	NA	NA	NA	6482	NA
Volta Volare DaVinci	Hybridelectric	General aviation/ recreational aircraft	NA	2+2	2017	24,000	160	NA	NA	NA

[92]. In other words, it is the selection of the rate of energy consumption from each available source at each moment. This management can be done in real-time or pre-defined before a mission, depending on the selected control methodology. When optimization techniques are used to optimize the onboard energy management, it becomes optimal and can be referred to as optimal energy management. When integrated in the aircraft design methodology, the optimal energy management is used in each step of the aircraft sizing loop, minimizing the block fuel for the entire mission. This will lead, owing to synergies, to a fuel efficiency improvement of the final hybrid aircraft design, potentially making hybrid aircraft a more attractive option to decrease fuel consumption and emissions in aviation [91,93–96].

Despite the theoretical advantages, the integration of onboard energy management in the design exercise is not yet in widespread use. Most of the design exercises involving hybrid aircraft published so far fix, *a priori*, the power split between the fuel power and electric power and not changing these values during the optimization run. A notable exception is the work of [94], where a rule-based controller regulating the power split is active during the simulation. The conclusions of *supra* point that "[…], more advanced energy management strategies need to be developed." In a second research study by the same authors [97], dynamic programming (DP) algorithms were used to optimize the energy management of an aircraft in an approach very similar to the one in the study by Pinto Leite and Voskuijl [98], although with relevant differences (e.g., the effect of variable weight in performance was not taken into account).

In the study by Pinto Leite and Voskuijl [98], no attempt is made for integrating optimal energy management in the design exercise. Instead, another step toward this goal is taken: an improved method to determine the optimal energy management for hybrid aircraft is presented and studied. Similar methods were presented in previous works [97,99,100], but the models used in these studies include neither the effect of weight reduction during the mission nor energy recovery. The weight reduction detail in particular becomes very relevant when considering long missions [101]. The study of Pinto Leite and Voskuijl [98] is also an attempt to investigate which are the main factors influencing optimal energy management for a fixed-wing general aviation aircraft.

Besides directly improving the fuel economy of a hybrid aircraft during preliminary design, at least one more reason exists to study optimal energy management in this context: the development of nonoptimal controllers for use in operational environments. For most real-life problems, the deployment of global-optimum real-time controllers is simply not possible. This is an effect of several characteristics of optimal controllers (e.g., the associated computational burden and the lack of robustness to certain types of external disturbances) that limit their use in operational context. To avoid these limitations, nonoptimal controllers are usually developed and tuned by comparing their performance against optimal controllers in high-fidelity simulation environments [102–104]. This application of optimal energy management is briefly addressed in the study by Pinto Leite and Voskuijl [98], as the presented methodology can also be promptly used for this purpose. Some remarks regarding the application of optimal controllers to hybrid aircraft will also be given.

Hybrid aircraft, unlike hybrid cars, are a relatively new topic, and more research needs to be done to make them a viable and efficient mean of transportation. However, some promising steps are being taken: companies specialized in small/general aviation aircraft developed or are developing fully electric aircraft and hybrid electric aircraft [105]. Some of the technologies developed for general aviation will, hopefully, be exported to airliners, the largest segment of the market.

Despite all the hopes surrounding hybrid aircraft, they face some major difficulties in the fight for market share. The most important one is likely to be the enormous cost of new technology development and certification, very difficult to overcome, as the expected gains in fuel efficiency are quite small. Aircraft manufacturers are not able to finance the development cost of a new aircraft unless it delivers substantial reductions in fuel consumption.

Two main hybrid electric architectures used for propulsion purposes: the serial architecture and the parallel architecture [102,106]. The main difference between both is the way the electric and the fuel-powered systems are connected: in the parallel case, this is done using a mechanical coupling mechanism (e.g., a gearbox), while in the serial case, the connection is done through a central electrical bus. In the latter case, the fuel-powered system (e.g., an internal combustion [IC] engine) is connected to a generator to provide energy to the bus. Both architectures have pros and cons: in the parallel architecture, complicated mechanical coupling mechanisms must be designed and maintained, while in the serial architecture, the complexity is in the design of a large and redundant electrical system. The choice between architectures is usually made very early in the design phase, and for small aircraft, the serial architecture is usually favored to take advantage of the high-specific power of electric motors. It is important to note that the aircraft used in the study by Pinto Leite and Voskuijl [98] was designed according to a serial architecture, but the proposed method is applicable to both serial and parallel architectures. The authors would also like to note that these are not the only possible architectures, and several variants can be found in between the two mentioned architectures.

To test the novel proposed methods, the preliminary design data of an aircraft already under construction are used. This aircraft is the hybrid version of the Panthera aircraft, manufactured by Pipistrel, known as Panthera hybrid. The original aircraft is a four-seat general aviation aircraft, and the hybrid version will use the same airframe, but the powertrain will be a hybrid-electric system with serial architecture [107]. The Panthera hybrid is powered by an electric motor rated to 300 kW (150 kW continuous). The motor is powered by two electricity sources: Li-ion batteries and a generator powered by a turbocharged piston engine. For redundancy, both the electric motor and the generator are dual winding machines connected to dual inverters. The IC engine powering the generator provides up to 100 kW in continuous operation [108].

The batteries of the aircraft are still under development thus approximate data were used in this study. The batteries are considered to store up to 20 kWh of energy and weigh about 100 kg. The maximum capacity of the fuel tanks is 210 L. Based on the density of gasoline, one can conclude that the batteries store approximately 1% of the total energy onboard despite having approximately 63% of the total energy storage mass (fuel tank mass not taken into account).

As already described, having a hybrid aircraft creates a whole new range of problems to be investigated, especially in terms of energy control. In the case of conventional aircraft, there is only one source of energy, fuel, and the rate of fuel consumption is fully defined by the desired engine performance at a given instant [109]. That is not the case with hybrid aircraft, where more than one source of energy is available, and the rate of consumption of each source must be selected to yield the desired engine(s) performance. The selection of the energy consumption rates for an hybrid aircraft is thus an underdetermined problem, opening a new degree of freedom that can be used for optimization [89,91,94,99,100].

In most hybrid aircraft designs, it is assumed that there is an automatic controller selecting the amount of energy to be drained from each available source at each moment (i.e., in real time). This controller can be programmed in many ways and with many objectives such as improving take-off performance or decreasing noise during approach. In the specific case under study, the controller has the objective of decreasing total fuel consumption and does this by selecting the throttle position of the IC engine powering the generator ("generator throttle"), while the pilot takes care of the electric motor driving the propeller (i.e., the "normal" throttle).

These automatic controllers envisioned for hybrid aircraft can have different levels of complexity and share several similarities with the ones used in hybrid cars, as they serve the same propose: onboard energy management. But, unlike cars, the trajectory of an aircraft is usually much better known, *a priori*, i.e., before the trip actually begins. This opens the door for new types of high-performance controllers, of types not generally used in the car industry.

The best possible controllers are called optimal controllers [110]. In a comparison with conventional controllers, optimal controllers are computationally heavier and require an *a priori* knowledge of all the external disturbances applied to the system. The flight trajectory is usually defined before the flight, and thanks to the development of low-cost and low-weight powerful computers, these issues became straightforward. Even in the case when optimal controllers are not used in real time during the mission, they are still of utmost importance; they are used to quantify the performance of a real-time controller. This is regularly done when studying the performance of controllers for hybrid cars [102–104].

A major difference between optimizing hybrid cars and hybrid aircraft is the high correlation between weight and performance in the latter case. This effect is usually neglected when designing a controller for cars but becomes important when dealing with aircraft. Addressing this effect in energy management was only done in qualitative ways [91], and one of the objectives of this work is to address it in a quantitative way. To conclude the remarks on the optimal energy management problem, all optimizations require an objective function (a *target* quantity to be minimized or maximized). As already noted, in this work, the objective is to minimize the total fuel consumption. Other metrics may be relevant for this problem, such as the minimization of total energy, minimization of total energy cost, or more advanced metrics [111].

The study by Pinto Leite and Voskuijl presents the detailed methodology of the novel approach [98]. Based on the study, it is concluded that the method can be used during conceptual and preliminary design phases and uses a DP open source solver together with an aircraft performance model. This model includes effects such as weight variations and in-flight energy recovery during the descent. It was shown that optimal controls only yield small improvements when compared with nonoptimal results. A sensitivity study showed that the battery recharge efficiency can have a large impact on the control history despite having a small effect on the final fuel consumption (when optimal solutions are compared). Regarding shortcomings of the presented method, it is important to summarize that it is only applicable "as-is" during the design and development phases of the aircraft. To use such a methodology in operational environments, it should be complemented by alternative methods, such as optimal control solvers robust to external disturbances.

4.4   HYBRID TRAINS AND RAILWAYS

Hybrid trains are used all over the world. In May 2003, *JR East* started test runs using a KiYa E991 "NE Train" ("New Energy Train") railcar, testing the system performance in cold regions (see Figure 4.11). The design had two 65-kW fuel cells and six hydrogen tanks under the floor, with a lithium-ion battery on the roof. The test train was capable of 100 km/h (60 mph) with a range of 50–100 km (30–60 mi) between hydrogen refills. Research was underway into the use of regenerative braking to recharge the test train's batteries, intending to increase the range further. JR had stated that it hoped to introduce the train into scheduled local service during the summer of 2007. Technology tested on this train was incorporated in the KiHa E200 diesel/battery railcars entering service in 2007.

The first JR Freight Class HD300 shunting locomotive was delivered from Toshiba on 30 March 2010 [12]. The new locomotive uses lithium-ion batteries and is designed to reduce exhaust emissions by at least 30%–40% and noise levels by at least 10 dB compared with existing Class DE10 diesel locomotives. Japan's first

FIGURE 4.11   The experimental "NE Train" [112].

hybrid train with significant energy storage is the KiHa E200, with roof-mounted lithium-ion batteries.

The new Autorail à grande capacité (AGC or high-capacity railcar) built by the Canadian company Bombardier for service in France is diesel/electric motors, using 1,500 or 25,000 V on different rail systems. It was tested in Rotterdam, the Netherlands, with Railfeeding, a Genesse and Wyoming company. The First Hybrid Evaluating locomotive was designed by rail research center MATRAI in 1999 and built in 2000. It was a G12 locomotive upgraded with batteries, a 200 kW diesel generator and 4 AC motors. Indian railway launched one of its kind CNG-Diesel hybrid trains in January 2015. The train has a 1,400 hp engine which uses fumigation technology [112–117].

In the US, General Electric made a locomotive with $Na-NiCl_2$ battery storage. They expect ≥10% fuel economy. Variant diesel electric locomotive include the Green Goat and Green Kid switching/yard engines built by Canada's Railpower Technologies, with lead acid (Pba) batteries and 1,000–2,000 hp electric motors, and a new clean burning ~160 hp diesel generator. No fuel is wasted for idling—~60%–85% of the time for these type of locomotives. It is unclear if regenerative braking is used, but in principle, it is easily utilized.

Since these engines typically need extra weight for traction purposes, anyway the battery pack's weight is a negligible penalty. The diesel generator and batteries are normally built on an existing "retired" "yard" locomotive's frame. The existing motors and running gear are all rebuilt and reused. Fuel savings of 40%–60% and up to 80% pollution reductions are claimed over a "typical" older switching/yard engine. The advantages hybrid cars have for frequent starts and stops and idle periods apply to typical switching yard use "Green Goat" locomotives have been purchased by Canadian Pacific Railway, BNSF Railway, Kansas City Southern Railway, and Union Pacific Railroad among others [112–117].

4.4.1 SOLAR-POWERED TRAIN SYSTEM

Railway presents a low rolling resistance option that would be beneficial for planned journeys and stops. PV panels were tested as APUs (Auxiliary Power Unit) on Italian rolling stock under EU project PVTRAIN. Direct feed to DC grids avoids losses through DC to AC conversion. DC grids are only to be found in electric-powered transport: railways, trams, and trolleybuses. Conversion of DC from PV panels to grid AC was estimated to cause around 3% of the electricity being wasted; PVTRAIN concluded that the most interest for PV in rail transport was on freight cars where onboard electrical power would allow new functionality:

1. GPS or other positioning devices, so as to improve its use in fleet management and efficiency.
2. Electric locks, a video monitor, and a remote control system for cars with sliding doors, so as to reduce the risk of robbery for valuable goods.
3. Antilock braking system brakes, which would raise the maximum velocity of freight cars to 160 km/h, improving productivity.

The Kismaros–Királyrét narrow-gauge line near Budapest has built a solar-powered railcar called "Vili." With a maximum speed of 25 km/h, "Vili" is driven by two 7 kW motors capable of regenerative braking and powered by 9.9 m^2 of PV panels. Electricity is stored in onboard batteries [29]. In addition to onboard solar panels, there is the possibility to use stationary (off-board) panels to generate electricity specifically for use in transport [74,75,78,98].

A few pilot projects have also been built in the framework of the "Heliotram" project, such as the tram depots in Hannover Leinhausen and Geneva (Bachet-de-Pesay). The 150 kW$_p$ Geneva site injected 600 V DC directly into the tram/trolleybus electricity network provided about 1% of the electricity used by the Geneva transport network at its opening in 1999. On December 16, 2017, a fully solar-powered train was launched in New South Wales, Australia. The train is powered using onboard solar panels and onboard rechargeable batteries. It holds a capacity for 100 seated passengers for a 3 km journey. Recently, Imperial College London and the environmental charity 10:10 have announced the renewable traction power project to investigate using track-side solar panels to power trains. Meanwhile, Indian railways announced their intention to use onboard PV to run air-conditioning systems in railway coaches. Also, Indian Railways announced that it is to conduct a trial run by the end of May 2016. It hopes that an average of 90,800 L of diesel per train will be saved on an annual basis, which in turn results in reduction of 239 tons of CO_2 [5,113–120].

The invention by Dearborn [119] provides a means by which rail (railroad) transportation operators can generate megawatts of carbon-free electrical power from the space over rail tracks and right-of-way without buying or leasing significant amounts of sun accessible space or power transmission right of way. Public rail transportation is becoming increasingly important in the effort to lower carbon emissions from automobile usage. Public rail transportation is migrating from diesel-electric locomotive to electric module unit vehicles as they operate more efficiently, create less noise, and are less dependent on carbon-based energy. Electrified rail systems depend on large quantities of electric power, much of which is derived from coal, natural gas, or carbon-based energy sources generating tens of thousands of tons of carbon dioxide and other GHGs.

Periods of hot weather or peak electrical demand strain electrical producing capacities or cause reductions in service. Population growth and increasing urban densification will require more power generation plants, typically dependent on carbon-based sources of energy. U.S. Department of Energy (DOE) projections state that carbon-free RE generating plants such as wind, solar, nuclear, or other will not come on line fast enough to meet this growing demand.

The U.S. DOE regulates the generation and distribution of electrical power. The U.S. Department of Transportation (DOT) and the Federal Railroad Administration regulate rail and public transportation. The current body of regulations and assumptions of these and state-level regulations do not address or expressly permit the concept embodied in the preferred invention. Global warming, the rising cost of carbon-based energy, rapid advances in solar PV technology, and large-scale automated manufacturing are rapidly changing the way we view energy, transportation, and the way we live. Research, analysis, and design work embodied herein relate directly to these recent developments.

A system of solar PV canopy modules positioned over rail track scan provide a method by which rail system operators can become carbon neutral or make some portions of those systems carbon neutral and, in the process, offer rail operators an increased level of control over their carbon footprint and cost of traction power. Just as no one or two of the developments noted above (see background) make the case for favoring this invention; no single or pair of benefits make the case favoring modular solar PV canopy over rail power generation systems. The combination of changing environmental conditions and social values together with various benefits of this invention provides the foundation for its considerations and deployment. Example of one benefit: energy production from an urban or regional solar PV canopy over rail right of way during periods of peak demand coincides with the canopy's periods of peak power output. During periods of peak demand, a carbon-neutral modular solar PV canopy system over railroad tracks or right of way can produce 40%–50% more electrical power than required for traction power of said rail vehicles. This surplus energy going into the local grid will reduce the need for carbon-based peak generating plants and the probability of reduced electrical supply service levels.

This invention and the long-term fixed cost of energy (levelized cost of energy) offered by the same would allow an electrified rail system to become carbon neutral faster than the local power grid, assist the local power grid supplier in reducing its carbon footprint, fix its average annual cost of rail traction power for 25–30 years, insulate itself from uncontrolled market-driven pricing variations of carbon-based energy, produce local jobs in design, construction, installation, and maintenance in clean and sustainable technology, and make a greater contribution to a nation's energy independence.

4.4.2 HYBRID ELECTRIC RAILWAY

For future locomotive procurement, both hybrid and gas-powered trains provide the potential for major gains in fuel efficiency and pollution reduction, with developments and trials of these technologies already underway. Electrification of rail lines offers reduction in emissions when the electricity is sourced from cleaner sources than diesel. Due to the current Australian electricity generation mix and higher efficiency of electric locomotives, they already have the ability to reduce emissions than standard diesel-powered locomotives. Electric-powered locomotives have an advantage over diesel-powered locomotives in that they are cheaper to buy and cheaper to run. However, rollout is limited by the high capital costs it takes to electrify rail lines. One issue with diesel-powered locomotives is that kinetic energy created from breaking is wasted and overall efficiency is reduced. Despite being available for the previous few decades, regenerative breaking is a technology that can be utilized to reduce the amount of energy required from the electricity lines to power a train. In electric railways, electricity produced from regenerative breaking is fed back into the supply system. Where high capital costs make the electrification of long-distant freight routes unfeasible, hybrid locomotives look to be the rail industry's solution to reducing emissions. Hybrid locomotives use a combination of diesel and electricity stored in onboard batteries to power their loads. Hybrid technology allows an onboard rechargeable ESS to be placed between the power source and the traction system connected to the wheels, which allows surplus energy to be collected through regenerative braking. The energy storage device uses batteries, super capacitors, and flywheels to increase locomotives efficiency.

General Electric is one company developing hybrid technology for locomotives. GE states that these locomotives will deliver an additional 15% reduction in fuel consumption and a 50% reduction in emissions. Due to the longevity of locomotives, newer more efficient locomotives will be slowly rolled out within the industry. However, higher oil prices as well as government incentives may fasten entire fleet rollout. The procurement of new locomotives brings with it a host of technological developments that deliver fuel consumption savings of 8%–10%. This results in substantial returns for assets where fuel is 60% of the lifecycle cost. Where new locomotives are compliant with U.S. emission standards, dramatic improvements in emissions of other pollutants can also be expected. The very high capital cost of locomotives means procurement is relatively insensitive to policy. Healthy demand for rail services from new and existing routes can underpin these large purchases [112–117].

A hybrid train is a locomotive, railcar, or train that uses an onboard rechargeable energy storage system, placed between the power source (often a diesel engine prime mover) and the traction transmission system connected to the wheels. Since most diesel locomotives are diesel-electric, they have all the components of a series hybrid transmission except the storage battery, making this a relatively simple prospect. Surplus energy from the power source, or energy derived from regenerative braking, charges the storage system. During acceleration, stored energy is directed to the transmission system, boosting that is available from the main power source. In existing designs, the storage system can be electric traction batteries, or a flywheel. The energy source is diesel, liquefied petroleum gas, or hydrogen (for fuel cells), and transmission is direct mechanical, electric, or hydrostatic [112–117].

Diesel electric locomotives may have most of what they need for regenerative braking since they might already use dynamic braking. This uses the traction motors as generators to absorb much of the train's energy, but without a way to store the generated electricity, it is simply dumped into the atmosphere as heat with large rooftop resistor banks and cooling fans. Using a storage system means that a nonfully electric train can use regenerative (as opposed to merely dynamic) braking and even shut down the main power source while idling or stationary. Reducing energy consumption provides environmental benefits and economic savings. A smaller scale version of the concept is found in hybrid automobiles, such as the Chevrolet Volt.

The Patton Motor Car, manufactured by Patton Motor Company, was a gas-electric hybrid system, although the term hybrid was not yet in use. William H. Patton filed for a patent on February 25, 1889; the drawings on his patent application resemble later descriptions of his first prototype [1]. Patton built a tram car that was in experimental service in Pullman, Illinois, in 1891 and a small Patton locomotive was sold to a street railway company in Cedar Falls, Iowa, in 1897. The latter used a

2-cylinder, 25 hp gasoline engine to drive a 220 volt generator that served to charge the 200-Ampere hour 100-cell lead acid battery in parallel with the traction motors. The engine ran at constant speed, with a shunt-wound generator that also served as an electric starter motor. A conventional series-parallel controller was used for the two 35 hp traction motors that drove the wheels of the locomotive.

The term mixed drive train came to be used at the turn of twentieth century. The Pieper system was applied to Belgian (Vicinal tramway) and French (Compagnie des Chemins de Fer de Grande Banlieue) railcars as early as 1911. The Thomas system, manufactured by Thomas Transmission Ltd. of England, which is similar in design to the mechanical part of the Hybrid Synergy Drive, was used in the United Kingdom and tested in New Zealand in a NZR (New Zealand Railways department) RM class railcar [112–117].

In 1986, Czechoslovak locomotive manufacturer ČKD built a prototype hybrid shunting locomotive termed the DA 600. The locomotive was powered a 190-kW diesel engine and four electric motors, with a maximum overall power 360 kW powered from batteries. The batteries were recharged while the diesel engine was running, by regenerative braking or from external electric power [9]. After tests on the Railway test circuit Velim and some minor tweaks, the locomotive was lent to the Olomouc train depot and successfully operated there for ten years. Czechoslovak socialist economics failed to start mass production, mainly because of a lack of proper battery manufacturing capacities. As shown in Figure 4.12, a Swedish Rc locomotive was the first series locomotive that used thyristors with DC motors [116].

The *Hybrid Electric Train (HET)* is a HET built by the Department of Science and Technology's Metals Industry Research and Development Center. It is the first train crafted and designed locally by Filipino engineers with parts imported from abroad. It was officially turned over to the Philippine National Railways on June 20, 2019.

FIGURE 4.12 The Swedish Rc locomotive was the first series locomotive that used thyristors with DC motors [116].

The development of the HET is a project of the Metals Industry Research and Development Center (MIRDC) of the Philippine Department of Science and Technology in partnership with the Philippine National Railways.

The HET project began in 2012, and the designing of the train commenced in the following year. The bidding process took place in 2013, with the manufacturing taking place from 2014 to 2015. The project was introduced to the public in June 2016. The train was developed by ten Filipino engineers and technicians from MIRDC led by head engineer Pablo Acuin. The HET is a hybrid electric vehicle powered by electricity and diesel. It has 260 lead acid batteries which are used to run the train and operate its automatic doors, air-conditioning, and CCTV systems [5]. It can be converted to utilize lithium battery. It also makes use of regenerative braking technology.

The trainset has five nonarticulated cars measuring 12 m (39 ft) long, 2.85 m (9.4 ft) wide, and 4.432 m (14.54 ft) high with one double-sliding doors on each side of each car, which can carry 220 passengers each at 50 kph (31 mph). Four of its cars serve as passenger coaches with one having a driver's cabin, while the other car not mentioned solely serves as a generator car that also includes a driver's cabin. The train also has an automatic stop safety feature that would activate in an event of a strong earthquake The MIRDC contracted local bus and truck manufacturer Fil-Asia Automotive and Industries Corp. to build the train. Fil-Asia in turn outsourced the motor, chassis, engine, motor, axle, and wheels from outside of the Philippines to be able to manufacture it.

BNSF and project partners are developing and will soon begin testing a battery-electric high-horsepower road locomotive (the type that moves freight trains from Point A to Point B). BNSF and other railroads have tested low-horsepower battery-electric locomotives in rail yards for years, but mainly for switching freight cars. In 2018, BNSF and Wabtec (formerly GE Transportation) joined forces to begin developing a 100% battery-electric road locomotive prototype that works with conventional diesel locomotives to make a battery-electric hybrid consist. (Consist refers to when two or more locomotives are coupled together.) Performance testing of the hybrid is expected to begin in the late 2020. The project is being supported by a grant from the California Air Resources Board as part of its Zero- and Near Zero-Emission Freight Facilities program. Once all the equipment and support systems are in place, the plan is to run tests between Stockton and Barstow, California—about 350 miles.

Once fully developed, the battery-electric locomotive will provide environmental benefits and fuel savings for the entire locomotive consist. While in the rail yard, the consist will shut down or idle the other locomotives (when possible) and use the battery-electric locomotive to reduce local emissions and noise. Once on the road, the locomotive consist will work behind the scenes to determine the best way to use the battery power. Thanks to this capability, the consist could also choose to "graze" on battery power when the train is cruising through open landscape, saving hundreds of gallons of diesel. Over the next few years, BNSF and Wabtec expect to learn much about how to build, configure, operate, and maintain a battery-electric locomotive. Like the transition from steam to diesel-electric locomotives, it will take years to support an all battery-electric fleet [112–117].

4.4.3   ENERGY STORAGE TECHNOLOGY FOR HYBRID ELECTRIC RAILWAY

Ogasa [118] examined the application of energy storage technologies for electric railway vehicles, in specific with hybrid electric railway vehicles.

Regenerative braking of railway electric vehicles is effective when other powering vehicles, in other words electrical load, exist near the regenerating train on the same electrified line. So, at early mornings and late nights, or in the low-density district lines, regeneration cancellation phenomenon often occurs and the regenerative brake force cannot be operated in accordance with the commanded value. The new high-performance energy storage devices tackle the issues of energy storage and reuse technologies for on-ground and on-vehicle situations. A hybrid energy source is one effective solution. The study by Ogasa [118] examined the application of energy storage technologies for electric railway vehicles, in specific with hybrid electric railway vehicles. Specifically the study examines the effective use of regenerating braking and the expected electricity storage technology in future for further energy conservation in the electric railway field.

The conclusions of this study were as follows: The study clarifies the purpose of mounting an ESS for electric railway vehicles supplied from contact wire is clarified: i.e., effective utilization of regenerative braking and reuse of absorbed energy and, additionally, partial contact-wire-less operation. Demand power and energy specification to contact-wire/onboard storage hybrid system is described. To absorb the surplus energy alone and not to return to the main feeder, power of several hundred kilowatts and energy of several mega joules are needed. And for partial contact-wire-less operation, power of several hundred kilowatts and energy of more than 10 MJ are needed. This difference is caused by the duration of powering and by overcoming running resistance with the auxiliary energy consumed. A comparison of developments of contact-wire/onboard storage hybrid vehicles together with the features of electric storage device is made. There are many projects in the world, starting from the beginning of the 21st century.

As an example of a power circuit and control strategy, a contact-wire/battery hybrid LRV (Light rail vehicle) "Hi-tram" developed by the RTRI (Railway Technical Research Institute) in Japan was presented. From the on-commercial-line trial operation, the LRV achieved an energy savings of over 30% compared to an inverter-fed regenerative tram, with the distance of contact-wireless operation of 25.8 km and a maximum speed up to 40 km/h; the quick charging of 60-seconds duration by a battery charging current of 1,000 A ensures over 4 km running with continuous maximum auxiliary power.

The above-mentioned type of hybrid technology has proven itself through operation between sections with different types of electricity such as different DC voltages and nonelectrified, or interoperation with different regulated tracks such as railway main lines and on-street lines. Hybrid-controlled electric railway vehicle driven cooperatively using contact wire and energy stored onboard is expected to proliferate as the next-generation standard. This technical field is progressing further and advancing rapidly.

Liu and Li [120] reviewed energy storage devices in electrified railway systems. As mentioned above, as a large energy consumer, the railway systems in many countries have been electrified gradually for the purposes of performance improvement and emission reduction. With the widespread utilization of energy-saving technologies such as regenerative braking techniques, and in support of the full electrification of railway systems in a wide range of application conditions, ESSs have come to play an essential role. Liu and Li [120] reviewed some recent developments in railway ESSs and presented a comprehensive comparison of various ESS technologies. The advantages and disadvantages of each ESS technology have been analyzed. Both flywheels and EDLCs have been commercially utilized as both stationary and onboard ESSs in tram and LRV systems. With the rapid development of Ni-MH and Li-ion battery technologies, the application of BESSs (battery energy storage system) in railway systems has received substantial attention in recent years. In addition to their high energy density, significant improvements in power density and cycle life mean that BESSs have great potential for the future. Although HFCs have been commercially utilized in road vehicles (cars and buses), they still have not been widely used in railway systems due to the limitations of low power density, low cycle efficiency, and high investment cost. SMESes are commercially utilized in the electrical grid for power-quality control and grid stabilization due to their fast response capability, but their low cost efficiency and high safety requirements (very low temperature and high magnetic operational environment) are major obstacles to their application in railway systems. By combining the distinctive advantages of different energy-storage technologies in a single solution, HESSs may have a greater potential for railway applications in the future. This paper has demonstrated that ESSs can not only improve energy efficiency and power quality but also bring considerable economic returns in railway applications. Finally, the foremost functionalities of the railway ESSs are presented together with possible solutions proposed from the academic arena and current practice in the railway industry. In addition, the challenges and future trends of ESSs in the railway industry are briefly discussed.

REFERENCES

1. Badin, F., Scordia, J., Trigui, R. et al., (2006). Hybrid electric vehicles energy consumption decrease according to drive train architecture, energy management and vehicle use. *IET Hybrid and Electric Vehicles Conference*: 213–223, Doi: 10.1049/cp:20060610.
2. Singh, K.V., Bansal, H.O. and Singh, D. (2019). A comprehensive review on hybrid electric vehicles: Architectures and components. *Journal of Modern Transportation* 27: 77–107, Doi: 10.1007/s40534-019-0184-3.
3. Thomas, C.S. (2009). Transportation options in a carbon-constrained world: Hybrids, plug-in hybrids, biofuels, fuel cell electric vehicles, and battery electric vehicles. *International Journal of Hydrogen Energy* 34: 9279–9296, Doi: 10.1016/j.ijhydene.2009.09.058. CrossRefGoogle Scholar.
4. Shah, Y.T. (2021). *Hybrid Power-Generations, Storage and Grids*. CRC Press, New York.
5. Shah, Y.T. (2020). *Modular Systems for Energy Usage Management*. CRC Press, New York.
6. Electric boat. (2020). Wikipedia, The free encyclopedia, last visited 21 August 2020.
7. Solar Boat. (2019). Wikipedia, The free encyclopedia, last visited 5 January 2019.
8. List of Solar Powered Boats. (2019). Wikipedia, The free encyclopedia, last visited 10 July 2019.

9. Eco Marine Power. (2020). Wikipedia, The free encyclopedia, last visited 21 May 2020.
10. Hybrid ferry. (2020). Wikipedia, The free encyclopedia, last visited 9 May 2020.
11. EnergySail-Wind & Solar Power for Low Emission Shipping. (2018). Wind-assisted propulsion device.-pathway to decarbonizing shipping. *ZERO Emissions*. A website report by eco marine power.
12. Hybrid marine power-low emission and cost effective solutions for shipping. (2015). A website report by eco marine power.
13. Mofor, L., Nuttall, P. and Newell, A. (2015). Renewable energy options for shipping technology brief. *Irena report*, Abu Dhabi.
14. Roh, G., Kim, H., Jeon, H. and Yoon, K. (2019). Fuel consumption and CO_2 emission reductions of ships powered by a fuel-cell-based hybrid power source. *Journal of Marine Science and Engineering*, 7: 230, Doi: 10.3390/jmse7070230. www.mdpi.com/journal/jmse, an MDPI publication.
15. MARPOL 73/78. (2020). Wikipedia, The free encyclopedia, last visited 6 August 2020.
16. Michel, M.R., Sorensen, A.J. and Vartdal, B.J. (2016). Hybrid marine power plants model validation with strategic loading. *IFAC-PapersOnLine* 49(23): 400–407.
17. UNCTAD. (2014). *Review of Maritime Transport 2014*. United Nations, New York.
18. Review of maritime transport-2008 a report by UNCTAD. (2008). United Nations, New York.
19. IMO. (2009). MEPC 59—second IMO green house emission study. A report by International Maritime Organization.
20. Corbett, J.J., Wang, H. and Winebrake, J.J. (2009). The effectiveness and costs of speed reductions on emissions from international shipping. *Transportation Research Part D: Transport and Environment* 14 (8): 593–598.
21. Corbett, J.J., Wang, C., Winebrake, J.J. and Green, E. (2007, January 11). Allocation and forecasting of global ship emissions. *A Report Prepared for the Clean Air Task Force*, Boston, MA.
22. Haynes, J. (2015, August). The hour of power, hybrid marine technology and green ports, *Reprinted with Permission from the August 2015 edition of Maritime Reporter* – www.marinelink.com, Meritime reporter and engineering news.
23. Hanley, S. (2018, September 3). Rolls Royce electric debuts SAVe energy battery propulsion system for ships. A website report in Clean Technico.
24. Johnson, E. (2019, July 3). Growing interest in energy storage for maritime applications. A website report www.pv-magazine.com/.../growing-interest-in-energy-storage-for-maritime-ap.
25. Research Partnership Aims to Enable Zero Emission Ships—Rolls-Royce. (2017, November 29). A website report www.rolls-royce.com/.../29-11-2017-research-partnership-aims-to-enable-zero.
26. DNV-GL. (2015). Ship rules for classification. Chapter 2 Propulsion, power generation and auxiliary systems. DNV-GL.
27. Seth, A. (2015, July 8). Rolls-Royce, EVP Electrical, automation and control—commercial marine, said: GE modular naval vessel electrification in Brest.
28. GE powers US Navy's. (2020, June 11). 1st full electric power and propulsion ship. A website report by GE, www.Ge.com.
29. Dedes, E., Hudson, D. and Turnock, S. (2012). Assessing the potential of hybrid energy technology to reduce exhaust emissions from global shipping. *Energy Policy* 40: 204–21840, Doi: 10.1016/j.enpol.2011.09.046.
30. Lloyds Maritime Information Services. (2007). Lloyd's Fairplay Ship Database.
31. Mohamed, M.R., Sharkh, S.M. and Walsh, F.C. (2009). Redox flow batteries for hybrid electric vehicles: Progress and challenges. In: *Vehicle Power and Propulsion Conference (VPPC)*. IEEE, Dearborn, MI, 551–557.

32. Alvarez, R., Schlienger, P. and Weilenmann, M. (2010). Effect of hybrid system battery performance on determining hybrid electric vehicles in real-world conditions. *Energy Policy* 38(11): 6919–6925, Doi: 10.1016/j.enpol.2010.07.008.

33. Fontaras, G., Pistikopoulos, P. and Samaras, Z. (2008). Experimental evaluation of hybrid vehicle fuel economy and pollutant emissions over real-world simulation driving cycles. *Atmospheric Environment* 42: 4023–4035.

34. Kyrtatos, N. (1993). *Marine Diesel Engines Design and Operation Aspects*. Symetria, Athens.

35. Klein Woud, H. and Stapersma, D. (2002). *Design of Propulsion and Electric Power Generation Systems*. IMarEST publications, London.

36. MAN Diesel. (2009). *Emission Control MAN B&W Two-Stroke Diesel Engines*. MAN B&W Deisel A/S, Copenhagen.

37. Baker, J. (2008). New technology and possible advances in energy storage. *Energy Policy* 36: 4368–4373.

38. Chalk, S.G. and Miller, J.F. (2006). Key challenges and recent progress in batteries, fuel cells and hydrogen storage for clean energy systems. *Journal of Power Sources* 159: 73–80.

39. Divya, K.C. and Ostergaard, J. (2009). Battery energy storage technology for power systems—an overview. *Electric Power Systems Research* 79: 511–520.

40. Linden, D. and Reddy, T.B. (2002). *Handbook of Batteries*, 3rd ed. McGraw-Hill Handbooks, New York, NY.

41. Galloway, R.C. and Haslam, S. (1999). ZEBRA electric vehicle battery: Power energy improvements. *Journal of Power Sources* 80: 164–170.

42. Sudworth, J. (2001). The Sodium/nickel chloride (ZEBRA) battery. *Journal of Power Sources* 100: 143–149.

43. Ponce de Leon, C., Frias-Ferrer, A., Gonzalez-Garcia, J., Szanto, D.A. and Walsh, F.C. (2006). Redox flow cells forenergy conversion. *Journal of Power Sources* 160: 716–732.

44. Neglur, S. and Ferdowsi, M. (2009). Effect of battery capacity on the performance of plug-in hybrid electic vehicles. In: *Vehicle Power and Propulsion Conference (VPPC)*. IEEE, Dearborn, MI, 649–654, Doi: 10.1109/VPPC.2009.5289789. Corpus ID: 38100493.

45. Hall, W.J. (2010). Assessment of CO_2 and priority pollutant reduction by installation of shoreside power. *Resources, Conservation and Recycling* 54: 462–467; IMO. (2009). MEPC 59—Second IMO Green House Emission Study. International.

46. Denholm, P. and Kulcinski, G.L. (2004). Life cycle energy requirements and greenhouse gas emissions from large scale energy storage systems. *Energy Conversion and Management* 45: 2153–2172, Doi: 10.1016/j.enconman.2003.10.014.

47. Caravella, T., Austell, C., Brady-Alvarez, C. and Elsaiah, S. (2019). Hybrid Marine microgrids: A quest for future onboard integrated marine power system. An Intech open access paper, Doi: 10.5772/Intechopen.89004.

48. ABB. 2011. The step forward: Onboard DC Grid. Report.

49. Prenc, R., Cuculic, A. and Baumgartner, I. (2016). Advantages of using a DC power system on board ship. *Pomorski zbornik* 52: 83–97.

50. Elsaiah, S. and Mitra J. (2015). A method for minimum loss reconfiguration of radial distribution systems. *Proceedings of the IEEE Power and Energy Society General Meeting and Exposition*, Denver, CO.

51. Elsaiah, S., Benidris, M. and Mitra, J. (2014). Reliability improvement of power distribution system through feeder reconfiguration. *Proceedings of the 13th IEEE International Conference on Probabilistic Methods Applied to Power Systems, PMAPS*, Durham, North Carolina.

52. Elsaiah, S., Benidris, M. and Mitra J. (2016). A method for reliability improvement of microgrids. *Proceedings of the IEEE 19th Power System Computation Conference*, Genoa.

53. Sulligoi, G., Tessarolo, A., Benucci, V., Barret, M., Rebora, A. and Taffone, A. (2010). Modeling, simulation and experimental validation of a generation system for medium-voltage DC integrated power systems. *IEEE Transactions on Industry Applications* 46: 1304–1310.

54. Wei J. and Yu Z. (2011). Load sharing techniques in hybrid power systems for DC microgrids. *Proceedings of the Asia- Pacific Power and Energy Engineering Conference*, IEEE, IEEE computer society, Washington, DC, 1–4.

55. Shariatzadeh, F., Zamora, R. and Srivastava, A. (2011). Real time implementation of microgrid reconfiguration. *Proceedings of the North American Power Symposium*, IEEE, Boston, MA, 1–6, Doi: 10.1109/NAPS.2011.6025181.

56. Ahamad, N.B., Othman, M., Vasquez, J.C., Guerrero, J.M. and Su, C.-L. (2018). Optimal sizing and performance evaluation of renewable energy based microgrid for future seaports. *Proceedings of International Conference on Industrial Technology (ICIT)*, IEEE, Lyon, 1–6, ISBN: 9781509059492 (eBook), 9781509059508 (Print).

57. Hari Kumar, R. and Ushakumari, S. (2015). A two stage algorithm for optimal management of isolated microgrid. *Proceedings of International Conference on Signal Processing, Informatics, Communication and Energy*, IEEE, 1–5.

58. Su, C.-L., Lan, C.-K., Chou, T.-C. and Chen, C.-J. (2014). Design of a multi-agent system for shipboard power systems restoration. *Proceedings of the 50th Industrial and Commercial Power Systems Technical Conference*, IEEE/IAS, Fort Worth, TX.

59. Jiang, Y., Jiang, J. and Zhang Y. (2012). A novel fuzzy multiobjective model using adaptive genetic algorithm based on cloud theory for service restoration of shipboard power systems. *IEEE Transactions on Power Systems* 27(2): 612–620.

60. Butler-Purry, K.L. and Sarma N.D.R. (2004). Self- healing reconfiguration for restoration of naval shipboard power systems. *IEEE Transactions on Power Systems* 19(2): 754–762.

61. Grigoryev, A.V., Malyshev, S.M. and Zaynullin, R.R. (2017). Unified ship power grids with alternators and DC power distribution. *Proceedings of the IEEE International Conference on Industrial Engineering, Applications and Manufacturing (ICIEAM)*, IEEE, Chelyabinsk, 1–3.

62. Guerrero, J.M., Jin, Z., Liu, W. et al., (2016, June 28–30). Shipboard microgrids: Maritime islanded power systems technologies. *PCIM Asia, 2016. PCIM Asia 2016*, Shanghai, 1–8.

63. Jin, Z., Savaghebi, M., Vasquez, J.C., Meng, L. and Guerrero, J.M. (2016). Maritime DC microgrids: A combination of microgrid technologies and maritime onboard power system for future ships. In: *Proceedings of the IEEE 8th International Power Electronics and Motion Control Conference (IPEMC- ECCE Asia)*, IEEE, Piscataway, NJ, 1–6.

64. Shekhar, A., Ramírez-Elizondo, L. and Bauer, P. (2017). DC Microgrid Islands on ships. In: *Proceedings of the IEEE Second International Conference on DC Microgrids (ICDCM)*, Nurenberg.

65. Brady-Alvarez, C. and Elsaiah, S. (2018). A review of renewable energy resources and their applications. In: Scundurra G., (ed). *Focus on Renewable Energy Sources*. Nova Science Publishers Inc., New York, 131–156. ISBN: 978-1-53613-802-3.

66. Jin, Z., Sulligoi, G., Cuzner, R., Meng, L., Vasquez, J.C. and Guerrero, J.M. (2016). Next- generation ship-board DC power system. *IEEE Electrification Magazine*, 45–57.

67. Othman, M., Anvari-Moghaddam, A. and Guerrero, J. (2017). Hybrid shipboard microgrids: System architectures and energy management aspects. *Proceedings of the IEEE 43rd Annual Conference of the IEEE Industrial Electronics Society, (IECON)*, IEEE, Piscataway, NJ, 1–6.

68. Elsaiah, S., Cai, N., Benidris, M. and Mitra, J. (2015). A fast economic power dispatch method for power system planning studies. *IET Generation Transmission and Distribution* 9(5): 417–426. London, UK.

69. Elsaiah, S., Benidris, M. and Mitra, J. (2014). An analytical approach for placement and sizing of distributed generators on power distribution system. *IET Generation Transmission and Distribution* 8(6): 1039–1049.

70. Tang, W., Dickie, R., Roman, D., Robu, V. and Flynn, D. (2019). Optimization of hybrid energy systems for maritime vessels. *The Journal of Engineering*: 4516–4521, Doi: 10.1049/joe.2018.8232.

71. Tang, W., Dickie, R., Roman, D., Robu, V. and Flynn, D. (2019). Optimisation of hybrid energy systems for maritime vessels. *The 9th International Conference on Power Electronics, Machines and Drives (PEMD 2018)*. *The Journal of Engineering* 2019(17): 4516–4521, IET Journals, eISSN 2051-3305, Doi: 10.1049/joe.2018.8232, www.ietdl. org.

72. Shepard, J. (2017, May 30). Hybrid power module is a first for the marine industry. A website report.

73. van Biert, L., Godjevac, M. and Visser K. (2016, September 30). A review of fuel cell systems for maritime applications. *Journal of Power Sources* 327: 345–364, Doi: 10.1016/j.jpowsour.2016.07.007.

74. Kuhn, H., Falter, C. and Sizmann, A. (2011). Renewable energy perspectives for aviation. *Conference: 3. Conference of the Council of European Aerospace Societies (CEAS) and 21. Italian Association of Aeronautics and Astronautics (AIDAA) Congress*, At Venice.

75. Hydrogen Powered Aircraft. (2020). Wikipedia, The free encyclopedia, last visited 7 February 2020.

76. Unmanned Aerial Vehicle. (2020). Wikipedia, The free encyclopedia, last visited 11 September 2020.

77. Solar Impulse. (2019). Wikipedia, The free encyclopedia, last visited 24 June 2019.

78. Electric Aircraft. (2019). Wikipedia, The free encyclopedia, last visited 12 July 2019.

79. Solar-Powered Aircraft Developments Solar One. (2020). Wikipedia, The free encyclopedia, last visited 4 September 2020.

80. Secretariat ICAO. (2019). Electric, hybrid, and hydrogen aircraft – state of play. Chapter Four Climate Change Mitigation: Technology and Operations: 124–130.

81. Pipistrel Taurus. (2020). Wikipedia, The free encyclopedia, last visited 22 June, 2020.

82. Rosero, J.A., Ortega, J.A., Aldabas, E. and Romeral, L. (2007). Moving towards a more electric aircraft. *IEEE Aerospace and Electronic Systems Magazine* 22(3): 3–9.

83. Sarlioglu, B. and Morris, C.T. (2015). More electric aircraft: Review, challenges, and opportunities for commercial transport aircraft. *IEEE Transactions on Transportation Electrification* 1(1): 54–64.

84. Mavris, D., Chakraborty, I., Garcia, E., Perullo, C., and Trawick, D. (2015, December). Onboard Energy Management. *Encyclopedia of Aerospace Engineering*: 1–10. Doi: 10.1002/9780470686652.eae1029.

85. Sehra, A.K. and Whitlow, W. (2004). Propulsion and power for 21st century aviation. *Progress in Aerospace Sciences* 40(4/5): 199–235.

86. Gohardani, A.S., Doulgeris, G. and Singh, R. (2011). Challenges of future aircraft propulsion: A review of distributed propulsion technology and its potential application for the all electric commercial aircraft. *Progress in Aerospace Sciences* 47(5): 369–391.

87. Pornet, C. and Isikveren, A. (2015). Conceptual design of hybrid-electric transport aircraft. *Progress in Aerospace Sciences* 79: 114–135.

88. Pornet, C., Gologan, C., Vratny, P.C., Seitz, A., Schmitz, O., Isikveren, A.T. and Hornung, M. (2013). Methodology for sizing and performance assessment of hybrid energy aircraft. *2013 Aviation Technology, Integration, and Operations Conference*, American Institute of Aeronautics and Astronautics.

89. Harmon, F.G., Frank, A.A. and Chattot, J.-J. (2006). Conceptual design and simulation of a small hybrid-electric unmanned aerial vehicle. *Journal of Aircraft* 43(5): 1490–1498.
90. Voskuijl, M., van Bogaert, J. and Rao, A.G. (2017). Analysis and design of hybrid electric regional turboprop aircraft. *CEAS Aeronautical Journal* 9(1): 15–25.
91. Perullo, C. and Mavris, D. (2014). A review of hybrid-electric energy management and its inclusion in vehicle sizing. *Aircraft Engineering and Aerospace Technology* 86(6): 550–557.
92. Serrao, L., Onori, S. and Rizzoni, G. (2011). A comparative analysis of energy management strategies for hybrid electric vehicles. *Journal of Dynamic Systems, Measurement, and Control* 133(3).
93. Perullo, C.A., Trawick, D., Clifton, W., Tai, J.C.M., and Mavris, D.N. (June 16–20, 2014). "Development of a Suite of Hybrid Electric Propulsion Modeling Elements Using NPSS." *Proceedings of the ASME Turbo Expo 2014: Turbine Technical Conference and Exposition. Volume 1A: Aircraft Engine; Fans and Blowers.* Düsseldorf, Germany. V01AT01A042. ASME. Doi: 10.1115/GT2014-27047.
94. Donateo, T., Ficarella, A. and Spedicato, L. (June, 2018). A method to analyze and optimize hybrid electric architectures applied to unmanned aerial vehicles. Aircraft Engineering and Aerospace Technology. Doi: 10.1108/AEAT-11-2016-0202.R1.
95. Donateo, T., Pascalis, C.L.D. and Ficarella, A. (2019). Synergy effects in electric and hybrid electric aircraft. *Aerospace* 6(3): 32.
96. Ficarella, A., Pascalis, C.L.D. and Donateo, T. (2018). Exploiting the synergy between aircraft architecture and electric power system in unmanned aerial vehicle through many-objective optimization. *International Journal of Sustainable Aviation* 4(3/4): 247.
97. Donateo, T., Ficarella, A. and Spedicato, L. "Applying Dynamic Programming Algorithms to the Energy Management of Hybrid Electric Aircraft." *Proceedings of the ASME Turbo Expo 2018: Turbomachinery Technical Conference and Exposition. Volume 3: Coal, Biomass, and Alternative Fuels; Cycle Innovations; Electric Power; Industrial and Cogeneration; Organic Rankine Cycle Power Systems.* Oslo, Norway. June 11–15, 2018. V003T06A015. ASME. https://doi.org/10.1115/GT2018-76500.
98. Pinto Leite, J.P.S. and Voskuijl, M. (2020). Optimal energy management for hybrid-electric aircraft. *Aircraft Engineering and Aerospace Technology* 92(6): 851–861, Doi: 10.1108/AEAT-03-2019-0046.
99. Bongermino, E., Mastrorocco, F., Tomaselli, M., Monopoli, V.G. and Naso, D. (2017). Model and energy management system for a parallel hybrid electric unmanned aerial vehicle. *2017 IEEE 26th International Symposium on Industrial Electronics (ISIE),* IEEE. Edinburgh.
100. Bradley, T., Moffitt, B., Parekh, D., Fuller, T. and Mavris, D. (2009). Energy management for fuel cell powered hybrid-electric aircraft. *7th International Energy Conversion Engineering Conference,* American Institute of Aeronautics and Astronautics, Denver, CO.
101. Hepperle, M. (2012). *Electric Flight-Potential and Limitations.* German Aerospace Centre, Institute of Aerodynamics and Flow Technology. NATO/OTAN report by R&T organization.
102. Guzzella, L. and Sciarretta, A. (2013). *Vehicle Propulsion Systems, SpringerLink: Bücher.* Springer, Heidelberg.
103. Sinoquet, D., Rousseau, G. and Milhau, Y. (2009). Design optimization and optimal control for hybrid vehicles. *Optimization and Engineering* 12(1/2): 199–213.
104. Liu, J. and Peng, H. (2008). Modeling and control of a power-split hybrid vehicle. *IEEE Transactions on Control Systems Technology* 16(6): 1242–1251.

105. Brelje, B. and Martins, J. (2018). Electric, hybrid, and turboelectric fixed-wing aircraft: A review of concepts, models, and design approaches. *Progress in Aerospace Sciences* 104.

106. Wall, T.J. and Meyer, R. (2017). A survey of hybrid electric propulsion for aircraft. *53rd AIAA/SAE/ASEE Joint Propulsion Conference*, American Institute of Aeronautics and Astronautics, Atlanta, Georgia.

107. MAHEPA consortium. (2017). D1.1: Concept of modular architecture for hybrid electric propulsion of aircraft, Technical report, MAHEPA project.

108. MAHEPA consortium. (2017). D2.1 Performance and energy efficiency trade-off study, Internal Document – Private Communication, MAHEPA project.

109. Roskam, J. and Lan, C.-T.E. (1997). *Airplane Aerodynamics and Performance*. DAR Corporation. Published by Springer Nature, New York.

110. Betts, J.T. (1998). Survey of numerical methods for trajectory optimization. *Journal of Guidance, Control, and Dynamics* 21(2): 193–207.

111. Pornet, C., Kaiser, S. and Gologan, C. (2014). Cost-based flight technique optimization for hybrid energy aircraft. *Aircraft Engineering and Aerospace Technology* 86(6): 591–598.

112. Hybrid Train. (2020). Wikipedia, The free encyclopedia, last visited 1 July 2020.

113. Modular and Mobile Photovoltaics on Wheels | SEMI.ORG. (2016). A website report prod7. semi.org/en/modular-and-mobile-photovoltaics-wheels.

114. Hybrid Electric Vehicles. (2019). Wikipedia, The free encyclopedia, last visited 8 July 2019.

115. Railway Electrification System. (2020). Wikipedia, The free encyclopedia, last visited 31 August 2020.

116. Electric Locomotive. (2020). Wikipedia, The free encyclopedia, last visited 25 August 2020.

117. Electric-Diesel Locomotive. (2020). Wikipedia, The free encyclopedia, last visited 3 September 2020.

118. Ogasa, M. (2010, April 20). Application of energy storage technologies for electric railway vehicles examples with hybrid electric railway vehicles special issue review paper / special issue on energy saving technologies on electric railways in Japan. *Transactions on Electrical and Electronic Engineering*, Doi: 10.1002/tee.20534.

119. Dearborn, D.D. (2010, August 12). Modular solar photovoltaic canopy system for development of rail vehicle traction power. US20100200041A1 (6 February 2019).

120. Liu, X. and Li, K. Energy storage devices in electrified railway systems: A review. *Transportation Safety and Environment* 2(3): 183–201, Doi: 10.1093/tse/tdaa016.

5 Hybrid Energy Systems for Coal Industry

5.1 INTRODUCTION

While coal industry has come under severe pressures in recent years, coal is still one of the major sources for energy and chemicals. Major criticism against the use of coal is the amount of CO_2 emission and other pollutants that are created by its use. There are several strategies adopted to combat this issue. All of these involve less and more efficient use of coal. The strategy also involves, as pointed out in my previous book, the conversion of CO_2 to power and fuel. The efficiency on the use of coal can also be increased by hybrid processes of converting waste heat into power. While the use of coal for power is on the decline, the use of coal for chemicals continues to increase. The use of renewable energy in coal industry has many positives, particularly when it is used in the hybrid form. In this chapter, we describe the use of hybrid energy and processes in the coal industry for decarbonization to the extent possible.

Hybrid energy is sometimes used in the mining of coal. The major efforts in coal industry are, however, in the use of coal to produce, power, generate synthetic gaseous and liquid fuels, and generate chemicals. In each application, hybrid energy in the form of coal-biomass, coal-solar, coal-nuclear can be used to reduce the generation of carbon dioxide during the conversion processes. As pointed out in my previous books [1–5], coal-biomass hybrid is the most widely examined subject for the production of power, fuels, and chemicals. Other hybrid processes such coal-solar and coal-natural gas for power production are also described in my previous books [1–5]. In this chapter, we examine several aspects of coal-biomass hybrid system.

Mining operations are often done in remote areas. A growing number of mining companies are incorporating on-site or locally produced renewable energy into their operations—these include alternate power solutions such as solar PV, wind, geothermal, and hydropower. According to Black & Veatch estimates, if an average remote project with a power demand of 5 MW generated 15% of its energy supply through wind, it could save 10% in energy costs and significant reduction in CO_2 emission. In some cases, a hybrid solution that pairs solar with energy storage (batteries) and diesel generators can be cost-effective for mining companies. This is particularly true for mines that have one or more peaks in demand throughout a day, as opposed to a fairly flat load profile. Often, microgrids—stand-alone, small-scale electrical grids with their own power system—are a viable solution for remote areas, with the option to be connected to larger grids for increased resilience. This adds an additional level of flexibility and sustainability, which can improve financial performance and environmental compliance [6]. The most compelling reason for use of hybrid energy systems in mining operation is the drop in price of renewable energy. Hybrid renewable energy solutions can help mining operations reduce costs, improve reliability, and become more sustainable, and are worth investigating [6]. One barrier is the length of power purchase agreement (PPA) mining company can make with the renewable energy provider. PPA of 10 years is desirable.

5.2 COAL-BASED HYBRID POWER PLANTS

The combustion of coal to generate power results in an excessive amount of carbon dioxide emission. There are seven hybrid energy or hybrid processes methods to reduce carbon emission: (a) cocombustion of coal and biomass, (b) cofiring of coal and natural gas, (c) coal-solar hybrid power generation, (d) hybrid process of coal combustion with carbon capture and storage (CCS) technology, (e) hybrid process for coal combustion with subsequent conversion of carbon dioxide to power by fuel cell technology or conversion of carbon dioxide to diesel fuel, (f) combined cycle of power generation to improve conversion efficiency, and (g) hybrid processes which convert waste heat to power or use waste heaty for other useful purposes like industrial heat need. Here we briefly examine these processes.

5.2.1 COCOMBUSTION OF COAL AND BIOMASS

Multifuel combustion using coal and biomass or waste is the easiest way to reduce carbon dioxide emission. The cocombustion of coal and biomass can be done in a variety of different ways such as (a) separate combustion with mixing of heat, (b) separate feed preparation with combined combustion, (c) separate feed preparation with multistage combustion, etc. The postcombustion residual products for cocombustion will also be different from the ones used in separate combustion. All of these issues are discussed in great details in my previous books [1–5]. There are numerous commercial processes already in existence for coal-biomass (waste) cocombustion and they are also described in my previous book. The literature has indicated that a mixture of 70% coal, 30% biomass makes the process carbon neutral in its whole cycle. Thus, this is a very effective way to reduce carbon emission during coal combustion. Since there are different types of coal and biomass, choice of suitable combination is important for cocombustion [1–5].

5.2.2 COFIRING COAL-NATURAL GAS

Both coal- and natural gas-based power generation technologies are vital in powering many of the world's developed and emerging economies. Both are used widely to provide secure uninterrupted electricity, needed to ensure that economies and societies develop and prosper. In some countries, coal provides much of the power, in others, natural gas dominates. However, there are many instances where the national energy mix comprises combinations of the two. Each brings its own well-documented advantages and disadvantages. Recent years have seen a growing interest in ways that these two fuel sources might be combined in an environmentally acceptable and cost-effective manner.

Changing market conditions are forcing many power plant operators to evaluate and implement alternate modes of operation such that their plants remain capable of dispatching electricity in an efficient and cost-effective manner. Important tools in meeting these criteria are fuel and operational flexibility. There are various ways for existing coal-fired power plants to achieve this. For example, as noted in the first part of this report, there is the potential for some to incorporate solar energy. However, another possibility for existing coal-fired may be cofiring, replacing a percentage of the coal feed with natural gas and burning them together. Within this context, cofiring is the combustion of two different fuels simultaneously to produce heat in a steam generator—it is often implemented in coal-fired power plants using natural gas, or sometimes fuel oil [4,7,8].

In some economies, coal-fired plants are facing increasing competition from natural gas and renewables. Furthermore, environmental legislation is being tightened, aimed at reducing permissible levels of SO_2, NO_x, particulates, and CO_2. Thus, many plant operators face the dilemma of whether to invest in emission control equipment or withdraw their plant from service. Many coal-fired power plants in, for example, the USA, currently face this situation. There are several possible options open to plant operators: (a) decommissioning of the coal-fired facility; (b) complete conversion from coal firing to natural gas; (c) using gas to reduce plant emissions (for example, reburning for NOx control); or (d) a switch to coal-natural gas cofiring.

A number of utilities are considering converting some of their coal plants to gas cofiring, such that they can operate on a mix of the two. Importantly, some are evaluating systems whereby the ratio of coal:gas can be changed, providing a degree of flexibility in terms of fuel supply. In the USA, a number of plant operators are investigating the test firing of natural gas to determine the long-term feasibility of either full conversion or dual-fuel firing. Others are undertaking feasibility studies to evaluate the possible ramifications of cofiring. The cost of generating electricity from coal or gas can be similar. Even slight changes in fuel price can result in significant swings in production costs, and this can create market opportunities for utilities that have both gas- and coal-fired assets. Cofiring can be a possible option, allowing pricing and market conditions to drive the fuel choice and mix (89). Substituting some coal input with gas is considered to be a low-risk option, allowing utilities to better meet changing market requirements.

In the coming years, natural gas is forecast to continue partially replacing coal for power generation in some major economies. Thus, the operating advantage will go to utilities with diversified fleets capable of switching between coal and gas as the market price of each fluctuates. This will be of particular advantage during periods of flat electricity growth, such as that experienced in the USA in recent years. Adding gas to coal-fired plants offers utilities the possibility of rapid response to changes in load demand and deep cycling capability, but retains the ability to fire low cost coal. In economies where electricity demand fluctuates, a power plant that can cycle quickly to meet peaks and troughs in demand, and also ramp down during periods of low demand, is more likely to be profitable. However, most coal-fired units can only operate as low as 30%–35% load and still sustain good combustion, restricting the plant's ability to cycle. Furthermore, coal plants can be slow to cycle up to full load—this can take 12 hours or more to ramp up to load from a cold start [9]. A plant capable of switching to gas at low loads and take load down even further, then switch back to coal at higher loads, could be at an advantage over the competition.

As noted, in the US coal-fired generating sector, cofiring natural gas is one option being considered. Cofiring offers increased fuel flexibility, and potentially, this can provide significant fuel and operational savings. Cofiring introduces a second main fuel and can also offer the possibility of using lower cost/grade types of coal, generating even greater savings. According to Mills [7,8], there are good environmental and economic reasons for cofiring. It mitigates CO_2 emissions from coal fired boilers; it reduces emissions of SO_2, NO_x, and particulates; it provides a mechanism of generation and sale of cost-effective dispatchable electricity and it has public acceptance. Cofiring also save money in competitive markets through fuel diversity; it maintains coal supply and employment in coal industry and it allows companies to get credits for early voluntary greenhouse gas abatement measures. Cofiring in existing boilers may be the least-cost way for coal-burning utilities to improve fuel flexibility.

5.2.2.1   Options for Natural Gas Addition

Potentially, there are a number of ways in which natural gas can be combined with coal. Some replace a portion of the main coal feed whereas others are more focused on minimizing plant emissions. Mills [7,8] has suggested three options: (a) preheating coal prior to combustion, (b) cofiring of coal and natural gas, and (c) addition of natural gas for reburning.

Preheating coal using gas-fired burners prior to combustion has been shown to help release fuel nitrogen and other volatiles, destroying NO_x precursors, and thus reducing NO_x formation. In operation, a concentrated pulverized coal stream enters a preheating chamber where flue gas from natural gas combustion is used to rapidly heat the coal up to about 820°C prior to complete combustion in the PC burner. This converts coal-derived nitrogen compounds to molecular nitrogen and avoids the need for posttreatment of NO_x after combustion [10].

Cofiring of coal and natural gas can allow much higher amounts of gas to be fed into an existing boiler and burned simultaneously with the coal feed. The process entails modifying the boiler to make it capable of using combinations of coal and gas. Depending on the individual plant and operational requirements, there are a number of possible ways to reconfigure the existing unit such that cofiring becomes possible. The simplest option is often to retrofit existing oil-fired ignitors with natural gas equivalents which typically allows for a maximum gas-firing capability of 10%–20%. If the unit is equipped with oil-fired warm-up guns, these can be replaced with natural gas-fired units—this increases potential natural gas capability to ~30%–50%. Should a greater level be required, gas firing needs to be incorporated into the main burner system [11]. Oil and gas igniters are deployed widely in coal-fired power plants, used to generate the heat needed to safely ignite coal, the main boiler fuel. The igniter fulfils three basic functions: (a) furnace warm-up, (b) ignition (light-off) of the main fuel, and (c) load stabilization. Of the different igniter types, Class 1 units provide the maximum flexibility in pulverized coal applications [12]. For continuous service applications, the igniter fuel can account for a significant portion of total energy [12].

Various types of burner assemblies are offered commercially, some designed specifically for cofiring. For example, Breen Energy Solutions of the USA has developed a proprietary system known as dual orifice cofiring [13]. This was developed as an effective means for handling variable gas input to a coal-fired boiler. Breen Energy Solutions has also developed a system known as enhanced gas cofiring based on its dual orifice technology that aims to replace up to 35% of coal energy input with natural gas. This is achieved using a Class 1 dual-fuel outlet igniter coupled with a high volume, annular, secondary gas supply. Other US manufacturers also produce dual-fuel burners. For example, Texas-based Forney and Storm Technologies have developed a proprietary system known as the Eagle Air burner. This is a wall-fired boiler dual-fuel burner that operates on coal or natural gas. Available in a range of capacities, it can operate on 100% coal or natural gas and incorporates multiple zones of secondary air that allow for combustion and NO_x tuning [7,8]. Replacing a percentage of coal feed with gas will clearly help reduce overall plant emissions of SO_2, NO_x, particulates, mercury, and CO2. Breen Energy Solutions [13] claims that replacement of 35% of coal feed using their cofiring system can reduce SO2/SO3 emissions by up to 35%, NO_x emissions by 45%, particulates by 35%, mercury by 35%, and CO_2 by 20% [7,8].

Finally, reburning technology was originally developed for NO_x combustion control, primarily on coal-fired furnaces. It is a staged fuel approach that uses the entire volume of a furnace, rather than the control of NO_x production/destruction within the flame envelope. Reburn is a three-stage combustion process that takes place in primary, reburn, and burn-out zones. In the primary zone, pulverized coal is fired through conventional or low-NO_x burners operating at low excess air. A second fuel injection is made in a region of the boiler after the coal combustion, creating a fuel-rich reaction zone (the reburn zone). Here reactive radical species are produced from the natural gas that reacts chemically with the NO_x produced in the primary zone, reducing it to molecular nitrogen. The partial combustion of the natural gas in this reburn zone results in high levels of CO. A final addition of overfire air, creating the burn-out zone, completes the overall combustion process. Natural gas reburn can reduce NO_x emissions by up to 70%. Some technological advances have increased the effectiveness of reburning, such as dual-fuel orifice cofiring technology. Alongside this, Breen Energy Solutions has also developed a system known as fuel lean gas reburning which can replace conventional selective noncatalytic burning technique [14]. More details on this system are given by Mills [7,8]. Reburning systems can provide a means for incorporating a sizeable amount of natural gas into an existing coal-fired power plant, thus providing a number of environmental and operational benefits [7,8].

5.2.3 COAL-SOLAR HYBRID FOR POWER AND FUELS

While in previous book and here we have defined the term "hybrid energy" in very broad terms, often hybrid energy or hybrid power are often colocated generation facilities. For example, photovoltaic (PV) solar cells might be added to a combined-cycle gas turbine (CCGT) plant. Clearly, these solar assets generate electricity, but this is fed into the grid independently of the gas-fired plant. Under this type of arrangement, the solar facility may serve to diversify the economic interests of the plant's owner or reduce the overall environmental footprint of the site, but the PV and CCGT are not as tightly integrated as they might first appear. Similarly, cogeneration or addition of thermoelectric generator to the waste heat from CCGT plant described in my earlier book is not always closely integrated hybrid energy system. India plans to install a significant amount of solar PV generating capacity, with some new facilities being located at existing coal-fired power plants. Both will generate electricity that will be fed to the grid independently of the other. Although these two technologies will share a site and some assets such as grid connection, they will operate largely as independent units and not as integrated hybrids [7,8].

A limited number of coal-solar projects are true hybrids. These operate under an entirely cooperative arrangement where the two sources of energy are harnessed to create separate but parallel steam paths. These paths later converge to feed a shared steam-driven turbine and generate electricity as a combined force. This form of hybrid technology integrates these two disparate forms of power so that they combine the individual benefits of each. This approach can replace a portion of coal demand by substituting its energy contribution via input from a solar field. During daylight operation, solar energy can be used to reduce coal consumption (coal-reducing mode). As solar radiation decreases during the latter part of the day, the coal contribution can be increased, allowing the plant's boiler to always operate at full load. When solar radiation increases again, the process is reversed, with solar input gradually reducing that of coal. Alternatively, input from the solar field can be used to produce additional steam that can be fed through the turbine, increasing electricity output (solar boost). Whichever mode is adopted, the design and integration of the solar field into the conventional system are critical for the proper functioning of a hybrid plant. In principle, this form of hybrid technology can be applied to any form of conventional thermal (coal-, gas-, oil-, or biomass-fired) power plant, either existing or new build [7,8].

Solar energy is usually harvested in one of two ways. The first is via conventional PV cells that convert solar radiation directly into electricity. The second is solar thermal, usually in the form of concentrated solar power (CSP), where radiation is used to produce heat. These systems generally rely on a series of lenses or mirrors that automatically track the movement of the sun. They focus a large area of sunlight into a small concentrated beam that can be used as a heat source for a conventional thermal power plant. In all types of systems, a working fluid (such as high-temperature oil or, increasingly, molten salts) is heated by the concentrated sunlight, then used to raise steam that is fed into a conventional steam turbine/generator. In addition to the solar collection system, a stand-alone CSP plant (*not* hybridized) will also require many of the systems and components, such as steam turbine/generator, found in a conventional power plant. For a coal-solar hybrid, most of these will already be in place and available as part of the coal-fired plant, greatly reducing the cost of each unit of electricity generated [7,8].

5.2.3.1   Advantages of Coal-Solar Hybridization

The concept of combining coal and solar-based systems may appear counterintuitive. However, the idea has gained traction as hybridization is often considered more effective, reliable, and less expensive than relying solely on PV-based generation [15].

Although renewable energy systems such as solar are attractive on environmental grounds, they face two principal challenges, namely, their inherent variability, and that they usually require some form of support or subsidy to be considered economically viable. Potentially, a coal-solar hybrid is an elegant solution. Such a partnership can offer the environmental benefits of solar power, but with the advantage of shared costs of major plant equipment. This significantly lowers the lifetime cost of energy of the solar component. Combining these two types of generation within a single power cycle capitalizes on the strengths of both. Importantly, developers consider that, from a grid operator perspective, a coal-solar (and gas-solar) hybrid can be considered fully dispatchable [16].

There are a number of reasons why power utilities might find coal-solar hybrids more attractive than PV. There is no requirement for new turbines, grid connection, and other major systems as most of these are already in place. Furthermore, particularly when running in boost mode, the overall system operates in a manner familiar to utilities. The cost of electricity from a hybrid could be between a third and a half that from a stand-alone CSP plant—this is mainly because there is little extra cost apart from the addition of the solar field [17,18]. A further advantage of CSP over PV systems is that the land area required per unit of electricity generated is normally much less. For example, solar supplier AREVA's steam generators are used in various CSP applications around the world—they are claimed to generate between 1.5 and 2.6 times more peak energy per hectare of land than PV.

As solar projects tend to generate most electricity on hot days when there is likely to be high demand from air conditioners, adding solar to conventional power plants could reduce the requirement for peaking power plants. This could cut peak electricity costs, given that such plants are only brought on line at times of peak demand and deliver the most expensive electricity. As there is more inertia in the system, output from a solar-thermal-based plant will fluctuate less than an equivalent PV facility. This minimizes troughs and spikes that result from changing weather conditions, reducing the extent of intermittency. Thus, under appropriate conditions, unlike PV, a coal-solar hybrid is considered to be dispatchable.

Many thermal power plants often have a degree of flexibility in the amount and pressure of steam that can be integrated, so it should be possible for such variables to be optimized for the utilities' specific needs [15]. Where legislation imposes limits on greenhouse gas emissions, hybridization may offer a cheaper option to CCS. This could also apply with other types of emissions.

Thus, depending on the particular circumstances, according to Mills [7,8] the main advantages cited for coal-solar hybridization [16,19–22] are as follows:

1. the higher initial investment is balanced by reduced fuel consumption or increased power output;
2. combining the two technologies allows "greening" of existing coal-fired power assets;
3. hybridization can provide both dispatchable peaking and base load power to the grid at all times. CSP coupled with conventional thermal capacity (with or without thermal storage) can offer that capability;
4. hybrid technologies could help meet renewable portfolio standards and CO_2 emissions reduction goals at a lower capital cost than deployment of stand-alone solar plants. Capex is less for the same capacity;

5. siting solar technology at an existing fossil fuel plant site can shorten project development timelines and reduce transmission and interconnection costs;

6. solar thermal augmentation can lower coal demand, reducing plant emissions and fuel costs per MWh generated;

7. solar augmentation can boost plant output during times of peak demand. According to US studies carried out by EPRI, potentially, a solar trough system could provide 20% of the energy required for a steam cycle;

8. hybridization will reduce the level of coal and ash handling, reducing load on components such as fabric filters, pulverizing mills, and ash crushers. It could also avoid the requirement to upgrade fabric filters or electrostatic precipitators;

9. solar input could provide some level of mitigation against difficult coal contracts, such as wet coal, fines, and variable coal quality; the majority of solar plant components could be sourced locally, helping boost local economies;

10. rapid deployment—depending on size and configuration, hybrid plants could be completed in less than two years from notice to proceed;

11. hybridization could be used to extend the lifespan of existing thermal facilities—for example, where regulatory changes require a coal-fired plant to reduce emissions or face closure;

12. hybridization could avoid certain limitations and restrictions applied to new greenfield site projects; and

13. hybrid plants will benefit from the general cost reductions that CSP technology is achieving. Many of these will also be directly applicable to hybrid plants.

5.2.3.2 Disadvantages of Coal-Solar Hybridization

Although coal-solar hybridization can provide some benefits, there are obvious criteria that must be met [7,8,21,23]:

1. the location must receive good solar intensity for extended periods, both on a daily and yearly basis. This is not always the case (Figure 5.5);

2. a suitable area of land close to the existing thermal power plant is required. It must meet certain criteria in terms of hectares available, topography, and issues such as shading;

3. the land will no longer be available for other purposes such as agriculture;

4. a solar add-on will require capital investment;

5. there will be additional costs for operation and maintenance of the solar component, such as mirror washing; and

6. the scale of most coal-solar hybrid projects has so far been low, mainly because these have been retrofits at existing coal-fired plants. Practical issues have tended to limit the solar contribution to ~5%. A new power plant, designed and built based on the hybrid concept from the outset, could possibly accommodate up to 30%–40% solar share.

Each potential project, however, brings its own combination of advantages and disadvantages.

A number of solar collection systems are available commercially, some more effective than others. Systems that use two-axis tracking to concentrate sunlight onto a single point receiver (tower and dish systems) are usually more efficient than linear focus systems. Where they form part of a CSP plant, they can operate at higher temperatures and hence generate power more efficiently. However, they are also more complex to construct. The four main types of solar collection systems are (a) parabolic trough systems, (b) linear Fresnel systems, (c) power tower systems, and (d) parabolic dishes. Coal-solar technology has been under consideration and development for some years. The world's first true coal-solar hybrid power project was located at the Cameo Generating Station in Colorado, USA—the Colorado Integrated Solar Project. It was undertaken as part of the state's Innovative Clean Technology Program, an initiative designed to test promising new technologies that had the potential to reduce greenhouse gas emissions and produce other environmental improvements. The system at the Cameo plant was put into operation in 2010 [7,8].

Xcel Energy and Abengoa Solar partnered on this US$4.5 million demonstration that used sun-tracking parabolic trough technology to supplement the use of coal. The main aim was to demonstrate the potential for integrating solar power into large-scale coal-fired power plants to increase plant efficiency, reduce the amount of coal required, and hence reduce conventional plant emissions and CO_2. It was also to test the commercial viability of combining the two technologies. A 2.6-ha solar field housed eight rows of 150-m-long parabolic solar troughs. In operation, this arrangement concentrated solar radiation onto a line of receiver heat-collecting elements filled with a mineral oil-based heat transfer fluid. This was heated to about 300°C before feeding to a heat exchanger where it preheated feedwater (to about 200°C) supplied to one of the Cameo plant's two 49-MW coal-fired units. Coal-solar hybrid technology has also been investigated in a number of other countries that maintain major coal-fired power sectors, and significant work has been undertaken in, for example, South Africa, China, and several European countries. Parts of Chile have significant potential for the application of solar power. Yearly total direct normal irradiation levels are >3,000 kWh/m² in most of the country, with >3,500 kWh/m² to the north in the Atacama Desert, close to much of the mining sector. As part of increasing electricity supply in the region, a 5-MW coal-solar hybrid project is being developed by Engie and Solar Power at the existing 320-MW Mejillones coal-fired power plant. A new coal-solar project is also underway in India. In a recent development, NTPC (formerly the National Thermal Power Corporation Limited) announced the start of a coal-solar hybrid project (the Integrated Solar Thermal Hybrid Plant) to be developed at its Dadri power plant. This will be the first Indian project to use solar energy to heat boiler feedwater with the aim of increasing plant efficiency and reducing coal demand. To save costs and manpower, the project will feature the robotic dry cleaning of the solar panels [7,8].

Another hybrid method to use solar power is to incorporate electricity produced by PV systems with coal and/or biomass-based systems. There are also projects under way that propose to combine solar power (in the form of CSP) with existing thermal (gas- and coal-based) power stations. Efforts to develop such coal-solar hybrids are being pursued in several parts of the world that have the appropriate weather conditions. An advantage of adding a solar thermal module to an existing coal-fired power plant is that much of the necessary infrastructure (steam cycle, etc.) and plant requirements already exist. This can make the economics more attractive than those of a stand-alone solar thermal generating unit. Solar energy input can be harnessed by parabolic troughs, compact linear Fresnel reflectors (CLFRs), or power towers. As part of a hybrid system, these raise steam that is fed into a power plant, reducing the amount of coal (or gas) required. Such thermal hybrid projects may be the most cost-effective option for large-scale use of solar energy. According to US studies carried out by EPRI, potentially, a solar trough system could provide 20% of the energy required for a steam cycle. The IEA GHG R&D Program has recently examined the integration of solar energy technologies with CCS-equipped plants [24].

Globally, there are around twenty hybrid solar thermal plants being developed, some based on gas and some on coal. In the USA, a major demonstration project (using parabolic trough solar collectors) was undertaken at Unit 2 of Xcel Energy's coal-fired Cameo Generating Station. Main aims were to decrease coal use, increase plant efficiency, lower CO_2 emissions, and to test the commercial viability of combining the two technologies. The solar component was deemed to have operated satisfactorily—coal use was lower, and overall SO_2, NO_x, and CO_2 emissions reduced. During operation, the system produced the equivalent of one MW (of the plant's 49 MW) from solar power. In Chile, GDF Suez and German renewable energy company Solar Power Group are developing a 5 MW concentrated thermal solar power plant. The facility will provide steam to the 150 MW Mejillones coal-fired plant in the northern part of the country. This pilot plant was scheduled to become operational in 2012. In Australia, the utility CS Energy is building a 44 MW solar thermal add-on to its 750 MW coal-fired supercritical Kogan Creek plant in Queensland. This US$110 million project (the Kogan Creek Solar Boost Project) is using CLFR technology. This is the largest solar project in the Southern Hemisphere, the world's largest hybrid integrated solar steam/coal-fired power plant augmentation project, and the world's largest linear Fresnel reflector solar CSP installation. The add-on allowed increased electricity production and avoided an estimated 35,600 tCO_2. A second Australian CSP project (using a Linear Fresnel reflector system with a total mirror surface of 18,500 m^2) is at Macquarie Generation's 2 GW coal-fired Liddell Power Station in New South Wales. This incorporates a 9.3 MW capacity Novatec solar boiler to generate steam that is fed into the existing plant, helping reduce coal requirements and plant emissions. The reduction in coal used cuts CO_2 emissions by 5,000 t/y [7,8].

A scoping study examining possible ways of harnessing solar power with brown coal (within an Australian context) was completed (90) in 2012. This study examined ways of combining input from a CSP facility with particular emphasis on coal gasification. There is increasing interest in using concentrated radiation from CSP to drive directly thermochemical reactions in this manner. Direct conversion of coal to liquids is also possible. The study examined the technical and economic viability of various options for combining brown coal and concentrated solar and identified the following possible approaches: (a) use of solar process heat; (b) low temperature supercritical water gasification of brown coal in a linear concentrator; direct coal to liquids reactions in a linear concentrator; gasification within a high-temperature solar-heated molten salt tank; and (c) high-temperature supercritical water gasification using a tower or dish concentrator; entrained flow or fluidized bed gasification using a tower concentrator [7,8].

Such solar-driven technologies are still at the R&D stage and it is expected that associated costs will decrease over time. Advantageously, their input into a coal-based process should reduce CO_2 emissions by effectively replacing part of the coal feed. Concentrating solar thermal technologies has been proven on a large utility scale, although so far, they have only been employed commercially to heat working fluids for power generation. However, the principle of driving high-temperature endothermic reactions with concentrated solar energy is well established in the R&D phase. Conversion of hydrocarbons such as brown coal using solar heat appears to be technically feasible via a number of possible routes [7,8].

A major challenge is to increase the solar share of coal-solar hybrids. To date, input from the solar component has tended to be limited, often as a consequence of their application to retrofit applications on ageing plants as opposed to new build. There is some consensus that the sector needs bigger hybrid projects based on highly efficient, newly built coal-fired plants; this would provide more scope for improved efficiency and better economics. The projects developed in Chile and India will provide useful data and operational experience, hopefully encouraging further uptake of the technology. Technology development continues, aimed at reducing system costs, increasing the efficiency of solar-to-electricity conversion and minimizing the environmental footprint. Some argues that a much greater solar share (possibly up to 30%–40%) could be realistic [7,8].

5.2.4 ROLE OF WIND ENERGY

A major drawback with wind and solar power is their intermittency. Consequently, times of peak output may not correspond with periods of high electricity demand, and vice versa. At times, there can be significant amounts of surplus unwanted energy available, particularly from wind farms. One approach is to use it to surplus energy to electrolyze water, producing hydrogen and oxygen. Both gases have the potential to be component parts of hybrid systems, and there are various schemes where the hydrogen could be fed into the syngas from a gasification system, used in fuel cells, used directly as a transport fuel, or stored for later combustion in gas turbines to generate electricity. Similarly, oxygen could be stored, used for a number of industrial applications, or fed to a coal/biomass gasifier or an oxyfuel combustion plant to generate electricity. Different concepts and schemes combining gasification, intermittent renewables, and electrolysis are currently being examined. Some also aim to incorporate some form of CCS. Some hybrid systems are at early stages in their development or have been undertaken at a very small capacity. Ongoing improvements in gasifier and electrolyzer design are likely to encourage further development of such systems for energy production. Where hydrogen and/or oxygen production forms part of such schemes, the reduction in the cost of electricity supplied by renewable energy sources such as wind would also be beneficial as it would reduce the cost of electrolysis. A number of ways to integrate electricity generated by wind power with coal use have also been proposed; the direct coupling of wind- and coal-based power generation has been examined in a number of US and European studies and reported in a previous Clean Coal Centre report [25,26]. A number of schemes for incorporating wind-generated electricity as part of coal and biomass-based processes are also discussed.

5.2.5   CARBON CAPTURE FROM BIOMASS AND COFIRED PLANTS

Numerous CCS development programs and projects are under way around the world [27–32], many focused on coal-fueled processes. The three main routes being pursued are precombustion capture, postcombustion capture, and oxyfuel firing technology; the status of many of these was reported in a recent Clean Coal Centre report [27,30]. The potential of using CCS on biomass- and biomass/coal-fueled plants is also being examined. Substituting part of the coal feed to a coal-fueled process with biomass will help reduce overall CO_2 emissions, although clearly, if carbon capture technology is also deployed, the impact will be even greater. Where a process is fired only on sustainably produced biomass, the addition of CCS could be expected to produce a net overall removal of CO_2 from the atmosphere, making the process "carbon negative." This combination (Bioenergy with CCS, or BECCS) is claimed to be the only large-scale technology currently capable of achieving net negative emissions. Potentially, BECCS could be applied to any process where CO_2 forms a significant proportion of the flue gas stream. However, in the case of forestry-supplied biomass, there are concerns that it might be impossible to meet demand using only residues, as opposed to increased tree harvesting [28]. For example, Smolker and Ernsting suggest that a 50 MW biomass-fueled power plant would require woodchips/pellets made from ~500 kt/y of wood. Since biomass transportation is difficult, coal-biomass hybrid plant makes more sense with local use of biomass.

In 2012, the Zero-Emissions Platform and the European Biofuels Technology Platform published (via the Joint Task Force for Bio-CCS) a report considering the future for biomass with CCS within a European context. This concluded that in order to minimize CO_2 emissions, effective carbon-negative solutions (such as BECCS) will be required. The report suggested that within Europe, by 2050, Bio-CCS could potentially remove 800 Mt/y of CO_2 from the atmosphere through the use of sustainable biomass. This is equivalent to >50% of current emissions from the EU power sector and would be in addition to any emissions reductions achieved by replacing fossil fuels with biomass. Combining biomass with CCS could produce a global saving of up to 10 Gt/y of negative CO_2 emissions by 2050 [31], equivalent to a third of all current global energy-related emissions.

CCS in general is considered to have significant potential, and bodies such as the IPCC, IEA, and UNIDO have flagged up its importance for both coal and biomass-fueled systems. A recent IEA Technology Roadmap set out a vision for CCS in industrial applications up to 2050 and examined its potential for a number of applications, including biomass conversion. It concluded that the application of CCS to biomass processes is likely to start modestly; by 2020, some 14 Mt/y of CO_2 could be captured, mostly from ethanol and hydrogen production [32]. Both North America and China are expected to play key roles in CCS deployment in biomass conversion.

5.2.6   CONVERSION OF CARBON DIOXIDE TO POWER BY
FUEL CELL TECHNOLOGY OR TO DIESEL FUEL

Both of these conversion processes are described in details in my previous book [4]. Exxon-Mobil and Fuel Cell Energy companies are jointly developing "Sure Source application" technology to convert CO_2 to power. In a typical Sure Source application, clean natural gas is combined with ambient air for the fuel cell power generation process. Due to the unique capabilities of the carbonate fuel cell technology, the exhaust flue gas of coal or gas-fired power plants is directed to the air intake of the fuel cells when configured for carbon capture. The fuel cells act as a carbon purification membrane, transferring CO_2 from the air stream, where it is very dilute, to the fuel exhaust stream, where it is concentrated, allowing the CO_2 to be easily and affordably captured, chilled, and compressed for industrial use or sequestration. In the process, fuel cell also generates additional power. More details are given in my previous book [4].

   CO_2 can also be converted to diesel fuel by a commercially viable CO_2-to-Fuel technology with the recent announcement, by the Audi-Sunfire-Climeworks consortium of companies, operating in Germany and Switzerland, of a pilot project for producing diesel fuel from CO_2 and H_2O and renewable energy sources, such as wind or solar or hydropower. The CO_2 supply can be obtained mainly from a combustion plant supplemented with some CO_2 captured directly from the air. A pilot plant in Dresden is destined in the months ahead to produce around 160 L/day of the synthetic diesel, which they have dubbed "blue crude." This process is described in detail in my previous book [4].

5.2.7   COMBINED CYCLE TO IMPROVE EFFICIENCY

Combined cycle technology allows a power plant to generate 50% more electricity from its fuel than it could with a single-cycle power system. In a two-on-one combined cycle system, two combustion turbine generators work in conjunction with two heat-recovery steam generators and a steam turbine generator. In the first cycle, natural gas or diesel gas is burned to directly power two gas turbine generators that produce electricity. Some industry like TVA (Tennessee Valley Authority) also has one-on-one combined cycle units where there is a single heat-recovery steam generator and a single steam turbine generator.

   The hot exhaust normally lost during this process is captured for the second cycle, where it used to boil water into steam in the heat recovery steam generators. The steam spins an additional turbine generator to produce more electricity. The steam is then condensed back to water and recycled. This process is widely used in the industry and widely publicized in the literature [1–5].

5.2.8   CONVERSION OF WASTE HEAT TO POWER
OR ADDITIONAL INDUSTRIAL USE

Waste heat can converted to power by numerous other techniques like thermoelectricity, thermophotovoltaic, etc., or it can be used for numerous industrial heat need through hybrid cogeneration processes. These are well described in my previous book [4].

5.3   COAL-BIOMASS COGASIFICATION

In order to reduce carbon dioxide emission, cogasification of coal and biomass has become important. Combining both coal and biomass and cogasifying can also provide a number of benefits. Although there are many commercial-scale plants that gasify coal and numerous small units that gasify biomass, relatively few have so far cogasified combinations of the two [33]. However, interest in cogasification has been growing, as adding biomass to a coal-fueled facility can be advantageous. As mentioned before, 70–30 coal-biomass mixture results in carbon neutral gasification over its entire cycle. Gasifying biomass in a large coal gasifier can achieve higher efficiency and improve process economics through economies of scale. The biomass may be a no-cost waste. Hence it may be possible to take advantage of a lower cost feedstock [33]; cogasifying with coal may help smooth out the seasonal fluctuations in the availability of biomass, and so on. Furthermore, replacing a portion of the coal feed with biomass may enable a coal plant to obtain credits for the use of a renewable fuel.

Coal has been gasified commercially for many years and the various reactions involved in the process have been well examined and described. Similarly, gasification of many biomass materials has been studied. However, the chemical processes involved when cogasifying coal-biomass mixtures are less well understood, and investigations continue [for instance, 35]. Whereas coal comprises mainly carbon, biomass is often a mixture of complex compounds such as cellulose, hemicelluloses, lignin, extractives, and minerals [34]. When gasified, minor oxidation and major pyrolysis occur, and as the temperature increases, a series of pyrolysis reactions take place: (a) below 125°C, drying occurs; (b) between 125°C and 500°C, hemicelluloses, cellulose, and lignin (partially) decomposes; and (c) above 500°C, remaining lignin degrades. During gasification, all of the volatiles and some tars are thermally cracked and broken down into simple gaseous products. Remaining tars (and some alkali minerals) exit with the product gases.

Because of their different chemical and physical properties, a number of issues may arise when cogasifying coal and biomass, and these can present some technical challenges. Biomass has a higher hydrogen/carbon ratio and oxygen content than coal, making it more reactive under less severe (lower temperature and pressure) gasification conditions. This lower severity can lead to tar formation that can constitute a major challenge for gas clean-up. Trace elements present in coal and biomass may also be different, requiring alternative gas clean-up methods for each feedstock. Whereas char and tars are of greater importance for biomass gasification, species such as sulfur and heavy metals are likely to be of greater concern with coal. Biomass tends to have as much higher metallic ion content than coal. This can result in higher levels of acidic species capable of producing highly corrosive products. However, because of their lower sulfur and ash contents, biomass materials generally produce cleaner syngas than coal. Because of such issues, it may be more economically attractive and less technically challenging to cogasify biomass with coal [36].

In terms of gasifier operating conditions, there are a number of similarities, but also some significant differences between biomass and coal. The minimum temperature required for coal gasification is ~900°C and that for biomass is generally in the range 800°C–900°C. Thus, the temperature required for the complete thermal gasification of biomass may be similar to that of coal [33]. However, dissimilarities attributable to differences in fuel properties may include as follows: (a) unlike coal particles, some biomass materials are very fibrous in nature; (b) biomass is a more reactive fuel than coal; (c) biomass ash has a comparatively low melting point and under some circumstances, in the molten state, can be very aggressive; and (d) at low gasification temperatures, biomass can produce high tar levels in the product gas.

Advantageously, studies have detected the existence of synergetic effects when cogasifying coal and biomass. However, there appears to be no firm consensus on the precise nature of the chemical reactions occurring and a number of possible reactions and processes have been suggested. Under some circumstances, the addition of biomass to a coal feed has been found to increase both the rate and extent of gasification reactions occurring—reaction rate has sometimes been observed to double [37]. This may be a consequence of the catalytic activity of char formed, or the high oxygen content of the biomass and investigations are continuing. From a technical standpoint, biomass gasification can be improved in the presence of coal as tar content in the product gas is often significantly reduced [38,39]. Furthermore, certain types of biomass ash have been found to catalyze coal gasification [34].

However, while cogasification is advantageous from a chemical point of view, there can be technical challenges associated with the gasification stage as well as a number of upstream and downstream processes. Upstream, the uniformity, size, and characteristics (such as moisture content) of the coal and biomass particles are of importance and various pretreatments may be required prior to feeding. While upstream processing influences material handling, the choice of gasifier operating parameters (temperature, gasifying agent, and catalysts) determines product gas composition and quality. As biomass decomposes at a lower temperature than coal, different reactors suited to the particular feedstock mixture may be required [35]. The choice of feedstocks and the type of gasifier and its operating parameters not only influence product gas composition they also dictate the amount and types of impurities to be handled downstream. Some downstream processes may require modification when biomass is added to a coal gasification process. For instance, alkalis stemming from biomass may result in corrosion problems in downstream pipes and components.

Over the course of several decades, studies into the cogasification of coal and biomass have been carried out, although R&D efforts are still continuing. Many investigations have been undertaken using different types of gasifier and specific types of coal and biomass in North America, Europe, and the Asia-Pacific region, although most have been at a small scale. Despite these efforts, some of the chemical pathways involved are still unclear and practical operating experience, particularly on a large scale, remains limited. Various programs are addressing these issues. For instance, entrained flow cogasification is being actively pursued in several parts of the world. In the USA, the entrained flow gasification of a variety of coal-biomass combinations is being investigated. Work is examining operating parameters, analysis gasification products, and assessing gas clean-up systems [34,40]. Similar themes are being studied elsewhere. For instance, for some years, ECN (Energy Research Center of the Netherlands) in The Netherlands has been examining both entrained flow and fluidized bed cogasification [41]. Elsewhere, work has not been limited to the use of hard coals and the potential for cogasifying biomass and lignite has also been examined [38]. Although historically, a large body of work has already been generated, R&D programs focused on cogasification are continuing, motivated mainly by combinations of environmental and economic concerns. Some projects are examining cogasification as a means for the large-scale production of hydrogen. Here, the critical challenges and major R&D needs have been identified as further development of biomass/coal cofed gasifiers, the need to reduce process costs and improve overall system efficiency, the removal/control of feedstock impurities, and the minimization of process carbon footprint [42]. Thus, there are still some technical uncertainties associated with the process, largely because of the limited experience with cogasification at a commercial scale. For instance, Weiland and others [43] and Brar and others [34] noted: (a) the technology for feeding mixed feedstocks via pressurized dry feed systems is not mature; particle size and shape are critical factors for specific feeder types; (b) gasifier performance—more operational data are required on reaction kinetics, material interactions, and product effects; detailed models have not yet been developed or validated; (c) in-depth studies and practical evaluation of some combinations of coals and biomass types is lacking; and (d) process optimization requires further more detailed examination.

In the case of cogasification of forest residues with coal, there are still issues that have not yet been fully explored. For instance, the tar evolution profile needs further investigation in order to tailor gasifier design, plus determine operating conditions and tar removal systems. In addition, further studies are required to determine the influence of particle size and specific energy consumption for forest residues. This information will be critical to performing sustainability and economic analysis of the size reduction process [34]. Similar issues remain with some other types of biomass. The advantages and disadvantages of cogasification are described in Table 5.1. More complete details on all issues regarding coal-biomass cogasification are also described in great details in my previous books [1–5]. The nature of syngas produced from cogasification depends on coal/biomass ratio, gasifier temperature profile, catalyst, and air-to-steam ratio. Cogasification also requires downstream processing for tar removal, alkali removal, particulates removal, and ammonia removal. Special techniques may have to adopted to remove tar depending on the nature and extent of tar. The composition of Syngas produced from cogasification depends on the ratio of coal and biomass, air to steam ratio, gasifier temperature profile, nature of coal and biomass, and catalyst used. Pressure of the reactor also affects the Syngas composition. These issues along with existing commercial processes are also described in detail in my previous book [1].

5.4   HYBRID POWER BY IGCC PLANTS

Biomass-derived syngas gas can be used for direct combustion in power generation/ cogeneration plants, although most such facilities are of small scale. Alternatively, it can be gasified in a stand-alone gasifier and the resultant syngas injected into the combustion zone of a coal-fired power plant. A major advantage of this technique is that biomass does not enter the plant's boiler, thus avoiding any issues with unwanted

TABLE 5.1
Main Advantages of Cogasifying Biomass with Coal [23,25,26,36,39,44]

Disadvantages of Stand-Alone Biomass Gasification	Advantages of Cogasification with Coal
Separate, stand-alone gasification can be expensive	Reduction of specific investment costs. Cogasification has the potential to reduce investment cost and improve efficiency compared to separate gasification facilities
Capacity of biomass plants sometimes limited to relatively small scale	Improved efficiency by allowing operation in a larger system (compared to biomass-only gasification facilities)
Seasonal variations in biomass supply. Overall supply may be limited. Mainly limited to small-scale processes and small-capacity plants	Coal is easily available in many countries. Larger scale possible; improved process economics; adding biomass to existing coal-fired processes provides economies of scale and helps smooth out seasonal fluctuations in biomass availability
Some feedstocks unsuitable for stand-alone gasification	Some biomass sources are unsuitable for alternative utilization hence would remain unused (as a waste). There may be associated disposal costs. Cogasification may provide a cost-effective disposal route
Low gasification temperature	Biomass has higher reaction rate than coal; synergistic improvement when cogasified with coal
Tar production may be excessive	Bio-ash can act as catalyst for coal gasification. Cogasification usually produces higher overall efficiency than individual feedstocks. Higher gas yield
	Little/no tar production; most tars converted to gas at high temperature
Low heating value gas	Higher heating value gas produced
High CO_2 content	CO_2 converted to CO at high temperatures
Fluidization quality (in fluidized bed gasifiers)	Improved fluidization quality
Anomalistic shape	Coke as bed particles—abrasion reduced and power saved
Inert bed particle (sand) can cause problems of abrasion	
Biomass is carbon-neutral but a limited resource	CO_2 emissions—although not carbon-neutral, coal is a huge global energy resource. Biomass is seen as carbon-neutral and renewable. Combining the two reduces overall CO_2 emissions
Biomass harvesting and pretreatment can be expensive	Cogasification in IGCC plant—compared to coal firing alone: • power output increases; • emissions of SO_2 and NO_x decrease; • process efficiency improves slightly; • less energy required for H_2S removal; • less water required for FGD scrubbers Coutilization would reduce competition with land use for food production. Also, large releases of CO_2 from standing biomass and/or soil carbon when new lands are cultivated would be avoided

species such as alkalis. It also avoids any impact on the properties or quality of fly and bottom ash, often sold commercially. Examples of commercial plants using this technology have been examined earlier. To date, these have been limited to conventional pulverized coal boilers. However, syngas generated by gasifying biomass also has the potential for use in integrated gasification combined cycle (IGCC) plants. IGCC employs a combined cycle format that incorporates a gas turbine fired on syngas; heat is recovered from the turbine's exhaust gas and used to generate steam to drive a steam turbine. Typically, more of the power produced by an IGCC plant (~60%–70%) comes from the gas turbine. Coal-based IGCC plant configurations and operation have been examined in detail in a number of IEA Clean Coal Centre reports [for instance, 45,46], and are not therefore considered in detail here. Several technology variants are offered commercially and a number of plants are in commercial operation; many others have been proposed and several are close to completion. However, largely because of a combination of prevailing market conditions and economic issues, a number of other projects have been cancelled or delayed.

Most coal-fueled IGCC plants use oxygen-blown technology. This has a number of fundamental advantages over air-blown systems that include smaller gasifier size (and hence lower cost), the heating value of the cooled/purified syngas is higher, syngas volume is about half that for an air-blown unit for the same amount of gasification energy (thus gas handling and clean-up require smaller units), and smaller heat exchangers are required to recover sensible heat from the syngas prior to clean-up. Technical disadvantages include the energy needed for air separation and the higher degree of plant integration required. Auxiliary power consumption in an air-blown system is estimated to be ~8%, compared with 10%–15% for oxygen-blown systems.

As part of their commercial operations, a number of existing coal-fueled IGCC plants cogasify combinations of coal and biomass. In addition, various studies have examined the potential advantages and disadvantages associated with the use of such fuel combinations [for instance, 36]. These suggest that using biomass in an existing coal-based IGCC can provide a number of benefits. Long and Wang [36] modeled the impacts of adding up to 50% (wt) biomass on the performance of a 250 MW coal-fueled IGCC. This was the highest proportion thermodynamically feasible, and also the one that produced the most significant results. It was suggested that, initially, plant efficiency would increase as 10% biomass was added. At 30%, it would decrease but increase again when biomass addition reached 50%. The variation of efficiency was minor, only within one percentage point between 38% and 39% [36]. The results suggested that system efficiency, emissions, and power output are not linear functions with biomass blending ratio. The authors concluded that by cogasifying biomass with coal in this manner: (a) net power output would increase; (b) emissions would decrease; (c) efficiency would improve slightly.

Adding biomass to the feed reduced plant auxiliary power requirements as overall pollutant (such as SO_2) levels were lower, reducing syngas clean-up requirements. Water requirements for the particle scrubber were also reduced, suggesting that higher levels of biomass produce less fly ash and slag output. There was also a positive impact on CO_2 emissions. Other investigations have reached similar conclusions.

Some studies have modelled cogasification as the basis for clean electricity generation combined with hydrogen production. For instance, work carried out at Babes-Bolyai University in Romania examined combining coal with various types of biomass and solid wastes in an IGCC-based system. This was capable of producing varying amounts of electricity and hydrogen, and was coupled with 90% carbon capture. The process modelled was based on a Siemens gasifier and generated between 400 and 500 MW net electricity, with a flexible output of 0–200 MWth hydrogen. Evaluations were undertaken for gasifying (80:20) combinations of coal with sawdust, wheat straw, corn stalks, MSW (Municipal Solid Waste), waste paper, sewage sludge, and meat and bone meal.

Unlike most other IGCC + CCS evaluations, where hydrogen is only used to generate power in a combined cycle gas turbine plant, this concept proposed purifying part of the hydrogen-rich gas coming from acid gas removal system and exporting it either for direct use (chemicals/liquid fuels production, or fuel cell use) or for temporary storage. With the latter, hydrogen could then be converted to electricity at times of high power demand, taking advantage of high electricity prices. It could also help smooth out variations in supply from the growing level of intermittent renewables being deployed in many countries. Such a concept could provide a high degree of operational flexibility as it would allow the plant to operate solely on base-load, irrespective of electricity demand, offering significant technical and economic advantages. Overall, the process would capture 90% of the feedstock carbon, enhance security of supply by fuel diversification, and improve plant flexibility in terms of varying the energy vectors according to the instant grid demand. Such flexible plant operation would offer some advantages within the context of the emerging hydrogen market. In the initial stages of the developing the hydrogen economy, plants could operate predominantly in electricity production mode, with low or even zero hydrogen output. However, as hydrogen applications increase, more plant energy could be delivered in the form of hydrogen [47].

5.4.1 Commercial Cogasification IGCC Plants

5.4.1.1 ELCOGAS IGCC Plant, Puertollano, Spain

Puertollano is the location of Spain's 335 MWe coal/petcoke-fuelled IGCC plant. This was built by ELCOGAS, a consortium of eight European utilities and three technology suppliers, set up in 1992 to develop the project. The plant uses a Krupps-Koppers PRENFLO single-stage, oxygen-blown entrained flow gasifier with dry pulverized coal feed. Fuel is transported pneumatically to the gasifier using nitrogen as carrier gas. Steam/nitrogen is used as moderator, fed through four horizontally arranged burners located in the lower part of the gasifier. The reaction chamber has a membrane wall with an integral cooling system that produces pressurized steam. Gasification takes place at a pressure of 2.5 MPa, at a temperature of 1,200°C–1,600°C [33]. The plant started commercial operation in 1996 with natural gas, but has operated with syngas since 1998.

The plant's basic fuel is local high ash (~40%) subbituminous coal blended in equal proportion with high sulfur (5.5%) petcoke from the Puertollano REPSOL refinery. At full operational capacity, the plant burns 700 kt/y of mixed fuel.

Although designed to operate primarily on this fuel combination, a number of cogasification trials have been carried out using different biomass feedstocks. A dedicated preparation system was used to feed up to 10% (wt) of materials that included olive wastes, almond shells, waste wood, and vineyard and grape wastes. These were cogasified with the coal/petcoke; total operating time under these conditions was more than 1,100 hours.

The plant also successfully cogasified Meat and Bone Meal (MBM – 1% and 4.5%). In this case, around 2% limestone was added to the feed to capture chlorine and avoid high-temperature chemical corrosion. There were no observable operational differences in the behavior of the fuel preparation and sluicing systems. During gasification, there were no effects with the 1% test, although at 4.5%, there was an appreciable reduction in the fouling values, which was coincident with a limestone reduction. There was also a progressive increase in chlorate level in the wet scrubber system. However, there were no changes in hazardous emissions.

More recently (during 2007 and 2008), further cogasification trials were undertaken using olive oil waste (orujillo). Initial tests were undertaken using 1%–2%, although this was subsequently increased to 4%. The appropriate amount of orujillo was mixed with limestone and introduced into the gasifier with a 50% mixture of coal/petcoke. The limestone/orujillo mixture tended to cake in the storage hopper so was fed manually. However, overall, no significant operational problems were encountered.

5.4.1.2   The Willem Alexander IGCC Plant, Buggenum, The Netherlands

Developed initially by Demkolec BV, initial syngas trials were carried out in 1994–1995. In 2001, the plant was bought by NUON for operation as a commercial base load station. Based on Shell gasification technology, it can gasify a range of coal types as well as petroleum coke. Gross power output is 284 MWe (156 MW from the gas turbine and 128 MW from the steam turbine). The plant generally operates at a net efficiency of ~43% LHV (Lower Heating Value).

Around 2,000 t/d of imported coal is used, usually cogasified with biomass. The Dutch Coal Covenant requires that CO_2 emissions from the plant are reduced by 200 kt/y—the equivalent of ~35 MWe from biomass, or ~30% (wt) of biomass in the coal feed. Eventually, up to 50% biomass may be used, generating 60 MWe of "green power." Biomass is cogasified with coal in a Shell single-stage upflow entrained flow, oxygen-blown, dry feed gasifier. This is of the membrane walled type and operates at a temperature of 1,500°C and a pressure of 2.8 MPa. Cleaned syngas is diluted with nitrogen and steam to achieve the required specification for the gas turbine. The gasifier has a design capacity of ~4.0 million m³/d of syngas. In order to maximize thermal efficiency, the cycle uses full integration with extraction of air from the gas turbine compressor for the Air Products ASU (Air Separation Unit) [45].

A program of biomass cogasification began in 2001–2002. Testing included 34 types of biomass that included wood, chicken litter, municipal sewage sludge, grape seed, palm-pits, cacao meal, sunflower pits, paper/plastic residue, and paper mill wastes. Up to 30% (wt) were used. However, in order to maintain the required power output, biomass addition was limited to 10%–15% [33,48,49]. Biomass feedstocks currently used include dried sewage sludge, chicken litter, and saw-dust. Ideally, particle size is limited to ~1 mm, and moisture content to ~5%. Initial cogasification operations threw up a number of technical issues although these were systematically resolved. The successful program undertaken at Buggenum has con-firmed that a wide range of secondary fuels can be cogasified at relatively high ratios on a continual basis for significant periods of time. More details on this process are given in my previous books [1–3,5].

5.4.1.3 Polk IGCC Plant, Florida, USA

Tampa Electric's 250 MWe Polk plant was built with support from the U.S. Department of Energy (DOE) as part of its Clean Coal Technology Program. Since 1996, the facility has been operating using a 2000 t/d oxygen-blown GE gasifier. Coal for the process is stored on-site in two 5,000 ton capacity silos. The gasifier gener-ates syngas with a CV of ~9.3 MJ/m^3; this is cooled to 750°C–800°C, then cleaned. Particulates and HCl are removed using a water scrubber, and sulfur species removed via COS hydrolysis. H$_2$S removal is by means of an MDEA (Methyl diethanolamine)-based acid gas removal system, with sulfur produced in a Claus unit. Since start-up, the plant's reliability and availability have increased steadily. According to the DOE, the Polk facility is now one of the world's cleanest coal-fired power plants. More than 98% of SO$_2$ and 90% of NO$_x$ are captured. Particulate emissions are also very low.

The plant also evaluated a combination of syngas clean-up and CO$_2$ capture (the *Warm Gas Cleanup (WGC) and CCS Demonstration*). A DOE cooperative agree-ment was put in place in 2010 to demonstrate RTI's (Research Triangle International) WGC sulfur removal technology which cleans syngas at elevated temperatures. This was followed by the announcement of DOE support for the addition of CCS to the existing WGC project. The new development involved the addition of a shift reactor and syngas cooling systems [1,2,3,5,25,26].

Cogasification tests using eucalyptus and bahia grass were carried out between 2001 and 2004. The eucalyptus trials used a total of 8.8 tons of coarsely ground material (moisture content ~47%, ash content 5.3%). Felled trees were reduced in size using a hammer mill and trommel screen. This produced small particles capable of passing through the pumps and screens of the plant's slurry feed system. The ground material was introduced into the process via a stirred recycle tank and mixed with water. It was blended with the normal coal (and petcoke) mixture to form a slurry that was fed to the gasifier. The biomass comprised 1.2% of the plant's fuel input. During the test, plant performance was unaffected. In 2004, further testing examined the addition of 5% bahia grass to the coal feed. A total of 50 tons were used. Again, apart from some difficulties in material handling, there was no significant impact on syngas quality or emissions. It was confirmed that up to 5% biomass addition was technically feasible. However, the amounts used and the length of the test periods were relatively short; hence, the trials gave little indi-cation of what issues might arise if a greater proportion of biomass was cogasified on a continuous basis. Such a move would require the installation of dedicated biomass handling and preparation systems. More details on this process are also given in my previous books [1–3,5].

5.4.2 OTHER COGASIFICATION PROJECTS AND PROPOSALS

Historically, the cogasification of various biomass and wastes with coal has been demonstrated using several types of gasification technology. As described above, some of them generated power from produced syngas. Some, however, used syngas to produce chemicals. Two of the original ones were in Germany. At Berrenrath, during the 1980s and 1990s, lignite was cogasified successfully with several waste materials (MSW pellets, dried sewage sludge, and loaded cokes) in a 600 t/d pressurized circulating fluidized-bed High-Temperature Winkler gasifier. Biomass materials were fed without problem by the existing plant solids handling and feeding systems, and during all trials, gasifier operation was virtually trouble-free. In the case of MSW, up to 50% was gasified successfully. There were no appreciable changes in gasification temperature or syngas output and composition [33]. Syngas produced was used for methanol production.

Cogasification was also carried out successfully at the Schwarze Pumpe plant in Germany where a range of coals, biomass, and wastes were gasified in the site's three gasifiers (operated in an integrated manner). These comprised an oxygen-blown FDV rotating grate unit, an oxygen-blown GSP entrained flow gasifier, and an oxygen-blown BGL slagging gasifier [27]. Between 1995 and 2002, wastes such as demolition wood, sewage sludge, plastics, and household waste were processed. Different combinations of materials were regularly used and at times, up to 80% waste and 20% lignite were cogasified. Syngas was supplied to an on-site methanol production unit.

In recent years, there have been several coal/biomass IGCC projects proposed. Two of these were in Rotterdam in The Netherlands. CGEN (Compugen Ltd.) NV proposed the development of a 450 MW IGCC plant (fueled on coal and biomass) that could also produce hydrogen. Foster Wheeler has reportedly undertaken a feasibility study. CO_2 from the plant would be captured and stored in a depleted oil and gas field. In 2011, the project was reportedly at the feasibility stage. Essent and Shell also proposed the construction of a 1,000 MW IGCC plant. This would use Shell gasification technology and be fueled on coal and biomass. Most of the CO_2 produced would be captured and a number of onshore and offshore storage options, such as depleted oil and gas fields, were investigated. A proposed start-up date of 2016 was originally suggested. However, the project is currently on hold [1,2,3,5,25–27].

5.5 LIQUID SYNTHETIC FUELS BY COGASIFICATION

Gasification is the foundation for converting coal and other feedstocks to gasoline, diesel, and jet fuel. Two main processes have been demonstrated at full commercial scale for coal-to-liquids, namely, Fischer-Tropsch (FT) and methanol-to-gasoline (MTG). Both start with coal gasification and follow this stage with further downstream processing. The global potential for producing liquid fuels from coal and biomass combinations is substantial, and various studies have examined the possible scale of production for different countries. Most conclude that producing transport fuels from coal and biomass is best achieved by cogasifying, followed by either FT or MTG synthesis—these two appear to offer the most viable and economic options.

Various studies agree that the production of transport fuels via cogasification of coal and biomass appears to offer the most economical option [38] and efforts are under way to determine the most effective way forward. Department of Energy has funded several projects to convert coal-biomass mixture to liquids (CBTL (Coal-biomass to liquids) process). Cogasification of coal and biomass and the production of synthetic fuels are also being addressed by the Department of Defense. Feasibility studies have examined the possibility of producing 100,000 bbl/d of jet fuel from coal and biomass (as woody biomass, switchgrass, and corn stover) using entrained flow gasification and FT technology, with a net carbon footprint 20% less than that of jet fuel produced from petroleum. It was determined that FT diesel could be produced at the target CO_2 level by cogasifying coal with only a modest amount of biomass (between 10% and 15%) [33]. Other US studies undertaken by NETL (National Energy Technology Laboratory) investigated the use of coal/biomass mixtures to generate electricity and coproduce zero-sulfur diesel via low temperature FT technology. Both bituminous and subbituminous coals were evaluated. Switchgrass (15%) was selected as a representative type of biomass for CBTL facilities. The plant designs envisaged the production of 50,000 bbl/d of FT liquids, comprising 34,000 bbl/d of FT diesel, with the balance consisting of FT naphtha [50]. Another innovative method for producing liquid transportation fuels from domestic coal and biomass by CBTL process used novel FT catalyst developed by Chevron. The Southern Research Institute developed a process based on the highly selective conversion of coal/biomass-derived syngas to gasoline and diesel. This approach should eliminate the typical FT product upgrading and refining steps and enhance the ability of CBTL processes to compete with petroleum-based processes. CBTL process (with 85–15 wt % mixture) was also examined as a statewide process in West Virginia using forestry biomass [51]. The production of liquid fuel by CBTL process using FT technology was also pursued in Europe. For instance, Spanish studies investigated the cogasification of different combinations of coal, biomass, and petcoke [52]. Similar studies aimed at producing liquid fuels via gasification and FT technology have been undertaken elsewhere using different fuel combinations.

In Turkey, as part of the TUBITAK 1007 research program, a four-year project (the Trijen Project) investigated the cogasification of local coal and biomass for the production of FT liquids. The main aim was the development of an economically viable process for the production of liquid fuels from domestic lignite and biomass [53]. Liquid fuel production focused on the low temperature (180°C–250°C) FT process, using a slurry phase reactor and novel Fe-based catalysts [54]. The feasibility of cogasifying coal and biomass was also investigated by Sasol in South Africa using a Sasol-Lurgi fixed bed dry bottom gasifier. Biomass evaluated comprised bark and bark/wood pulp combinations. It was concluded that such blends could be gasified successfully although further investigations were required in areas concerning operating stability, high gasifier outlet temperature, lower carbon conversion, and higher CO_2 conversion [33].

Baliban et al. [55] introduced a hybrid coal, biomass, and natural gas to liquids (CBGTL) process that can produce transportation fuels in ratios consistent with current U.S. transportation fuel demands. Using the principles of the H2Car process, an almost-100% feedstock carbon conversion was attained using hydrogen produced from a carbon or noncarbon source and the reverse water-gas-shift reaction. Seven novel process alternatives that illustrate the effect of feedstock, hydrogen source, and light gas treatment on the process were considered. A complete process description is presented for each section of the CBGTL process including syngas generation, syngas treatment, hydrocarbon generation, hydrocarbon upgrading, and hydrogen generation. Novel mathematical models for biomass and coal gasification are developed to model the nonequilibrium effluent conditions using a stoichiometry-based method. Input–output relationships were derived for all vapor-phase components, char, and tar through a nonlinear parameter estimation optimization model based on the experimental results of multiple case studies. Two distinct Fischer–Tropsch temperatures and a detailed upgrading section based on a Bechtel design were used to produce the proper effluent composition to correctly match the desired ratio of gasoline, diesel, and kerosene. Steady-state process simulation results based on Aspen Plus were presented for the seven process alternatives with a detailed economic analysis performed using the Aspen Process Economic Analyzer and unit cost functions obtained from literature. Based on the appropriate refinery margins for gasoline, diesel, and kerosene, the price at which the CBGTL process becomes competitive with current petroleum-based processes was calculated. This break-even oil price was derived for all seven process flowsheets, and the sensitivity analysis with respect to hydrogen price, electricity price, and electrolyzer capital cost, was presented.

5.6   HYBRID ENERGY SYSTEMS FOR COAL TO CHEMICALS

The major users of coal are China, US, India, Australia, and Russia. As a major part of China's industrial sector, the output value of the coal-based chemical industry is around 132.2 billion US dollars ($), accounting for 3.7% of the gross industrial output value in 2014. Meanwhile, the related CO_2 emission is about 495 million tons, occupying 7.2% of the total industrial carbon emission. With the full production capacity of coal-based SNG (Synthetic Natural Gas), oil, olefin, ethylene glycol, ammonia, and methanol, the related carbon emission may increase to 1.7 billion tons by 2030, and the carbon emission intensity may reach 8.95 kg CO_2/$ in 2030. This has forced China to implement strategy for carbon reduction; a low-carbon strategy is vital for the sustainable development of the coal-based chemical industry. There is a more than 900 million tons CO_2 emission reduction by using a hybrid energy system for global coal to chemicals in 2020, with China's contribution to carbon emission reduction at 23%.

In China, the emission source is mainly located in the northwestern part of China, the middle-eastern area of China, and the eastern coast, which in total account for 65% of total CO_2 emission. Not surprisingly, the distribution of CO_2 emission is in line with the geographical distribution of energy resources. Taking into consideration the energy resource distribution and energy consumption patterns, four large-scale low-carbon integrated systems are proposed in China [56]: (a) integration of solar energy, wind energy, and natural gas with coal in the northwestern part; (b) integration of biomass with coal in the east-central area; (c) integration of natural gas, wind, and biomass with coal in the northern area; and (d) integration of nuclear energy, wind energy, and coal along the eastern coast.

Based on the geographical distribution of energy resources and coal-based chemical industry planning in China, a comparison of CO_2 emission between the conventional coal system and hybrid system is conducted by assuming the utilization rate of the hybrid system at 80% in 2030. Both carbon emission and emission intensity in 2030 will reach 22% and 57% less than that in 2020. According to the World Gasification Database released by the DOE, there are numerous coal chemical plants existing or under construction in coal-rich countries such as the US, India, South Africa, Indonesia, and Australia, and the major coal gasification projects include coal-based synthesis ammonia, oil, and SNG. Thus, there is huge potential to develop low-carbon hybrid energy systems in these countries as well [56].

According to the U.S. Energy Information Administration data, coal resources in the US are mainly distributed in Appalachia in the east, the mid-west, and west, which account for 22.6%, 28.1%, and 49.3% of the already explored reserves, respectively. The nuclear plants in the US are distributed in the eastern coastal area and central eastern area; the solar power plants are widely distributed in the eastern coastal and southwestern areas. According to the U.S. wind resources investigation data published by the National Renewable Energy Laboratory (NREL) and the American Wind Energy Association, the wind resources of the US are mainly distributed in the mid-west and the Pacific, and Atlantic coastal areas. Biomass resources are mainly distributed in the northwestern and mid-west areas. The geographical compatibility of coal resources and low-carbon clean energy makes it possible for the implementation of a hybrid energy system. In fact the major clean energy labs in the US all have made techno-economic evaluations of the production of chemicals from hybrid energy systems. Thus, the following implementation of the hybrid energy system can be considered [56]: (a) coal-nuclear or coal-solar integration for the production of fuels/chemicals in the eastern area; (b) a coal-wind or coal-nuclear system for the production of fuels/chemicals in the mid-western plain area, and coal/biomass-wind system for the production of fuels/chemicals in northern mid-western area; (c) coal-biomass gasification for the production of fuels/chemicals in the northwestern area, and solar-biomass system for the production of fuels/chemicals in the southwestern area.

Chen et al. [56] presented the process flowchart of a comprehensive hybrid system for various fuel/chemical productions. In this system, nuclear/wind/solar-assisted water electrolysis is applied to produce hydrogen [57–60] in order to adjust the H/C ratio of syngas, hence to eliminate the water-gas shift process and to constrain the carbon emission. Besides this, biomass can be gasified to produce fuel and chemicals and the amount of CO_2 absorbed during biomass growth is almost equivalent to the amount of CO_2 produced during the combustion process. Thus, it is possible to achieve zero-carbon emission in a biomass-based system.

5.6.1 Nuclear-Coal Integration System

According to the Gasification and Syngas Technologies Council Worldwide Syngas Database, chemical production shared 45% of coal gasification products in 2008, and from 2008 to 2010, 22% of new gasifier additions were intended for chemical production [61]. However, the principal issue of current coal chemical projects is the resultant tremendous CO_2 emission. Carbon emission of coal-based chemical industries mainly includes indirect CO_2 emission from heat/electricity supply and direct CO_2 emission from the coal chemical conversion process, e.g., water-gas shift reaction.

Notably, carbon emission of coal to carbide or coking process is 80% lower than for other processes since they only involve indirect carbon emission from the heat/electricity supply. Other coal chemical industries always include the water-gas shift unit to increase the H/C ratio of the syngas that comes from the coal gasification unit by consuming a part of C. During this process, more than 50% of the carbon turns into CO_2 and is emitted into the atmosphere.

It is clear that either clean power or hydrogen supply is the key to CO_2 emission reduction in the coal chemical industries. Thus, it will be an effective and reasonable solution to integrate nuclear/renewable energy with coal for low-carbon fuel and chemicals production, i.e., hybrid energy systems. Nuclear/renewable energy can supply heat and electricity for low-/high-temperature water electrolysis processes to produce clean hydrogen, after which the hydrogen will be mixed with the syngas from the coal gasification unit to adjust the H/C ratio for downstream chemicals synthesis processes (FT-synthesis, methanol to olefin, methanation process, etc.). In this case, the water-gas shift unit can be eliminated and the CO_2 emission significantly reduced.

Chen et al. [56,57] proposed a nuclear-coal hybrid energy system as a potential solution to reduce CO_2 emission from coal conversion. A high-carbon energy such as coal is integrated effectively with a low-carbon energy such as nuclear in a flexible and optimized manner, which is able to generate the chemicals and fuels with low carbon dioxide emissions. The nuclear-coal hybrid energy system is presented in this study for the detailed analysis. In this case, the carbon resource required by the fuel syntheses and chemical production processes is mainly provided by coal while the hydrogen resource is derived from nuclear energy. Such integration can not only lead to a good balance between carbon and hydrogen but also improve both energy and carbon efficiencies. More importantly, a significantly lower CO_2 emission intensity is achieved. A systematic techno-economic model is established, and a scenario analysis is carried out of the hybrid system to assess the economic competitiveness based on the considerations of various types of externalities. It is found that with the rising carbon tax and coal price as well as the decreasing cost of nuclear energy, the hybrid energy system can become more and more economically competitive with the conventional option, which make it a potential viable solution for the future carbon-constrained world.

A comparison can be made between a nuclear-assisted hybrid system and a conventional coal-based system in terms of system economic performance [61]. Referring to the current coal price ($111/ton) and nuclear power generation price ($41.7/MWh), the fuel/chemical costs via hybrid energy systems are 4%–38% higher than those of conventional coal-based systems but will approach the present coal-based chemical production cost when the nuclear electricity price falls to $31.7/MWh. In most of the hybrid energy systems, hydrogen cost accounts for more than 60% of the total fuel/chemicals production cost. For instance, the proportion of renewable hydrogen cost is up to 83% in the nuclear-assisted ammonia production system due to its high demand for hydrogen. However, renewable hydrogen cost accounts for just 32% in a nuclear integrated coal to ethylene glycol system due to the long and complex process consuming large amounts of utilities (power and steam).

5.6.2   WIND/SOLAR-COAL INTEGRATION SYSTEM

Renewable hydrogen cost is also the key component of a wind/solar integrated coal to chemicals system. The current hydrogen production cost from renewable energy (wind/ solar)-assisted water electrolysis technology [59–63] is still 1.8–2.8 times the cost of a coal-based hydrogen production process but will further decline to $0.69/Nm^3 and $1.6/Nm^3, respectively, if considering the lowest wind and solar electricity price level. Referring to the present wind ($70/MWh) and solar ($100/MWh) electricity price, ammonia cost via the hybrid energy system is 2.4 and 3.4 times the cost of conventional coal-based system due to its high hydrogen demand (e.g., renewable hydrogen cost accounts for 88% and 92% of total ammonia cost). Therefore, the more hydrogen is required per unit product, the higher its production cost is compared with the conventional coal-based chemical production system. Even if adopting the lowest price of wind power and solar power, most chemical production costs in the hybrid energy system are still higher than those in the coal-based system except the wind integrated coal to ethylene glycol system, where the production cost is slightly ($7%) lower than in the coal-based system.

In terms of economic feasibility, based on the price fluctuation of fuel chemicals in the last 2 years and minimum acceptable rate of return (12%), the current wind/solar energy to ethylene glycol system can achieve relatively better economic performance due to its high added value. For the wind/solar energy to olefin system to achieve economic feasibility, the price of wind power needs to be lower than $50/MWh while that of solar power needs to be lower than $38/MWh, which is over 24% less than the lowest assumed solar power price level assumed by Chen et al. [56]. For the wind/ solar energy to traditional bulk chemicals (methanol/synthetic ammonia/SNG) system, it is difficult to achieve economic feasibility because of their low added values.

Renewable energy power generation technology has been developing by leaps and bounds in recent years, and accordingly, the wind/solar power generation costs have fallen sharply. Therefore, the wind/solar energy integrated coal to high value-added chemicals system is still promising.

5.6.3   BIOMASS-COAL INTEGRATION SYSTEM

In the major food-producing areas, which are rich in straw and other biomass resources, utilization of biomass energy is an effective way to relieve the pressures of shortage of conventional energy resources and serious environment pollution. High value-added fuel and chemicals production via a biomass gasification process based on $31.7/ton of biomass raw material, the cost of ethylene glycol and methanol production is equivalent to the cost of a coal-based system, and the cost of biomass-based synthetic ammonia/oil/SNG/olefin is 5%–10% lower that of the coal-based system [64–66].

To conclude, the current cost of hybrid systems to produce fuel/chemicals is more expensive than the conventional coal-based systems from an economic point of view. However, a break-even point will appear for some areas, depending on the market price of products. Taking into consideration cheap nuclear electricity price in the long run along with the introduction of a carbon tax, the hybrid system shows the potential to become economically competitive. For the wind/solar energy integration system, it is economically feasible to produce high value-added chemicals (olefin, ethylene glycol) rather than traditional bulk chemicals. With the falling prices of renewable energy due to the rapid development of renewable energy generation technology, the integration system will have a bright future. The economic performance of biofuel and biochemicals appears competitive with the conventional system given the low biomass price [56].

5.6.4 CARBON TAX IMPACT ON THE ECONOMIC COMPETITIVENESS OF HYBRID ENERGY SYSTEM

As one of the most cost-effective tools of emission reduction, carbon tax has attracted considerable attention from international organizations and governments. Most of the Nordic countries began to levy a carbon tax or related energy tax in the 1990s, and Australia began to levy a carbon tax of $23/ton CO_2e on major domestic polluters in 2012. In 2013, China launched its "pilot emission trading scheme" in seven provinces and cities, and the carbon trading price fluctuates between $2 and $20/ton CO_2e [67].

According to the State and Trends of Carbon Pricing Report released by the World Bank, the carbon prices in most nations or regions are still generally low, ranging from $5/ton CO_2e to $15/ton CO_2e. Only in some Nordic countries where carbon taxes are implemented do carbon prices can reach more $40–80/ton in 2020 and $50–100/ton in 2030 [68]. Based on this literature, Chen et al. [56] set carbon price between $5 and $50/ton CO_2e. Taking the methanol production system as an example, to achieve the same production cost as that of the traditional coal to methanol system based on coal price of $79.4/ton, the nuclear power and wind power prices in the hybrid system need to range between $23.4–$47.1/MWh and $17.5–$35.1/MWh, respectively, and the solar power price needs to be lower than 23.3$/MWh. Therefore, when the carbon tax is higher than $26.2/ton CO_2e, the nuclear energy-integrated coal to methanol system will become competitive; when the carbon tax is higher than $11.4/ton CO_2e, the wind energy-integrated coal to methanol system will become competitive. Due to the high price of solar PV power generation, the solar energy-integrated coal to methanol system is at a competitive disadvantage in the short term. The price of solar PV power is, however, rapidly declining [56].

5.6.5 CARBON-NEUTRAL CYCLE VIA CO_2 CAPTURE AND CONVERSION SYSTEM

In the process of using fossil resources for fuel and chemicals production, it is essential to minimize the occurrence of the reaction C/CO_2 via a hybrid energy system. On the other hand, capture and transformation of CO_2 into useful chemicals or fuels will be another major opportunity to achieve a carbon-neutral cycle. The utilization of CO_2 as a feedstock to produce methanol [69,70], gasoline [71,72], and other chemicals has attracted wide attention. Sandia National Laboratories has developed the direct solar thermolysis of H_2O/CO_2 reactor and analyzed the feasibility of liquid hydrocarbon fuel and chemicals production through such technology. Moreover, extensive studies on the process design, techno-economic analysis, and life cycle assessment [73,74] of renewable energy-integrated CO_2 conversion to chemicals systems have been put forward. This indicates that by implementing a renewable hydrogen and CO_2 hydrogenation process for methanol, the life cycle of greenhouse gas emissions alone can be reduced by 86% compared with using conventional petroleum-based fuels. Chen et al. [56] also analyzed energy efficiency and economic feasibility of nuclear- or solar-assisted CO_2 hydrogenation for a methanol production system. The overall energy efficiency of the carbon-neutral process is 21.9%–73% lower than that of the conventional coal to methanol system, but the production cost is 92%–134% higher given the \$35/ton CO_2 capture cost. Thus, the CO_2 and renewable hydrogen conversion to produce a fuel and chemicals system is promising due to its carbon-neutral characteristic. However, there are also many challenges in this field, such as the development of low-temperature and high-efficiency CO_2 hydrogenation process and the application of high performance, low cost, and environment friendly CO_2 capture technologies [75,76]. Furthermore, a huge concern is the lower economic competitiveness due to its more expensive renewable hydrogen consumption relative to the conventional CO hydrogenation process.

Chen et al. [56] concluded that various proposed hybrid systems in their study combining coal with nuclear and renewable energies have significant potentials for reducing CO_2 emission. The systems should, however, be chosen based on geographical advantages. The feasible development of low-temperature and high-efficiency CO_2 hydrogenation process, and the application of high-performance, low-cost, and environment-friendly renewable and nuclear energy resources fit well in most coal-rich countries, especially China and the US. It will also be economically practical as the rapid development of renewable energy technologies will greatly reduce the cost of their power generation. As a result, the hybrid strategy can lead to a great reduction in carbon emission.

5.7 NOVEL HYBRID PROCESSES COMBINING COAL/BIOMASS TO CHEMICALS AND HYDROGEN PRODUCTION

Globally, around a third of total primary energy consumption is for transportation fuels and chemicals. Syngas is a versatile building block for the chemical industry and major uses include ammonia synthesis for fertilizer manufacture, generation of hydrogen for oil refining, and methanol production. Liquid fuels are produced from coal in considerable quantities by Sasol in South Africa. Most current syngas production is from coal-based plants, although in the future, syngas from biomass and biomass/coal combinations is likely to become a key intermediate in the production of some renewable fuels and chemicals.

Other applications for syngas include the hydroformylation (oxo-synthesis) of olefins. Here, olefins are reacted with syngas in the presence of homogeneous catalysts (often Rhodium-based) to form aldehydes and alcohols. The most important oxo-products are in the range C3-C19. These are converted to alcohols, carboxylic acids, aldol condensation products, and primary amines. Other applications of syngas have relatively small markets or the processes involved are still in early stages of development. These include the production of mixed alcohols, CO (for acetic acid and phosgene production), and aromatics [77].

In 2009, a consortium comprising Polish companies ZAK and PKE announced plans to develop such a polygeneration facility (with CCS) at the Kedzierzyn Chemical Plant in Poland. The proposed hybrid complex would enable the simultaneous generation of clean electrical power, heat, and syngas to be used by an adjacent methanol plant, as well as the capture and storage of CO_2 emitted (~2.5 Mt/y) by the production process. A proportion of the CO_2 would be utilized in the manufacture of fertilizers and plastics, as well as chemical raw materials, urea, and hydrogen. It may also be used later for the production of synthetic fuels. It was expected that the combination of coal/biomass gasification technology coupled with CCS and a source of renewable energy will potentially produce negative carbon emissions. It was anticipated that the Kedzierzyn complex will produce 1.55 billion m^3/y of syngas and generate a total of 300 MWe (gross electricity production will be 2.4 TWh/y). Around 137 MW of thermal energy will also be produced as well up to 550 kt/y of methanol. Fuel for the plant will be 2 Mt/y of coal plus 0.25 Mt/y of biomass.

In several parts of the world, concepts are being developed that propose combining several technologies to form effective, cost-effective energy-producing packages. Through such combinations, it may prove possible to capitalize on the respective individual strengths of the individual technologies. Some propose combining the cogasification of coal and biomass with the electrical output from intermittent renewable energy sources (predominantly wind power). Off-peak or excess electricity that would otherwise be wasted would be used to generate hydrogen and oxygen via the electrolysis of water. The resultant oxygen could be fed to either a gasifier, an oxyfuel combustion power plant, or directed to some other industrial application. Similarly, hydrogen could be stored and used in a variety of ways. Some projects have focused on gasifying combinations of coal and biomass, whereas others have concentrated on using coal or biomass alone; with both, woody biomass has frequently been selected as the biomass of choice. A number of promising concepts are briefly described below.

5.7.1 NREL HYBRID CONCEPTS (USA) (GASIFICATION/
COGASIFICATION + ELECTROLYSIS)

NREL examined a range of concepts and issues associated with the gasification of biomass (and potentially, cogasification) applications and also several that propose to combine these with electricity from renewable energy sources. A number of US studies have suggested using electrolyzers to produce hydrogen and oxygen from wind-generated electricity—for instance, the NREL/Xcel Energy wind-to-hydrogen (Wind2H2) demonstration project in Colorado. Biomass gasification or cogasification coupled with wind power/electrolysis has the potential to convert unwanted electricity into useful energy and could provide dispatchable electricity for local utilities.

Both directly and indirectly heated gasification-based hybrid systems were examined. Both required a source of oxygen, currently produced using conventional ASUs. Under certain circumstances, oxygen from electrolysis could form a viable alternative, with the added benefit of producing a pure hydrogen stream. Initial research suggests that from both technical and economic perspectives, such hybrid options hold promise [78]. However, in the case of pressurized gasification plants, replacing an ASU with an alternative system would remove the source of nitrogen used for pressurized feeding, necessitating another source of inert pressurization.

The indirect system proposed featured a two-stage fluidized-bed process where the heat needed for reaction was produced by burning char in a separate chamber to heat sand. This was then circulated through the reaction chamber to drive the reaction kinetics. A directly heated gasifier typically had a single combustion/reaction chamber and burns a small portion of the fuel feed to create heat. It was suggested that such hybrid systems would allow switching between fuel production and electricity production, based on grid demand. This would be accomplished by routing some or all of the syngas from the gasifier to a gas turbine instead of to fuel-production reactors. In addition to power production, the use of surplus electricity to heat the gasifier was investigated. Excess or low-value electricity generated during periods of low demand could be used to heat the gasifier reaction chamber. As the temperature of the gasifier increased, the proportions of syngas and char change. Adding heat energy would produce additional syngas, increasing plant efficiency. To optimize economics, the ideal plant would continuously adjust both feed use and fuel production.

Gasification and pyrolysis plants usually also require an external source of power for operation, and the use of renewable sources has been proposed. However, how to achieve the successful direct coupling of the process with intermittent sources such as wind power is rarely addressed. Although the commercial use of wind power is now well established in the USA, as elsewhere, the major drawbacks are often location and intermittent output; at times, this can be zero and at other times, surplus electricity may be produced. Since none of the current electricity storage systems are yet available for utility-scale operation, the use of surplus electricty for electrolysis plants is very attractive.

Key factors affecting the viability of producing oxygen and hydrogen via electrolysis are the value of the end products and the price of electricity supplied to the electrolyzers. The proposed hybrid systems would run associated electrolyzers intermittently. Thus, surplus or low-value electricity would be used to produce oxygen (and hydrogen) for use by the gasifier or stored for later use. During periods of peak electricity demand, rather than feeding the electrolyzers, the stored oxygen could be used to produce syngas in the gasifier. However, replacement of a single ASU supplying oxygen would require multiple electrolyzers. It is estimated that a 2,000 t/d fluidized bed gasifier would require about 27,800 kg/h of oxygen. Based on the technology currently available, this would require around 160 electrolyzers running at full capacity. Modern wind turbines use power electronics that permit variable speed constant frequency operation; this improves annual energy production, limits drive-train torque, and supplies quality power to the electric utility grid. NREL studies suggest that if the wind turbine was supplying only an electrolyzer load, deleting the grid-quality requirement would significantly reduce the capital and operation and maintenance costs of the turbine's electrical generating system, as well as reduce process energy losses.

With the aim of investigating hybrid hydrogen production systems, other NREL studies have examined how to combine gasification with renewable energy sources. These have included concepts for combining wind power and biomass gasification, combined electrolysis and biomass gasification, and combined coal and biomass cogasification with CCS. Each focused on the coproduction of electricity and fuel. Thus, various gasification pathways for hydrogen production and how they could be hybridized to support renewable electricity generation were considered. Several concepts were identified with the potential to increase efficiency and reliability, and reduce the cost of hydrogen production, or improve the sustainability of hydrogen production from nonrenewable resources. Those with the highest potential were direct wind and wind-electrolyzer combinations coupled with biomass gasification (indirectly heated gasifier) producing both electricity and hydrogen. An electrolyzer (replacing the ASU) coupled with a directly heated fluidized-bed biomass gasifier was selected for the coproduction of fuel and power. Both systems proposed to overcome wind intermittency by feeding stored hydrogen to a gas turbine (used for peaking duties), and/or absorbing excess renewable power (in the form of hydrogen) during periods of low demand.

It was determined that the direct gasification concept was unlikely to be cost-competitive in the near future, largely because of high electrolyzer costs. However, the various systems examined do not require significant technology breakthroughs and could become cost-competitive in the near term [78]. The work may be expanded to cover pathways such as hybrid coal gasification with CCS as well as the production of other fuels such as DME (Dimethyl Ether) and FT liquids.

5.7.2   CRL ENERGY, NEW ZEALAND (COAL/BIOMASS COGASIFICATION + ELECTROLYSIS)

In New Zealand, CRL Energy is developing a technology package that aims to combine oxygen-blown fluidized bed cogasification of indigenous lignites/subbituminous coals and woody biomass with intermittent renewable energy technologies (primarily wind, but possibly wave power later). The country has plentiful intermittent renewable energy resources and weather conditions and a long coastline well suited to generating electricity from wind. Conditions are favorable for much greater deployment of wind power in the country. The current total installed capacity is 622 MW; this generates 4%–5% of the country's total electricity. However, forecasts suggest that by 2030, there will have been a six-fold increase in wind generated electricity. Numerous new wind projects are proposed or in development [79].

Cogasification of indigenous coals and renewable biomass would provide an opportunity for effectively converting the latter into fuel gases using fluidized bed gasification technology. Advantageously, New Zealand lignites are very reactive— chars are generally very reactive and undergo conversion at low temperatures [80,81]. Furthermore, New Zealand lignites usually have ash fusion temperatures in excess of

1,100°C, hence should not cause problems of bed agglomeration. Potentially, syngas from cogasification can be used in several ways that include the production, via FT technology, of low carbon footprint chemicals and synfuels.

It is proposed that cogasification would be integrated with the high-efficiency electrolysis of water to produce hydrogen and oxygen; the latter would be fed to the gasifier. Hydrogen could be used to enrich the product gas or employed as a means for storing excess off-peak renewable electricity. It also has the potential for direct use as a transport fuel or for electricity generation using fuel cells. CRL's initial focus has been on the development of an air-blown gasification process. The gasifier and its associated syngas clean-up line have operated successfully for more than 2,000 hours. Steady plant operation (with air and steam) has generated good quality syngas using suitably sized indigenous coals alone or combined with 30% woody biomass. Steam injection slightly increases the hydrogen concentration of the syngas.

Particulates are removed from the syngas in two stages using a high efficiency cyclone (~95% capture) followed by a venturi scrubber that quenches the syngas and removes any remaining particulates, tars, and condensables. Only trace quantities of the latter two have been detected, indicating that a high-level volatile breakdown is achieved in the gasifier freeboard section. Syngas can be further cleaned using a hypochlorite and caustic soda wash, reducing sulfur content to very low levels. If the syngas is destined for hydrogen or electricity production, it could be water-gas-shifted. Alternatively, it could be fed to a separation plant to extract further hydrogen to add to the electrolysis stream for direct use as a fuel, or directed to a FT process. However, in order for this integrated cofueled coproduction process to be viable, considerable quantities of low-cost electricity must be available. There must also be a suitable market for the hydrogen produced. As the country's intermittent wind power capacity increases in the future, the CRL process may become suitable for load balancing. The plant would consume electricity when prices were low. When they were high, hydrogen from the process could be used to fire turbines to generate electricity.

When water-gas-shifted (WGS), syngas composition from the air-blown gasifier is typically around 20.8% hydrogen, 6.6% CO, 20.1% CO_2, 1.9% CH_4, and 51.2% nitrogen. The large quantity of nitrogen present in the syngas highlights a major advantage of gasifying with oxygen rather than air—by using oxygen, nitrogen can be virtually eliminated. Consequently, CRL is now focusing on oxygen-blown fluidized bed gasification. A newly developed gasifier operates at ambient pressure, has a maximum temperature of 1,150°C, and can accommodate a variety of feedstocks that include different coals, alone, or with high proportions of woody biomass. Furthermore, the unit can be operated with pure oxygen, air, or combinations of the two. The H_2:CO ratio from the gasifier (without WGS promotion) is generally 1.2:1. In order to upgrade it to 2:1, an additional 0.8 mole of H_2 is required per mole of CO.

In order to avoid the expense and energy penalty associated with a cryogenic ASU, CRL plans to supply the oxygen required for gasification using a modular electrolysis plant designed by Industrial Research Limited of New Zealand. This is being developed as part of a program to investigate production technologies suitable for cost-effective production of hydrogen from small-scale wind and other renewable resources [82]. The concept of combining cogasification with renewable-powered electrolysis offers the prospect of using the country's coal, biomass and intermittent renewable resources for near carbon neutral production of hydrogen, synthetic hydrocarbons and biofuels [80,81]. It offers a real possibility for building on the respective strengths of each resource.

5.7.3 OTHER HYBRID PROJECTS FOR CHEMICALS AND HYDROGEN

In the USA, the Leighty Foundation and partners have investigated the cogasification of coal and biomass combined with electrolysis. This examined a large-scale, long-distance transmission system for hydrogen produced by electrolysis using wind-generated electricity. Oxygen produced from the process would be supplied to an oxygen-blown gasifier, either gasifying coal alone or cogasifying coal and biomass. As it is uneconomic for oxygen to be piped over large distances, it was determined that the electrolyzers should be in close proximity to the gasification plant.

Advanced Alternative Energy Corporation (AAEC) of Lawrence, Kansas, has conceptualized an energy system that offers to combine cleaner coal-derived power with that produced from municipal wastes and urban and agricultural biomass [83]. The AAEC system would call up such solid fuels to back up wind when insufficient wind was available. The system has not yet been constructed, and as the process is still in the process of patenting, full details have not yet been released. In 1993, a patent was secured for the biomass portion of the system termed the Sequential Grates System; further R&D has since been carried out. In operation, electricity would be generated primarily through wind turbines until the wind began to drop, at which time system operators would dispatch the biomass burning furnaces to make up the shortfall. If biomass supply became inadequate, this would be backed up with coal. Thus, the concept would comprise a plant that could be fired on coal, biomass, or municipal solid waste, all of which would be transformed into clean electric power or potentially, various types of biofuels. Overall system efficiency could be high. The technology could have application in both developing and developed nations. It is claimed to be clean, highly efficient, low cost, modular, scalable, and expandable, and provide waste disposal and energy efficiency [84]. Similar concepts are also implemented in Switzerland, Italy, and Denmark. However, in their studies, only biomass gasification was used.

In China, cogasification is accompanied by wind-generated electricity to be used to electrolyze water, generating oxygen, and hydrogen. Oxygen is employed as gasification agent for the gasification of coal to produce syngas, with hydrogen used to adjust the proportion of carbon and hydrogen in the desulfurized syngas. The suitably adjusted syngas is then used for the production of methanol [85]. Other Chinese studies have examined the possibility of producing methanol, SNG, and ethylene glycol by combining coal gasification with wind power and electrolysis. In each case, the system proposes the use of wind power to provide electricity to an electrolyzer. The oxygen generated is fed to the coal gasifier and the hydrogen mixed with the syngas produced to adjust the H_2:CO to an appropriate ratio suitable for methanol synthesis. Use of an electrolyzer avoids the requirement of an ASU for oxygen production and reduces the WGS process. The combination of these technologies is claimed to reduce raw material requirements and reduce associated CO_2 emissions [86–88]. These systems could be particularly attractive in the regions of China with abundant wind and coal resources.

REFERENCES

1. Shah, Y.T. (2016). *Energy and Fuel Systems Integration*. CRC Press, New York.
2. Shah Y.T. (2019). *Modular Systems for Energy and Fuel Recovery and Conversion*. CRC Press, New York.
3. Shah Y.T. (2020). *Modular Systems for Energy Usage Management*. CRC Press, New York.
4. Shah Y.T. (2021). *Hybrid Power, Generation, Storage and Grids*. CRC Press, New York.
5. Shah Y.T. (2017). *Chemical Energy from Natural and Synthetic Gas*. CRC Press, New York.
6. Renewable Energy, Hybrid Solutions Can Power the Mining of Tomorrow. A website report 10 October 2017.
7. Mills, S. (2017, September). Combining solar power with coal-fired power plants, or cofiring natural gas, IEA Clean Coal Centre London SW15 2SH CCC/279.www.iea-coal.org.
8. Mills, S. (2018, June). Combining solar power with coal-fired power plants, or cofiring natural gas. *Clean Energy* 2(1): 1–9, Doi: 10.1093/ce/zky004.
9. Gossard, S. (2015, June 18). Coal-to-gas plant conversions in the U.S. *Power Engineering* 119(6): 3 p.
10. Bryan, B., Nester, S., Rabovitser, J. and Wohadlo, S (2005, December). Methane de-NOX for utility boilers. Final Report: Reporting period: January 11, 2000–October 31, 2005. Cooperative agreement: DE-FC26-00NT40752. Submitting organization: Gas Technology Institute, Des Plaines, IL, 117 p.
11. Reinhart, B., Shah, A., Dittus, M., Nowling, U. and Slettehaugh, R. (2012, December 12) A case study on coal to natural gas fuel switch. *Power-Gen International 2012*, Orlando, FL, 6 p.
12. Parent, R. and Czarniecki, P.E. (2016). Orlando utilities commission ignites shift to fuel diversity. *[Forney] Second Quarter 2016 Newsletter – Combined Cycle Focus*. Forney Corporation, Addison, TX, 2 p. (2016).
13. Breen Energy Solutions. (2014). Coal/natural gas cofiring solutions. Breen Energy Solutions, Carnegie, PA, 4 p. Available at: http://breenes.com/wp-content/uploads/2014/10/Coal-and-Natural-Gas-CoFiring-Solutions1.pdf.
14. Liss, W.E. (2016). Using natural gas to improve air quality. *Advances in Natural Gas Utilization and Production Workshop*, 31 March–1 April 2016, Yibin, 23 p.
15. Kho, J. (2012, May 7). Solar with fossil fuels: Partner or competitor? *PV Magazine* 5: 3 p.
16. Appleyard, D. (2015, December 9). CSP hybrids: Optimizing renewable steam production. *Power Engineering International* 23(12): 3 p.
17. Siros, F. (2014). Hybridization of thermal plants is a great driver to increase the CSP share in the global energy mix. *IEA Technology Roadmap, 1st Workshop*, 3–4 February 2014, International Energy Agency, Paris, 4 p.
18. Zhu, Q. (2015, March). *High-Efficiency Power Generation – Review of Alternative Systems*. CCC/247, IEA Clean Coal Centre, London, 120 p.
19. EPRI. (2010, April 30). Solar augmented steam cycles for coal plants: Conceptual design study. Product ID: 1018648. Electric Power Research Institute, Palo Alto, CA, 2 p.

20. IT Power. (2012, May). CSP can contribute significantly to Australia's energy needs. Realising the potential for concentrating solar power in Australia. Produced for The Australian Solar Institute (ASI). IT Power (Australia) Pty Ltd, Canberra, 273 p.

21. Rajpaul, V. (2014, April 8). Concentrating solar power in Eskom. *CSP Today*: 2014, 13 p.

22. Roos, T. (2015). Solar thermal augmentation of coal-fired power stations. Council for Scientific and Industrial Research (CSIR), South Africa, 18 p. Available at: http://www.fossilfuel.co.za/conferences/2015/Independent-Power-Generation-in-SA/Day-2/Session-3/01Thomas-Roos.pdf.

23. Stancich, R. (2010, November 29). Solar power group: Cleaning, not greening, conventional energy. *New Energy Update*. FCBI Energy Ltd., London, 2 p.

24. IEA GHG. (2012, March). Integration of solar energy technologies with CCS; a preliminary study. Report 2012/TRI. IEA GHG R&D Programme, Cheltenham, 40 p.

25. Mills, S.J. (2011). *Integrating Intermittent Renewable Energy Technologies with Coal-Fired Power Plant*. CCC/189, IEA Clean Coal Centre, London.

26. Mills, S. (2013). *Combining Renewable Energy with Coal*. CCC/223, ISBN 978-92-9029-543-3 September 2013 © IEA Clean Coal Centre, IEA, London.

27. Mills, S.J. (2010, November). *Prospects for Coal, CCTs and CCS in the European Union*. CCC/173, IEA Clean Coal Centre, London, 77 p.

28. Smolker, R. and Ernsting, A. (2012, October 5). BECCS (Bioenergy with Carbon Capture and Storage): Climate saviour or dangerous hype? *Biofuelwatch Journal* 25 p., a website report by Biofuelwatch (2012).

29. Pathway for readying the next generation of affordable clean energy technology — Carbon Capture, Utilization, and Storage (CCUS) (2012). *Technology Readiness Coal Assessment Research Program —Analysis of Active Research Portfolio*. US DOE, Department of Fossil Energy, Washington, DC.

30. Mills, S. (2012). *Coal-Fired CCS Demonstration Plants*. CCC/207, ISBN: 978-92-9029-527-3, IEA Clean Coal Center, 11 p, 1 October 2012.

31. IEA GHG/Ecofys. (2011, August). New Ecofys study for the IEA Greenhouse Gas R&D Programme. *Large Global Potential for Negative CO_2 Emissions through Biomass Linked with Carbon Dioxide Capture and Storage*. IEA GHG R&D Programme, Cheltenham. Available at: http://www.ecofys.com/en/press/new-ecofys-study-for-the-iea-greenhouse-gas-rd-programme/ www.ieaghg.org

32. ZEP. (2012). *Biomass with CO_2 capture and Storage (Bio-CCS). The way forward for Europe*. Available at: http://bellona.org/ccs/uploads/tx_weccontentelements/filedownload/EBTP__ZEP_Report_Bio-CCS_The_Way_Forward.pdf, European Technology Platform for Zero Emission Fossil Fuel Power Plants, 32 p., a website report by ZEP (2012).

33. Fernando, R. (2009, November). *Cogasification and Indirect Cofiring of Coal and Biomass*. CCC/158, IEA Clean Coal Centre, London, 37 p.

34. Brar, J.S., Singh, K., Wang, J. and Kumar, S. (2012). Cogasification of coal and biomass: A review. *International Journal of Forestry Research* 2012 (2012): 10 p, Article 363058.

35. NETL. (2010, January). Development of secure and clean energy technologies: Cogasification. US DOE, National Energy Technology Laboratory, Reaction Chemistry & Engineering Group, Morgantown, WV, 16 p. Available at: http://www.netl.doe.gov/publications/proceedings/10/gfe/Dirk%20Link_TourDay2.pdf.

36. Long, H.A. and Wang, T. (2011). Case studies for biomass/coal cogasification in IGCC applications. *ASME Turbo Expo 2011*, 6–10 June 2001, Vancouver, 15 p.

37. NETL. (2010). Cogasification—researchers combine coal and biomass. *Netlog* 10: 2–3. (July 2008)

38. Vreugdenhil, B.J. (2009). Co gasification of biomass and lignite. *International Pittsburgh Coal Conference*, 20–23 September 2009, Pittsburgh, PA, 18 p.
39. Bi, J. (2005). Cogasification of biomass and coal in fluidised bed. *The 2nd Biomass-Asia Workshop*, 13–15 December 2005, Bangkok, 25 p.
40. Laumb, J. (2011). Biomass gasification in entrained-flow systems. *Renewable Energy Council Meeting*, Bismarck, ND, 24 March 2011, 16 p.
41. Carbo, M., Kalivodova, J., Cieplik, M., van der Drift, B., Zwart, R. and Kiel, J. (2012). Entrained flow gasification of coal/torrefied woody biomass blends. Efficient carbon footprint reduction. *5th International Freiberg Conference on IGCL & XtL Technologies*, Leipzig, 21 May 2012, 23 p.
42. EERE. (2009, January). Hydrogen production. Overview of technology options. US Department of Energy, The FreedomCAR and Fuel Partnership, Washington, DC, 16 p. Available at: http://www1.eere.energy.gov/hydrogenandfuelcells/pdfs/h2_tech_road-map.pdf.
43. Weiland, N., Means, N. and Morreale, B. (2010). Kinetics of coal/biomass cogasification. *NETL 2010 Workshop on Multiphase Flow Science*, 4–6 May 2010, Pittsburgh, PA, 28 p.
44. Hanssen, J.E. and Hagen, E.F. (2006, June). Prospects for hydrogen from biomass. IEA Hydrogen Implementing Agreement. Annex 16. Subtask B. Final Report, 70 p. Available at: http://ieahia.org/pdfs/finalreports/Task16BFinal.pdf.
45. Mills, S.J. (2006, June). *Coal Gasification and IGCC in Europe*. CCC/113, IEA Clean Coal Centre, London, 37 p.
46. Henderson, C. (2008, December). *Future Developments in IGCC*. CCC/143, IEA Clean Coal Centre, London, 45 p.
47. Cormos, C.-C. (2012). Hydrogen and power co-generation based on coal and biomass/solid wastes cogasification with carbon capture and storage. *International Journal of Hydrogen Energy* 37(2012): 5637–5648.
48. van der Drift, B. (2010). Biomass gasification for second generation biofuels. *Gasification, the Development of Gasification as a Key Technology Contributor to Future Clean Coal Power Generation Conference*, 19–20 April, London, 18 p.
49. van der Drift, A., Boerrigter, H., Coda, B., Cieplik, M.K. and Hemmes, K. (2004, April). Entrained flow gasification of biomass. Ash behaviour, feeding issues, and system analyses. Report ECN-C-04–039, Energy Research Centre of the Netherlands (ECN), Petten, 58 p.
50. Tarka, T.J. (2012, May). Production of zero sulfur diesel fuel from domestic coal: Configurational options to reduce environmental impact. NETL/DOE-2012/1542, National Energy Technology Laboratory, Pittsburgh, PA, 94 p.
51. Wang, J. and McNeel, J. (2009, 10 July). *Assessments of Coal/Biomass to Liquid Fuels in West Virginia*. Division of Forestry and Natural Resources, Biomaterials and Wood Utilization Research Center, Morgantown, WV, 56 p.
52. Fermoso, J., Arias, B., Gil, M.V., Plaza, M.G., Pevida, C., Pis, J.J. and Rubiera, F. (2010). Cogasification of different rank coals with biomass and petroleum coke in a high-pressure reactor for H(2)-rich gas production. *Bioresource Technology* 101(9): 3230–3235.
53. Akgun, F., Caglayan, E., Durak, Y., Gul, O., Olgun, H., Sarioglan, A. and Unlu, N. (2009, 12–13 November). Liquid fuel production from biomass and coal blends. Cogasification of coal, biomass and waste. *Flexgas International Workshop*, CIEMAT, Madrid, 43 p.
54. Ziypak, M. (2011). Development of gasification and activities in Turkish Coal Enterprises. *28th Annual International Pittsburgh Coal Conference*, 12–15 September 2011, Pittsburgh, PA, 58 p.

55. Baliban, R.C., Elia, J.A. and Floudas, C.A. (2010). Toward novel hybrid biomass, coal, and natural gas processes for satisfying current transportation fuel demands, 1: Process alternatives, gasification modeling, process simulation, and economic analysis *Industrial & Engineering Chemistry Research* 49(16): 7343–7370, Doi: 10.1021/ie100063y.
56. Chen, Q., Lv, M., Gu, Y., Yang, X., Tang, Z., Sun, Y. and Jiang, M. (2018, April 18). Hybrid energy system for a coal-based chemical industry. *Joule* 2: 607–620.
57. Sanz-Bermejo, J., Munˉ oz-Antoˊ n, J., Gonzalez-Aguilar, J. and Romero, M. (2014). Optimal integration of a solid-oxide electrolyser cell into a direct steam generation solar power plant for zero-emission hydrogen production. *Applied Energy* 131: 238–247.
58. Herring, J.S., O'Brien, J.E., Stoots, C.M., Hawkes, G.L., Hartvigsen, J.J. and Shahnam, M. (2007). Progress in high-temperature electrolysis for hydrogen production using planar SOFC technology. *International Journal of Hydrogen Energy* 32: 440–450.
59. Abbasi, T. and Abbasi, S.A. (2011). 'Renewable' hydrogen: Prospects and challenges. *Renewable and Sustainable Energy Reviews* 15: 3034–3040.
60. Menanteau, P., Queˊ meˊ reˊ, M.M., Duigou, A.L. and Bastard, S.L. (2011). An economic analysis of the production of hydrogen from wind-generated electricity for use in transport applications. *Energy Policy* 39: 2957–2965.
61. Chen, Q., Tang, Z., Lei, Y., Sun, Y. and Jiang, M. (2015). Feasibility analysis of nuclear-coal hybrid energy systems from the perspective of low-carbon development. *Applied Energy* 158: 619–630, Doi: 10.1016/j.apenergy.2015.
62. Khan, M.J. and Iqbal, M.T. (2009). Analysis of a small wind-hydrogen stand-alone hybrid energy system. *Applied Energy* 86: 2429–2442.
63. Olateju, B., Monds, J. and Kumar, A. (2014). Large scale hydrogen production from wind energy for the upgrading of bitumen from oil sands. *Applied Energy* 118: 48–56.
64. Tijmensen, M.J.A., Faaij, A.P.C., Hamelinck, C.N. and Hardeveld, M.R.M.V. (2002). Exploration of the possibilities for production of Fischer Tropsch liquids and power via biomass gasification. *Biomass Bioenergy* 23: 129–152.
65. Hamelinck, C.N., Faaij, A.P.C., Uil, H.D. and Boerrigter, H. (2004). Production of FT transportation fuels from biomass; technical options, process analysis and optimisation, and development potential. *Energy* 29: 1743–1771.
66. Sansaniwal, S.K., Rosen, M.A. and Tyagi, S.K. (2017). Global challenges in the sustainable development of biomass gasification: An overview. *Renewable and Sustainable Energy Reviews* 80: 23–43.
67. Dong, H., Dai, H., Geng, Y., Fujita, T., Liu, Z., Xie, Y., Wu, R., Fujii, M., Masui, T. and Tang, L. (2017). Exploring impact of carbon tax on China's CO_2 reductions and provincial disparities. *Renewable and Sustainable Energy Reviews* 77: 596–603.
68. Zechter, R., Kerr, T., Kossoy, A., Peszko, G., Oppermann, K. and Ramstein, C. (2016). *State and Trends of Carbon Pricing*. World Bank Group, 140. Washington, DC.
69. Goeppert, A., Olah, G.A. and Surya Prakash, G.K. (2018). Chapter 3.26. Toward a sustainable carbon cycle: The methanol economy. In: B. Toˉ roˉ k and T. Dransfield, (eds) *Green Chemistry*, Elsevier, 919–962. Netherland.
70. Arena, F., Mezzatesta, G., Zafarana, G., Trunfio, G., Frusteri, F. and Spadaro, L. (2013). Effects of oxide carriers on surface functionality and process performance of the Cu-ZnO system in the synthesis of methanol via CO_2 hydrogenation. *Journal of Catalysis* 300: 141–151.
71. Wei, J., Ge, Q., Yao, R., Wen, Z., Fang, C., Guo, L., Xu, H. and Sun, J. (2017). Directly converting CO_2 into a gasoline fuel. *Nature Communications* 8: 15174.
72. Gao, P., Li, S., Bu, X. et al., (2017). Direct conversion of CO_2 into liquid fuels with high selectivity over a bifunctional catalyst. *Nature Chemistry* 9: 1019–1024.

73. Matzen, M. and Demirel, Y. (2016). Methanol and dimethyl ether from renewable hydrogen and carbon dioxide: Alternative fuels production and life-cycle assessment. *Journal of Cleaner Production* 139: 1068–1077.

74. Martín, M. and Grossmann, I.E. (2017). Towards zero CO_2 emissions in the production of methanol from switchgrass. CO_2 to methanol. *Computers & Chemical Engineering* 105: 308–316.

75. Koytsoumpa, E.I., Bergins, C. and Kakaras, E. (2018). The CO_2 economy: Review of CO_2 capture and reuse technologies. *Journal of Supercritical Fluids* 132: 3–16.

76. Song, C., Liu, Q., Ji, N., Deng, S., Zhao, J., Li, Y., Song, Y. and Li, H. (2018). Alternative pathways for efficient CO_2 capture by hybrid processes—a review. *Renewable and Sustainable Energy Reviews* 82: 215–231.

77. Boerrigter, H. and Rauch, R. (2005). Review of applications of gases from biomass gasification ECN-RX-06-066. In: Knoef, H.A.M. (ed), *Handbook of Biomass Gasification*, Chapter 10, Biomass Technology Group, ECN, Enschede, 33 p.

78. Dean, J., Braun, R., Munoz, D., Penev, M. and Kinchin, C. (2010, January). Analysis of hybrid hydrogen systems final report, Technical Report NREL/TP-560-46934, NREL, Golden, CO.

79. Wind power in New Zealand. (2020). Wikipedia, The free encyclopedia, last visited 8 September 2020.

80. Levi, T. (2011). Fuel production from coal and biomass in New Zealand. *Advanced Biofuels Research Network, Science Symposium*, 28–29 November 2011, Wellington, 29 p.

81. Levi, T.P., Gardiner, A.I., Whitney, R., Iwasaki, Y., Pang, S., Xu, Q. and Clemens, A. (2010). Coal: Biomass gasification – a pathway for new technology development of oxygen blown cofired gasification with integrated electrolysis. *27th Annual International Pittsburgh Coal Conference*, 11–14 October, Istanbul.

82. Whitney, R.S., Levi, T.P. and Gardiner A.I. (2011). A technology package utilising coal, biomass and intermittent renewable electricity. *IEA Clean Coal Centre Fifth International Conference on Clean Coal Technologies*, 9–12 May 2011, Zaragoza, 11 p.

83. Blevins, L. (2013, August 8). Advanced Alternative Energy Corporation, personal communication, Lawrence, KS.

84. AAEC. (2013, August). Advanced Alternative Energy Corporation. Lawrence, KS, Available at: http://www.aaecorp.com/.

85. Weidong, G. and Weidou, N. (2011). Method for preparing methanol by directly using wind power of non-grid-connection on a large scale. Chinese patent. Publication dates 27 May 2009–30 November 2011. Assigned IPC Subclass: C07C.

86. Gu, W. and Yan Z. (2009). Research on the wind/coal multi-energy system. *World Non-Grid-Connected Wind Power and Energy Conference—WNWEC 2009*, 24–26 September 2009, Nanjing, 5 p.

87. Gu, W. and Yan Z. (2010). Research on integrated system of large-scale non-grid-connected wind power and coal-to-SNG. *World Non-Grid-Connected Wind Power and Energy Conference—WNWEC 2010*, 5–7 November 2010, Nanjing, 4 p.

88. Ni, W., Chen, Z., Gu, W. and Yan Z. (2009). Green integrated system: Non-grid-connected wind power and coal-to-methanol. *World Non-Grid-Connected Wind Power and Energy Conference 2009*, 24–26 September, Nanjing, 4 p.

89. Nowling, U. (2016). *Utilities explore dual fuel as low-cost option to full gas conversion.* Available from: http://bv.com/Home/news/solutions/energy/utilities-explore-dual-fuel-as-low-cost-option. Kansas, USA, Black & Veatch, 2 pp (2016).

90. Lovegrove, K. and McDonald, J. (2012). *Background paper on solar conversion of brown coal.* A0102, Australia, O'Connor, IT Power (Australia) Pty Limited, 66 pp (Dec 2012).

6 Hybrid Energy Systems for Nuclear Industry

6.1 INTRODUCTION

The U.S. energy system, like many others, is evolving to better meet environmental constraints and at the same time continuing to provide secure, reliable, and affordable energy services to the economy. In practice, this has led to an increased interest in producing low-carbon electricity in the power sector and utilizing domestically sourced alternatives to imported petroleum in the transportation sector. The use of fossil fuels for heating and cooling needs is also reduced by improving energy efficiency and replacing fossil fuels with renewable sources through hybrid energy systems [1–9,10]. As shown in this chapter, nuclear energy can also play an important role toward this decarbonization effort. The development of small modular nuclear reactors further helps improvement of thermal efficiency of nuclear power plants.

In the electric power industry, significant capacity additions of variable renewable energy systems such as wind and photovoltaic (PV) power are likely to continue. These changes, although beneficial in terms of greenhouse gas (GHG) emission reductions and improved fuel diversity, in some cases have led to a need for additional operating reserves and other ancillary services [11]. Continued integration of these variable renewable resources drives the need for flexible generation to accommodate fluctuations in supply and demand. Such load-following flexible facilities typically are used as intermediate or peaking plants utilized for a relatively small number of hours during times of high net demand (Net demand, as defined herein, is the output the grid requires from an individual generator to make supply and demand equal in the generator's balancing area). High net demand can occur when demand for electricity is high and/or variable production is low and vice versa. Thus, under the current paradigm, a large amount of capital equipment (and, ultimately, investment capital) is not being utilized near its capacity when demand is lacking. In many cases, the equipment could be out of use during the majority of the year.

At the same time, energy use by industrial processes (e.g., major chemical manufacturing and minerals conversion industries) is large in scale and diverse in the proportions and types of energy services required. Overall, approximately 40% of the energy used in these industries is provided by fossil-fired heaters, 43% by steam systems, and 17% from electrical inputs, though each industry differs [12]. Additionally, changes—including advanced informatics, energy management systems, and forecasting—are enabling new innovation in integrated plant design [13,14] and power system operations [15]. These innovations can be utilized to design new types of hybrid energy systems which (a) allow the use of traditionally base-load systems to generate economical load-following power, (b) improve grid flexibility (grid flexibility is the ability of an electric system's conventional generators to vary output and respond to the variability and uncertainty of the net load [16]) and allow for multiple types of ancillary services, (c) produce additional commodities such as fuels for the transportation sector and chemicals production, and (d) improve energy conversion efficiency.

A strategy to achieve these goals is hybrid operation of nuclear reactors coupled with renewable energy technologies and industrial processes in a single facility which has the potential to provide secure, reliable, and affordable, low-carbon energy services. In this chapter, we examine various facets of this strategy in detail and apply nuclear-renewable hybrid energy systems (N-R HESs) for some industrial processes. As outlined by Suman [8], the N-R HES can offer several advantages:

1. GHG emission can be reduced and the of global warming can be mitigated.
2. Helps in making renewable energy highly competitive.
3. Renewable energy can have high grid penetration by upgrading the grid infrastructure to provide grid-scale energy storage and dispatch.
4. Advanced integration via smart control and heat management technologies will allow increased energy conversion efficiency.
5. Economical, highly reliable supply of electricity in addition of other supplementary services to the grid.
6. Hybrid energy systems are capable of producing biofuel, synfuel, or hydrogen.
7. Overcomes the unease and psychological fear associated with nuclear power.

Against these benefits, there are challenges such as economics and financial issues, safety and security concerns, scale and commercialization, materials input requirements, intermittency of renewable sources and land foot print, waste disposal, and other ecological issues such as water requirements, research needs for compatibility and technological barriers and public perception that need to be resolved. Recent developments of small modular reactors (SMRs) help mitigating some of these challenges.

6.2   DIVERSITY OF HYBRID ENERGY SYSTEMS

As indicated in my previous two books [17,18] and in Chapter 1, the term hybrid energy system is used to describe various concepts. As an example, a long history of work exists on small, decentralized hybrid energy systems which utilize multiple generation sources, often with storage, to provide electricity to remote populations. As pointed out by Ruth et al. [4], this includes concept proposals [19], analyses of technical challenges and opportunities [20], feasibility studies [21–23], and cost-benefit analyses [24]. Single-energy-source-centralized generation facilities that provide multiple services (e.g., electricity, heating, cooling, water) also have been referred to as hybrid energy systems, and research has been exploring those concepts from various standpoints [25]. Cogeneration (or combined heat and power (CHP)) optimizing both design and output in accordance with technical constraints and market signals [26] also can be termed as hybrid energy system. Cogeneration may include, as indicated in my previous book, conversion of waste heat to additional power by thermoelectricity or CO_2 to additional power by fuel cell. Use of cofuels like coal and biomass is also considered as an integrated or hybrid energy system. The degree of hybridization, however, can significantly vary in these cases.

To a lesser extent, larger, hybridized electric generation facilities which use fossil fuels in combination with renewables are also reported. Kang et al. [27] developed a generalized computational framework to determine optimal operation procedures of an integrated system consisting of a coal-fired power station, a temperature-swing absorption carbon-capture facility powered by a natural gas combustion turbine, and a wind farm. Kieffer et al. [28] proposed using the term flex-fuel poly-generation systems for multifeed, multiproduct energy systems. Researchers developed techniques to measure and optimize cost, sustainability, and resilience of such systems. Phadkee et al. [29] performed an economic and technical feasibility assessment on a system consisting of a coal gasification combined-cycle power plant equipped with carbon capture, a wind plant, and the option for a fuel production or hydrogen/carbon monoxide gas mixture (referred to as syngas) storage facility. An integrated system which includes a syngas storage facility or fuel production plant increased utilization of capital and reduced the levelized cost of electricity. Cherry et al. [30] corroborated the results of Phadkee in an evaluation of the technical and economic benefits of hybrid systems that integrate chemical and fuels synthesis plants with wind power to help ameliorate wind power intermittency. As pointed out by Ruth et al. [4], work has also been done to assess the feasibility and added utility of a hybrid energy system in which solar-generated steam is injected into fossil power cycles, such as that described by Turchi and Ma [31]. Analysis indicated that a hybridized design of a gas turbine with a concentrated solar power (CSP) system could produce electricity more efficiently and dispatchably than either system could produce alone [32].

Some researchers have extended the definition of hybrid energy systems to include systems with components coupled across the electrical grid joined by hybrid or integrated processes. Forsberg [33] explores several integrated system solutions for the larger U.S. energy system, which include combining nuclear, fossil, and renewable energy sources to sustainably create electricity and transportation fuels. Cherry et al. [30] concluded that the amount of excess capacity in the power-generation systems could be cost-competitively converted into chemicals and fuels, thus replacing one-third or more of all foreign oil foreign oil imports into the United States. Accounting for electrical and thermal energy management, Garcia et al. [34] modeled and predicted the ability to load-follow increasing amounts of variable energy integration on the grid in association with battery storage and chemical production (methanol) [34,35].

As defined in my previous books, we use the term "hybrid energy systems" here in its broadest interpretation. We define hybrid energy system as a single facility which takes two or more energy resources as inputs and produces two or more products, with at least one being an energy commodity such as electricity or transportation fuel with or without storage. These systems comprise two or more energy-conversion subsystems that are traditionally separate or isolated. In hybrid systems, they are physically coupled to produce outputs by dynamically integrating energy and materials flows among energy production and delivery systems. While some hybrid systems are fully integrated, others are loosely coupled. As pointed out by several Idaho National Laboratory/National Renewable Energy Laboratory (INL/NREL) reports [2–5], the components of a hybrid energy system can be organized and interconnected in a number of different ways. Some hybrid energy system requires coupling "behind" the electrical transmission bus, where all subsystems within the hybrid energy system share the same interconnection so that the grid is exposed to a single, highly dynamic, and responsive system. Several coupling opportunities exist to link the energy conversion subsystems. As pointed out by Ruth et al. [4], Dixit [7], and Suman [8], these include thermal, electricity, chemical, hydrogen, and mechanical and information interconnections. Additionally, data transfer between subsystems is essential to a functioning hybrid energy system. While functionally different, all of these systems are a part of hybrid energy systems.

An important feature of this type of umbrella system is that it produces several products. These products can include electricity, hydrogen, substitute natural gas, hot process gases or steam, and transportation fuels for merchant or captive use. Additionally, this type of system might be designed to produce many nonenergy outputs whose production is energy-intensive, such as chemical feedstocks for fertilizer, polymers, plastics, and textiles; hydrogen from syngas, potable water from desalination of seawater and brines; minerals from geothermal brines; and CO_2 for enhanced oil recovery or as a heat-transport medium. The inclusion of such products allows a broader range of operations and products to maximize overall system performance and profitability.

Ruth et al. [2–5] in their excellent set of reports point out that because the hybrid energy systems operate dynamically, large nominally base-load power plants could be used flexibly within hybrid systems to produce the electricity necessary to mitigate times of either high demand or low production from variable sources. Increased efficiency is achieved by matching electrical output to demand while utilizing excess generation capacity for other purposes when it is available. Hybrid energy systems provide an alternative approach to systems optimization, leveraging previously untapped attributes of energy production, increasing efficiency, and offering a greater return on investment [2–5].

6.3   NUCLEAR-RENEWABLE HYBRID ENERGY SYSTEMS

Integrating nuclear energy and renewable energy into a single hybrid energy system, coupled through informatics linkages, is one way to enable the nuclear plant to run at high capacity while simultaneously addressing the need for flexibility of generation rates and producing energy services, ancillary services, and low-carbon coproducts. Because of their large capital costs and low fuel costs, nuclear power plants require a high load or capacity factor to be economically viable (i.e., they need to be run as many hours annually as possible) [36]. Transient reactor operation can also increase costs of nuclear facility operation by accelerating wear on various nuclear system components [37–39]. As pointed out by Ruth et al. [2–5], a future hybrid nuclear–renewable energy facility, incorporating an appropriate industrial process, presents opportunities to produce revenue from a variety of product streams and avoid capital inefficiencies of underutilized capacity. Such coupling allows the system to respond to market signals, diverting electricity to spot or ancillary service markets or internal industrial processes. It relies upon advanced informatics and market data to choose which activity is most profitable. This type of operation has become feasible as a result of several factors: increased demand for low-CO_2 energy generation; the advent of a new generation of small modular nuclear reactors having energy output similar to the needs of a large chemical process complex; and increased data generation, collection, and utilization capabilities to make continuous online operations optimization possible. Furthermore, such a system might reduce the integration costs (integration costs are the costs associated with providing reliability necessary to accommodate variable renewable electricity (VRE) sources onto the grid) of VRE sources by providing firming power to such sources within the hybrid system and additional grid flexibility to sources outside the system. Although fully integrated hybrid systems such as these have not yet been demonstrated, the component technologies are mature. It is expected that nuclear-renewable hybrid systems proposed by Ruth et al. [2–5] and others [7,8] will continue to operate component technologies similarly to how they have been operated independently. Thus, key technical issues will involve nature of interconnections and additional system issues due to the added complexity of integration.

Numerous renewable energy sources can be used as inputs to an N-R HES: wind, solar, hydroelectric, biomass (such as forest and agricultural residues as well as purpose-grown energy crops and algae), geothermal, and marine technologies (wave or tidal). These are just as varied and numerous as the inputs currently used for existing single input, single output plants, and each resource has different geographic, economic, and environmental considerations, making some far more practical than others. The different energy resources, possible products, and coupling modes allow for many combinations. Depending upon various factors such as geographical, economical, desired form of output, etc., different renewable energy sources may be coupled to form an N-R HES; Table 6.1 mentions few of the possible integrated systems [4,8,40,41].

There are more design and scale options available for hybrid systems than ever before because recent developments in the nuclear industry have resulted in options for smaller reactors. The advent of SMRs allows for nuclear reactors capacities as low as 10 MW while maintaining favorable economics. Renewable facilities are scalable and range from very small to large capacities. Many wind farms are over 100 MW and at least

---

TABLE 6.1
Examples of Nuclear and Renewables in a Hybrid Energy System [4,8]

Resources	Coupling Mode	Storage Mode	Products
Nuclear and wind energy	Electrical	Hydrogen	Electricity, hydrogen
Nuclear and biomass	Thermal	Chemical	Electricity, biofuels
Nuclear and CSP	Thermal	Thermal	Electricity, heat
Nuclear, wind energy, and natural gas	Electrical and thermal	Chemical	Electricity, chemical products (e.g., ethylene), diesel fuel
Nuclear, geothermal	Thermal	Thermal	Electricity. Heat

one U.S. wind farm over 1 GW capacity and many solar power stations are over 100 MW with the largest over 250 MW nominal capacity [42]. Renewable resources—such as electricity produced by wind turbines and PVs—are characterized by variability of generation. High penetration of those sources requires a flexible grid, and consequently, other generators such as nuclear–renewable hybrid energy systems that can provide outputs at the rates necessary to meet demand. SMR also allows the proximity of nuclear reactor to industry which is often required to transfer heat without significant loss.

There is a growing body of literature on the economics and business cases for N-R HESs. Cherry [43] analyzed the technical and economic performance of an N-R HES that produces methanol from natural gas. Methanol can be used as a fuel or precursor for other fuels using heat from a nuclear facility during nonpeak hours for electricity. The resultant cost of methanol from a hybrid facility was 10% higher than a conventional, nonhybrid facility; however, cost externalities such as reducing GHG emissions, utilizing resources more efficiently to extend their lifetimes, and producing vehicle fuels domestically were not included in his estimate. In further analysis, Cherry [30] assessed the potential for the state of Wyoming to upgrade coal and wind resources to obtain higher values using N-R HESs. One particular system design uses electricity from a nuclear facility to balance variability of wind-generated electricity. Excess heat from the nuclear facility is used as an energy input for production of gasoline from coal resources (via methanol). That study found that the coal-gasoline process is competitive with conventional methanol-to-gasoline processes, earning a 12% IRR (Internal Rate of Return) with a gasoline wholesale price of $2.13/gallon.

Garcia [34,35] explored the feasibility of an N-R HES that uses the nuclear facility and storage to balance variability of wind-generated electricity. Excess heat from the nuclear facility is used in a chemical plant complex. The focus was specifically on dynamic response and its value instead of time-averaged output as used for other economic analyses. The proposed hybrid energy systems become more profitable than conventional configurations at 20% wind-electricity penetrations on the grid with greater profitability at higher penetrations. As pointed out by Ruth et al. [4], at 40% wind-electricity penetration the additional return is 4% higher than the conventional configuration. Bragg-Sitton [44] presents the technical and economic value associated with the hybridization of SMR architectures that are dispatched in concert with wind energy produced by the system. Conventional systems are compared to integrated hybrid energy systems producing hydrogen (via high temperature electrolysis) and methanol in addition to electricity. This study found that, when wind energy penetration exceeds about 30% of the total electrical power generation, the internal rate of return of the integrated hybrid energy systems is higher than that of the conventional systems.

Nuclear-solar hybrid systems have been examined in numerous studies. Keller [40] uses a nuclear reactor to heat a compressed working fluid that is expanded within turbines rotating compressors of both reactor plant and air pressurizing plant. Heat exchangers are used to extract low-grade heat from the working fluid and transfer to the moisture removal equipment located in downstream side of air compressing plant. In addition, intercooler heat exchangers are used to further cool the working fluid prior to entry into the compressors, thereby reducing compressor power needs.

A regenerative heat exchanger is used to preheat the working fluid before its reentry into the reactor by extracting low-grade heat from the working fluid discharged from the turbines. In another few studies [45,46], a new type of hybrid nuclear-solar power plant having small modular nuclear reactor and concentrated solar-thermal plant is analyzed. In this proposed hybrid energy system, the solar heat is transferred to nuclear steam to raise its temperature. Continuous superheating is provided through thermal energy storage. The results from design point calculations show that solar superheating has the potential to increase nuclear plant electric efficiency significantly, pushing it to around 37.5%. Bartev et al. [47] proposed a solar-nuclear hybrid power plant (US Patent US20150096299A1). The disclosure relates to a solar-nuclear hybrid power system that combines a solar energy loop with a nuclear energy loop. The solar energy loop can attain higher temperatures compared to the nuclear energy loop and transfer that heat energy to steam. This solar-nuclear hybrid/cogeneration plant can have improved operation and efficiency, compared to a nuclear plant. Method of operating the hybrid plant with the nuclear power generation and solar power generation system operating in tandem are also disclosed. Heat transfer fluids/media are used to facilitate the integration of the solar and nuclear power generation systems. Many more disclosures are included in the patent.

In developing N-R HESs, there are several challenges for both renewable energy and nuclear energy that need to be overcome [8]. For renewable energy, issues such as scalability, commercialization, and timeline are important. Scalability refers to the capability of a system or source to accommodate surge in production or number of users without a penalty in cost, performance, reliability, or functionality. Material input requirements for renewable energy are also important. Renewable energy sources are not only about high initial capital investment but also resources and materials for the sustained growth of its infrastructure. Nondispatchable renewable sources like solar and wind are intermittent and require large land footprint. Because of their intermittent nature, their compatibility with nuclear energy and the grid and resulting technological barriers are also challenging. Finally, ecological impacts of renewable systems need to be properly considered. For nuclear energy, economic and financial issues can be important for developing countries. The use of SMR, however, requires lower cost and provides more flexibility. The safety and security of nuclear reactors is always an issue. To this end, public perception and social acceptance are very important. Once again SMR lessens the risks and improves perception. The disposal of nuclear waste is also important. Again, SMR reduces this requirement. Finally, more advanced fourth generation nuclear reactors used in the hybrid system still require some additional research [8].

6.4 NATURE OF INTERACTIONS IN COMPONENTS OF N-RES HYBRID ENERGY SYSTEMS

Ruth et al. [4], Dixit [7] and Suman [8] point out that there are numerous ways components of N-R HES can be coupled to produce desired energy products. Although the conversion from energy resource to energy product can be done using many different processes, the forms of energy are limited and therefore provide a convenient way to structure the discussion. Each section identifies coupling interconnections, discusses opportunities for their use, and proposes work necessary before the interconnection can be used for N-R HESs.

It is expected that proposed nuclear-renewable hybrid systems will continue to operate component technologies similarly to how they have been operated independently. Thus, key technical issues will involve interconnections and few additional system issues due to the complexity of integration [4,8]. Thermal interconnections are at the heart of nuclear-renewable energy system in order to efficiently use the generated heat during low demand of electricity. Thermal energy storage reservoirs may be needed to soften the imposition of rapid transitions on the thermal hydraulics systems. Additionally, new remote flow monitoring system, control valves and pumps for high-temperature thermal hydraulic fluids, gases, molten salts, liquid metals, and ultrasupercritical steam will be needed. High-temperature heat circulation in aggressive environments associated with high-temperature steam, molten salts, and gases requires validation of metallurgy and possibly new heat-exchanger fabrication techniques [48]. Research and development (R&D) is also required to make hybrid energy systems respond rapidly, efficiently, and safely to electricity market signals. Successful electricity interaction will need development of advanced, interconnected sensing, and informatics systems which identify all the needs and provide information to the control systems, thus enabling control systems to optimize profitability. In addition, advanced power electronics are necessary. They need to be of low cost, highly responsive, durable, and are able to switch between multiple uses without disruption to operations [4]. Production of hydrogen is also crucial for future energy systems since it has many possible uses. High-temperature electrolysis (HTE) might provide a better interface for combining nuclear with renewables in a hybrid energy system because wind and solar PV systems generate electricity directly. This allows for the possibility of utilizing heat from a nuclear reactor and most or all of the electricity from renewable sources. This is further illustrated in case study outlined in Section 6.7 [1] and by Orhan et al. [49]. Few other studies [50,51] have also been reported on nuclear energy-based integrated system for hydrogen production and their thermodynamic modeling has shown good efficiency. The details on several options for interactions are outlined by Ruth et al. [4], Dixit [7] and Suman [8]. The following discussion on various types of interactions largely summarizes the excellent report of Ruth et al. [4].

6.4.1   TIGHTLY COUPLED N-R HES FOR POWER AND HEAT

In this architecture, nuclear and renewable generation sources and industrial processes would all be linked and cocontrolled behind the electricity bus, such that there would only be a single connection to the grid, as shown in Figure 6.1. The closely coupled system would likely be managed by a single financial entity to optimize profitability for the integrated system.

6.4.2   THERMAL INTERCONNECTIONS OF COMPONENTS
OF N-RES HYBRID ENERGY SYSTEMS

This architecture would thermally integrate subsystems and tightly couple them to the industrial processes, but the nuclear and renewable electrical subsystems could have more than one connection to the same grid balancing area and would not need to be

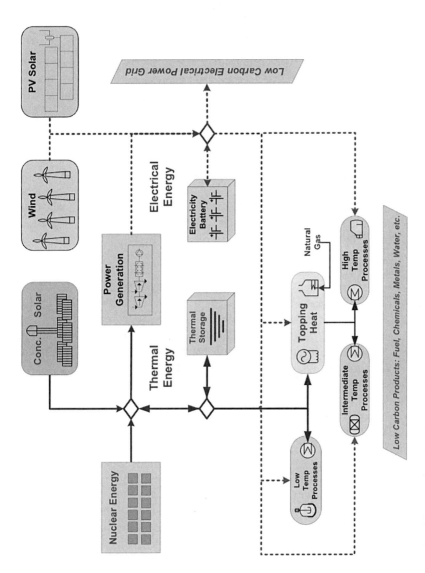

FIGURE 6.1 General architecture for a tightly coupled N-R HES, where the generation sources are integrated behind a single connection point to the grid and are managed by a single financial entity [52]. (Credit: Idaho National Laboratory.)

co-located, but would be cocontrolled to provide energy and ancillary services to the grid. The thermally integrated subsystems would need to meet industrial process requirements considering the required heat quality, the heat losses to the environment along the heat delivery system, and the required exclusion zone around the nuclear plant. These systems would likely be managed by a single financial entity (see Figure 6.2).

As pointed out by Ruth et al. [4], Suman [8] and Dixit [7] a key motive for N-R HESs is the efficient alternative use of the heat generated when it is not needed for electric power production due to low net demand conditions. Heat from nuclear reactors is a key focus point; however, renewable sources such as solar energy in CSP systems, biomass, and geothermal have the similar issues and opportunities [53]. These concepts also would apply to coal-fired power plants and natural gas combined-cycle plants [54]. Technologies that enable heat utilization in an industrial process—instead of reducing reactor output or releasing the energy through cooling—can create new revenue streams.

Shared use of nuclear reactor thermal energy is not a new concept. Nuclear heat is currently used for combined power generation and district heating in Europe [55,56]. The proposed load-following behavior in a system that incorporates a greater percentage of variable power generation, however, requires systems that are more complex than district heating. This is due to timing (when the heat is available), time scales (required response rate), and the large amount of excess heat. Industrial processes potentially can be designed to absorb the heat at time scales more closely aligned with heat availability.

The range of dynamic apportionment between power production and process-oriented heat use must be considered for selection and design of the nuclear subsystem. This includes analyses of heat versus electrical tariff structures, the range of electrical versus heat demands of the industrial process, and the ramping limits of the nuclear system. It is likely that small- and medium-sized reactors will have several technical or economic advantages in different markets/installations, and modular designs also could allow phased expansion of the hybrid system, or hedge against times of contraction. Operational optimization techniques for industrial CHP [57,58] can be transferrable to nuclear-renewable hybrid systems, but techniques to optimize designs need to be developed.

Many industrial processes requiring large thermal inputs could be well-suited for coupling in a nuclear-renewable hybrid system. For example, steam in a power cycle could be diverted to energy storage or an industrial user prior to the final condensing turbine. As outlined by Ruth et al. [4] Table 6.2 classifies temperature ranges corresponding to industrial process reaction mechanisms and identifies potential heat sources.

Suman [8] and Dixit [7] point out that many other industrial processes could utilize heat from a nuclear reactor. In the realm of petroleum production, lower/intermediate temperature heat is widely used in hot water extraction [59,60] and steam-assisted gravity drainage [61] processes for heavy crude and oil sand bitumen extraction (300°C–350°C steam). Also, petroleum refineries use 300°C–500°C steam to refine crude oil into asphalt, fuels, and distillate products. This steam often is generated by combustion of residual coke and vacuum bottom residuals, which typically contain high levels of sulfur (up to 10% by weight) and metals, and can result in toxic air emissions [62]. In the future, the organic kerogen in oil shale can be depolymerized and converted to crude oil and combustible gases by thermally retorting the shale over the temperature range of 350°C–500°C [63,64].

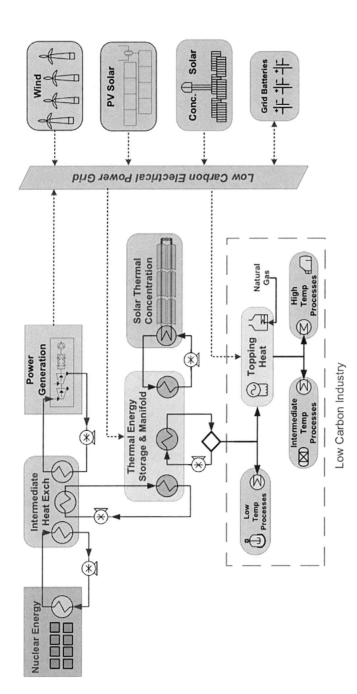

FIGURE 6.2 General architecture for a thermally coupled N-R HES, where the nuclear and renewable generation sources are cocontrolled and managed by a single financial entity but may not be colocated [52]. (Credit: Idaho National Laboratory.)

TABLE 6.2

Heat Sources and Applications Organized by Operating Temperature Range [4]

Temp. Range	Mechanism	Examples	Potential Heat Sources
High (1,000°C –1,500°C)	Metal refining; heterogeneous gas-solid reactions; high temperature gas phase reactions	Coal gasification; steel; cement and glass manufacturing; steam superheating	Combustion of natural gas or coal; electric arc; high-temperature plasma generation; concentrating solar power
Higher/intermediate (700°C–950°C)	Multi-bond scissioning; hydrogen abstraction reactions	Steam methane reforming, cracking of natural gas liquids to ethylene and propylene; biomass gasification; high temperature steam electrolysis	High-temperature gas-cooled nuclear reactor
Lower/intermediate (350°C–600°C)	Devolatilization endothermic reactions, organic compound pyrolysis	Distillation, cracking, and reforming of petroleum heavy end products; biomass pyrolysis; in-situ oil shale retorting	Molten salt reactor; liquid metal cooled nuclear reactor; biomass combustion
Low (50°C–320°C)	Saturated steam production; sensible heating	Many chemical processes; biomass torrefaction; water desalination; district Heating	Light water nuclear reactors; geothermal sources

Information drawn from Refs. [37,43–46].

Biofuel production can use process heat for a variety of purposes. Low-temperature heat can be used for purposes including feedstock drying and thermal torrefaction [65]. Temperatures in the range of 350°C–500°C [66,67] are considered ideal for biomass decomposition and pyrolysis, which converts biomass into a bio-crude that, after hydrotreating, is fungible with petroleum crude [68]. Integration opportunities could exist to store excess energy in the form of biofuels and to combust these fuels in times of high demand. Higher temperature chemical production processes which typically utilize fossil-fired process heat, such as natural gas reforming [69] (850°C), biomass gasification [70] (800°C–1,000°C), and coal gasification [71] (1,000°C–1,200°C), also could be coupled with nuclear-renewable hybrid systems. Heat recuperation, topping combustion, and alternative heat generation by electrically powered plasma generators or induction heating effectively can achieve higher temperatures of these processes. As pointed out by Ruth et al. [4], Suman [8] and Dixit [7] this will necessitate process reactor designs or plant layouts different from those currently used with conventional fossil-fired process heating.

A large number of different designs for small (10–300 MWe) and medium (< 700 MWe) nuclear reactors are in development around the world; many are still at the conceptual stage. These reactors are more versatile than traditional large reactors of 1,000–1,700 MWe (roughly 3,000–5,000 MW thermal) that were designed to capture economies of scale operating as base-load plants. The thermal demand of a large chemical plant is in the range of a few hundred megawatts, so one or even a few small- or medium-sized reactors can be matched to industrial-scale process plants to make a single operating complex. The seven U.S. designs under development in 2011 are representative and are summarized in Table 6.3.

All nuclear systems deliver their heat to a primary coolant (e.g., water, a molten metal mixture, helium) flowing in a closed loop. The primary coolant transfers heat to a secondary coolant through an intermediate heat exchanger isolating the power block and chemical process from each other and mitigating the potential of radioactive contamination entering the chemical process [73]. In a hybrid energy system, the secondary coolant then can then be dynamically apportioned between the power generation block, a thermal energy storage buffer, and an industrial process. One example of this is shown in Figure 6.2.

Ruth et al. [4], Suman [8] and Dixit [7] point out that the selection of the secondary heat transfer medium depends on various factors, including the outlet temperature of the primary coolant, the power generation cycle of choice, and the nature of the thermal coupling with the industrial process. If high-temperature heat is available, combinations of Brayton and Rankine cycles can be considered to attain electrical generation efficiencies approaching 50% [74,75]. The possibility of dual external thermal hydraulic loops that independently serve the power generation block and the industrial process could be considered as well.

Thermal interconnections with the dynamics and scale necessary for nuclear-renewable hybrid systems require R&D before they can be implemented. That research could start with optimization of the hybrid configuration using both static process models and dynamic system predictive models to understand system response to variable and uncertain grid demand and on-site variable energy production. Multiphysics transient behavior modeling can evaluate nuclear reactor choices and help establish reactor design and operating requirements that are driven by technical needs as well as probabilistic risk assessment during reactor licensing. Technical and economic evaluations should consider the costs and benefits of design alternatives for the nuclear reactor, external thermal heat transfer loops, the power block, and the energy storage buffer.

TABLE 6.3
Thermal Characteristics of U.S. SMRs under Development [4,72]

Reactor Class/Name	Manufacturer	Max Heat Delivery Temperature (°C)	Thermal Capacity (MW)
Light water reactors			
NuScale	NuScale Power Inc.	≈300	165
Westinghouse SMR	Westinghouse	310	800
mPower	Babcock and Wilcox	320	500
Liquid metal-cooled reactors			
PRISM	GE-Hitachi	485	471
Hyperion power module	Gen4 Energy, Inc.	500	70
Gas-cooled reactors			
GT-MHR	General Atomics	750	350
Energy multiplier module	General Atomics	850	500

The heat-to-power conversion system for an N-R HES should incorporate turbine technology and power generation blocks that are designed to respond to rapid shifts in power demand. Power turbines and electrical power generators that vary power output in accordance with dynamic demand cycles have already been developed for nuclear power plants [76,77]. Alternatively, smaller, parallel turbines and generation sets that can be independently ramped could be developed and operated such that they perform load-following. This concept embraces the philosophy of some small-modular reactors that match individual reactors modules with small power turbine units.

Thermal energy storage reservoirs might be needed to soften the imposition of rapid transitions on the thermal hydraulics systems. Energy storage technologies being developed for CSP are likely applicable to nuclear hybrid energy systems. However, storage required for a nuclear reactor could be hundreds of megawatt-hours, which is far greater than the thermal storage reservoirs currently under development for CSP [78,79]; hence, new systems would have to be developed to that scale.

R&D can help make effective use of relatively low-temperature heat from light water nuclear reactors in the petrochemical and industrial manufacturing processes. The R&D could focus on efficient methods to boost the temperature of steam or hot gases from 200°C–300°C to 500°C–900°C. Vapor compression or electrical heating systems could be effective for this purpose. Topping heat also can be provided by the process user through heat recuperation, electrical heating, or fossil-fuel combustion. Another option is chemical heat pumps. In most designs, temperature boosting should be done close to the point of use to avoid high-temperature heat transfer materials costs and significant heat loss. Additionally, new remote flow monitoring and control valves and pumps for high-temperature thermal hydraulic fluids, gases, molten salts, liquid metals, and ultra-supercritical steam will be needed. High-temperature heat circulation for aggressive environments associated with high-temperature steam, molten salts, and gases requires validation of metallurgy and possibly new heat-exchanger fabrication techniques.

6.4.3 Electricity Interconnections of Components of N-RES Hybrid Energy Systems

This configuration would be electrically coupled to industrial energy users but there would be no direct thermal coupling of subsystems. This design would allow management of the electricity produced within the system (e.g., from the nuclear plant or from renewable electricity generation) prior to the grid connection. Although there would not be a direct coupling of thermal energy to the industrial processes, the system could include electrical to thermal energy conversion equipment to provide thermal energy input to the industrial process(es). Such an option may allow for potential retrofit of existing generation facilities with fewer regulatory challenges. These systems could have more than one connection point to the grid but would likely be managed by one financial entity (see Figure 6.3). Molten salt reservoirs such as those currently being used to store concentrated solar energy, or a mass of firebrick similar to heat recuperators used by the steel manufacturing industry, may provide thermal energy storage for the heat that can be generated from electricity. In principle, electrical-to-thermal energy conversion would be economical when the cost of producing heat by these systems drops below the cost of producing heat

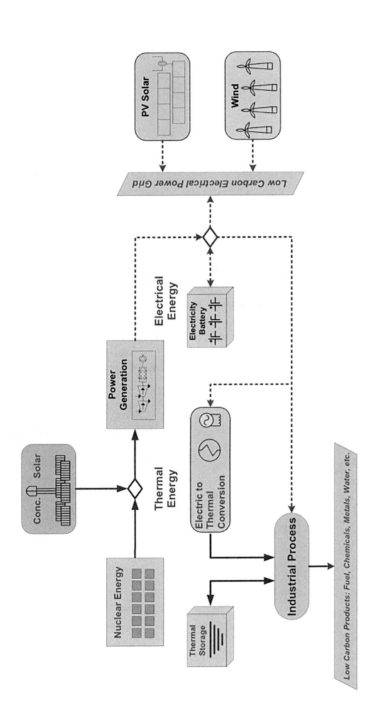

FIGURE 6.3 General architecture for a loosely coupled (electricity-only) N-R HES, where the generation sources are only electrically connected to the industrial process. Note that electrical to thermal energy conversion systems may be included to provide thermal energy to some processes [52]. (Credit: Idaho National Laboratory.)

from traditional combustion-fired process heaters. The type and quality of heat must match the industrial heat user technical specifications.

As pointed out by Ruth et al. [4], Suman [8] and Dixit [7] electricity is a key interconnection because it is both valuable on an existing market and useful internally. Thus, while designing and operating a hybrid energy system, one must consider both internal uses for electricity and its dynamic market value. With sufficient operational flexibility, facility operators can respond to market signals and choose whether to utilize electricity internally or divert energy to the market. In times of high net demand, ancillary services or spot-market electricity prices can become elevated, incentivizing a diversion of energy from an on-site industrial process toward maximizing electrical output. In times of low net demand, electricity prices are reduced and a hybrid energy system can be producing excess electricity. To increase revenue, it might become more profitable for the operator to divert energy from nuclear and variable renewable energy production to industrial processes.

Use of electricity internally will not only be dependent upon momentary market demands but also ramp rate requirements of equipment using the electricity; thus, adjusting the electricity use within the hybrid system will require both short- and long-term planning. In addition, excess electricity in hybrid systems also could be stored using batteries, pumped hydropower, or compressed air energy technologies and dispensed when it is economically attractive to do so. System costs and round-trip efficiency affect the economics of storage options.

Ruth et al. [4], Suman [8] and Dixit [7] point out that R&D is necessary to allow hybrid systems to respond rapidly, efficiently, and safely to electricity market signals. This includes development of advanced, interconnected sensing, and informatics systems that identify all needs and provide information to the control system, thus enabling control systems to optimize profitability. In addition, advanced power electronics are necessary. They need to be low cost, highly responsive, durable, and able to switch between multiple uses without disruption to operations [80].

There are unique concerns for financing a hybrid nuclear-renewable system using electrical interconnections. Nuclear power plants are a significant capital investment and historically have required the long-term certainty of a return on investment to attract capital [81,82]. This certainty can be granted by a public utility commission representing a large group of ratepayers, which determines an appropriate retail electricity rate and provides a guaranteed market for the plant operator. Current legal and regulatory frameworks do not have established methodologies to valuate—on behalf of ratepayers—a system that (a) in addition to electricity produces an industrial product not sold to ratepayers, (b) transfers production dynamically to maximize profits, (c) cannot accurately predict its long-term operation schedule, and (d) requires purchase and construction of components/systems shared by multiple processes—not all of which provide a service to ratepayers. Instead of the long-term certainty resulting from a ratemaking process, plant operators might have to agree to power-purchase agreements which could have shorter contract lengths or less profitable terms. This could make the cost of capital more expensive to operators or discourage investment. Analysis of market redesign solutions could help mitigate project financing issues. Cochran et al. [15] argue that electric markets in their current form do not always assign appropriate value to plants which provide grid flexibility. Establishing market mechanisms which properly reward the flexibility that a power plant provides to the grid might reduce barriers to entry of an N-R HES.

6.4.4 CHEMICAL INTERCONNECTIONS OF COMPONENTS OF N-RES HYBRID ENERGY SYSTEMS

Figure 6.4 shows a generalized, tightly coupled N-R HES. N-R HESs comprise a nuclear reactor, at least one renewable energy source, power conversion system, energy and thermal storage, and a chemical process using both electricity and thermal energy for making numerous types of chemical products. The detailed components can vary with the situation. The N-R HES hierarchical controller controls all aspects of the system within the boundary.

Recognition of the central role for chemical intermediates can expand the role of hybrid energy systems in the chemical industry [4,7,8]. Nuclear plants can be designed to generate heat to produce chemical products such as syngas, high purity hydrogen, and other key chemicals that can then be transported to industrial processes. Syngas is produced by reforming natural gas with steam [70], or by gasifying coal or biomass and separating gas diluents and impurities to produce a clean mixture of hydrogen and carbon monoxide [72]. Both processes require thermal energy that is produced by burning up to 65% of the carbonaceous feedstock, resulting in carbon dioxide emissions. Nuclear reactors can supply both the process heat and steam necessary to carry out these reactions. High-temperature, gas-cooled nuclear reactors can provide superheated helium that could replace the burners in the steam reforming process [83]. The steam produced by a light water reactor can also significantly reduce combustion requirements with changes to the reforming process [30,43].

Syngas and hydrogen can be used as fuel for a gas turbine to produce power in the manner of an integrated-gasification/combined-cycle facility. Alternatively, syngas and hydrogen can be converted into commodity chemicals and products, fertilizer, and synthetic fuels. Methanol is a noteworthy chemical intermediate that is used to produce key chemicals such as formaldehyde, acetic acid, ethylene glycol, vinyl acetate, and olefins [84]. Methanol also can be converted into a fungible motor gasoline substitute by the methanol-to-gasoline process [85] or to produce biofuels via trans-esterification of fatty oils [86].

Fischer-Tropsch catalysis and refining process to produce diesel is a second route to producing synthetic fuels in hybrid energy systems. This technology has been advanced by several energy companies and catalyst companies [87,88]. Various unit operations in the product upgrading and refining section of the plant can utilize the steam or heat provided by a nuclear reactor connection [89]. Some case studies for the high-temperature, gas-cooled reactor have been recently completed for steady-state, cogeneration operations in which favorable return on investment cases for synthetic fuels and chemical production was demonstrated [65,90]. Multiple design and technology improvements are necessary before implementing hybrid energy systems with chemical interconnections. Because chemical manufacturing plants are generally designed to run at nearly constant operation, new design schemes that are resilient to time-varying electrical and thermal inputs are needed. Designs might require dual steam sources (e.g., a nuclear facility and a supporting natural gas-fired boiler). As pointed out by Ruth et al. [4], heat integration issues might also require new reactor designs that improve heat transfer into the chemical reactor vessels. Heat recuperation schemes could require modification, resulting in new designs of

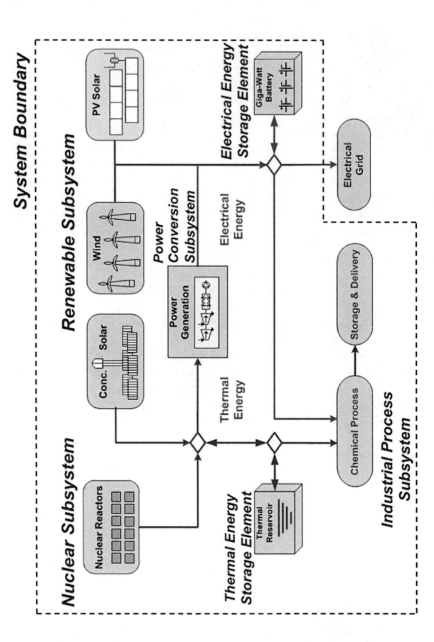

FIGURE 6.4 Generalized N-R HES showing the system boundary and linkage to the grid [3].

heat exchangers and gas production equipment. Induction heating for electrically induced plasma arc heating of gaseous inputs and reacting flows could to be considered. Materials qualification will likely be necessary for these new chemical reactors. In-plant power generation also could be considered in hybrid process plants. New micro-turbines or reciprocating internal combustion engines can burn syngas. Fuel cells that burn hydrogen, methane, or vaporized methanol also are becoming commercially viable [89]. Auxiliary power production using stored chemical energy could help smooth transitions in nuclear and wind-power conversions.

6.4.5 HYDROGEN INTERCONNECTIONS OF COMPONENTS OF N-RES HYBRID ENERGY SYSTEMS

Ruth et al. [4] points out that hydrogen is a special case of chemical coupling of energy systems. Much work has been done to develop hydrogen production technologies, and hydrogen is a common feedstock in chemical and industrial processes. Hydrogen also offers an important possibility for storing energy. Although it is currently produced primarily from natural gas via steam methane reforming [91], hydrogen created using nuclear reactor heat (and electricity) has been researched intently in recent years (see, e.g., National Academies Press [92]; Herring et al. [74]; Forsberg [93]). Hydrogen can be produced in an N-R HES by three primary means: thermo-chemical (T-C) cycles, electrolytic processes, and gasification or pyrolysis of biomass or coal-biomass mixtures. Recent progress in high temperature electrolysis for hydrogen production using planar SOFC technology is reviewed by Herring et al. [149].

T-C cycles produce hydrogen through a series of chemical reactions that extract hydrogen and oxygen from water, requiring heat at temperatures of between $750°C$ and $1,000°C$ [94,95]. The heat can be provided by employing high-temperature nuclear reactors, CSP technology, or the combination of a lower temperature nuclear reactor with CSP. R&D needs vary by the T-C cycle. Lower temperature sources might be used instead if temperature-boosting technology in the desired temperature range is developed. Development priorities for the hybrid sulfur process include techniques to control the gas dynamics of sulfur dioxide (SO_2) and improvements to the design of the acid decomposer. For the sulfur-iodine process, priorities include a better understanding of the fundamental kinetics, catalyst development, and improved product/by-product separation operations [96].

Electrolysis is a means by which electrical power is used to separate water into its constituents, hydrogen and oxygen. Of particular interest is HTE, which increases the electrical efficiency of hydrogen production by running at temperatures ranging from $100°C$ to $850°C$, consuming heat in addition to electricity [97]. HTE might provide a better interface for combining nuclear with renewables in a hybrid energy system because wind and solar PV systems generate electricity directly. This allows for the possibility of utilizing heat from a nuclear reactor and most or all of the electricity from renewable sources. Work on dynamic options could supplement current work on steady-state HTE [96].

As shown in Chapter 11, hydrogen can also be produced by using nuclear-renewable heat for gasification and pyrolysis of biomass or coal-biomass mixtures. More details on this are given in Chapter 11.

Hydrogen has many possible end uses. It can be sold as merchant hydrogen for applications including transportation fuel and as a feedstock for other industrial chemical processes. Annually, more than 50 million metric tons of hydrogen are used globally, with most used for petroleum refining, ammonia production, and methanol production. That quantity is expected to grow in the near term as the market share of heavier crude oils, such as those from oil sands, increases. Refining sulfur-rich heavy crude oils requires more desulfurization, a process which requires hydrogen [91]. Hydrogen is also used for upgrading some biofuels. Oils produced from biomass via fast pyrolysis require hydrogen for hydrotreating and hydrocracking to convert large hydrocarbon molecules into naphtha and diesel range products [98].

Hydrogen is also under consideration as an energy-carrier for transportation. It can be utilized efficiently in fuel cell electric vehicles (FCEVs), thus helping to make FCEVs a viable option for the future [99]. Because hydrogen is an effective energy carrier and transportation fuels are not required at constant rates, a hybrid energy system producing hydrogen for fuels could potentially produce hydrogen when demand for electricity is lower than potential production. Hydrogen also can be used for energy storage for the power grid. It can be produced when the net demand is low, stored on-site, and used later in a fuel cell system or combusted to provide additional electricity and heat. If this option is selected, the hybrid system gains flexibility by being able to generate electricity when the bidding price of electrical power is high. However, additional capital costs for hydrogen storage and fuel cells will need to be overcome.

Additional issues are raised by siting on-site production of hydrogen at a nuclear site. First, hydrogen has its own set of safety codes, standards, best practices, and regulations. Second, the presence of a volatile flammable substance invokes more rigorous scrutiny and application of 10 CFR 50 and 52 nuclear power regulations [100,101]. It may be necessary to manufacture the hydrogen away from the nuclear reactor, outside the exclusion area boundary. This poses additional security problems in addition to heat loss concerns and additional concerns pertaining to potential loss of heat sink.

6.4.6 MECHANICAL INTERCONNECTIONS OF COMPONENTS OF N-RES HYBRID ENERGY SYSTEMS

Ruth et al. [4] and Suman [8] point out that mechanical interactions of components of N-RES can occur in a number of different ways. Rotational energy from a turbine in a hybridized system could be transferred directly to a work-performing machine such as a pump or compressor. This may require the custom design of mechanical coupling and gearing to achieve proper transfer of power and torque. A flywheel system could also be employed to store mechanical energy. While flywheels do not typically store large amounts of energy, they are able to absorb or release energy at high rates, which may be advantageous in certain industrial processes, or to quickly smooth out large fluctuations in electrical load if a power electronic converter is also coupled to the flywheel. Mechanical energy could also be used as a supplement or in tandem with electric or gas motors.

Ruth et al. [4] also point out that direct mechanical interconnections will need to compete with electric motors. The energy that provides shaft work to directly drive an end-use pump or compressor could be used to generate electricity. The choice between these alternatives is based largely on the relative costs of power and fuel (which includes byproduct purge streams), CO_2 emission penalties, the need for variable speed operation (for which turbines are more suitable than motors), and maintenance costs (which are greater for turbines). Because a significant electric motor is needed to start the end-use device, that capital cost is expended whether the primary drive is electric or mechanical. Thus, mechanical drives are likely to be more expensive than purely electric drives. Additionally, the current trend is toward electric motors in applications traditionally powered by gas or steam turbines [102]. However, mechanical interconnections can be beneficial in special cases such as direct-drive emergency-backup pumps or compressors.

6.4.7 INFORMATION INTERCONNECTIONS OF COMPONENTS OF N-RES HYBRID ENERGY SYSTEMS

The ready availability of information on the status of the electrical grid and each of the plants in a hybrid system is critical to realize the advantages discussed in this chapter. In the absence of such continuous monitoring, a hybrid system behaves like an ordinary market interaction of several buyers and sellers each responding to the price signals they can gather. With near-instantaneous measurement, collection, and distribution of comprehensive information, they can operate as an integrated and optimized entity.

Ruth et al. [4] point out that informatics enables two distinct capabilities. The first is for business and production planning with each subsystem providing information on production plans, needs, and associated prices. Having this information, the overall plan for the hybrid energy system is refined and subsystem operation adjusted. The timeframe of this planning can range from year-ahead plans for major maintenance to daily production plans based on weather forecasts. Optimization tools similar to those used by petroleum refineries and for supply-chain management must be developed to optimize the product slate; thus, maximizing profits and meeting constraints such as production guarantees [103]. Supply-chain management tools that are specifically designed to support facilities meeting rapid dynamics of the electrical grid also are likely to be necessary.

The second level of interaction is online monitoring of process performance and, perhaps more importantly, identification, tracking, and control of operational upsets. These upsets, in mild cases, simply are periods of reduced production for which each plant would like forewarning to prepare accordingly. In severe cases—or mild cases where the process units interact in self-reinforcing fashion and become severe— temporary plant shutdowns or damage to equipment might result. Instantaneous access to system information to diagnose and respond to such conditions in coordinated fashion could be an important safety feature.

6.4.8 SYSTEM-LEVEL CONSIDERATIONS FOR N-R HES DEVELOPMENT

There are system-level considerations in the development of N-R HES. The overall system will require appropriate nuclear reactors (preferable SMR with high temperature capacities), power generation, renewable energy sources, an industrial process and a storage device. The nuclear reactor would provide baseload heat and power without direct emission of GHGs. The nuclear system should operate at a high capacity factor to cover capital and operating costs. The reactor(s) would also perform more efficiently and maintenance costs would be minimized if operated near steady-state design conditions. As pointed out by Ruth et al. [4], nuclear-generated heat would be apportioned to the industrial process and storage, to the power generation system, and to fuels production (such as hydrogen) based on net loads and optimum earnings. The steam turbine in the power generation subsystem would convert thermal energy generated by the nuclear reactor into electrical power. This would be a flexible generator since power can be ramped up or down depending on the amount of steam used. The remaining thermal energy will either be used for industrial process or stored. Steam turbines' large mass provides significant rotational inertia, and together with the synchronous generators they drive, they could help support grid frequency stability. The renewable source(s) would provide near-zero marginal cost energy (heat and/ or power) without direct emission of GHGs. While in principle both dispatchable and nondispatchable renewable sources can be used, more often solar and wind (nondispatchable) sources will be used which may not provide large amounts of power as needed to follow grid load. Electricity and heat from renewable energy sources could also be used by the industrial process, fuels production, or stored. When coupled within an N-R HES, the industrial process would receive heat and/or power from the nuclear reactor(s) and the renewable energy source(s) as needed or as available. The system would use that energy to produce high-value products or fuels that would provide another income stream for the N-R HES. When heat from the nuclear reactor is diverted to power production, the heat needed by the industrial process could be provided by stored thermal energy or derived from another clean energy source such as a biomass boiler when constant operation of the industrial process is necessary or desired. The overall objective is to economically optimize the system between power and heat need of the industrial process and grid power requirement with storage as a buffer in this optimization process. The balance between energy generated by nuclear and renewable sources will depend on the prevailing situation with the overall setup.

Unlike CHP systems, the goal of N-R HES is to transfer as much low-carbon energy to the industrial process as possible. N-R HESs are essentially a cooptimization approach to support grid reliability and stability and to support industrial production, providing power generation and thermal energy to industry while maximizing profitability and minimizing GHG emissions. With the advent of SMRs and Concentrating Solar Thermal Power (CSP) systems, the potential exists to apply these to CHP applications in the traditional manner where heat generation is located in proximity to the industrial process. DOE (Department of Energy) is currently examining the market for nuclear and renewable sources heat applications [104]. Electrical storage options include batteries and flywheels. Thermal storage options include both liquid (e.g., molten salt) and solid (e.g., firebrick) forms. Chemical storage could include hydrogen production carried out by thermally assisted electrolysis. Heat removed from storage could be used either directly in the industrial process or to generate power.

As pointed out by Ruth et al. [4], the overall system can operate in a number of different ways. Tightly coupled and thermally coupled N-R HES concepts would require a dual heat delivery system and the controls necessary to apportion heat between power production, a given industrial process, or fuels production. Similarly, the electrical output would be apportioned between the grid, the coupled industrial process, or fuels production. If necessary, power would be drawn from the grid and combined with the heat and/or electricity delivered from within the hybrid system to operate the industrial process. In the thermally coupled case, the renewable subsystem could be loosely coupled and operated in close coordination with the nuclear subsystem via the grid balancing area. Thermal energy generators (e.g., nuclear reactor and CSP) could supply heat, steam, and power to the manufacturing industry or power to the grid, apportioned to maximize earnings. These systems could operate as dynamic cogeneration plants, adjusting output to meet grid needs and to maintain economic operation of the overall plant. By comparison, traditional nuclear power plants typically connect to the grid alone. Interaction between generators is variously managed by independent system operators, Regional Transmission Organizations, utilities, cooperatives, Federal systems, etc., depending on the location and level.

Ruth et al. [4] points out that successfully developed and managed N-R HES could (a) reduce the cost and volatility of energy production, particularly by helping balance electricity supplies from variable renewable sources; (b) provide dispatchable, carbon-free electricity generation for the grid, with little to no impact on the nuclear reactor core, fuels, and heat transfer loops; (c) provide a second customer for nuclear heat which will improve thermal efficiency and reduce the impact of cost of overall capital equipment; (d) provide greater grid support than variable renewable sources alone; (e) reduce the carbon footprint of the industrial sector; and (f) reduce energy system impact on fresh water resources when using excess thermal or electrical energy to produce potable water, and by coupling low temperature heat rejection to an industrial heat user rather than relying on a cooling tower to condense the power cycle water.

As with many complex systems, technical, economic, environmental, and social aspects of system integration should be considered to justify the added complexity. Above all, an N-R HES requires deliberate project selection and optimization so that customer needs are met and the project's profitability justifies the capital expense. Many opportunities are likely to exist for such hybrid energy systems, and many selection criteria are warranted—ranging from project economics to sustainability issues and from national policy to development risks [105]. These criteria will also be affected by whether a hybrid facility is located on a site with an existing nuclear plant.

Each project development team is likely to have its own set of criteria and weighting factors to determine the most promising options for a given situation. Initial heuristics that simplify estimates of key criteria can assist project developers in narrowing the selections to a number manageable for further analysis. Model-centric tools that allow for rapid screening based on resource availability and cost data, environmental conditions and constraints, project goals, and market conditions can help steer technology choices and conceptual hybrid energy system configuration.

The successful choice of N-R HES for given situation will require conceptual design and cost-benefit analysis of various options of hybrid energy systems. Specific challenges for analyzing hybrid systems include construction and operational lifetimes of overall systems, identifying timelines for upgrade and renovations and system adjustments for evolving market opportunities.

Project economics will be driven by the dynamic variation of market values for different products; current tools and techniques for design and operation do not encompass the needs of hybrid systems. In design, new tools could improve understanding of tradeoffs, including those between storage capacity (of heat or electricity) and system flexibility/response. New tools also could aid design of systems that are stable and controllable during start-up, shut-down, and process interruptions.

Since the hybrid systems may require large upfront capital investment and more likely involve SMR, modular approach to the overall system development would be very desirable. This approach would allow staged capacity construction and evolutionary development so that certain subsystems (or portions thereof) are built initially, and subsystems are added or expanded during operation when demand increases. Risk mitigation is also likely to involve improved control systems and reduced complexity. Thorough life-cycle assessments of hybrid energy system options can be used to quantify environmental effects. Metrics should include both emissions and resource utilization. A means for allocating GHG emissions and energy use between operational systems with varying capacity factors will need to be developed to improve the accuracy of the life-cycle assessments. Other factors to be considered include (a) reduction in reuse, reclamation and efficiency derived from linking multiple systems; (b) environment consequences of materials requiring mining and refining; and (c) storage, security, potential reuse, and eventual disposition of used nuclear fuel.

Finally, cross-sectoral issues that can be raised due to linkages of complex processes from multiple industries should be addressed. Those issues include (a) communication of technical aspects between groups; (b) reconciliation of regulatory, design, and operational standards between independent industries. Some examples are (a) the distance required between nuclear reactor and industrial operation for proper safety, (b) proper control systems for linking processes from multiple industries, and (c) technique for coordination of multiple linked but independent systems with differing dynamic timescales. The last issue may require new supervisory monitoring and control system for the overall N-R HES system. More details on different types of interconnections for N-RES hybrid systems are outlined in an excellent report by Ruth et al. [4], Suman [8] and Dixit [7].

6.5 INDUSTRIAL APPLICATIONS OF N-R HES

According to DOE report [1], in 2014, the U.S. electrical power generation capacity exceeded 1,068 GW [106]. In this same year, U.S. electricity consumption totaled 3,900 billion kWh in sales to end-users [107], or about 450 GW on a continuous output basis. This indicates that a significant amount of the overall power generation capacity (about 60%) is idle for substantial periods during the year. Depending on the season of the year and diurnal use patterns, the location and type of power generation facilities, and the disposition of hydro and variable electricity generation sources, a significant percentage of the power generation capacity could, in principle, be directed part-time to industrial processes. Initial estimates indicate that about one-third of the current power generation resources could be redirected to manufacturing and fuels production [108]. Hybrid energy systems look to expand thermal and power generation to industrial manufacturing and fuels production with better overall capacity utilization. With the build-out of renewable power, hybrid systems could offer an alternative off-take for baseload nuclear plants that are optimally operated at their name plate capacity. Additionally, hybrid systems may offer another option for managing the electricity that will be produced by renewable energy sources.

TABLE 6.4

Annual Energy Use of the Six Largest U.S. Industrial Users of Energy in Exajoules (EJ) [113]

	EJ Electrical	EJ Steam Systems	EJ Fired Heaters	EJ Total by Industry
Chemicals	1.1	1.7	1.3	4.1
Petroleum refining	0.3	1.1	2.2	3.6
Forest products	0.8	2.6	0.2	3.6
Iron and steel mills	0.2	0.1	1.5	1.8
Food and beverage	0.3	0.6	0.3	1.2
Mining	0.7	0.03	0.2	0.9
Total (EJ)	3.4	6.1	5.7	15.2

Market opportunities to apply a large amount of energy on a variable basis have been presented in recent publications by researchers at the Massachusetts Institute of Technology, INL, and the NREL [109–112]. A breakdown of the energy use of the top six industrial energy users is summarized in Table 6.4. Other industries with large energy use include inorganic minerals production (cement, phosphates, sodium carbonates, silica, etc.), textiles, glass, and computers and electronics. Prime opportunities for using N-R HES include industrial plants that have high steam duties and processes that require low-to-intermediate temperature heat. SMRs are an appropriate size to service some of these plants and could be located near these end-users to service them.

The benefits of SMRs potentially include reduced manufacturing costs, safety advantages, incremental scalability, reduced land use, and, in the case of high-temperature reactors, reduced water usage [114]. Some designs provide passively safe, gravity-driven, natural circulation through the primary core. Some designs are submerged in the coolant pool to provide long term to permanent emergency cooling and some are below ground to be more tolerant of physical damage by earthquakes or tornados. Multiple barriers are designed-in to prevent the release of radiation should an accident occur and could potentially allow an SMR to be located adjacent to an industrial user, and in closer proximity to population centers [115].

Flexible operations could also enable production of clean water from saline or compromised water sources for use in power plant, industrial, or even community services. Water desalination is one option for flexible operation of an existing plant or for future SMRs [116]. The IAEA (International Atomic Energy Agency) has an on-going program to address the issues related to the use of nuclear energy for desalination of alternative sources of water, including wastewater from munic-ipalities, agricultural run-off, brackish groundwater, or seawater [117]. A recent study of an N-R HES indicates fresh water may be efficiently and cost-effectively produced from brackish water in the Southwest US when future demands exceed fresh water availability [118]. Clean-up of water displaced from deep saline aqui-fers by future CO_2 injection into these reservoirs to sequester the carbon is another potential application. Low-rank coal drying could release significant by-product water that could be cleaned and used for other purposes.

New energy systems might also be integrated with process heat applications [119]. This could, for example, reduce the water cooling requirements of SMRs that are located in proximity of the heat application. Such cases could operate like a traditional CHP system where the SMR is located near the industrial heat user, such as for inorganic minerals concentration and drying, or for distiller grain drying and for distillation in corn-ethanol plants. Other less apparent industrial uses might include paper pulp operations (~10–20 MWt is typical), food processing plants (~5–20 MWt is typical), and chemical plants (e.g., methanol distillation), with a typical plant using 100 MWe and 90 MWt) [120].

The reactor design could be optimized for a particular industrial service, considering the process heating requirements relative to scale, peak temperature, steam quality, time-of-use, overall conversion efficiency, and other factors. In some cases, thermal energy storage in a steam accumulator or a molten salt or liquid metal tank could be used to buffer the thermal/electrical energy available from the N-R HES or from the grid. Technical assessment of the scale, duty cycles, and associated costs of thermal energy storage buffers is needed for future DOE or industrial consideration of possible hybrid configurations. Hybridization options will vary in accordance with regional resources, industry, electricity, and financial markets.

DOE report [1] suggests extensive opportunities for N-R HES. These opportunities are illustrated DOE reported in Figure 6.5 [1]. The report indicates that the design basis for these depends on case-specific industrial user technical requirements and economic drivers associated with these or other options. Systems should be tailored to regional resources and markets to dynamically optimize the use of thermal and electrical energy. Definition, prioritization, and analysis of key options based on pertinent figures of merit are necessary to identify energy systems that have the greatest likelihood for success.

Heat delivery to meet end-user requirements depends on the physical properties and temperature of the heat transfer fluid and the design of the thermal hydraulics system. Temperature and distance comparisons are indicative of the challenges of distributing heat at a specific temperature (and for a particular working fluid and heat exchanger) from a nuclear reactor to an industrial user. The distances shown are an initial approximation of the distances that various heat levels can be economically circulated for heat deposition at the industrial user site. Actual distances will vary depending on design and economics of the system. Direct (or tight) coupling with the industrial user can include purely thermal energy, purely electrical energy, or a combination of the two. Thermal, electrical (e.g., Compressed Air Energy Storage, Battery Storage), and chemical-derived energy can be stored and then delivered to industry when needed.

DOE reports by Nelson et al. [124,125] included technical and economic evaluation of the use of nuclear heat for cogeneration applications under the Next Generation Nuclear Plant (NGNP) program. The NGNP Alliance with industry continues to develop heat application markets for the high temperature gas reactor (HTGR) with an outlet of approximately 750°C helium based on current code-qualified materials. The results of these studies indicated one or more of the nominal 600 MWt plants could supply the quality and quantity of heat needed for steam methane reforming to produce hydrogen, ammonia, and ammonia-based derivatives (ammonia-based

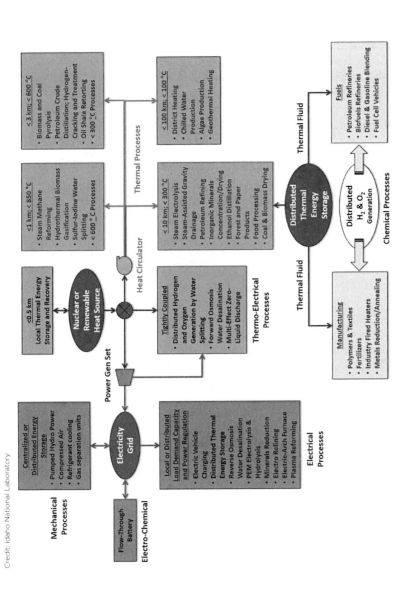

FIGURE 6.5 Summary of potential N-R HES applications indicating energy conversion varieties possible and their appurtenant processes [121,122]. Heat transfer distances are only an approximation based on preliminary calculation under the NGNP Program [123]. Arrows indicate energy flows. Color is only intended for graphical rendering.

fertilizers, nitric acid, urea), synthetic chemicals, and nonconventional fuels (from oil sands, oil shale, coal, and natural gas). Although the capital investment for a nuclear plant is comparatively higher than traditional fossil-fired heat sources and steam boilers, the analysis indicated that heat produced from nuclear reactors could favorably compete if fossil fuels combustion took into account estimated externality costs of GHG emissions. The potential use of N-R HES for applications in the electrical power market with coordinated synthesis of methanol from natural gas and hydrogen production through high temperature steam electrolysis is also examined [126,127]. The value proposition and technical integration challenges of nuclear and renewable energy in hybrid systems are also being evaluated by joint efforts among DOE, INL, NREL, and industry [119]. Three case studies using N-R HES examined by DOE, NREL, and INL are described in Section 6.7.

6.6 TOOLS REQUIRED FOR SUCCESSFUL N-R HES APPLICATIONS

Nuclear hybrid energy systems are currently in the early stages of development. Several of the identified development needs correspond to technology needs for flexible and resilient energy structures. Currently, DOE-NE and DOE-EERE are jointly supporting analysis of N-R HESs based on specific regional opportunities led by government, industry, and university to discuss pathways to a low-carbon energy economy, including the role of nuclear energy and relevant technology development program needs [128–130]. An initial gap analysis of technology development needs was completed to identify the necessary tools, experimental facilities, and component testing that is needed to develop hybrid systems that could be marketable in the relatively near term. The following is a summary of the most important development subjects relevant to N-R HES [119] as reported by DOE [1].

6.6.1 DYNAMIC MODELING TOOLS FOR N-R HES IMPACT ASSESSMENT, DESIGN OPTIMIZATION, AND NUCLEAR REACTOR DESIGN STUDIES

The modeling and simulation tools are required to account for the time variability, dynamic interaction of unit operations, and transient phenomena that may impact hardware performance and control of hybrid energy systems. These tools are being developed jointly by DOE national laboratories and universities [131–134]. The tools are needed to understand the case-specific interactions with the electrical grid, behavior of the nuclear reactor and associated heat transfer systems, behavior of power cycle ramp rates, energy storage and recovery rates, and chemical plant reaction response to changes in heat transfer and material feeds. The major objective of the modeling and simulation is the economic and technical optimizations which would require multiphysics, dynamic model with the tools for multiobjective optimization. The simulation must help decisions such as the choice of an appropriate nuclear reactor, its design and operating conditions, risk assessment and its interactions with industrial processes. Optimal configuration of N-R HES that will minimize the cost of electricity production under the defined constrain a tightly coupled system has on the capability of the N-R HES to meet demand needs to be developed. These constraints play a fundamental role in enabling the economic evaluation framework by monetizing the ability of the N-R HES to help cope with electricity demand volatility. A framework to assess and optimize the value of energy utilization by considering the value of ancillary services and other figures of merit that are important for grid operations, reliability, and resiliency, as well as other societal priorities such as environmental impacts on air, water, and land is being developed by DOE in cooperation with the industries and universities [135].

Dynamic modeling tools are also important for nuclear reactor design studies under transient conditions. In order to accommodate flexible operations required for hybrid process, design requirements for SMR which include reactor size (i.e., size of SMR and number of modules), heat delivery systems, and safety considerations that derive from direct coupling of subsystems must be provided by modeling and simulation. An evaluation of the nuclear reactor choices based on the thermal energy output and operating characteristics of the reactors for specific N-R HES applications should also be a part of these activities. The reactor design must also include verification of future advanced SMR reactors for rapid transient operation capabilities that meet Nuclear Regulatory Commission (NRC) design certification requirements. Furthermore, thermal energy buffers may be required when maneuvering heat from power generation to an industrial process [119].

6.6.2 Thermal Hydraulics and Electricity Interconnections

As pointed out earlier, N-R HES can be operated in thermal or electricity interconnections mode. It was also pointed out that each will require different types of component interactions. DOE report [1] points out that the thermal energy interconnections for hybrid energy systems will require new heat exchanger and heat circulation designs. This will include heat transfer fluids, heat delivery systems, and heat deposition systems. Relative to N-R HES, dynamic heat delivery ramp-up and ramp-down rates will need to be characterized for industrial systems and storage options. These new heat exchanger designs are necessary to dynamically and safely apportion heat between the power generation and heat application users. In particular, an intermediate heat exchanger that can rapidly maneuver heat for power production and industrial users without significant impact on reactor operations needs to be developed for each of the possible N-R HES reactor types. This work should include the development and testing of high-temperature thermal fluid circulators and new control valves, as well as new materials and heat exchanger designs for increased thermal and mechanical cycling.

The requirement that reactor outlet temperature and reactor inlet temperature must be reliably maintained within operating specifications during thermal interconnections may necessitate a thermal buffer such as a steam accumulator or heat sink to ensure stable reactor operations [119]. It may also require heat exchanger fabrication, and prototype testing [109]. Electrical heating can be used to heat a thermal storage reservoir when the cost drops below that of natural gas heating [136]. Excess electricity can similarly be used to chill water or other fluids that can be cooled and stored and used by an industrial plant or building for cooling loads during the day. The use of commercial ice storage systems which exist to cool commercial buildings [137,138] should be tested for this purpose. University of Texas-Austin is considering this concept for energy storage [139]. Electrical-to-thermal conversion systems can deliver heat at very high temperatures [140].

When N-R HES is operated under electricity interconnections, the design and the operation of N-R HES system must consider both internal uses for electricity and its dynamic market value. With sufficient operational flexibility, a hybrid system can respond to market signals and choose the most profitable use of its electric power and thermal energy. Application of real-time digital simulators can help evaluate the electric grid and provide automatic or supervisory control updates. DOE report [1] points out that as the grid incorporates more distributed power generation, N-R HESs potentially offer a new paradigm for "micro-industrial grids." SMRs, including both light water and high temperature reactors, may find increased opportunities for applications with industrial complexes. This may lead to independence from the grid with potential benefits of more reliable power and heat delivery. The modularity of SMRs can allow for staggered refueling which could create a nearly constant heat source for process heat applications. If the entire nuclear plant shuts down, natural gas-based back-up supplement process should be used [141].

6.6.3 Power Generation and Storage Systems

DOE report indicates [1] that the heat-to-power conversion system for an N-R HES must incorporate turbine technology and power generation blocks that are designed to respond to rapid shifts in power demand. Technical issues pertaining to transient response needs of the turbine and generator set are outlined in early efforts to model the transient behavior of these unit operations [145]. Nuclear power has the capability to load follow based on the French and German experiences [142]. The load-following capabilities of nuclear plants are on the order of coal-fired plants but below that of open-cycle gas turbines. For example, the maximum ramp-up rate of a typical European nuclear reactor ranges from 1% to 5%/minute, while an open-cycle gas turbine can ramp up at a rate up to 20%/minute and a combined-cycle gas turbine power plant can ramp up at a rate of approximately 5%–10%/minute. The current fleet of U.S. nuclear reactors was designed to operate close to their name plate capacity, however, and because of a different control rod configuration than used in typical European plants, their ramp rates are lower. They are usually ramped up or down at a percent or two per hour. These limitations are due more to the heat source than to the turbine [143]. DOE report [1] indicates that research activities in this area will need to consider new Brayton power cycles (e.g., super-critical CO_2 power cycles), or a combination of Brayton and Rankine cycles that allow heat to be extracted from various stages in the power cycle. Smaller parallel turbines and generation sets that can be independently ramped may also best accommodate load-following power generation. This concept embraces the design philosophy of some SMRs for which individual reactors modules could be matched with small power turbine units [1].

According to DOE report [1], thermal energy storage reservoirs will likely be needed to soften the imposition of rapid transitions on the thermal hydraulic systems of nuclear systems. Energy storage technologies that are being developed for CSP systems may also be applicable to nuclear hybrid energy systems as long as scale of the thermal reservoir is tailored to the needs of the hybrid energy system. Technical and economic assessments are needed to arrive at an understanding of the costs and benefits of capital investments that support different temporal scales—from minutes, to hours, to days, to seasonal—of storage operation for both electricity and thermal supply and demand. Results to date indicate that production of the nonelectric product is often economically more compelling [145], but further analysis is needed to understand the value of providing grid stability, and the real value of clean electricity in potential carbon-constrained markets. Projections for the cost of future nuclear SMRs also need to be developed for comparison to natural gas combined cycle (NGCC) with CCS (Carbon Capture and Sequestration) [1].

6.6.4   CONTROL, SAFETY, SECURITY, AND LICENSING

DOE [1] points out that N-R HES will require control systems that co-process input signals from the electric grid, the processing plants, and the nuclear plant. System diagnostics and prognostics for intelligent control of large complex systems are becoming possible with faster computer processors and artificial intelligence. In addition, methods to detect and manage cyber security issues will need to be further developed and implemented [1]. Operator control room environments need to be developed to protect it from human factors. Hybrid plant complexity will require careful development of the panels, alarms, and supervisory decisions that may need to be made to maintain stable, safe, and secure operations for the dynamic operating conditions. The present DOE efforts for licensing of advanced nuclear SMR need to be expanded for N-R HES systems. Industry input from both nuclear reactor developers and industrial users, as well as input from the NRC, is needed for this technology area [1].

6.7   CASE STUDIES

INL, NREL, and DOE developed three case studies to illustrate the value of N-R HES. Two case studies, west Texas synthetic gasoline production and Arizona desalination plant, were carried out for tightly coupled N-R HES and hydrogen production plant was examined with thermal-only interaction plant.

6.7.1   CASE STUDIES 1 AND 2: WEST TEXAS SYNTHETIC GASOLINE AND ARIZONA DESALINATION PLANT

Ruth et al. [2–4] point out that tightly coupled N-R HESs are an option that can generate zero-carbon, dispatchable electricity and provide zero-carbon energy for industrial processes at a lower cost than alternatives. Tightly coupled N-R HESs are defined as systems that are managed by a single entity and link a nuclear reactor that generates heat, a thermal power cycle for heat-to-electricity conversion, at least one renewable energy source, and an industrial process that uses thermal and/or electrical energy. They produce multiple products and can dynamically vary the amount of each produced. Because of that flexibility, N-R HESs are potentially advantageous over traditional technologies that produce a single product and use a minimal number of energy sources. For example, an N-R HES can produce electricity at times when its value is higher than that of the industrial product and produce the industrial product when its value is higher than that of electricity. An N-R HES can also produce both electricity and the industrial product during hours when electricity services such as regulation or flexibility reserves are valuable but the value of electrical energy alone is lower than that of the industrial product. Potential benefits of N-R HESs include the ability to generate dispatchable, flexible, zero-carbon electricity that can support the grid's resource adequacy requirements; zero-carbon energy sources for industry; real inertia to support the grid; and alleviation of the impacts of price suppression [2–4].

The first case study of Texas-synthetic gasoline scenario examined by Ruth et al. [3] included four subsystems: a nuclear reactor, a thermal power cycle, a wind power plant, and synthetic gasoline production technology. The second case study of Arizona-desalination scenario [3] included four subsystems: a nuclear reactor, thermal power cycle, PV, and a desalination plant. The analysis focused on the economics of the N-R HESs and how they compare to other options, including configurations without all the subsystems in each N-R HES and alternatives where the energy is provided by natural gas. This analysis was built upon a previous analysis performed by the INL that focused on dynamic operability and performance of the same two N-R HES options and treated financial performance as a secondary objective [2]. INL's team found that both N-R HESs can potentially meet operational requirements necessary to provide flexibility and that each is projected to be profitable [2]. In this subsequent economic study, Ruth et al. [3] assumed that the configurations would be operable and focused on identifying the financially optimal configurations and operating schemes. The study tested five hypotheses regarding the potential benefits of the N-R HES in each of the two case studies [2–4]:

1. The N-R HES configurations analyzed have the potential to be profitable to investors and are likely to be more profitable than uncoupled configurations.
2. Using nuclear-generated heat in an N-R HES can economically reduce GHG emissions from industry. If a cost of carbon is included in the economic analyses, the N-R HES will have a lower cost than competing uncoupled natural gas configurations.
3. N-R HESs can support resource adequacy for the electric grid while maximizing production of a more profitable industrial product if the market structures incentivize that option.
4. N-R HESs will be more profitable than uncoupled configurations because they can produce electricity when its price is high and the industrial product when the price of electricity is low.
5. The internal flexibility of N-R HESs makes them beneficial as a hedge against changing and uncertain future prices.

The study used REopt, the NREL-developed energy planning platform, to optimize the design (i.e., identify the optimal subsystem configurations) and operation of each N-R HES. REopt's optimization was set to maximize the N-R HES's net present value (NPV) over the analysis period based on capital costs, operating costs, conversion efficiencies, product prices, and financial assumptions. Capital, fixed operating costs, and efficiencies for the nuclear reactor, thermal power cycle, renewable electricity generation, and the industrial process are from other published reports. The study did not include a cost of carbon in the base case; however, the study priced carbon in some sensitivity cases. The study sets the REopt inputs for the price of electrical energy and reserve products to marginal costs generated using the PLEXOS production cost model. PLEXOS performs a security-constrained, least-cost economic dispatch of generation assets to identify the least-cost option to serve the loads subject to operational constraints on a generation mix with a high penetration of VRE generation. The study included a capacity payment for configurations selling electricity to the grid during the highest load hours. The study estimated water and gasoline product prices based on a water supply curve developed from proposed projects and published gasoline price projections, respectively.

The analysis showed that for both cases analyzed, industrial processes using nuclear-generated energy can be more profitable if a cost of carbon is included. The Texas-synthetic gasoline scenario's configuration with only the nuclear reactor and industrial process has an NPV of $3,699 million, an IRR of 25%, and an NPV/TCI ratio of 1.48. Replacing the nuclear reactor with a natural gas boiler reduces the NPV to $3,600 million and does not change the IRR but it increases the NPV/TCI ratio to 1.55. Although the natural gas-heated system has less profit (lower NPV), it is likely to be more attractive to investors who are concerned about the uncertainty of recovering their capital investment due to the higher NPV/TCI ratio. The U.S. government provides four sets of social costs of carbon for analyses for regulatory analyses because the cost of climate-related impacts is uncertain. For most analysis, the study used the 2035 value with a 3% discount rate—$61/metric ton CO_2. The natural gas-heated synthetic gasoline configuration emits 281,000 metric tons of CO_2 annually. If that configuration is assessed a $61/metric ton CO_2, its NPV is $3,520 million and its NPV/TCI ratio is 1.45; thus, the nuclear-industrial process is both more profitable and may be more attractive to investors at that cost of carbon.

Using an NGCC to produce the electric power for desalination resulted in emissions of 153,000 metric tons CO_2 annually. Using the $61/metric ton CO_2 cost of carbon, the NPV for the nuclear-powered process is slightly higher than the NGCC-powered process—$3.16 billion as compared to $3.08 billion; however, at 48% the IRR for the nuclear-powered process was lower than that of the NGCC-powered process, which was 60%. Thus, a high cost of carbon may be necessary to make the nuclear-desalination configuration more attractive than one powered by an NGCC.

The study supported hypothesis #3 in the Arizona-desalination scenario but not in the Texas-synthetic gasoline scenario under the base case parameters and the capacity payments used in this analysis. The analysis of the Texas-synthetic gasoline N-R HES did not identify any condition where either a capacity payment or high electrical energy prices result in selling electricity to support electricity resource adequacy instead of maximizing production of synthetic gasoline. This is because synthetic gasoline is a more valuable product than electricity, and electricity generation requires a thermal power cycle, which is an additional investment. The capital cost for the thermal power cycle (and the opportunity costs of not generating gasoline during hours when the N-R HES generates electricity instead of producing gasoline) disincentivizes the nuclear-synthetic gasoline configuration from participating in the capacity markets.

The Arizona-desalination system is likely to optimally generate sufficient electricity to receive the capacity payment while maximizing production of water (a more profitable product) the majority of the year with a \$50/kilowatt-year (kW-y) capacity payment, provided the water price is between \$1.15/thousand gallons (gal) and \$3.50/thousand gal (i.e., high enough for the desalination unit to be profitable, but not so high that the opportunity cost of not producing water is greater than the value of electricity, including the capacity payment). At higher capacity payments, the Arizona-desalination scenario is likely to optimally generate electricity to receive the capacity payments. The key difference between the two scenarios is that the desalination scenario inherently includes a thermal power cycle, so electricity production requires no additional capital cost.

The analysis did not support hypothesis #4. The analysis indicated that no configurations in the Texas-synthetic gasoline scenario realized sufficient value from the capacity payments to include the thermal power cycle necessary for a system capable of switching without decreasing its NPV; hence, the configuration with the maximum NPV is unlikely to select between products. The Arizona-desalination system only switches from water desalination to electricity production as necessary to receive capacity payments under the base-case parameters. Thus, the key value provided for electricity generation is the capacity payments. Without higher electricity prices, short periods of high-priced electricity are insufficient to incentivize N-R HESs to produce additional electricity because the opportunity cost of not producing the industrial product was too high.

The analysis partially supported hypothesis #5. N-R HESs have internal flexibility that makes them beneficial as a partial hedge against uncertain product prices. However, the benefits of using a combined system are limited by the additional capital cost of a flexible configuration when compared to a configuration with fewer subsystems; the increase in fixed operating costs over a configuration with fewer subsystems; and the reduction in capacity factors because some subsystems in the flexible configuration are not operated at all times.

In summary, the study showed that N-R HES configurations in the two cases analyzed are profitable and the key driver for each is the industrial product. For electricity production to increase each N-R HES's value, the price of electricity needs to be higher than considered in this analysis. In addition, the study found comanagement of the nuclear and renewable assets to be beneficial only when the VRE subsystem generates some of the electricity necessary to receive capacity payments while minimizing the reduction in production of the industrial product, as it does in the Arizona-desalination scenario. Under the limitations of this analysis, the study found that full hybridization did not improve the economics of these two scenarios; however, it may be beneficial under different market conditions or for other scenarios. The study suggested that additional analysis of the potential for tightly coupled nuclear and renewable energy generators to provide thermal energy for industrial processes and both thermal and electrical energy for hydrogen production would be useful, because those options introduce more potential thermal loads. In addition, improving analyses of markets may identify opportunities for N-R HESs that were not identified in this analysis [2–4].

6.7.2 Case Study 3: N-R HES for Hydrogen Production

Figure 6.6 illustrates the coupling of nuclear and wind to produce power in response to grid demand up to the maximum generation capacity of the combined resources, and to produce hydrogen by steam electrolysis when demand for electricity falls below the capacity of the system [144]. In this illustrative example examined by DOE [1], power generation was assumed to be the highest priority, even when the market may actually drive the systems to produce hydrogen any time the value of hydrogen is higher than electricity.

In general, a loosely coupled, electricity-only HES involves only power production with multiple generators supplying a single output of electricity to the electrical grid-as described earlier [119]. The "advanced hybrid energy system" defined within this case study falls into the category of a thermally coupled HES through the addition of time-varying hydrogen production when heat and electricity are transferred to the steam electrolysis plant, creating a second product output.

A transient physics-based model was developed by DOE [1] to address the technical feasibility of shifting electricity and heat to the electrolysis unit based on a representative wind farm operating in Wyoming, and a load profile representative of the Midwest [145,146]. The dynamic simulation for this case demonstrated that the advanced nuclear hybrid solution is physically capable of resolving the power generation variability introduced by wind turbines down to a minute-by-minute time scale. The electrical battery storage unit helped smooth the transients associated with power generation spin-up and spin-down. The electrolysis unit, with gas flow and heat recuperation, is also capable of being operated intermittently as functionally required. Physical testing of this modeled outcome is needed to confirm these conclusions.

In order to address the question regarding economic viability, the ratio of cash flows (computed as the profitability of the system as a percentage of the total wind power generation in the system) was calculated by DOE [1] for both the traditional and the advanced HES. The ratio of profitability (referred to as the Profitability Index, or PI) is plotted in Figure 6.7. When PI is less than 1.0, then the additional profit gained by operating the electrolysis unit does not justify the additional capital investment for this unit. In other words, the rate of return on investment for the advanced system would be less than construction and operation of only the traditional HES when PI < 1.

DOE report [1] points out that for the example considered, when the wind capacity exceeded 23%, the additional revenue for hydrogen production from excess generation that would otherwise be curtailed (assuming commodity prices of $2.50/ kg-$H_2$ at the plant gate and $0.12/kWh electricity) was sufficient to cover the cost of capital and operating costs for the electrolysis unit and the PI of the overall system is greater than 1.0. Further increases in the penetration of wind raised the percentage of time that electricity was dedicated to hydrogen production, resulting in higher PI. These results may not be specifically relevant in all power markets.

In this example, providing ancillary services to the grid—such as regulation-up or regulation-download, power stability control, or contingency reserve—was not factored into the economic analysis. Analysis of the material balance for

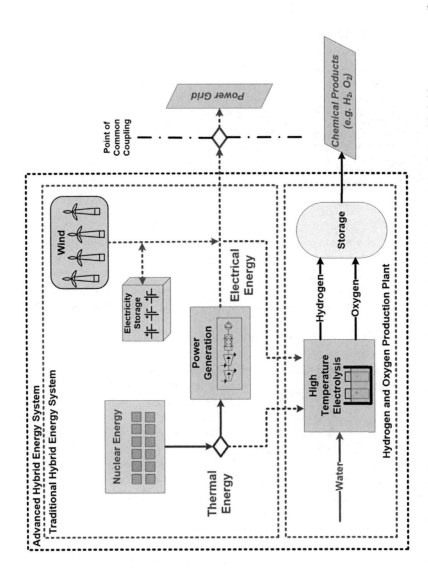

FIGURE 6.6 Possible configuration for a traditional hybrid energy system (yellow box) versus an advanced hybrid energy system (large box) [145]. (Credit: Idaho National Laboratory.)

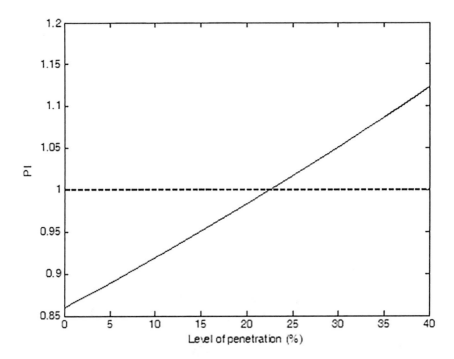

FIGURE 6.7 PI of advanced (thermally coupled) N-R HES as a function of the fraction of the grid demand met by variable renewable sources. (Credit: Idaho National Laboratory [1].)

steady-state steam electrolysis with electrical and thermal input versus hydrogen production by a conventional steam methane reforming process is summarized in Figure 6.8 [147]. Almost 5 tons of CO_2 is produced by steam methane reforming for each ton of hydrogen produced. Assuming a cost of $50/ton for CO_2 capture, compression, delivery, and permanent storage, this would add about $0.25 per kilogram of hydrogen produced from natural gas. With the hybrid system, the cost of hydrogen production by electrolysis depends on the cost of electricity. Hence, in cases where the price of electricity is low, for example, when renewable sources are at their low marginal cost of production, then the N-R HES hydrogen coproduct hybrid system may compete with steam methane reforming. The report [1] points out that an additional analysis is needed to understand the overall cost/trade-off benefits of the hybrid systems for realistic markets [1].

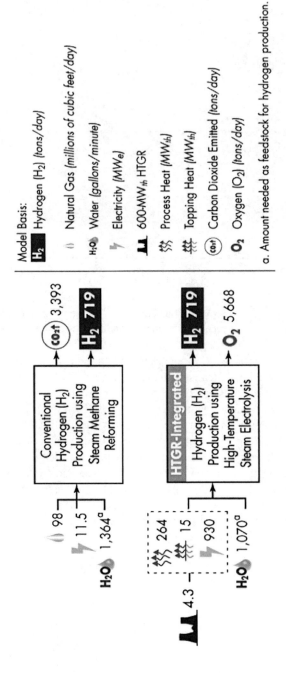

FIGURE 6.8 Mass and energy balance calculation results for the HTGR-integrated process for hydrogen production via the high temperature steam electrolysis process [148]. (Credit: Idaho National Laboratory.)

REFERENCES

1. Hybrid Nuclear-Renewable Energy Systems. (2015). Chapter 4: Advancing clean electric power technologies technology assessments. *Quadrennial Technology Review 2015*, DOE report, Washington, DC.
2. Ruth, M., Cutler, D., Flores-Espino, F., Stark, G. and Jenkin, T. (2016). The economic potential of three nuclear-renewable hybrid energy systems providing thermal energy to industry. Technical Report NREL/TP-6A50-66745 December 2016 Contract No. DE-AC36-08GO28308. NREL, Golden CO.
3. Ruth, M., Cutler, D., Flores-Espino, F., Stark, G., Jenkin, T., Simpkins, T. and Macknick, J. (2016). The economic potential of two nuclear-renewable hybrid energy systems. Technical Report NREL/TP-6A50-66073 August 2016 Contract No. DE-AC36-08GO28308. NREL, Golden, CO.
4. Ruth, M.F., Zinaman, O.R. and Antkowiak, M. (2014). Nuclear-renewable hybrid energy systems: Opportunities, interconnections, and needs. INL/JOU-14-33322, Prepared for the U.S. Department of Energy Office of Nuclear Energy Under DOE Idaho Operations Office Contract DE-AC07-05ID14517 February 2014 INL Laboratory, Idaho.
5. Antkowiak, M., Ruth, M., Boardman, R., Bragg-Sitton, S., Cherry, R. and Shunn, L. (2012). Summary report of the INL-JISEA workshop on nuclear hybrid energy systems. Technical Report NREL/TP-6A50-55650 July 2012 Contract No. DE-AC36-08GO28308. NREL, Golden, CO.
6. Ruth, M., Zinaman, O., Antkowiak, M., Boardman, R., Cherry, R. and Bazilian, M. (2014). Nuclear-renewable hybrid energy systems: Opportunities, interconnections, and needs. *Energy Conversion and Management* 78: 684–694, Doi: 10.1016/j.enconman.2013.11.030.
7. Dixit, S. (2018). Hybrid nuclear-renewable energy systems: A review. *Journal of Cleaner Production* 181: 166–177, Doi: 10.1016/j.jclepro.2018.01.2620959–6526/©.
8. Suman, S. (2018). Hybrid nuclear-renewable energy systems: A review. *Journal of Cleaner Production* 181: 166e177.
9. Nuclear–Renewable Hybrid Energy Systems for Decarbonized Energy Production and Cogeneration, (2019). IAEA Tecdoc Series IAEA-TECDOC-1885, IAEA, Vienna.
10. Garcia, H.E., Chen, J., Kim, J.S., McKellar, M.G., Deason, W.R. Vilim, R.B., Bragg-Sitton, S.M. and Boardman, R.D. (2015, April). *Nuclear Hybrid Energy Systems—Regional Studies: West Texas & Northeastern Arizona* INL/EXT-15-34503. Idaho National Laboratory INL ART Program Prepared for the U.S. Department of Energy Office of Nuclear Energy Under DOE Idaho Operations Office Contract DE-AC07-05ID14517.
11. Ela, E., Milligan, M. and Kirby, B. (2011). *Operating Reserves and Variable Generation.* NREL/TP-5500-51978, National Renewable Energy Laboratory, Golden, CO. http://www.nrel.gov/docs/fy11osti/51978.pdf. Accessed 28 August 2012.
12. Pellegrino, J., Margolis, N., Miller, M., Justiniano, J. and Thedki, A. (2004, December). Energy use, loss and opportunities analysis: U.S. Manufacturing and mining, Energetics, Inc. and E3M, Inc. for the U.S. Department of Energy, Industrial Technology Programs.
13. Wheeley, C., Mago, P.J. and Luck, R. (2012). Methodology to perform a combined heating and power systems feasibility study for industrial manufacturing facilities. *Distributed Generation & Alternative Energy Journal* 27(1): 8–32.
14. Robinson, P.J. and Luyben, W.L. (2011). Plantwide control of a hybrid integrated gasification combined cycle/methanol plant. *Industrial & Engineering Chemistry Research* 50(8): 4579–4594.
15. Cochran, J., Bird, L., Heeter, J. and Arent, D. (2012, April). *Integrating Variable Renewable Energy in Electric Power Markets: Best Practices from International Experience.* NREL/TP-6A00-53732. National Renewable Energy Laboratory, Golden, CO.

16. Denholm, P. and Hand, M. (2011, March). Grid flexibility and storage required to achieve very high penetration 39(3): 1817–1830, NREL Report No. JA-6A20-49400, Doi: 10.1016/j.enpol.2011.01.019. Accessed 28 March 2013.
17. Shah, Y.T. (2020). *Energy and Fuel Systems Integration*. CRC Press, New York.
18. Shah, Y.T. (2021). *Hybrid Power-Generation, Storage and Grids*. CRC Press, New York.
19. Borges Neto, M.R., Carvalho, P.C.M., Carioca, J.O.B. and Canafístula, F.J.F. (2010). Biogas/photovoltaic hybrid power system for decentralized energy supply of rural areas. *Energy Policy* 38(8): 4497–4506.
20. Coppez, G., Chowdhury, S. and Chowdhury, S.P. (2011, July 24–29). South African renewable energy hybrid power system storage needs, challenges and opportunities. *IEEE Power and Energy Society General Meeting*, a paper presentation referred in NREL/INL report on Nuclear-Renewable Hybrid Energy Systems: Opportunities, Interconnections, and Needs by Mark F. Ruth, Owen R. Zinaman, Mark Antkowiak. INL/JOU-14-33322 (2014).
21. Nixon, J.D., Dey, P.K. and Davies, P.A. The feasibility of hybrid-solar biomass power plants in India. *Energy*, Doi: 10.1016/j.energy.2012.07.058. Accessed 25 March 2013.
22. Rehman, S. and Al-Hadhrami, L.M. (2010). Study of a solar PV–diesel–battery hybrid power system for a remotely located population near Rafha, Saudi Arabia. *Energy* 35(12): 4986–4995.
23. Zoubeidi, O.M., Fardoun, A.A., Noura, H. and Nayar, C. (2012). Hybrid renewable energy system solution for remote areas in UAE. *Global Journal of Technology & Optimization* 3: 115–121.
24. Kaldellis, J.K. and Kavadia, K.A. (2007). Cost-benefit analysis of remote hybrid wind-diesel power stations: Case study Aegean Sea islands. *Energy Policy* 35(3): 1525–1538.
25. Rubio-Maya, C., Uche, J., Martínez, A. and Bayod, A. (2011). Design optimization of a polygeneration plant fuelled by natural gas and renewable energy sources. *Applied Energy* 88(2): 449–457.
26. Bourgeois, T.G., Helman, B. and Zalcman, F. (2003). Creating markets for combined heat and power and clean distributed generation in New York State. *Environmental Pollution* 123: 451–462.
27. Kang, C., Brandt, R. and Durlofsky, L. (2011). Optimal operation of an integrated energy system including fossil fuel power generation, CO_2 capture and wind. *Energy* 36(12): 6806–6820.
28. Kieffer, M., Brown, T. and Brown, R. (2019). Flex fuel polygeneration: Integrating renewable natural gas into Fischer-Tropsch synthesis, Version of Record: https://www.sciencedirect.com/science/article/pii/S0306261916302719 1f0677d1cf99126113605ae 0426225d3, an open access paper by Elsevier user licence https://www.elsevier.com/open-access/userlicense/1.0/.
29. Phadke, A., Goldman, C., Larson, D., Carr, T., Rath, L., Balash, P. and Yih-Huei, W. Advanced Coal Wind Hybrid: Economic Analysis, Ernest Orlando Lawrence Berkeley National Laboratory. LBNL-1248E, under contract No. DE-AC02-05CH11231. http://eetd.lbl.gov/ea/ems/reports/lbnl-1248e.pdf. Accessed 20 September 2012.
30. Cherry, R.S., Breckenridge, R.P., Boardman, R.D., Bell, D., Foulke, T. and Lichtenberger, J. (2012, November). *Preliminary Feasibility of Value-Added Products from Cogeneration and Hybrid Energy Systems in Wyoming*. INL/EXT-12-27249. Idaho National Laboratory, Idaho Falls, Idaho.
31. Turchi, C. and Ma, Z. (2011). *Gas Turbine/Solar Parabolic Trough Hybrid Design Using Molten Salt Heat Transfer Fluid*. NREL/CP-5500-52424. National Renewable Energy Laboratory, Golden, CO. http://www.nrel.gov/docs/fy11osti/52424.pdf. Accessed 26 March 2013.

32. Livshits, M. and Kribus, A. (2012). Solar hybrid steam injection gas turbine (STIG) cycle. *Solar Energy* 86: 190–199.

33. Forsberg, C. (2008). Sustainability by combining nuclear, fossil, and renewable sources. *Progress in Nuclear Energy* 51(1): 192–200.

34. Garcia, H., Mohanty, A. Lin, W.-C. and Cherry, R. (2013). Dynamic analysis of hybrid energy systems under flexible operation and variable renewable generation—Part I: Dynamic performance analysis. *Energy* 52: 1–16.

35. Garcia, H., Mohanty, A., Lin, W.-C. and Cherry, R. (2013). Dynamic analysis of hybrid energy systems under flexible operation and variable renewable generation—Part II: Dynamic cost analysis. *Energy* 52: 17–26.

36. Organization for Economic Co-Operation and Development (OECD). (2011, June). Nuclear energy agency report on, technical and economic aspects of load following with nuclear power plants. www.oecd-nea.org.

37. Maillot, V., Fissolo, A., Degallaix, G. and Degallaix, S. (2005). Thermal fatigue crack networks parameters and stability; An experimental study. *The International Journal of Solids and Structures* 42(2): 759–769.

38. Kwon, J.D., Woo, S.W., Lee, Y.S., Park, J.C. and Park, Y.W. (2000). Thermal aging and low cycle fatigue characteristics of CF8M in a nuclear reactor coolant system. *Fourth International Conference on Fracture and Strength of Solids.*

39. Eggerston, E.C., Kapulla, R., Fokken, J. and Prasser, H.M. (2011). Turbulent mixing and its effects on thermal fatigue in nuclear reactors. *World Academy of Science, Engineering and Technology* 76: 206–213.

40. Keller, M.F. (2011). *Hybrid integrated energy production processes.* US Patent US7,961,835B2 (June, 14, 2011).

41. Richard, E., Mead, B., Zlotnikov, E., Park, H., Us, N.J., Haders, D. and Nj, S. (2011). (12) United States Patent. Doi: 10.1145/634067.634234.

42. Energy Information Administration. (2013, October 10). Form EIA-860 detailed data set – final 2012 data. www.eia.gov/electricity/data/eia860/. Accessed 21 October 2013.

43. Cherry, R. Aumeier, S. and Boardman, R. (2012). Large hybrid energy systems for making low CO_2 load-following power and synthetic fuel. *Energy & Environmental Science* 5(2): 5489–5497.

44. Bragg-Sitton, S., Boardman, R., McKellar, M., Garcia, H., Wood, R., Sabharwall, P. and Rabitti, C. (2013, May). Value proposition for load-following small modular reactor hybrid energy systems. INL/EXT-13-29298, a report by INL, Idaho Falls, Idaho.

45. Borissova, A. (2015). Analysis and synthesis of a hybrid nuclear-solar power plant. *BgNS Transmission* 20: 58–61.

46. Popov, D. and Borissova, A. (2017). Innovative configuration of a hybrid nuclear-solar tower power plant. *Energy* 125: 736e746. Doi: 10.1016/j.energy.2017.02.147.

47. Bartev B. S., William A.A. and David L.K. (2013). Solar-nuclear hybrid power plant—Google Patents, US20150096299A1.

48. Adee, S. and Guizzo, E. (2010). Reactors redux. *IEEE Spectrum* 47: 25e32. Doi: 10.1109/MSPEC.2010.5520624.

49. Orhan, M.F., Dincer, I., Rosen, M.A. and Kanoglu, M. (2012). Integrated hydrogen production options based on renewable and nuclear energy sources. *Renewable and Sustainable Energy Reviews* 16: 6059e6082. Doi: 10.1016/j.rser.2012.06.008.

50. Al-Zareer, M., Dincer, I. and Rosen, M.A. (2017). Performance analysis of a super-critical water-cooled nuclear reactor integrated with a combined cycle, a Cu-Cl ther-mochemical cycle and a hydrogen compression system. *Applied Energy* 195: 646e658. Doi: 10.1016/J.APENERGY.2017.03.046.

51. Ozcan, H. and Dincer, I. (2016). Thermodynamic modeling of a nuclear energy based integrated system for hydrogen production and liquefaction. *Computers & Chemical Engineering* 90: 234e246.

52. Bragg-Sitton, S.M., Boardman, R. and Rabiti, C. et al., (2016, March). *Nuclear-Renewable Hybrid Energy Systems: 2016 Technology Development Program Plan.* INL/EXT-16-38165, Idaho National Laboratory, Idaho Falls.
53. Taibi, E., Gielen, D. and Bazilian, M. (2010). Renewable energy in industrial applications: An assessment of the 2050 potential. United Nations Industrial Development Organization (UNIDO). http://www.unido.org/fileadmin/user_media/Services/Energy_and_Climate_Change/Energy_Efficiency/CCS/Renewable_%20Energy_%20Assessment_%202050_%20Potential.pdf, Accessed 26 March 2013.
54. Wyoming Business Council. (2012). *Preliminary Feasibility of Value-Added Products from Cogeneration and Hybrid Energy Systems in Wyoming.* INL/EXT-12-27249, Idaho National Laboratory, Idaho Falls, Idaho. https://inlportal.inl.gov/portal/server.pt/document/116037/preliminary_feasibility_of_valueadded_products_from_cogeneration_and_hybrid_energy_systems_in_wyoming_pdf, accessed 26 March 2013.
55. Safa, H. (2012). Heat recovery from nuclear power plants. *International Journal of Electrical Power & Energy Systems* 42(1): 553–559.
56. Bergroth, N. (2012). Large-Scale Combined Heat and Power (CHP) Generation at Loviisa Nuclear Power Plant Unit 3. a website report.
57. Evins, R., Pointer, P. and Vaidyanathan, R. Optimisation for CHP and CCHP decision-making, Proc of Building Simulation 2011. *12th Conference of International Building Performance Simulation Association*, 14–16 November 2011, Sydney, 1335–1342.
58. Vasebi, A. Fesanghary, M. and Bathaee, S. (2007). Combined heat and power economic dispatch by harmony search algorithm. *International Journal of Electrical Power & Energy Systems* 29(10): 713–719.
59. Houlihan R. (1987). Recent enhancements in mined oil sands bitumen extraction technology. *The Journal of Canadian Petroleum Technology* 26(1): 91–96.
60. Hong, P.K., Cha, Z., Zhao, X., Cheng, C.-J. and Duyvesteyn, W. (2013). Extraction of bitumen and oil sands with hot water and pressure cycles. *Fuel Processing Technology*, 106: 460–467.
61. Oyeneyin, B., Bali, A. and Adom, E. (2012). Optimization of steam assisted gravity drainage (SAGD) for improved recovery from unconsolidated heavy oil reservoirs. In: A.O. Akii Ibhadode (ed.), *Advanced Materials Research: Advances in Materials and Systems Technologies III*, vol. 367, Trans Tech Publications, Inc., Stafa-Zurich. 403–412.
62. Wang, M.R. and Chang, K.C. (1991). Study on reduction of CO_2 and NO_x emissions in a pulsating combustor burning petroleum coke. *Energy* 15(5): 849–858.
63. Forsberg, C. (2009). Meeting U.S. liquid transport fuel needs with a nuclear hydrogen biomass system. *International Journal of Hydrogen Energy* 34(9): 4227–4236.
64. Nelson, L., Gandrik, A., McKellar, M., Patterson, M., Robertson, E. and Wood, R. (2011). Integration of high temperature gas-cooled reactors into selected industrial process applications. INL/EXT-11-23008. Idaho National Laboratory. www.inl.gov/technicalpublications/Documents/5163472.pdf, Accessed 26 March 2013.
65. Tumuluru, J.S., Hess, J.R., Boardman, R.D., Wright, C.T. and Westover, T.L. (2012). Formulation, pretreatment, and densification options to improve biomass specification for co-firing high percentages with coal. *Industrial Biotechnology* 8(3): 113–132.
66. Steele, P.H. (2009). Hydrocarbons production via biomass fast pyrolysis and hydrodeoxygenation, *AIChE Annual Meeting, Conference Proceeding*, Nashville, TN.
67. Trippe, F. (2010). Techno-economic analysis of fast pyrolysis as a process step with biomass-to-liquid fuel production. *Waste & Biomass Valorization* 1(4): 415–430.
68. Arbogast, S., Bellman, D. and Wykowski, J. (2012). Advanced bio-fuel from pyrolysis oil: The impact of economies of scale and use of existing logistics and processing capabilities. *Fuel Processing Technology* 104: 121–27.

69. Chen, W.H., Lin, M.R., Lu, J.J. and Leu, T.S. (2010). Thermodynamic analysis of hydrogen production from methane via autothermal reforming and partial oxidation followed by water gas shift reaction. *International Journal of Hydrogen Energy* 35(21): 11787–11797.

70. Higman, C. and van er Burgt, M. (2003). *Gasification*. Elsevier/Gulf Professional Pub., Boston, MA.

71. Bell, D.A., Towler, B.F. and Fan, M. (2011). *Coal Gasification and Its Applications*. William Andrew/Elsevier, Oxford/Burlington, MA.

72. International Atomic Energy Agency. Status of small and medium sized reactor designs. http://www.iaea.org/NuclearPower/Downloads/Technology/files/SMR-booklet.pdf. Accessed 26 March 2013, Brussels, Belgium.

73. World Nuclear Association (WNA). Nuclear power reactors. www.world-nuclear.org/info/inf32.html. Accessed 15 November 2012.

74. McKellar, M. (2010). *Power Cycles for the Generation of Electricity from a Next Generation Nuclear Plant*. INL/TEV-674. Idaho National Laboratory, Idaho Falls, Idaho.

75. McKellar, M. (2012). *Sensitivity of High Temperature Gas Reactor (HTGR) Heat and Power Production to Reactor Outlet Temperature (ROT), Economic Analysis*. INL/TEV-998. Idaho National Laboratory, Idaho Falls, Idaho.

76. Etchepareborda, A. and Flury, C. (2002). Multivariable robust control of an integrated nuclear power reactor. *Brazilian Journal of Chemical Engineering* 4(4): 441–447.

77. Adee, S. and Guizzo, E. (2010, August). Reactors redux. *IEEE Spectrum* 47(8): 25–32.

78. Dunn, R., Hearps, R.J. and Wright, M.N. (2012, February) Molten-salt power towers: Newly commercial concentration solar storage. *Proceedings of the IEEE, Special Issue, Addressing the Intermittency Challenge: Massive Energy Storage in a Sustainable Future* 100(2): 504–515.

79. Wagner, S.J. and Rubin, E. (2012). Economic implications of thermal energy storage for concentrated solar thermal power. *Renewable energy* 39: 1–15.

80. U.S. Department of Energy—Office of Electricity Delivery & Energy Reliability. Smart grid research & development multi-year program plan (MYPP) 2010–2014. September 2012 Update. http://energy.gov/sites/prod/files/SG_MYPP_2012%20Update.pdf. Accessed 2 June 2013.

81. Boarin, S., Locatelli, G., Mancini, M. and Ricoti, M.E. (2012). Financial case studies on small- and medium-size modular reactors. *Nuclear Technology* 178(2): 218–232.

82. Kuznetsov, V. (2008). Options of small and medium sized reactors (SMRs) to overcome loss of economies of scale and incorporate increased proliferation resistance and energy security, *Progress in Nuclear Energy* 50: 242–250.

83. Wood, R.A., Gandrik, A. and Boardman, R.D. (2010). *Sensitivity of Hydrogen Production Via Steam Methane Reforming (SMR) to High Temperature Gas Cooled Reactor (HTGR) Reactor Outlet Temperature (ROT) Economics Analysis*. INL/TEV-962, Idaho National Laboratory, Idaho Falls, Idaho.

84. Wender, I. (1984). Chemicals from methanol. *Catalysis Reviews. Science and Engineering* 26: 303–321.

85. Lee, S., Gogate, M. and Kulik, C. (1995). Methanol-to-gasoline vs. DME-to-gasoline, II process comparison and analysis. *Fuel Science and Technology International* 13: 1039–1057.

86. Marulanda, V.F. (2012). Biodiesel production by supercritical methanol transesterification: Process simulation and potential environmental impact assessment. *Journal of Cleaner Production* 33: 109–116.

87. Leckel, D. (2009). Diesel production from Fisher-Tropsch: The past, the present, and new concepts, *Energy & Fuel* 23: 2342–2358.

88. Schulz, H. (1999). Short history and present trends of Fischer-Tropsch synthesis. *Applied Catalysis A: General* 186(1–2): 3–12.

89. Neburchilov, V., Martin, J., Wang, H. and Zhang, J. (2007). A review of polymer electrolyte membranes for direct methanol fuel cells. *Journal of Power Sources* 169(2): 221–238.

90. Nelson, L., Gandrik, A., McKellar, M., Patterson, M. and Wood, R. (2012). *Integration of High Temperature Gas-Cooled Reactors into Industrial Process Applications*. INL/EXT-09-16942, Rev 2. Idaho National Laboratory, Idaho Falls, Idaho.

91. Suresh, B., Schlag, S., Kumamoto, T. and Ping, Y. (2010, July). CEH marketing research report: Hydrogen. In: *Chemical Economics Handbook* Vol. 743, 5000 A, IHS Markit, London, UK.

92. Board on Energy and Environmental Systems (BEES). (2008). *Review of DOE's Nuclear Energy Research and Development Program, Chapter 3*. National Academies Press (NAE), 31–46. http://www.nap.edu/openbook.php?record_id=11998&page=31. Accessed 26 March 2013.

93. Forsberg, C. (2009). Is hydrogen the future of nuclear energy? *Nuclear Technology* 166(1): 3–10.

94. Rosen, M.A. (1995). Energy and exergy analyses of electrolytic hydrogen production. *International Journal of Hydrogen Energy* 20(7): 547–53.

95. National Academy of Engineering (NAE). (2004). *The Hydrogen Economy: Opportunities, Costs, Barriers, and R&D Needs, Chapter 8*, a report by NAE, Washington, DC.

96. Varrin, R. Jr., Reifsneider, K., Sanborn, D., Irving, P. and Rolfson, G. (2009). *NGNP Hydrogen Technology Down-Selection. Results of the Independent Review Team (IRT) Evaluation*. Dominion Engineering, Inc, Reston, VA.

97. O'Brien, J.E. (2008). Thermodynamic considerations for thermal water splitting processes and high temperature electrolysis. *Proceedings of the 2008 International Mechanical Engineering Congress and Exposition*. IMECE2008-68880.

98. Wright, M., Daugaard, D., Satrio, J. and Brown, R. (2010). Techno-economic analysis of biomass fast pyrolysis to transportation fuels. *Fuel* 89(1): S2–S10.

99. Chu, S. and Majumdar, A. (2012). Opportunities and challenges for a sustainable energy future. *Nature* 488(16): 294–303.

100. Young, M. (1994). Evaluation of population density and distribution criteria in nuclear power plant siting, SAND-9300848, a report by Sandia Lab., Albuquerque, NM.

101. Nuclear Regulatory Commission (NRC). (2009). Safety evaluation report for an early site permit (ESP) at the Vogtle Electric Generating Plant (VEGP) ESP site. NTIS: NUREG1923.

102. Kleiner, F. and Kauffman, S. All electric driven refrigeration compressors in LNG plants offer advantages. Siemens 2005. www.energy.siemens.com/us/pool/hq/energy-topics/pdfs/en/oil-gas/1_All_electric_driven_refrigeration.pdf. Accessed 26 March 2013. Gastech (2005). Contact steve.kauffman@shell.com.

103. Seinfeld, J. and McBride, W. (1970). Optimization with multiple performance criteria. Application to minimization of parameter sensitivities in a refinery model. *Industrial & Engineering Chemistry Process Design and Development* 9(1): 53–57.

104. Office of Energy Efficiency & Renewable Energy. *Combined Heat and Power*. U.S. Department of Energy, [Online]. Available at: http://energy.gov/eere/amo/combined-heat-and-power. Accessed 16 February 2016.

105. Antkowiak, M., Ruth, M., Boardman, R., Bragg-Sitton, S., Cherry, R. and Shunn, L. (2012, July). *Summary Report of the INL-JISEA Workshop on Nuclear Hybrid Energy Systems*. NREL/TP-6A50-55650. National Renewable Energy Laboratory, Golden, CO.

106. Energy Information Administration. (2016, February 16). Electric power annual with data for 2014. http://www.eia.gov/electricity/annual/html/epa_04_02_a.html.

107. Lawrence Livermore National Laboratory. Estimated Energy Use in 2014. LLNL-MI-410527. https://flowcharts.llnl.gov/content/assets/images/energy/us/Energy_US_2014.png.

108. Cherry, R.S., Aumeier, S.E. and Boardman, R.D. (2012). Large hybrid energy systems for making low CO_2 load-following power and synthetic fuel. *Energy & Environmental Science* 5: 5489–5497. http://pubs.rsc.org/en/Content/ArticleHtml/2012/EE/c1ee02731j

109. Ruth, M.F., Zinaman, O.R., Antkowiak, M., Boardman, R.D., Cherry, R.S. and Bazilian, M.D. (2014). Nuclear-renewable hybrid energy systems: Opportunities, interconnections, and needs. *Energy Conversion and Management* 78: 684–694. http://www.sciencedirect.com/science/article/pii/S0196890413007516.

110. Forsberg, C.W. (2013). Hybrid systems to address seasonal mismatches between electricity production and demand in nuclear renewable electrical grids. *Energy Policy* 62: 333–341. http://www.sciencedirect.com/science/article/pii/S0301421513007003.

111. Antkowiak, M., Ruth, M., Boardman, R., Bragg-Sitton, S., Cherry, R. and Shunn, L. (2012). *Summary Report of the INL-JISEA Workshop on Nuclear Hybrid Energy Systems*. INL/EXT-12-26551. Idaho National Laboratory, Idaho Falls, ID NREL/TP-6A50-55650. National Renewable Energy Laboratory, Golden, CO. http://www.nrel.gov/docs/fy12osti/55650.pdf.

112. Ruth, M., Antkowiak, M. and Gossett, S. (2011). *Nuclear and Renewable Energy Synergies Workshop: Report of Proceedings*. NREL/TP-6A30-52256. National Renewable Energy Laboratory, Golden, CO. http://www.nrel.gov/docs/fy12osti/52256.pdf.

113. Pellegrino, J., Margolis, N., Miller, M., Justiniano, M. and Thedki, A. (2004, December) Energy use, loss, and opportunities analysis: U.S. manufacturing and mining. Energetics, Inc. and E3M, Inc. for the U.S. Department of Energy, Industrial Programs. https://www1.eere.energy.gov/manufacturing/intensiveprocesses/pdfs/energy_use_loss_opportunities_analysis.pdf.

114. Carelli, M.D. and Ingersoll, D.T. (2015). *Handbook of Small Modular Nuclear Reactors*. Woodhead Publishing Series in Energy Number 64, Elsevier, ISBN 978-0-85709-851-1 (print); ISBN 978-0-85709-853-5 (online). http://store.elsevier.com/Handbook-of-Small-Modular-NuclearReactors/Mario-D-Carelli/isbn-9780857098511/.

115. Advanced Reactors Information System (ARIS). (2012, September) *Status of Small and Medium Sized Reactor Designs: A Supplement to the IAEA*. International Atomic Energy Agency. http://aris.iaea.org.

116. Khamis, I. and Kavvadias, K.C. (2012, July). Trends and challenges toward efficient water management in nuclear power plants. *Nuclear Engineering and Design* 248: 48–54. http://www.sciencedirect.com/science/article/pii/S0029549312001732.

117. International Atomic Energy Agency. (2007). IAEA-TECDOC 1524 – status of nuclear desalination in IAEA member states. http://www-pub.iaea.org/mtcd/publications/pdf/te_1524_web.pdf; International Atomic Energy Agency. (2007). IAEA-TECDOC 1561 – economics of nuclear desalination: New developments and site specific studies. Vienna. http://www-pub.iaea.org/MTCD/Publications/PDF/te_1561_web.pdf.

118. Garcia, H.E., Chen, J., Kim, J.S. et al., (2016, February). *Dynamic Performance Analysis of Two Regional Nuclear Hybrid Energy Systems*. Accepted for publication in Energy. (For reference contact Humberto.Garcia@inl.gov).

119. Bragg-Sitton, S.M., Boardman, R., Rabiti, C. et al., (2016, March). *Nuclear-Renewable Hybrid Energy Systems: 2016 Technology Development Program Plan*. INL/EXT-16-38165. Idaho National Laboratory, Idaho Falls. (For reference contact Richard.Boardman@inl.gov).

120. Bragg Sitton S., Boardman, R., Wood, R., Garcia, H., McKellar, M., Sabharwall, P. and Rabiti, C. (2013, May) *Value Proposition for Load-Following Small Modular Hybrid Energy Systems*. INL/EXT-13-29298. (For reference contact Richard.Boardman@inl.gov).

121. Nelson, L., Gandrik, A., McKellar, M., Patterson, M., Robertson, E., Wood, R. and Maio, V. (2011, September) *Integration of High Temperature Gas-Cooled Reactors into Industrial Process Applications.* INL/EXT-09-16942 Rev. 3. Idaho National Laboratory, Idaho Falls, Idaho. https://inldigitallibrary.inl.gov/sti/4374066.pdf.

122. Nelson, L., Gandrik, A., McKellar, M., Patterson, M., Robertson, E. and Wood, R. (2011, August). *Integration of High Temperature Gas-Cooled Reactors into Selected Industrial Process Applications,* INL/EXT-11-23008. Idaho National Laboratory. https://art.inl.gov/NGNP/INL%20Documents/Year%202011/Integration%20of%20 High%20Temperature%20Gas-Cooled%20Reactors%20into%20Selected%20 Industrial%20Process%20Applications.pdf.

123. McKellar, M. (2011, September). *An Analysis of Fluids for the Transport of Heat with HTGR-integrated Steam Assisted Gravity Drainage.* TEV-1351, rev 0. Idaho National Laboratory. (For reference contact Michael.McKellar@inl.gov).

124. Nelson, L., Gandrik, A., McKellar, M., Patterson, M., Robertson, E., Wood, R. and Maio, V. (2011, September). *Integration of High Temperature Gas-Cooled Reactors into Industrial Process Applications.* INL/EXT-09-16942 Rev. 3. Idaho National Laboratory, Idaho Falls, Idaho. https://inldigitallibrary.inl.gov/sti/4374066.pdf.

125. Nelson, L., Gandrik, A., McKellar, M., Patterson, M., Robertson, E. and Wood, R. (2011, August). *Integration of High Temperature Gas-Cooled Reactors into Selected Industrial Process Applications.* INL/EXT-11-23008. Idaho National Laboratory, https://art.inl.gov/NGNP/INL%20Documents/Year%202011/Integration%20of%20 High%20Temperature%20Gas-Cooled%20Reactors%20into%20Selected%20 Industrial%20Process%20Applications.pdf.

126. Bragg Sitton S., Boardman, R., Wood, R., Garcia, H., McKellar, M., Sabharwall, P. and Rabiti, C. (2013, May). *Value Proposition for Load-Following Small Modular Hybrid Energy Systems.* INL/EXT-13-29298. (For reference contact Richard.Boardman@inl.gov).

127. Bragg-Sitton S., Boardman, R., Cherry, R., Deason, W. and McKellar, M. (2014, March). *An Analysis of Methanol and Hydrogen Production via High-Temperature Electrolysis Using the Sodium Cooled Advanced Fast Reactor.* INL/EXT-14-31642. https://inldigitallibrary.inl.gov/sti/6236303.pdf.

128. Bragg-Sitton S., Boardman, R., Ruth, M., Zinaman, O., Forsberg, C. and Collins, J. (2014, August). *Integrated Nuclear-Renewable Energy Systems: Foundational Workshop Report.* INL/EXT-14-32857; NREL/TP-6A20-62778. https://mitei.mit.edu/ system/files/20150526INL-EXT-1432857IESWorkshopSummary.pdf. NREL, Golsen CO.

129. Forsberg, C. (2015, August). *Strategies for a Low-Carbon Electricity Grid With Full Use of Nuclear, Wind and Solar Capacity to Minimize Total Costs.* Massachusetts Institute of Technology, Cambridge, MA, MIT-ANP-162. http://mitei. mit.edu/system/files/ANP162%20Zero-Carbon%20Nuclear%20Renewable%20 Grid%5B1%5D%5B1%5D_1.pdf. MIT, Cambridge, MA.

130. Forsberg, C. and Golay, M. (2015, May 26–27). *Low-Carbon Energy Economy Workshop.* Massachusetts Institute of Technology, Cambridge MA. https://mitei.mit. edu/may–26-2015. MIT, Cambridge, MA.

131. Garcia, H.E., Chen, J., Kim J.S. et al., Dynamic performance analysis of two regional nuclear hybrid energy systems. *Energy,* submitted October 2015, under review. (For reference contact Richard.Boardman@inl.gov).

132. Chen, J. and Garcia, H.E. (2016). Economic optimization of operations for hybrid energy systems under variable markets. INL/JOU-15-36620, a report by INL under Contract DE-AC07-05ID14517 INL, Idaho Falls, Idaho, (September, 2016).

133. Chen, J., Garcia, H.E., Kim, J.S. and Bragg-Sitton, S.M. Operations Optimization of Nuclear Energy Systems. INL/JOU-15-36619, a report by INL, Idaho Falls, Idaho under Contract DE-AC07-05ID14517 (Aug., 2016).

134. Kim, J.S., Chen, J. and Garcia, H.E. Modeling, control, and dynamic performance analysis of a reverse osmosis desalination plant integrated within hybrid energy systems. *Energy*, submitted September 2015, under review, (For reference contact JongSuk. Kim@inl.gov).

135. C. Rabiti, Kinoshita, R., Kim, J., Deason, W., Bragg-Sitton, S., Boardman, R. and Garcia, H. (2015, September). *Status on the Development of a Modeling and Simulation Framework for the Economic Assessment of Nuclear Hybrid Energy Systems.* INL/EXT-15-36451. Idaho National Laboratory. https://inldigitallibrary.inl.gov/sti/6799582.pdf. INL, Idaho Falls, Idaho.

136. Stack, D.C. and Forsberg, C. (2015, June 7–11). Improving nuclear system economics using firebrick resistance-heated energy storage (FIRES) *American Nuclear Society Annual Meeting*, San Antonio, TX. https://mitei.mit.edu/system/files/20150526-3-FIRES.pdf.

137. Ice Bank Energy Storage. CALMAC, Fair Lawn, NJ. http://www.calmac.com/icebank-energy-storage-model-c.

138. Ice Storage Design and Application. (2009). TRANE, Ingersoll Rand. https://www.trane.com/content/dam/Trane/Commercial/global/productssystems/education-training/continuing-education-gbci-aia-pdh/Ice-Storage-Design-and-Application/Trane_ENL_Ice_Storage_Design.pdf. Ingersall-Rand, Swords, Ireland.

139. Akshay, S. (2013). Economic forecasting and optimization in a smart grid built environment. Masters Report. https://repositories.lib.utexas.edu/handle/2152/22436.

140. Forsberg, C. (2015, August). *Strategies for a Low-Carbon Electricity Grid with Full Use of Nuclear, Wind and Solar Capacity to Minimize Total Costs.* Massachusetts Institute of Technology, Cambridge, MA, MIT-ANP-162. http://mitei.mit.edu/system/files/ANP162%20Zero-Carbon%20Nuclear%20Renewable%20Grid%5B1%5D%5B1%5D_1.pdf.,MIT, Cambridge, MA.

141. Ingersoll, D., Houghton, Z., Bromm, R., Desportes, C., McKellar, M. and Boardman, R. (2014, August 24–28). Extending nuclear energy to non-electrical applications. *19th Pacific Basin Nuclear Conference*, PBNC2014-209, Vancouver BC. https://inldigitallibrary.inl.gov/sti/6303857.pdf.

142. Lokhov, A., Cameron, R., Cometto, M. and Ludwig, H. (2011, June). *Technical and Economic Aspects of Load following with Nuclear Power Plants.* Nuclear Energy Agency, Organization for Economic Cooperation and Development. http://www.oecd-nea.org/ndd/reports/2011/loadfollowing-npp.pdf. A report by NEA-OECD.

143. Keppler, J., Cometto, M. and Cameron, R. (2012). *Nuclear Energy and Renewables: System Effects in Low-carbon Electricity Systems.* NEA No. 7056. Nuclear Energy Agency & Organization for Economic Co-operation and Development, Paris. https://www.oecd-nea.org/ndd/pubs/2012/7056-system-effects.pdf.

144. Bragg Sitton S., Boardman, R., Wood, R., Garcia, H., McKellar, M., Sabharwall, P. and Rabiti, C. (2013, May). *Value Proposition for Load-Following Small Modular Hybrid Energy Systems.* INL/EXT-13-29298. (For reference contact Richard.Boardman@inl.gov).

145. Garcia, H.E., Chen, J., Kim, J.S., McKellar, M.G., Deason, W.R., Vilim, R.B., Bragg-Sitton, S.M. and Boardman, R.D. (2015, April). *Nuclear Hybrid Energy Systems Regional Studies: West Texas & North eastern Arizona.* INL/EXT-15-34503. http://www.osti.gov/scitech/biblio/1236837-nuclearhybrid-energy-systems-regional-studies-west-texas-amp-northeastern-arizona.

146. Garcia, H.E., Mohanty, A., Lin, W.C. and Cherry, R.S. (2013). Dynamic analysis of hybrid energy systems under flexible operation and variable renewable generation – part II: Dynamic cost analysis. *Energy* 52: 17–26. http://www.sciencedirect.com/science/article/pii/S0360544212008936.

147. O'Brien, J. E. (2012, April). Review of the potential of nuclear hydrogen for addressing energy security and climate change. *Nuclear Technology* 178(1): 55–65, http://www.ans.org/pubs/journals/nt/a_13547.

148. Nelson, L., Gandrik, A., McKellar, M., Patterson, M., Robertson, E., Wood, R. and Maio, V. (2011, September). *Integration of High Temperature Gas-Cooled Reactors into Industrial Process Applications*. INL/EXT-09-16942 Rev. 3. Idaho National Laboratory, https://inldigitallibrary.inl.gov/sti/4374066.pdf.

149. Herring, J., O'Brien, J., Stoots, C. and Hawkes, G. (2006). Progress in high-temperature electrolysis for hydrogen production using Planar SOFC technology. *International Journal of Hydrogen Energy* 32(4): 440–450.

7 Hybrid Energy Systems for Manufacturing Industry

7.1 INTRODUCTION

Process heating systems are critical to the global manufacturing industry's ability to turn raw materials (such as oil, iron ore, trees, crops, etc.) into products (including plastics, metals, paper, and food). These systems use energy to generate, supply, transfer, contain, or recover heat energy. The manufacturing industry must increase the efficiency and reduce the energy utilization of these systems in order to be competitive, while reducing fossil fuel use and greenhouse gas (GHG) emissions [1,2].

The industrial sector was the third-largest source of direct U.S. GHG emissions in 2014 behind electricity generation and transportation and accounted for roughly 20% of total emissions [2]. The Energy Information Administration (EIA) projects that total U.S. energy consumption will grow to about 108 exajoules (1 EJ = 10^{18} J) or 102 quads (1 quad = 10^{15} British thermal units) in 2025, with nearly all of the growth coming from the industrial sector [2]. Energy consumption in the industrial sector is forecast to increase to 39.5 EJ (37.4 quads)—a 22% increase, exceeding 36% of total energy consumption in the United States. Therefore, it is imperative that industrial GHG emissions be considered in any strategy intent on achieving deep decarbonization of the energy sector as a whole.

It is important to note that unlike the transportation sector and electrical grid, energy use by industry often involves direct conversion of primary energy sources to thermal and electrical energy at the point of consumption. About 52% of U.S. industrial direct GHG emissions are the result of fuel combustion [3–6] to produce hot gases and steam for process heating, process reactions, and process evaporation, concentration, and drying. The heterogeneity and variations in scale of U.S. industry and the complexity of modern industrial firms' global supply chains are among the sector's unique challenges to minimizing its GHG emissions. A combination of varied strategies—such as energy efficiency, material efficiency (or less reliance of raw materials on fossil fuels), and switching to low-carbon fuels—can help reduce absolute industrial GHG emissions [7]. The concept of hybrid energy systems (HESs) and sometimes processes can be used in all three strategies.

The US remains the largest manufacturing country in the world, producing 21% of manufactured products on a global scale, and $1.7 trillion of value each year, 11.7% of US GDP, according to the National Association of Manufacturers. All of that manufacturing requires a significant amount of energy usage, particularly in electricity. According to a June 2014 report on renewable energy in manufacturing from the International Renewable Energy Agency, "Electricity accounts for around

20% of final energy use in manufacturing, and is used for the production of aluminum, equipment, and lighting and cooling in factories."

On a global scale, the energy-intensive sectors, namely iron and steel, chemical and petrochemical, non-ferrous metals, non-metallic minerals, and pulp and paper, are estimated to continue to use more than 75% of industrial energy use (including feedstock). However, they account for less than 5% of all manufacturing plants (30,000–60,000 plants globally). The global industry sector accounted for about a third of the global energy use and in 2009 when the sector's total final energy demand reached 128 exajoules (EJ). A total of 78 EJ of fuels were used to generate process heat. Another 9 EJ was used by blast furnaces and coke ovens for iron and steel production. Petrochemical feedstock use for the production of chemicals and polymers (together referred to as "materials" throughout this study) was about 16 EJ. The sector's remaining energy use was electricity demand (24 EJ) for various uses such as electrolysis, motor drives, cooling, or refrigeration [1,2,8,9].

Approximately 64% of the total final industrial energy use worldwide came from non-OECD (Organization of Economic Co-operation and Development) countries (81 EJ), the majority which are developing countries and economies in transition. Industrialized and high-income countries (i.e., OECD countries) used in total 47 EJ final energy (36%). Today, 91% of the sector energy use originates from fossil fuels, coal, petroleum products and natural gas accounting for 44%, 26%, and 21% of the total final energy use, respectively (excluding the demand for electricity and feedstock use). Renewable energy sources account for about 9% of the industrial energy use, which is mostly biomass and waste (the shares are the same in both OECD and non-OECD countries). The fuel mix varies substantially across different regions. While in the OECD Americas and Europe, natural gas accounts for at least 40% of the total fuel mix, in China and the OECD Pacific, 80% and 35%, respectively, of the total demand is met by coal. A few regions use a high share of renewable energy in their fuel mix, such as India (24%), Latin America (35%), and Africa (42%). In comparison, the share of renewable energy use in economies in transition and the Middle East is less than 1% [10].

Clean energy manufacturing involves the minimization of the energy and environmental impacts of the production, use, and disposal of manufactured goods, which range from fundamental commodities such as metals and chemicals to sophisticated final-use products such as automobiles and wind turbine blades. The manufacturing sector, a subset of the industrial sector, consumes 24 quads of primary energy annually in the United States—about 79% of total industrial energy use, as shown in Figure 7.1 [1]. Clean energy manufacturing can improve energy utilization and also yield economy-wide reductions in GHG emissions through changes in energy use enabled by the development of new materials and process technologies.

As pointed out by DOE report [1] this chapter examines the opportunities for improvements in energy and materials utilization with HESs and processes within three spaces:

1. Individual manufacturing processes and unit operations
2. Goods-producing facilities, including manufacturing business processes
3. Manufacturing supply chains and manufactured goods, including impacts from all phases of the product life cycle

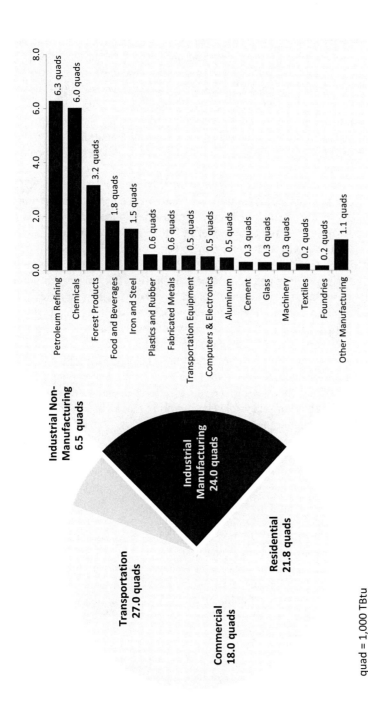

FIGURE 7.1 Manufacturing share of the nation's overall energy consumption and breakdown of manufacturing primary energy (including nonfuel feedstock energy) consumption by subsector [1,11].

These opportunities correspond to three levels of manufacturing system integration: manufacturing/unit operations, production/facility systems, and supply chain systems, as illustrated in Figure 7.2. In this chapter, in line with the scope of the book, emphasis is, however, placed on (a) how use of hybrid combined heat and power (CHP) process (cogeneration) and HESs can improve efficiency of fossil fuel based energy use in the manufacturing and industrial processes (generally this means the introduction of workable waste heat recovery processes), (b) how renewable technology infusion (through HESs and processes) reduce use of fossil fuel for heating and electricity needs of the manufacturing and industrial operations, and (c) how to reduce use of fossil fuel for raw materials used in industrial processes by the infusion of biofuels or biomass.

Since 2009, energy demand in the global industry sector has grown 2.7% per year (including demand as feedstock use). The share of energy consumption in non-OECD countries has also continued to grow, and as mentioned before, they account for more than two thirds of the total global in 2012. In most countries, industry is a key sector or the economy and it will remain so in the next decades as well since the demand for materials (e.g., steel, plastics, bricks) will continue to increase as a consequence of population and economic growth. In turn, this growth will have important effects on the sector's demand for energy. Increasing energy use will lead to the release of more carbon dioxide (CO_2) emissions, which are regarded as the main driver of climate change. Increasing concerns about climate change has created different policy responses. In 2011, the Sustainable Energy for All (SE4ALL) initiative was launched by the United Nations Secretary-General. It seeks commitment from all countries to meet three objectives by 2030: (a) ensuring universal access to modern energy services; (b) doubling the rate of improvement in energy efficiency; and (c) doubling the share of renewable energy in the global energy mix. The third objective is addressed by the International Renewable Energy Agency (IRENA) in its renewable energy roadmap—REmap 2030—to double the share of renewable energy in the global energy mix [12,13]. Given its large energy share, the industrial sector has an important role to play in meeting these targets. Chapas and Colwell [14] have examined industrial technologies program research plan for energy-intensive process industries.

A number of measures exist to reduce the industry sector's increasing demand for fossil fuels and the related CO_2 emissions. Among them, conservation of energy use by improving energy efficiency is the first step as it is cost-effective and various technologies exist which are suitable for different production processes [15,16]. According to Saygin, Patel, and Gielen [17], improving industrial energy efficiency by implementing best practice technologies could reduce total final industrial energy demand more than 25%. However, even more reductions in industrial energy demand would be required to meet ambitious targets in the long term. Improving efficiency also involves conversion of waste heat to power or other useful purposes. Thus cogeneration or CHP process is an inherent part of this strategy. CHP is inherently a HES.

Another measure is treatment of CO_2 generated by industry. One approach, carbon capture and storage (CCS) has not yet proven successful on a commercial scale, its costs for the industry sector are too high, and the capture processes (e.g., heat regeneration, compression) require additional energy [18]. Conversion of CO_2 to

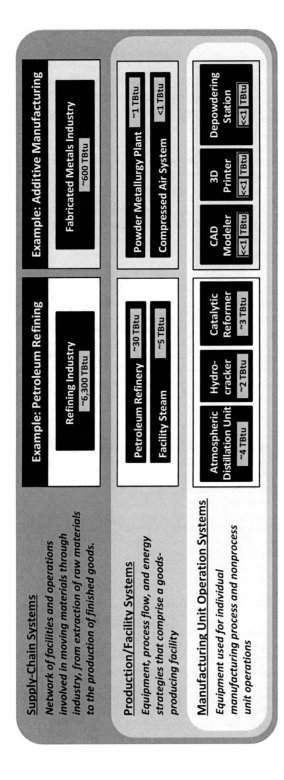

FIGURE 7.2 Levels of system integration in manufacturing: opportunities for energy savings occur at each system level. The energy usage estimates shown (small boxes) represent the typical annual energy consumption levels in the United States for a single industry, production facility, or piece of manufacturing equipment [1].

power by fuel cell or diesel fuel described in my previous book [19] has potential but more work is needed for their commercialization.

The third option is the infusion of renewable energy (in the form of HESs) which has so far not received much attention, especially compared to the power and transport sectors [20]. Suitable policies could also help the industrial sector to increase its share of renewable energy in the next decades as there are already many commercial-scale examples of biomass use technologies as fuel and feedstock, as well as solar thermal or geothermal energy use to provide process heat [21–23]. The share of renewable energy in 2009 was about 10%, and it has grown from around 5.5% in 1971 [12,13]. The share of renewable energy can continue to increase, as there is large potential to substitute the fossil fuel-based fuel use for process heat generation and feedstock use for materials production. With new climate policies, development in fossil fuel prices, and the introduction of efficient conversion technologies and learning from the deployment of renewable energy technologies could result in increased cost-competitiveness.

A technology roadmap specific to the industry sector, encompassing the technology and cost aspects of renewable energy technologies, as well as their respective potential, can provide valuable insights. Under the REmap umbrella, IRENA published its renewable energy roadmap for the manufacturing industry sector in June 2014 [12,13]. This working paper provides detailed analyses and background information to this roadmap. With this paper, IRENA also aims to support policy-makers in the development of effective policies to promote renewable energy technologies in the industry sector by identifying the most important subsectors and regions where renewable energy technologies can play a role [24,25].

7.2 METHODS FOR IMPROVING ENERGY EFFICIENCY BY HESs

Process heating and motor-driven systems collectively consume more than nine quads of end-use energy in the U.S. manufacturing sector. Continued technology maturation and improvements will drive technology uptake to reduce energy intensity and can narrow the gap between current energy use and practical minimum energy requirements, especially for major energy-intensive commodities. Transformational next-generation processes and technologies that are not bound by practical (energy and emissions) limitations of current processes, such as low-thermal-budget processes and next-generation motor-driven systems, can drive manufacturing energy reductions and expand capabilities of manufacturers [26–29].

7.2.1 PROCESS HEATING SYSTEMS (INCLUDING STEAM FOR UNIT OPERATIONS)

Process heating accounts for approximately 61% of manufacturing end-use energy use annually [1]. Energy for process heating is obtained from a combination of electricity, steam, and fuels such as natural gas, coal, biomass, and fuel oils. In 2010, process heating consumed approximately 330 TBtu of electricity, 2,290 TBtu of steam, and 4,590 TBtu of fuel [30]. Characteristics of Common Industrial Processes that require process heating [1] are illustrated in Table 7.1.

TABLE 7.1

Characteristics of Common Industrial Processes that Require Process Heating [1]

Process Heating Operation	Description/Example Applications	Typical Temperature Range (F)	Estimated (2010) U.S. Energy Use (TBtu)
Fluid heating, boiling, and distillation	Distillation, reforming, cracking, hydrotreating; chemicals production, food preparation	150–1,000°	3,015
Drying	Water and organic compound removal	200–700°	1,178
Metal smelting and melting	Ore smelting, steelmaking, and other metals production	800–3,000°	968
Calcining	Lime calcining	1,500–2,000°	395
Metal heat treating and reheating	Hardening, annealing, tempering	200–2,500°	203
Nonmetal melting	Glass, ceramics, and inorganics manufacturing	1,500–3,000°	199
Curing and forming	Polymer production, molding, extrusion	300–2,500°	109
Coking	Coke-making for iron and steel production	700–2,000°	88
Other	Preheating; catalysis, thermal oxidation, incineration, softening, and warming	200–3,000°	1,049
Total			7,204

Common process heating systems include equipment such as furnaces, heat exchangers, evaporators, kilns, and dryers. As shown in Table 7.1, three largest energy user processes are fluid heating, boiling, distillation; drying and metal smelting and melting [1]. Industrial process heating, which consumes more than 7,000 TBtu of energy annually [31], is used for fundamental materials transformations including heating, drying, curing, and phase change. Process heating systems are associated with significant thermal losses; nearly 36% of the total energy input to process heating is lost as waste heat [31]. The largest sources of waste heat for most industries are exhaust gases from burners, heat treating furnaces, dryers, and other equipment. Waste heat can also be released to liquids such as cooling water, heated wash water, boiler and blow-down water. Solid waste heat sources include hot products that are discharged after processing or after reactions are complete, hot by-products from processes or combustion of solid materials, and hot equipment surfaces. The quality of these heat sources varies. Industrial waste heat generally occurs in four forms: sensible heat of solids, liquids, and gases; latent heat contained in water vapor or other type of vapors and gases; radiation and convection from hot surfaces; and direct contact conduction (in a few instances). Waste heat losses are a major consideration in process heating, especially for higher temperatures process heating systems such as those used in steelmaking and glass melting. Losses can occur at walls, doors, and openings, and through the venting of hot flue and exhaust gases. Overall, energy losses from process heating systems total more than 2,500 TBtu annually [1,30]. The recovery and use of waste heat offer an opportunity to reutilize wasted heat for other purposes.

TABLE 7.2

RD&D Opportunities for Process Heating and Projected Energy Savings [1]

R&D Opportunity	Applications	Estimated Annual Energy Savings Opportunity (TBtu/y)	Estimated Annual Carbon Dioxide (CO_2) Emissions Savings Opportunity (Million Metric Tons [MMT]/y)
Advanced nonthermal water removal technologies	Drying and concentration	500	35
"Super boilers" (to produce steam with high efficiency, high reliability, and low footprint)	Steam production	350	20
Waste heat recovery systems	Crosscutting	260	25
Hybrid distillation	Distillation	240	20
New catalysts and reaction processes (to improve yields of conversion processes)	Catalysis and conversion	200	15
Lower-energy, high-temperature material processing (e.g., microwave heating)	Crosscutting	150	10
Advanced high-temperature materials for high-temperature processing	Crosscutting	150	10
Net-shape and near-net-shape design and manufacturing	Casting, rolling, forging, additive manufacturing, and powder metallurgy	140	10
Integrated manufacturing control systems	Crosscutting	130	10
Total		2,210	155

Novel processing techniques that involve lower temperature processing or fewer heating steps through process hybridization can also reduce energy consumption. Hybrid process heating systems that combine multiple forms of heat transfer (radiative, conductive, and/or convective methods) or multiple operations into a single piece of equipment (such as hybrid distillation systems) can reduce heating time, increase energy efficiency, and improve product quality. Key research, development, and deployment (RD&D) opportunities for energy and emissions savings in industrial process heating operations as outlined by DOE report [1] are summarized in Table 7.2. While the total energy savings opportunity (2,210 TBtu) is very large, only a portion of this opportunity is technically and economically feasible to capture, as discussed in the *Waste Heat Recovery Systems* Technology Assessment [1].

While every effort should be made to reduce waste heat losses (for example, by integrating advanced insulation techniques and selective heating technologies into process heating equipment), some heat losses are unavoidable. The recovery and reduction of waste heat generated in manufacturing systems offer an opportunity to reduce manufacturing energy use and associated emissions. Waste heat can be recycled either by redirecting the waste stream for use in other thermal processes (e.g., flue gases from a furnace could be used to pre-heat a lower-temperature drying oven) or by converting the waste heat to electricity in a process called waste heat-to-power. In some cases, the technologies needed to economically recover waste heat from hot gases, liquids, or solids are already available. However, industrial facilities often do not implement these technologies, based in part on technology issues (e.g., fouling, corrosion, and high maintenance requirements). According to U.S. EIA *Manufacturing Energy Consumption Survey* (MECS) data, approximately 6% of U.S. manufacturing facilities were using some type of waste heat recovery as of 2010 [1,32].

Industrial users demand equipment lifetimes of several years, low maintenance and cleaning requirements, and consistent and reliable performance over acceptable life. For low-temperature waste heat streams (i.e., less than 400°F), low heat transfer rates and large recovery equipment footprints are major barriers. For high-temperature waste heat streams (i.e., above 1,200° F), materials are needed that can withstand high-temperature gases that may be contaminated with particulate matter (PM) or corrosive chemicals [1,33]. DOE report [1] points out that in order to address these challenges, research and development in the following areas will require:

1. Antifouling technologies that can remove contaminants from waste heat streams or mitigate build-up of debris on heat exchanger surfaces, promoting long-term operation of heat recovery equipment, and avoiding service interruptions for cleaning
2. Advanced materials that can withstand high-temperature waste heat sources
3. Compact, low-cost heat exchangers to reduce the size or footprint of heat recovery equipment
4. Secondary heat recovery technologies to supplement and enhance the performance of primary waste heat recovery equipment
5. Heat recovery chillers that capture waste heat from chilled water systems
6. Integrated heat recovery technologies that combine heating elements with heat recovery equipment, eliminating the need for hot-air piping and external heat recovery equipment
7. Innovative condensing heat exchangers for gases containing high moisture levels and particulates, such as the waste streams discharged from paper and food production equipment
8. Liquid-to-liquid heat exchangers for heat recovery from wastewater that contains contaminants
9. Solid-state (e.g., thermoelectric) generators for electricity production from otherwise unusable waste heat streams, thermophotovoltaic (TPV) system, or piezoelectric system
10. Industrial heat pumps, including chemical heat pumps (e.g., adsorption/desorption and chemical looping reactions).

7.2.2 Motor-Driven Systems

Industrial machine and motor-driven systems include pumps, fans, compressors, air conditioners, refrigerators, forming and machining tools, robots, and materials processing and handling equipment. These systems account for 68% of manufacturing electricity consumption [1,32]. The majority of this energy is consumed in just three manufacturing sectors: chemicals, forest products, and food and beverage manufacturing. While electric motors have high efficiencies, end-use motor-driven systems have much lower system efficiencies, particularly for pumps, fans, compressed air, and materials processing equipment. As a result, overall machine-driven system losses total 1,470 TBtu annually [1,32]. The total energy uses for major categories of machine-driven systems in U.S. manufacturing are as follows: pumps (614), fans (291), compressed air (333), materials handling (175), materials processing (497), process cooling (212), and facility heating (241). The numbers in brackets are in TBtu units for 2010 [1]. Key energy-saving opportunities can be identified by focusing on opportunities to improve the motor system, rather than focusing solely on the motor. A 2004 study estimated the electricity savings opportunities from the use of available technologies on motor-driven systems. Only 13% of these opportunities were from the motors, while variable speed drive adoption accounted for an additional 25%, and improvements to applications would account for the remaining 62% [1].

7.2.3 Process Intensification

DOE report [1] points out that process intensification (PI) targets dramatic improvements in manufacturing and processing by rethinking existing operation schemes into ones that are both more precise and efficient. PI frequently involves combining separate unit operations such as reaction and separation into a single piece of equipment, resulting in a more efficient, cleaner, and economical manufacturing process. At the molecular level, PI technologies can significantly enhance mixing, which improves mass and heat transfer, reaction kinetics, yields, and specificity. These improvements translate into reductions in energy use, waste generation, environmental impact, and amount of equipment, and thereby minimize cost and risk in chemical manufacturing facilities. Table 7.3 shows energy reduction opportunities in eleven chemicals using PI technologies [1].

7.3 HYBRID ENERGY SYSTEMS WHICH INCLUDE WASTE HEAT RECOVERY AND CONVERSION

Waste heat losses arise both from equipment inefficiencies and from thermodynamic limitations on equipment and processes. For example, consider reverberatory furnaces frequently used in aluminum melting operations. Exhaust gases immediately leaving the furnace can have temperatures as high as 2,200°F–2,400°F [1,200°C–1,300°C]. Consequently, these gases have high heat content, carrying away as much as 60% of furnace energy inputs. Efforts can be made to design more energy-efficient reverberatory furnaces with better heat transfer and lower exhaust temperatures; however, the laws of thermodynamics place a lower limit on the temperature of exhaust

TABLE 7.3

2010 Production, Calculated Onsite Energy Consumption, and Energy Savings Potential for Eleven Chemicals [1]

Chemical	Annual Production (Million lbs/y)	Calculated Onsite Energy (TBtu/y)	Energy Reduction Opportunity (TBtu/y)
Ethanol	66,100	307	264
Ethylene	52,900	374	107
Ammonia	22,700	133	78
Benzene	13,300	104	67
Chlorine/sodium hydroxide	21,500/16,600	203	87
Nitrogen/oxygen	69,600/58,300	99	18
Ethylene dichloride	19,400	66	37
Propylene	31,100	42	11
Acetone	3,180	25	18
Ethylene oxide	5,880	11	4
Methanol	2,020	10	4
Total	382,000	1,370	695

gases. Since heat exchange involves energy transfer from a high-temperature source to a low-temperature sink, the combustion gas temperature must always exceed the molten aluminum temperature in order to facilitate aluminum melting. The gas temperature in the furnace will never decrease below the temperature of the molten aluminum, since this would violate the second law of thermodynamics. Therefore, the minimum possible temperature of combustion gases immediately exiting an aluminum reverberatory furnace corresponds to the aluminum pouring point temperature 1,200°F–1,380°F [650°C–750°C]. In this scenario, at least 40% of the energy input to the furnace is still lost as waste heat. Many examples of industrial waste heat and their possible uses are illustrated in Table 7.4. The Temperature Classification of Waste Heat Sources and Related Recovery Opportunity is illustrated in Table 7.5.

Recovering waste heat requires implementation of hybrid process like cogeneration or CHP. Industrial waste heat can be recovered via numerous methods. The heat can either be "reused" within the same process or transferred to another process through process hybridization. Ways of reusing heat locally include using combustion exhaust gases to preheat combustion air or feedwater in industrial boilers. By preheating the feedwater before it enters the boiler, the amount of energy required to heat the water to its final temperature is reduced. Alternately, the heat can be transferred to another process; for example, a heat exchanger could be used to transfer heat from combustion exhaust gases to hot air needed for a drying oven. In this manner, the recovered heat can replace fossil energy that would have otherwise been used in the oven. Such methods for recovering waste heat can help facilities significantly reduce their fossil fuel consumption, as well as reduce associated operating costs and pollutant emissions. As shown below, waste heat can also be converted to power by thermoelectric generator, TPV system, or piezoelectric system. Waste heat (particularly at low temperature) can also be used in heat pump.

TABLE 7.4
Examples of Waste Heat Sources and End Uses [34]

Waste Heat Sources

- Combustion exhausts:
  - Glass melting furnace
  - Cement kiln
  - Fume incinerator
  - Aluminum reverberatory furnace boiler
- Nuclear reactor:
  - Various applications of solar energy
  - Various applications of geothermal energy
  - Off-gases from internal combustion engine
- Process off gases:
  - Steel electric arc furnace
  - Aluminum reverberatory furnace
- Cooling water from: Furnaces
  - Air compressors
  - Internal combustion engines
- Conductive, convective, and radiative losses from equipment: Hall–Hèroult cellsa
- Conductive, convective, and radiative losses from heated products:
  - Hot cokes
  - Blast furnace slagsa

Uses for Waste Heat

- Combustion air preheating
- Boiler feedwater preheating
- Load preheating
- Power generation
- Steam generation for use in:
  - Power generation
  - Mechanical power
  - Process steam
- Space heating
- Water preheating
- Transfer to liquid or gaseous process streams
- Solar fuels
- Various applications of nuclear waste heat depending on temperature level
- Various applications of solar and geothermal energy depending on the temperature levels

TABLE 7.5
Temperature Classification of Waste Heat Sources and Related Recovery Opportunity [34]

Temp Range	Example Sources	Temp (°F)	Temp (°C)
Very high > 1,600°F	Electrical refractory furnace exhaust	2,900–4,500	1,600–2,700
High > 1,200°F (> 650°C)	Nickel refining furnace	2,500–3,000	1,370–1,650
	Steel electric arc furnace	2,500–3,000	1,370–1,650
	Basic oxygen furnace	2,200	1,200
	Aluminum reverberatory furnace	2,000–2,200	1,100–1,200
	Copper refining furnace	1,400–1,500	760–820
	Steel heating furnace	1,700–1,900	930–1,040
	Copper reverberatory furnace	1,650–2,000	900–1,090
	Hydrogen plants	1,200–1,800	650–980
	Fume incinerators	1,200–2,600	650–1,430
	Glass melting furnace	2,400–2,800	1,300–1,540
	Coke oven	1,200–1,800	650–1,000

(Continued)

TABLE 7.5 (*Continued*)
Temperature Classification of Waste Heat Sources and Related Recovery Opportunity [34]

Temp Range	Example Sources	Temp (°F)	Temp (°C)
	Iron cupola	1,500–1,800	820–980
Medium	Steam boiler exhaust	450–900	230–480
450°F–1,200°F	Gas turbine exhaust	700–1,000	370–540
(230°C–650°C)	Reciprocating engine exhaust	600–1,100	320–590
	Heat treating furnace	800–1,200	430–650
	Drying and baking ovens	450–1,100	230–590
	Cement kiln	840–1,150	450–620
Low < 450°F (< 230°C)	Exhaust gases exiting recovery devices in gas fired boilers, ethylene furnaces, etc.	150–450	70–230
	Process steam condensate	130–190	50–90
	Cooling water from: Furnace doors	90–130	30–50
	Annealing furnaces	150–450	70–230
	Air compressors	80–120	30–50
	Internal combustion engines	150–250	70–120
	Air-conditioning and refrigeration condensers	90–110	30–40
	Drying, baking, and curing ovens	200–450	90–230
Ultralow < 250°F	Hot processed liquids/ solids	90–450	30–230
	Kitchen, ventilation, fryer, condenser exhaust	90–130	30–45
	Cooling water from power plants	60–140	15–50
	Cooling water from air compressor	75–140	25–50

Industrial-scale energy systems integration provides a systems approach to optimize energy use at manufacturing facilities through technologies that can increase energy flexibility and reduce/recover/reuse waste energy, leading to reduced energy intensity, and narrowing the gap between current energy use and practical minimum energy requirements. Any option for waste heat recovery results in the overall system to be hybrid.

7.3.1 CHP Systems

CHP is the concurrent production of electricity or mechanical power and useful thermal energy from a single energy input, as shown in Figure 7.3. CHP technologies provide manufacturing facilities, commercial and institutional buildings, and communities with ways to reduce energy costs and emissions while also providing more resilient and reliable electric power and thermal energy. CHP systems combine the production of heat (for both heating and cooling) and electric power into one process,

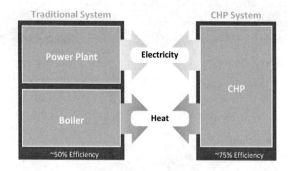

FIGURE 7.3 CHP systems produce thermal energy and electricity concurrently from the same energy input, and can therefore achieve higher system efficiencies than separate heat and power systems [1].

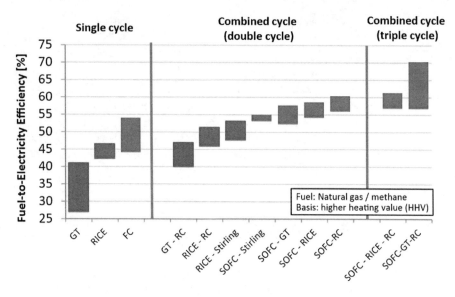

FIGURE 7.4 Theoretical efficiencies (electric generation only) for various CHP configurations, ranging from single-cycle systems to double- and triple-cycle systems that make use of multiple generation technologies. Efficiencies of up to 70% are theoretically possible [1].

using much less fuel than when heat and power are produced separately. CHP systems can achieve overall energy efficiencies of 75% or more [1] (see Figure 7.4) compared to separate production of heat and power, which collectively averages about 50% efficiency [1]. A recent executive order has set a national target of 40 gigawatts (GW) of additional CHP capacity by 2020 [1], an increase of nearly 50% above the current installed capacity of 83 GW in 2015 [1]; DOE (Department of Energy) analyses have identified R&D opportunities to increase the power-to-heat ratio of 1–10 MW CHP systems while maintaining the high overall system efficiencies of traditional thermally sized CHP systems. This would entail the development of ultra-high-efficiency

TABLE 7.6

Technical Potential and Energy and Cost Savings for High Power-to-Heat CHP Operation [1]

	Energy Benefits for High Power-to-Heat CHP Operation		
	Manufacturing Sector	Commercial/ Institutional Sector	Total
Incremental capacity potential (GW)*	4.7	45.1	52.9
Incremental annual primary energy savings (TBtu)**	144	1,160	1,310
User incremental energy cost savings ($ Millions)	$1,316	$8,660	$9,976

* Incremental CHP capacity based on a power-to-heat ratio of 1.5.

** Incremental primary energy savings based on a 33% average grid efficiency.

generation technologies. Existing CHP systems on average generate much more steam than electricity, with power-to-heat ratios [35] of individual systems as low as 0.1 but more commonly between 0.5 and 1, depending on the technology utilized [1]. If highly efficient CHP systems with a power-to-heat ratio of 1.5 were deployed, energy savings of up to 144 TBtu could be realized in the manufacturing sector, with economy-wide energy savings of 1,310 TBtu, as shown in Table 7.6. Major R&D opportunities for CHP exists in the areas: (a) packaging CHP for home and building-related systems; this involves packaging systems to avoid need for custom equipment design and onsite engineering expertise; (b) grid integration: technical solutions to enable grid interconnection, demand response, and ancillary services; (c) microgrid with CHP: small-scale autonomous energy grids with CHP generation, and possible facilitation of intermittent renewable sources, storage, energy efficiency measures, etc.; and (d) district energy with CHP: systems to enable use of rejected heat from CHP facilities to provide steam, hot, and chilled water to network buildings.

7.3.2 COGENERATION USING NUCLEAR HEAT

Excess nuclear heat from power generation (cogeneration) can be used for various industrial applications, such as seawater desalination, hydrogen production, district heating or cooling, the extraction of tertiary oil resources and process heat applications such as coal to liquids conversion and assistance in the synthesis of chemical feedstock. A large demand for nuclear energy for industrial applications is expected to grow rapidly on account of steadily increasing energy consumption, the finite availability of fossil fuels, and the increased sensitivity to the environmental impacts of fossil fuel combustion.

With increasing prices for conventional oil, unconventional oil resources are increasingly utilized to meet such growing demand, especially for transport. Nuclear energy offers a low-carbon alternative and has important potential advantages over other sources being considered for future energy. There are no technological impediments to extracting heat and steam from a nuclear power plant. This has been proven for low temperatures (<200°C) with nuclear-assisted district heating and desalination with an experience of approximately 750 reactor operation years from around 70 nuclear power plants. Detailed site-specific analyses are essential for determining the best energy option. The development of small and medium sized reactors would therefore be better suited for cogeneration and would facilitate non-electric applications of nuclear energy. The possibility of large-scale distribution systems for heat, steam, and electricity supplied from a central nuclear heat source (e.g., a multiproduct energy center) could attract and serve different kinds of consumers concentrated in industrial parks [34,36].

In the study by Peakman and Merk [37], the current market needs for high-temperature heat are considered based on UK industry requirements and work carried out in other studies regarding how industrial demand could change in the future. How these heat demands could be met via different nuclear reactor systems is also presented. The study found that found that of the current 300 TWh of industrial heat demand, around 35% of heat demand is at temperatures below 300°C and could be provided using all reactor types including conventional PWRs and BWRs; 35% falls into the range 300°C–500°C which could directly be provided from LMFRs (Liquid Metal Cooled Reactors), SCWRs (Super Critical Water Reactors), HTRs (High Temperature Reactors), VHTRs (Very High Temperature Reactors); 10% falls into the range 500°C–1,000°C, which could be provided using HTRs, VHTRs, and potentially some role using MSRs (Molten Salt Reactors) and GFRs (Gas Cooled Fast Reactors); and the remaining 20% of heat demand is for temperatures above 1,000°C, which to date no reactor has been demonstrated capable of operating at. For heat demand >1,000°C, it would be necessary to (a) continue to use fossil fuels; (b) use hydrogen or biofuels; or (c) artificially raise the outlet temperature of reactor systems, for example, via the use of heat pumps. The study inferred that, for wider industry, heat demand is dominated by temperatures below 500°C, implying that LMFRs and SCWRs, and to some extent, LWRs would be suitable candidates or MSRs, GFRs, HTRs, and VHTRs operating at lower temperatures. The use of small modular reactors (SMRs) for desalination is further illustrated in Chapter 10. The use of SMR for various other applications is also illustrated in my previous books [21,22,36,44,113]. Nuclear energy CHP hybrid process has significant future potential.

7.3.3 Options for Waste Heat to Power

Various options for recovering waste heat for power are illustrated in Table 7.7. The choice of method depends on temperature and source of heat [34].

7.3.3.1 Thermodynamic Cycles

The most frequently used system for power generation from waste heat involves using the heat to generate steam, which then drives a steam turbine. The traditional steam Rankine cycle [34,36] is the most efficient option for WHR (Waste Heat Recovery) from exhaust streams with temperatures about 650°F–700°F (340°C–370°C) [34,36]. It is used for waste heat from gas turbines, reciprocating engines, incinerators, and furnaces. At lower waste heat temperatures, steam cycles become less cost-effective,

TABLE 7.7

Options for Power Generation from Waste Heat [34]

Method	Temp. Range	Nature of Heat Source	Approximate Cost
Traditional steam cycle	M, H	Exhaust from gas turbines, reciprocating engines, incinerators, and furnaces.	$1,100–1,400/kW
Kalina cycle	L, M,	Gas turbine exhaust, boiler exhaust, and cement kilns	$1,100–1,500/kW
Organic rankine cycle	L, M	Gas turbine exhaust, boiler exhaust, heated water, and cement kilns	$1,500–3,500/kW
Thermoelectric generation	M, H	Not yet demonstrated in industrial applications	$20,000–300,000/kW
Piezoelectric generation	L	Not yet demonstrated in industrial applications	$10,000,000/kW
Thermal photovoltaic	M, H	Not yet demonstrated in industrial applications	N/A

Source: Johnson, I. and Choate, W., Waste heat recovery: Technology and opportunities in U.S. industry a report by U.S. department of energy-Industrial technologies program, Laurel, MD: prepared by BCS. Retrieved from http://www1.eere.energy.gov/manufacturing/intensiveprocesses/pdfs/waste_heat_recovery.pdf, 2008.

Note: Last three are the direct methods of power generation.

because low-pressure steam will require bulkier equipment. Moreover, low-temperature waste heat may not provide sufficient energy to superheat the steam, which is a requirement for preventing steam condensation and erosion of the turbine blades. Therefore, low-temperature heat recovery applications are better suited for organic Rankine cycle [34,36,38–46] or Kalina cycle [47,48], which uses fluids with lower boiling point temperatures compared to steam. Evaporator waste heat from process turbine condenser generates electricity for driving pump for grid [34]. Various cycles used to convert waste heat to power are described in Table 7.7. All the cycles mentioned in this table are described in detail in my previous books [34,36].

7.3.3.2 Thermoelectric Power

The basics of thermoelectric power are described in my previous books [19,34,36,49]. Most of the recent research activities on applications of thermoelectric power generation have been directed toward utilization of industrial waste heat [1,34]. Estimates of Waste Heat That Could Be Recovered with Thermoelectric Generation Technologies for Major Process Industries [34] are illustrated in Table 7.8. As mentioned earlier, vast amounts of heat are rejected from industry, manufacturing plants, and power utilities as gases or liquids at temperature which are too low to be used in conventional generating units (<450 K). In this large-scale application, thermoelectric power generators offer a potential alternative of electricity generation powered by waste heat energy that would contribute to solving the worldwide energy crisis and at the same time help reduce environmental global warming. In particular, the replacement of by-heat boiler and gas turbine by thermoelectric power generators makes it capable of largely reducing capital cost, increasing stability, saving energy source, and protecting environment. Recently, Min and Rowe [50,51] reported that in 1994 a research project sponsored by the Japanese New Energy and Technology Development Organization developed a series of WATT-X (Waste heat Alternative Thermoelectric Technology and X denoted power output in Watts) prototype generators. Basically, the generator consists of an array of modules sandwiched between hot and cold water-carrying channels. Some of the heat flux which is established by the hot and cold temperature difference between the hot and cold water flows is directly converted into electrical power. When operated using hot water at a temperature of approximately 90°C and cold flow at ambient Watt-100 generates 100 W at a power density approaching 80 kW/m^3. In this application, the system was scalable enabling 1.5 kW of electrical power to be generated [50,51].

Thermoelectric power generators have also been successfully applied in recovering waste heat from steel manufacturing plants. In this application, large amounts of cooling water are typically discharged at constant temperatures of around 90°C when used for cooling ingots in steel plants. When operating in its continuous steel casting mode, the furnace provides a steady-state source of convenient piped water which can be readily converted by thermoelectric power generators into electricity. It was reported by [50,51] that total electrical power of around 8 MW would be produced employing currently available modules fabricated using Bi_2Te_3 thermoelectric modules technology. Another application where thermoelectric power generators using waste heat energy have potential use is in industrial cogeneration systems [52–71]. For example, Yodovard et al. [56] assessed the potential of waste heat thermoelectric power generation for diesel cycle and gas turbine cogeneration in the manufacturing industrial sector in Thailand. The data from more than 27,000 factories from different sectors, namely, chemical product, food processing, oil refining, palm oil mills petrochemical, pulp and paper rice mills, sugar mills, and textiles, were used. It is reported that gas turbine and diesel cycle cogeneration systems produced electricity estimated at 33% and 40% of fuel input, respectively. The useful waste heat from stack exhaust of cogeneration systems was estimated at approximately 20% for a gas turbine and 10% for the diesel cycle. The corresponding net power generation is about 100 MW.

There are other novel approaches used to generate thermo electric power. Thermoelectric-based power generation system has been designed to be clamped onto the outer wall of a steam pipe. This can include a number of assemblies mounted on the sides of a pipe. Each assembly can include a hot block, an array of thermoelectric modules, and a cold block system. In this method, the hot block can create a thermal channel to the hot plates of the modules. The cold block can include a heat pipe onto which fins are attached [60–63]. The possibility of utilizing the heat from incinerated municipal solid waste has also been considered. For example, in Japan, the solid waste per capita is around 1 kg/day and the amount of energy in equivalent oil is estimated at 18 million kJ by the end of the 21st century. It was reported by [34,36,50,51,58,63] that an on-site experiment using a 60 W thermoelectric module installed near the boiler section of an incinerator plant achieved an estimated conversion efficiency of approximately 4.4%. The incinerator waste gas temperature varied between 823 K and 973 K, and with forced air-cooling on the cold side, an estimated conversion efficiency of approximately 4.5% was achieved. An analysis of a conceptual large-scale system burning 100 tons of solid waste during a 16 hour day indicated that around 426 kW could be delivered [34,36,50,51]. In the waste heat from incineration applications, the thermoelectric modules are typically placed on walls of the furnace's funnels. This construction can eliminate the by-heat furnace, gas turbine, and other appending parts of steam recycle [34,36,52–71].

Demonstrations of high-power thermoelectric WHR are necessary to prove that such systems can take advantage of economies of scale. Demonstrated power has typically been less than 1 kW, with notable demonstrations of 169 W generation from a cement kiln, 250 W generation from glass furnace exhaust, and 240 W generation from a steel carburizing furnace afterburner [34,36,67,69]. A 1 kW generator was mounted successfully into the Bradley fighting vehicle by a Department of Defense contractor based on a module from Hi-Z Technology [65], Inc. Two recent thermoelectric energy harvesting efforts have demonstrated power generation in excess of 1 kW. The first of these involved a high-power TEG installed over a continuous casting line. This system contained 896 Komatsu modules, 16 of which were used in the carburizing furnace demonstration, and it produced power on the order of 9 kW when exposed to the radiant heat of a 915°C slab. The other is the first plug-and-play TEG available for purchase, the 25 kW Alphabet Energy™ E1, which was announced in October 2014 [34,36,58,63].

Efficient and cost-effective thermoelectric energy generation systems depend on a number of module- and system-level design factors in addition to the selection of materials with good thermoelectric properties. At the module level, the choice of materials can be complicated by the fact that p- and n-type materials are both typically required. In addition, assembly of thermoelectric materials together with conductive leads and dielectric plates to form modules at an industrial scale is challenging. The different materials and their interfaces must be robust enough to tolerate thermal expansion and contraction, and modules must be designed to properly conduct heat through thermoelectric materials, maximizing temperature gradients and thus power generation. Module assembly quality requirements are high: a single bad connection makes a module useless due to serial electrical connections between the thermoelectric legs [34,36,58,63].

At the system level, the challenge of maximizing the temperature difference over the thermoelectric modules to maximize efficiency means that heat exchangers on the hot and cold sides must be well designed to maximize heat flux to the modules. Hot-side heat exchangers are particularly challenging to design in many cases in which they must tolerate high heat flux fluids that can be corrosive or contaminated with PM. Cost optimization of heat exchangers that transfer source heat to the cold side and also include heat transfer within the module is important. In high-temperature applications, heat exchanger materials and protective coatings impact costs.

Electrical design challenges include proper matching of resistance between thermoelectric modules and loads as well as efficiently inverting the DC power for use on the grid. Costs for electrically insulating plates (usually ceramic) must be reduced while maintaining good thermal conductivity at elevated temperatures. Electrical interconnects and other interfaces must be engineered to minimize electrical and thermal losses and to provide oxidation protection in order to maximize device and system efficiency and reliability. Using more corrosion-resistant heat exchangers would allow thermoelectric heat recovery from a more diverse range of industrial exhaust streams.

Studies to cooptimize the thermal and electrical properties of the whole TEG system while maintaining its mechanical integrity are also important [34,36,58,63]. Heat exchangers can present significant design challenges for applications with modest temperature gradients. Enabling high heat flux in these regimes requires more complex and highly engineered systems whose cost can significantly impact overall system cost. Progress must also be made in the area of automated assembly so that thermoelectric devices can be made in a reliable and cost-effective manner. Traditionally, metal interconnects are attached to ceramic insulating plates by using one of several available processes, such as soldering, thin film sputtering, and plating [34,36,50–71]. The thermoelectric legs are then soldered to these interconnects. More than 90% of the thermoelectric modules assembled today require some manual operations, such as attaching leads and visual inspections [34,36,50–71]. Some manufacturers have implemented automation systems for the assembly process, but in most cases, price points do not justify straying away from the standard pick and place machines used to assemble electronic components. Thin film thermoelectric modules offer an alternative to the manufacturing methods of conventional bulk materials because the p- and n-type materials can be sputtered onto separate wafers (using techniques from silicon microelectronics fabrication) that are then fused together, the added benefit of increasing the exposure of TEG technology [34,36,50–71].

The possibilities for installing a similar system to the one at the JFE Steel Corporation plant in Japan [34,36,67,69] and at the Nucor mini mill in Jewett, Texas, were explored. At 13–30 cm wide, the five continuous casting lines at the Nucor plant are not as wide as the 1.3–1.7 m slabs at the JFE plant, but assuming that the slab temperatures are the same, a similar amount of heat flux (on the order of 17 kW thermal/m^2) could be intercepted by placing 50 cm wide generators roughly 20 cm from the cast slabs. Assuming that the same level of thermoelectric generation per unit area that Kuroki et al. [69] discussed (1.13 kW electric/m^2) could be achieved over 14 m for each strand, 39 kW electric of thermoelectric power could be produced by 35 m^2 of generators (based on a 14 m × 0.5 m generator above each of five strands) at the Nucor plant. Under these conditions, using a nanobulk $Bi_{0.52}Sb_{1.48}Te_3$ TEG of a higher ZT value of 1 with a 15-year life span and a capacity factor of 60%, a modified version of LeBlanc's cost model [34,36,53,58] predicts an LCOE (Levelized cost of energy) of $0.31/kWh.

Besides the steel industry, any industry in which high-quality waste heat goes unused should be considered as a possible target for thermoelectric WHR. These potential targets include glass, aluminum, cement, and ethylene manufacturing, all discussed in a 2006 report by Hendricks and Choate [72]. This report also considered industrial and commercial boilers for their large aggregate waste heat, but these were found to lack high enough temperatures to make thermoelectric energy harvesting with current technology feasible. The most promising source the report found in terms of potential annual electricity generation was aluminum melting, which the authors determined could produce 1.4 TBtu/y by using thermoelectric materials with a ZT (dimensionless figure of merit) of 1.

In October 2014, Alphabet Energy, Inc. announced a TEG (Thermoelectric generator) product called the E1 that fits in a standard shipping container, connects to the exhaust pipe of a generator, and has a modular design so that thermoelectric components can be swapped out as materials improve. Alphabet Energy states that the E1 can produce 25 kW from the exhaust of a 1,000 kW generator, 66 implying an efficiency of about 2.5% based on the exhaust heat of such a generator [34,36]. If running constantly, the generator would produce roughly 219,000 kWh and would save 50,000 L of diesel fuel per year [34,36].

To obtain an estimate of the amounts of waste heat in each industrial sector, the DOE Advanced Manufacturing Office *Manufacturing Energy and Carbon Footprints* data [1,30] can be used. The potential of thermoelectric energy from manufacturing plant waste heat can be estimated with the knowledge of the fraction of the heat that can be recovered by the generation system and the efficiency value for those systems. The choice for the low end of the recoverable heat range was 10%, based on an estimate from Polcyn and Khaleel [73], and the high end of 25% was based on heat-recovery calculations for boiler exhaust from a study by Hill [64]. These estimates assume the thermoelectric generation efficiency to be 2.5%. This efficiency figure was chosen because module efficiencies of around 5% are seen in the sales literature of market modules for temperature differences around 200°C–250°C, and generally, only half of the temperature gradient across the TEG system is available for power conversion across the TEG material (whereas the other half is dissipated across the heat exchangers). In addition, this 2.5% figure matches the efficiency implied in early press related to the first published large-scale, off-the-shelf, exhaust-based thermoelectric generation system [34,36,52–55,74].

Based on this estimate, the thermoelectric recovery potential for U.S. manufacturing is about 1,880–4,701 GWh (6–16 TBtu). This is a conservative estimate based on the thermoelectric generation efficiencies of existing systems on the market and does not take future technology development. Table 7.8 presents estimate of Waste Heat That Could Be Recovered with Thermoelectric Generation Technologies for Major Process Industries.

7.3.3.3 Thermophotovoltaic Devices

Besides thermoelectric devices, TPV devices use the same physics as the solar panels (photovoltaics) typically used on rooftops—they both capture light in a semiconductor material that excites electrons, which then can flow out of the device as current. One major difference is that TPV devices can convert infrared light, which is the portion of the electromagnetic spectrum that is felt as heat directly into electricity. Currently, for TPVs to work efficiently, heat sources have to be in excess of 1,000°C. The glass and steel industries with their large, hot furnaces are low-hanging fruit best suited to TPV waste heat harvesting technology. For example, in the glass industry, furnaces must be maintained at not less than 1,575°C to keep glass in a molten state. In steel production, iron ore must first be heated in blast furnaces up to 2,300°C. Studies have shown that the technology implemented to capture industrial waste heat can pay for itself in at least 3–5 years, after which any additional electricity produced is essentially a free source of energy.

TABLE 7.8

Estimate of Waste Heat That Could Be Recovered with Thermoelectric Generation Technologies for Major Process Industries [34]

Manufacturing Process Industry	Process Heating Energy Use (TBtu/y) [105]	Process Heating Energy Losses (TBtu/y) [106]	Estimated Recoverable Heat Range (TBtu/y) [107]	Estimated Thermoelectric Potential (TBtu/y) [108]	Estimated Thermoelectric Potential (GWh/y) [109]
Petroleum refining	2,250	397	40–99	1–2	291–727
Chemicals	1,460	328	33–82	1–2	240–601
Forest products	980	701	70–175	2–4	513–1,280
Iron and steel	729	334	33–84	1–2	245–612
Food and beverage	518	293	29–73	1–2	215–537
Glass	161	88	9–22	0–1	64–161
Other manufacturing	1,110	426	43–107	1–3	312–780
All manufacturing	**7,200**	**2,570**	**257–642**	**6–16**	**1,880–4,700**

Source: Department of Energy, Innovating clean energy technologies in advanced manufacturing (Chapter 6), *Quadrennial Technology Review: An Assessment of Energy Technologies and Research Opportunities*, Department of Energy, Washington, DC.

In recent years, considerable research is being carried on to (a) extend the operating range of these TPV devices, solar energy directly to electricity. TPV generators can be used to convert total radiant energy into electricity. These systems involve a heat source, an emitter, a radiation filter, and a PV cell (like those used in solar panels). As the emitter is heated, it emits electromagnetic radiation. The PV cell converts this radiation to electrical energy. The filter is used to pass radiation at wavelengths that match the PV cell while reflecting remaining energy back to the emitter. These systems could potentially enable new methods for WHR. A small number of prototype systems have been built for small burner applications and in a helicopter gas turbine. The basics of TPV are described in my previous books [34,36]. With the use of nanomaterials, quantum effects, and some novel and clever design, the technology is being pushed to new boundaries.

7.3.3.4 Thermionic Devices

Thermionic devices operate similar to thermoelectric devices; however, whereas thermoelectric devices operate according to the Seebeck effect, thermionic devices operate via thermionic emission. In these systems, a temperature difference drives the flow of electrons through a vacuum from a metal to a metal oxide surface. One key disadvantage of these systems is that their applications have seen limited in use due to their low efficiencies and high cost. Most TE (Thermo electric) generation systems in use have efficiencies of 2%–5%; these have mainly been used to power instruments on spacecraft or in very remote locations. However, recent advances in nanotechnology have enabled advanced TE materials that might achieve conversion efficiencies of 15% or greater [34,36,75,76]. More details are given in my previous books [34,36,75,76].

7.3.3.5 Piezoelectric Devices

In principle, piezoelectric devices convert mechanical energy to electrical energy in the form of ambient vibrations. PEPG (Piezoelectric Power Generator) is an option for converting low-temperature waste heat (200°F–300°F or [100°C–150°C]) to electrical energy. A piezoelectric thin film membrane can take advantage of oscillatory gas expansion to create a voltage output. A recent study identified several technical challenges associated with PEPG technologies. PEPG technology has a very low (about 1%) efficiency. The device is also expensive ($10,000/W) and lacks long-term reliability and durability [34,36,77,78]. Although the conversion efficiency of PEPG technology is currently very low (1%), there may be opportunities to use PEPG cascading in which case efficiencies could reach about 10%. Other key issues are the costs of manufacturing piezoelectric devices and the design of heat exchangers to facilitate sufficient heat transfer rates across a relatively low-temperature difference [34,36,77,78].

7.3.3.6 Heat Pumps for Process Heat

Heat pumps convert energy from various sources into process heat. Heat sources could be air, river/lake/sea water, ground heat, or waste heat. Electricity input is required to operate the heat pump, so it is not fully a renewable energy but heat pumps can produce up to seven units of thermal energy from one unit of electricity input. For example, the European Commission Directive proposes that if seasonal performance factors (SPFs) of heat pumps are higher than the value of $1.15 \times 1/\eta$ (where η is the efficiency of power generation estimated based on Eurostat), then they can be considered as renewable energy [1,2,12,13,79]. SPF is determined based on temperature lift required, i.e., the difference between the temperature of the heat source and the temperature of the process heat, coefficient of performance (COP), and the efficiency of the heat pump. SPF is the ratio of heat delivered to energy consumed over the season.

Heat pumps can take heat from the environment or from waste heat streams and supply it to industrial applications without the need to burn any fuel. In applications where the pumping energy input is in the form of electricity produced from renewable energy sources, heat pumps are a fully renewable energy technology. Where the electricity is generated from fossil fuels, only part of the energy output of heat pumps can be regarded as renewable. So, for example, if the electricity comes from fossil fuel generation with an efficiency of 40%, the COP of the heat pump needs to be higher than 2.5 if the pump is to save primary energy and be considered as providing renewable heat. The amount of useful heat provided must be higher than the primary energy consumed. The COP is the ratio of useful output energy and useful input energy under standardized testing conditions.

The advantages of heat pump are (a) high COP in applications requiring a low temperature lift and/or operating in high ambient temperatures; (b) long annual operating time; (c) relatively low investment cost, due to large units and small distances between the heat source and heat sink; and (d) waste heat production and heat demand occur at the same time. Current disadvantages of industrial heat pumps are (a) lack of refrigerants in the relevant temperature range; (b) lack of experimental and demonstration plants; (c) user uncertainty about the reliability of heat pumps; and (d) lack of necessary knowledge among designers and consulting engineers about heat pump technologies and their application.

OECD countries have an important role to play in the potential deployment of heat pumps for industrial process heat. This reflects the fact that most OECD countries already have reliable electricity grids which deliver electricity at competitive prices. The high efficiency of electric industrial heat pumps makes this technology competitive with solar thermal technologies where electricity prices are low and solar radiation is less than optimum, conditions which describe many of the regions where OECD industrial production is located. Two other factors, the capital cost of the equipment and its performance, are also important in determining the competitiveness of heat pumps. Performance is expressed in terms of the number of units of energy the heat pump can move from the lower temperature of the source to the higher temperature needed, using one unit of electricity. In most normal operating conditions, the amount of electricity required is considerably less than the amount of heat provided, particularly in applications demanding relatively low-temperature process heat. The main thermodynamically limiting factor in the use of heat pumps for high temperature process heat, however, is that their performance decreases the greater the difference in temperature between the input source and the output demanded. So heat pumps are more efficient in delivering low-temperature process heat demands. And, air heat pumps are more efficient in warmer climates. This factor has been taken into account in analyzing the cost of the process heat in individual regions [1,2,12,13,79].

Heat pumps can already provide a competitive alternative to fossil fuels for low-temperature process heat in several regions. However, the competition for low-temperature renewable process heat production between heat pumps and solar thermal will be heavily dependent on regional and local conditions. In food processing industry, there are plenty of cases where heat pumps are being applied today. For example, in Japan, heat pumps are being used in brewing sake. India has been demonstrating the use of heat pumps in the dairy industry. In general, however, only a few heat pumps are installed in industry. Industrial heat pumps are typically used for space heating and cooling, simultaneous heating and cooling, refrigerating, low temperature steam production, cleaning, drying, evaporation, and distillation processes in various sectors. There are opportunities for heat pumps above 100°C as well. This could be done through mechanical vapor recompression by integrating excess heat to temperatures above 120°C and could also result in higher COP values. Furthermore, some heat pumps can deliver steam at 160°C [79]. Heat pumps have a potential in wastewater treatment facilities as well which are located on-site or close to most industrial plants. In anaerobic wastewater treatment process, the wastewater is usually heated by steam supplied by boiler where heat pumps can be used as an alternative.

7.4 ROLE OF BIOMASS SYSTEMS FOR INDUSTRIAL PROCESSES

As mentioned before, besides improving energy conversion efficiency by various hybrid processes outlined above, the reduction in fossil fuel consumption in industry can also be achieved by hybridization of fossil fuels with various renewable fuel sources. The most notable renewable sources for this purpose are biomass, solar thermal systems, geothermal energy, and wind energy. Industrial production processes operate across a wide temperature range. For example, while drying, washing, and heat treatment in the food industry and cleaning, dyeing and bleaching activities in the textile industry operate below 150°C, distillation processes, boilers, and reactors in the chemical industry operate above 250°C and temperatures are even higher for iron and steel production processes. While low (<150°C) and medium temperature (150°C–400°C) process heat is typically supplied via steam, high-temperature (>400°C) applications are provided in the form of direct heat (e.g., in cement kilns or in the iron and steel sector).

7.4.1 BIOMASS-BASED HYBRID SYSTEMS

Biomass is a very versatile material and it can be used in hybrid manner in the industry in a number of different ways. The versatility and substitution potential of biomass make this the top option for renewable energy in manufacturing. It can be used as a suitable replacement for fossil fuels for raw materials, fuel for localized energy production, and is a viable producer of low-, medium-, and high-temperature heat (Note: high-temperature heat applications make up more than two-thirds of the heat demand for total process). Additional factors pertaining to the economic viability of biomass include its reduced production costs, high energy density, shorter distance for transportation, and increased options for transportation methods [1,2,12,13,20,80–85].

Steam is typically generated by fossil fuels in steam boilers at high conversion efficiencies of about 90%. However, biomass can also be used to generate steam. Today typical sources are wood waste (e.g., bark, black liquor) used in the pulp and paper sector and charcoal use in small-scale blast furnaces [69]. Although biomass combustion for steam production is currently limited, there are large potential to provide low- and medium-temperature steam (<400°C) by fixed or fluidized bed boilers and CHP plants. High-temperature process heat can be provided by biomass gasification. Cofiring of biomass with coal is another option. The efficiency of bio-based steam generation from feedstocks such as rice husk, wood pellets, or wood chips is generally slightly lower (75%–90%) [80–82] than that of fossil fuels (85%–90%) [85]. The difference in efficiencies between bio-based gasifiers from wood, briquette, residues such as coconut shells (40%–50%) and fossil fuel fired furnaces, kilns, and stoves could be higher (50%–60%) [83,84].

Iron production requires the combustion of carbon-containing fuels to produce carbon monoxide which is reacted with ferrous oxide to produce iron and CO_2. Historically, iron was produced using charcoal exclusively as fuel. At the beginning of the 18th century, charcoal started to be substituted by coke. Coke is now by far the dominant fuel in iron and steel making, with at least 10 Gt of coke being consumed per ton of steel produced. Even so, significant amounts of pig iron are still successfully produced using charcoal. The use of electrochemical processes to produce iron ore, known as electrowinning, is currently in an early R&D phase. Aluminum is produced entirely by electrowinning and the approach is also used in the production of lead, copper, gold, silver, zinc, chromium, cobalt, manganese, and the rare-earth and alkali metals. Elector winning offers the possibility to produce iron without the use of carbonaceous fuels. If a technological breakthrough was to make the production of iron by electrowinning feasible, and if in future here were large quantities of low cost, low carbon electricity available, this would offer a route to the production of iron and steel with significantly reduced carbon emissions [1,2,12,13,20,80–85].

Carbon is also needed for the production of materials in the petrochemical sector, where it comprises around 75% of the total feedstock. The main alternative feedstock to fossil fuels in the petrochemical sector is likely to be biomass. But waste products, such as recycled plastics, can also substitute for some fossil fuel feedstock. Alternatively, organic materials such as cellulose fibers, coconut fibers, starch plastics, fiber boards, and paper foams can be produced which can directly substitute for petrochemical products in end-use applications. It is also possible to produce textile materials (mainly viscose and acetate) from wood pulp and as by-products from cotton processing. Replacement of petroleum fuels by biofuels is another possibility. Biomass availability and use is strongly dependent on regional conditions. Although biomass provides 8% of industry's final energy, in some regions, there is almost no biomass use in any industrial sector. In regions such as Latin America and Africa by contrast, biomass contributes around 30% of industry's final energy (International Energy Agency (IEA) statistics). Wide differences in use are also observed among different industrial sectors [24,25].

Biomass is used to a significant degree for industrial heat in the food and tobacco, paper pulp and printing, and wood and wood products sectors in most regions. By contrast, almost no process heat is produced from biomass in the iron and steel and nonmetallic mineral sectors, except in Brazil, or in the chemical and petrochemical, nonferrous metals, transport equipment, machinery, mining and quarrying, construction, or textile and leather sectors. The cement [86] and iron and steel sectors in Brazil use biomass for 34% and 40%, respectively, of the sectors' final energy consumption. The fact that such a high level of biomass contribution can be sustained in the two most energy intensive sectors in Brazil means that a similar level of contribution should also be technically feasible elsewhere. The limiting factors on the extension of biomass use in these two sectors are clearly therefore nontechnical ones. They may include resource availability, economics, and competition from other energy sources.

The estimates of the potential role of biomass in 2050 are strongly sensitive to the state of the markets for biomass trading among different regions. If there is no interregional trading of biomass, the potential contribution of biomass in industry is estimated to be 18.3 EJ/y; if there are liquid markets for interregional biomass trading, this contribution is estimated to be 30.3 EJ/y. Transporting biomass is unlikely to have a significant impact on overall emission reductions. A state-of-the-art coal-fired power plant with 46% efficiency cofiring pellets shipped by a 30 kiloton (kt) ship over 6,800 km would produce emissions of around 85 g of CO_2/kilowatt hour (kWh).

Using bio-coal5 shipped by an 80 kt ship over 11,000 km, the emissions would be reduced to 32 g of CO_2/kWh. By comparison, the same power plant using coal would emit 796 g of CO_2/kWh [1,2,12,13,20,80–85].

In the absence of interregional markets, the estimated marginal cost of biomass would be around USD 7/GJ of primary energy, mostly in the form of locally consumed residues and energy crops in Latin America, with a smaller level of local consumption in Africa. With liquid interregional markets, large volumes of biomass will be moved around the world, mostly into OECD countries (11 EJ) and some into the Chinese market (less than 1 EJ). Despite much higher levels of demand, the marginal cost would be around USD 7.5/GJ, assuming the exploitation of Africa's very large potential for energy crops, and significant use also of Asia's potential. It is clear from this analysis that creating tradable biomass commodities and allowing free trade from developing countries to industrialized ones will have a potentially positive impact on GHG emission reductions in industry.

Hybrid process cofiring coal and biomass can be good source for decarbonization. Significant quantities of biomass are already cofired with coal in conventional coal power plants. For example, the Amer 9 CHP power plant in the Netherlands, which produces 600 MW of electricity and 350 MW of heat, currently cofires 35% of biomass mostly in the form of wood pellets with 65% coal. The technological development of solid biomass fuels is likely to be directed at a scaling up in the energy density of the reprocessed biomass until it can be used without any modification on its own in coal-burning power plants, furnaces, and industrial process. Two main current forms of gaseous biofuels are biogas from anaerobic fermentation and producer gas or synthetic gas (syngas) from biomass gasification. Biomass gasification, although still only in an early commercial phase, offers good prospects for the use of biomass for process heat and power generation. Charcoal is used in blast furnaces and is widely used today as a fuel. World average charcoal production from 2001 to 2005 was around 43 Mt/y (equivalent to approximately 1.3 EJ/y). It has been expanding by around 2% a year in recent years. Most of this charcoal is used for cooking in developing countries. Around 37 million cubic meters (m^3) a year (2004 figures, equivalent to approximately 7.7 Mt), however, are used for iron-making particularly in small-scale blast furnaces in Brazil. Charcoal does not have the mechanical stability of coke, but it has similar chemical properties. A processed type of charcoal with better mechanical stability is under development. This "biocoal" could substitute for coke. Assuming the complete replacement of fossil fuels on a thermally equivalent basis, the production of 1 t of hot metal requires 0.725 t of charcoal produced from 3.6 t of wet wood. Charcoal produced in Minas Gerais costs about USD 200/t [87]. This is comparable with current coking coal industrial prices in nonsubsidized markets. So the economic impact on iron prices would be neutral [1,2,12,13].

The use of alternative fuels in the cement industry is a long established practice in many countries. It offers the opportunity to reduce production costs, to dispose of waste, and in some cases to reduce CO_2 emissions and fossil fuel use. Cement kilns are well-suited for waste combustion because of their high process temperature and because the clinker product and limestone feedstock act as gas-cleaning agents. Used tyres, wood, plastics, chemicals, treated municipal solid waste, and other types of waste are cocombusted in cement kilns in large quantities. Where fossil fuels are replaced with alternative fuels that would otherwise have been incinerated or land filled, this can contribute to lower overall CO_2 emissions. In a survey conducted by the World Business Council on Sustainable Development in 2006, participants reported 10% average use of alternative fuels, of which 30% was biomass [1,2,12,13]. European cement manufacturers derived 3% of their energy needs from waste fuels in 1990 and 17% in 2005 [1,2,12,13]. Cement producers in Belgium, France, Germany, the Netherlands, and Switzerland have reached substitution rates ranging from 35% to more than 80% of the total energy used. Some individual plants have achieved 100% substitution of fossil fuels with waste materials. Waste combustion in cement kilns also needs an advanced collection infrastructure and logistics (collection, separation, quality monitoring, etc.).

If waste materials are more generally to achieve widespread use in cement kilns at high substitution rates, tailored pretreatment and surveillance systems will be needed. Municipal solid waste, for example, needs to be screened and processed to obtain consistent calorific values and feed characteristics. A well-designed regulatory framework for waste management is an important factor in facilitating the use of waste. In developing countries, although interest is growing, alternative fuel use constitutes only 5% of cement industry fuel needs, compared to an average of 16% in the OECD. Bio-feedstocks are estimated to have the potential, based on the assumptions in GEA Scenario M, to supply 6.9 EJ/y of the petrochemical sector's energy needs in 2050. The achievement of this potential is likely, however, to be dependent on a number of factors, including the cost and availability of petrochemical feedstocks which will themselves be dependent on limitations in the refinery product mix and refinery product demand. Hybrid coal-biomass mixture also has future in this regard.

There are other opportunities for bio-based products. For example, bio-based polyethylene can be used as a substitute for polypropylene. Aromatics can also be produced from biomass feedstocks, particularly from lignin which is an important constituent of wood that may be produced in substantial amounts as a by-product if second-generation ethanol production takes off. This would offer new opportunities for the development of biorefineries. The most promising petrochemical bio-feedstocks other than bio-ethylene are polylactic acid (PLA) as a substitute for polyethylene terephthalate (PET) and polystyrene, polyhydroxy alkaonates as a substitute for high density polyethylene, and bio-polytrimethylene terephtalate (PTT), as a substitute for fossil-based PTT or nylon 6 [88]. Traditional fossil feedstocks can be substituted with bio-derived ones at a number of points in the petrochemical products production chain: (a) fossil feedstock can be substituted with a bio-based one (e.g., natural gas can be substituted with synthetic natural gas from biomass gasification and subsequent methanization); (b) petrochemical building blocks can be substituted (e.g., ethylene can be substituted with bio-ethylene); (c) traditional plastics can be substituted with a bio-based substitute (e.g., PET can be substituted with PLA); or (d) a petrochemically produced material can be substituted with a bio-based material with similar functional characteristics (e.g., plastic can be substituted with wood or nylon with silk) [1,2,12,13,20,80–88].

Worldwide plastics consumption amounts to approximately 245 Mt/y. Olefins (ethylene and propylene) are the most important feedstock. The steam-cracking of naphtha, ethane, and gas oil is the dominant production technology. Large amounts of aromatics are also produced from refinery streams. Worldwide steam-cracking accounts for approximately 3 EJ of final energy use and approximately 200 million tons of CO_2 emissions. This represents around 20% of the total final energy use and about 17% of the total CO_2 emissions from the chemical and petrochemical sector. A number of new technologies are being developed to manufacture olefins from natural gas, coal, and biomass. Only those based on biomass offer the potential to eliminate fossil fuel use and GHG emissions. Bio-based polymers are produced in three main ways: (a) by using natural polymers such as starch and cellulose which can be modified; (b) producing bio-based monomers by fermentation or conventional chemistry and polymerizing them, for example, to produce PLA; or (c) producing bio-based polymers directly in microorganisms or in genetically modified crops. The first of these three production methods is currently by far the most important, being involved for example in the use of starch in paper-making, in man-made cellulose fibers, and in the development of starch polymers. Much is expected of the future development of the second option, with the first large-scale plants currently coming into operation. Most chemical coproducts can be created from the basic chemical building blocks of sugars and alcohols [1,2,12,13,20,80–88].

7.5   ROLE OF GEOTHERMAL ENERGY

Geothermal heat (excluding geothermal source heat pumps) can be used as a source for low-temperature process heat applications. Today less than one percent of the total industrial heat use is provided from geothermal sources [89–97]. About half of the demand comes from the pulp and paper sector and the remainder from drying, evaporation, distillation, or washing applications in various other sectors [91,93]. In Iceland as an example, geothermal heat is typically used for fish drying [92]. Similarly, the drying of tomatoes is done with geothermal heat in Greece [93]. Geothermal heat can be directly applied to the industrial processes if the distance between the heat source and the end-user is sufficiently close [92,83]. Conventional deep geothermal heat-production technology for low-temperature heat applications offers the largest potential in all industry sectors with the exception of the chemical and petrochemical and the iron and steel sectors, where medium- and high-temperature process heat dominates the demand.

For industrial applications, geothermal energy in the temperature range below 150°C is used in the basic processes of preheating, washing, peeling and blanching, evaporation and distilling, sterilizing, drying and refrigeration. Some industrial applications of geothermal energy are (a) *preheating and heating*: geothermal energy can be effectively used to preheat boiler and other process-feed water in a wide range of industries. The geothermal resource can also be used to offload the boiler of some or all of preheating load; (b) *washing*: large amounts of low-temperature energy (35°C–90°C) is consumed in several industries for washing and clean-up. These industries include, food processing, (e.g., meat processing for scalding), soft drink production, poultry dressing, textile industry, and metal-fabricating industry; (c) *peeling and blanching*: in the typical peeling operation, the product is introduced into a hot bath, which may be caustic, and the skin or outer layer, after softening, is mechanically scrubbed or washed-off. Blanching operations are similar to peeling. Product is usually introduced into a blancher to inhibit enzyme action, provide produce coating, or for cooking; (d) *evaporation and distillation*: evaporation and distillation are basic operations in many processing plants to aid concentrating a product or separating products by distillation. Evaporators are commonly found in sugar processing, mint distilling, and organic liquor processes; (e) *sterilizing*: sterilizers are used extensively in a wide range of industries and include applications such as equipment sterilization for the canning and bottling industry [96].

One industrial technology that is well suited for geothermal energy is vapor recompression. This technology is briefly described below [91,94,98].

7.5.1 VAPOR RECOMPRESSION

Vapor recompression is a technology used in various industries in which low-pressure steam exiting from one part of a process (often a cooking or evaporation process) is directed to a compressor where the pressure of the steam is raised so that it can be used in a higher temperature/pressure part of the process. There are limitations in terms of the amount of pressure that can be added at the compressor (most applications are at compression ratios of less than 2:1). This translates into a minimum geothermal fluid temperature requirement of approximately 190°F. The attractiveness of recompression is that only that energy required to boost the steam to the higher pressure is added at the compressor. This amounts to only a fraction of the energy required to generate equivalent pressure steam at the same pressure with a boiler. In general, the compressor supplied steam costs (energy only) just 20% of boiler supplied steam. This number does not reflect the optimum configuration for a recompression design but only to illustrate the potential of the technology. It is potentially one method of capturing moderate pressure/temperature, steam-based, industrial applications with low-temperature geothermal water resources [89,91,94,96,98].

7.5.2 GEOTHERMAL HEAT FOR CHEMICAL INDUSTRY

Geothermal energy is also used for some low-temperature chemical manufacturing. Using information from the Brown et al.'s work [95], Rafferty [90,97] carried out an evaluation of 108 industrial processes to identify those processes most applicable to geothermal use. The individual processes were ranked first by the percentage of the heating requirements that were at or below 250°F and secondarily by the energy use per unit of production. This approach was based on the assumption that industries most likely to use geothermal would be those that could displace all or most of existing energy requirements and do so at temperatures closest to those commonly available from geothermal. Industrial processes are heavily dominated by steam as the heating medium, but there are a few that have significant hot-water use (acrylics, butyl rubber, malt beverages, and concrete). Most the industries with favorable heat use characteristics are in the plastics, rubber, chemical, and paper sectors. The amount of energy consumed in the industrial sector is substantial. For just the two top processes (Rayon and Acetate), according to Chemical and Engineering magazine, the total U.S. production in 2001 was 235,000,000 lbs. At an average of 40,000 Btu per pound of product, the energy consumed amounts to 9.4×10^{12} Btu— roughly equivalent to the annual energy supplied by all existing direct use projects in the US combined. Applications of geothermal energy for industry in Europe are described in detail by Popovska [89].

7.6 ROLE OF HYBRID SOLAR THERMAL ENERGY

Solar thermal systems have a significant potential for industrial process heating worldwide. While these systems are primarily used for low-temperature applications, new designs have been deployed to serve applications requiring up to 400°C. Small-scale plants and industries that are less energy-intensive (like the textile and food sectors) have a significant technical and economic potential for renewable energy through solar thermal systems. However, high initial capital cost is a barrier.

Solar thermal technologies can provide low- and medium-temperature process heat. In early 2014, there were around 130 solar thermal plants for industrial process heating worldwide, comprising a combined 93 megawatt-thermal (MWth) of total capacity [87]. Today most applications in the industry sector concern low-temperature heat generation from glazed and unglazed flat plate and evacuated tubular collectors [85]. For higher temperature process heat applications, solar concentrator technologies offer alternatives, such as parabolic trough concentrators, parabolic dishes (with fixed or moving focuses), or vacuum tube collectors with compound parabolic concentrators [99]. Industry sectors that employ solar thermal technologies in their processes are typically the food and beverages (drying, washing, pasteurizing processes) and the textile sectors (washing and bleaching processes) [88,90–92]. Heating make-up water for steam systems and for washing and cleaning in various other industry sectors is another example, e.g., chemical and petrochemical and pulp and paper sectors [89]. The conversion efficiencies depend on the annual solar yield of the region, the temperature of the process heat, and the type of collector (i.e., conversion factor, loss coefficients).

Solar thermal sources have the potential to produce 63 EJ of process heat for industry in 2050.

From the regional breakdown identified by IRENA [12,13,106], it is clear that most regions have the potential for the significant application of solar thermal systems for process heat production. OECD countries have a large potential due to their large industrial energy demand. Niches exist in several sectors in which part of the low-temperature energy demand can be economically supplied by solar thermal systems. In terms of the sectoral breakdown, the food and tobacco sector has almost half of the potential, with the balance well spread among other sectors. IRENA [12,13,106] reports that where good solar radiation is available, solar thermal technologies for industrial process heat are very close to break even. In areas with lower solar radiation, such as in central Europe, solar thermal solar energy is in the range of millions of EJ/y (e.g., 3.9 million EJ/y). This is significant compared to the current world total primary energy supply of 503 EJ in 2007 (IEA statistics).

Globally, industrial process heat accounts for more than two-thirds of total energy consumption in industry, and half of this process heat demand is at low to medium temperatures (< 400°C). Currently, approximately 40% of industrial primary energy consumption is covered by natural gas and approximately 41% by petroleum. This means that there is a technical potential to provide around 15 EJ of solar thermal heat by 2030 (around 10% of industrial energy demand) while the share of solar thermal deployed in the industrial sector could reach 33% [12,13,106]. Prime application areas for solar thermal systems are in the food, beverage, transport equipment, textile, machinery, and pulp and paper industries, where roughly 60% of the heating needs can be met by temperatures below 250°C. Smaller systems include absorption/adsorption chillers or other thermal chillers, and the drying of agricultural products.

An important barrier for the deployment of solar process heat is the structure of the industrial sector. The energy-intensive industries account for 75% of heat demand, but consist of only 30,000–60,000 plants. For larger industrial plants, integration into existing and optimized process heating streams, as well as the lack of familiarity with the technology, constitutes critical bottlenecks. The other 95% of the industrial plants are small- and medium-size enterprises. This means that solar process heat technologies need to be tailored to provide the specific energy demand needs at individual locations. Two smaller application areas are the use of solar thermal systems to drive absorption/adsorption chiller machines or other thermal chillers, and the active use of solar heating for drying agricultural products [12,13,21–23,100–106].

In industry, five sectors, transport equipment, machinery, mining and quarrying, food and tobacco, and textiles and leather, use a significant proportion of their process heat at temperatures lower than 400°C and are therefore likely to have a strong potential for solar thermal to meet their process heat needs. Solar radiation can be useful for washing, preheating of boiler feed water and space heating. The chemical sector has also a high potential for solar thermal, but generally on a very large scale. Cost reductions in concentrated solar power (CSP) technologies, combined with the growth in the production of chemicals in Africa and Middle East, suggest growing scope for the development of solar thermal applications in the chemical sector. The main barriers to the greater use of solar thermal in this sector are the scale of the area needed for solar collectors (2.3 EJ/y) below 100°C. If full potential is materialized, it could increase the total estimated potential for solar thermal in industrial applications in 2050 from 5.6 EJ/y to around 8 EJ/y.

Different solar technologies have different investment costs per unit of capacity and different levels of capability in terms of thermal output. The economic competitiveness of solar thermal energy in industry will be very positively affected by high carbon prices. Among renewable technologies, solar has an advantage over bioenergy as it is not exposed to feedstock price volatility. The main competitors for solar thermal systems are heat pumps. Heat pumps operate in similar low-temperature ranges. Solar thermal technologies will need to be implemented widely if they are to become common under all radiation conditions. To achieve the projected 5.6 EJ/y in 2050, the solar capacity needed by the industrial sector would be over 2,500 gigawatt hours (GWth) [12,13,21–23,100–106].

In general, there are three groups of solar thermal technologies that are useful for industrial process heat: solar air collectors, solar water systems, and solar concentrators. Solar air collectors are found primarily in the food processing industry to replace gas- or oil-based drying or to reduce food spoilage due to open-air drying. They can be built locally, and their cost depends on local building materials and labor. Conventional solar water systems, like flat-plate collectors (FPC) or evacuated tube collectors (ETC), are primarily used in residential applications, but they can readily be installed on industrial rooftops to provide heat demand of up to 125°C. More than one hundred systems exist around the world. They are heavily manufactured in countries such as Brazil, China, South Africa, and Turkey where the costs are three to ten times lower than in the United States or in European countries. A number of more advanced FPC and ETC designs are currently on the market and can generate temperatures of up to 250°C; however, they are also more expensive than conventional FPC and ETC. Solar concentrators include parabolic dish collectors, linear parabolic trough collectors, and linear Fresnel collectors. In India, local manufacturers sell mainly parabolic dish collectors that can generate temperatures of up to 400°C [12,13,21–23,100–106].

Most solar thermal systems for industrial process heat are small-scale pilot plants. Only a third of 140 projects has collector areas > 500 m², and the four larger projects (all FPC) account for 49% of the installed thermal capacity. The solar thermal plant opened in a copper mine in Chile in 2013 now accounts for 28% of installed capacity. Additionally, almost 80 parabolic dish collectors are used for community cooking in India, around 40 MWth in total. Countries with high sun hours (e.g., India, Mexico, and countries in the Middle East/the Arabian Peninsula) are seen as growth markets.

The costs of solar heat for industrial process heat strongly depend on process temperature level, demand continuity, project size, and the level of solar radiation of the site. For conventional FPC and ETC, investment system costs range between EUR 250 and 1,000/kW in Europe, and around EUR 200–300/kW in India, Turkey, South Africa, and Mexico. The energy costs for feasible solar thermal systems range from eurocents 2.5–8/kWh, and a European roadmap targets solar heat costs of eurocents 3–6/kWh [104]. For concentrated systems, heating costs are in the range of eurocents 6–9/kWh with a target of eurocents 4–7/kWh for concentrating systems by 2020 [104]. Concentrated systems include Parabolic Dish Collectors (developed and used in India) with costs ranging from USD 400 to 1,800/kW, Parabolic Trough Collectors with costs ranging from USD 600 to 2,000/kW, and Linear Fresnel collectors in the range of USD 1,200–1,800/kW. In comparison, the same technology is used in CSP plants with costs of around USD 34,000–6,000/kW [105]. A number of developments are pursued to reduce the costs of solar process heating systems. For conventional FPC and ETC, the use of polymers to replace steel and copper components is considered for future cost reductions, as well as more modular designs to allow for easier integration into industrial rooftops. For concentrating systems, the integration and optimization of solar process heating into existing and newly built industrial plants will be an essential technology improvement [12,13,21–23,100–106].

7.6.1 SOLAR COOLING

The chemical and petrochemical and food and tobacco sectors are the largest industrial users of process cooling. Most of the cooling in both sectors is currently done with electric chillers. The main alternative, especially in the chemical and petrochemical sector, is natural gas-powered absorption chillers. Data from the United States Energy Information Agency's MECS [1,30,31] indicate that process cooling accounts for 8.5% of the total power demand of the global chemical industry and for 12.5% of the global demand of the petrochemical industry. It is unlikely, however, that much of this demand can be met by solar cooling, given the very low temperatures required by chemical processes and the relatively high energy demands of individual facilities. These characteristics are difficult to meet with solar thermal systems, given the large areas of solar panel that would be needed to deliver them.

This leaves only one sector with a good potential for solar process cooling, the food and tobacco sector. According to the MECS, process cooling in the food and tobacco sector accounts for 27% of the sector's electricity demand, equivalent to 6% of the sector's total final energy demand. On this basis, the total process cooling demand for the food and tobacco sector is estimated in 2007 to have been less than 0.4 EJ/y worldwide. Although solar cooling can play an important part in niche applications in the industry, for example, in cooling greenhouses, it is unlikely to offer the potential to achieve significant savings in fossil fuel use or GHG emissions.

7.7 POTENTIAL OF RENEWABLE ENERGY TECHNOLOGIES FOR INDUSTRIAL ELECTRICITY USE

There are a number of electricity-intensive industrial production processes. These include the production of nonferrous metals, such as aluminum (~56 GJ/t), copper (~14 GJ/t), and zinc (24 GJ/t), as well as the chlor-alkali (12 GJ/t chlorine) process [107,108]. When the size of production plants is considered, primary aluminum smelters consume the largest quantities of electricity per plant (~14 PJ/y) [108–110]. It is therefore already common in the primary aluminum sector that smelters are located next to hydropower plants, which ensure the continuous supply of cheap electricity, e.g., in Iceland, Norway, or Brazil [111].

Electrification of production processes is another option if electricity is generated from renewable energy sources. Industrial production processes typically operate based on process heat, with the exception of a few processes, such as smelting or electrolysis. Some of these heat-based production processes can also operate via novel process routes running based on electricity. One example originates from Iceland where hydrogen is produced from water via electrolysis and is subsequently combined with CO_2 to produce bio-based methanol [112]. This process substitutes the fossil fuel-based steam reforming or partial oxidation process. However, since electrolysis is an electricity-intensive process, such transition is only possible in regions where electricity is cheap. Solar thermal technology can also provide an alternative to cooling processes in sectors, such as the food and tobacco sector [20,81].

At 20% of final energy use in manufacturing, the demand for electricity is only expected to grow. This is due in part to electrification of production processes. The decarbonization of the power sector is necessary, but other options for increasing the renewable potential of the electricity sector are possible. Some top solutions include the relocation of these industries to be within close proximity to renewable energy power plants and on-site renewable energy electricity generation. Technology development in the manufacturing industry and power sectors is necessary for effective deployment of increased electrification with renewable energy. Some of the unique ways these methods are implemented in commercial manufacturing include energy-efficient and self-sustaining facilities and plants that are fully powered with renewable energy sources such as wind turbines and solar power.

Manufacturers of tensile fabric structures for mining, construction, aviation, and other commercial industry applications are also participating in the transition toward a sustainable energy future. For example, fabric buildings for commercial manufacturing industries are made from highly resilient materials that last for many years in even the most extreme climates. They are equipped with proprietary environmental control and power systems to optimize energy use, provide stand-alone power generation, and insulate facilities to reduce energy loss.

7.8 REALIZABLE ECONOMIC POTENTIAL OF RENEWABLE ENERGY INTEGRATION

In order to evaluate realizable economical potential of renewable energy integration in industry sector, IRENA used two cases of realizable technical potential, namely, the "ambitious development scenario" (AmbD), which considers the temperature level of process heat and new/existing capacity as the main criteria, and the "accelerated development scenario" (AccD), which, in addition, considers technology development and sector-specific characteristics. Both scenarios, however, refer to realizable technical potential by accounting for the development and deployment of new and emerging renewable energy technologies, which are beyond the business as usual or market trends historically observed. The potential according to the AmbD scenario is higher than that of the AccD scenario [8].

Besides environmental impact of renewable energy, the second most important factor is the availability of renewable resources to meet the estimated economic potential. The annual total technical potential of solar energy ranges from 1,500 to 50,000 EJ worldwide [113]. In comparison to this large supply potential, the total estimated economic potential from the industry sector (3 EJ) is negligible, even when compared to regions where solar energy potential is low, such as Central and Eastern Europe (technical supply potential of 4–154 EJ). For geothermal energy, the technical potential for direct use is estimated to range between 10 and 312 EJ, with the lowest potential in OECD Europe (0.5–16 EJ) and the Middle East (5–21 EJ) [114]. Even with potential competition from heating for the residential sector and the utilization of these resources for power generation, no limitations are foreseen from deploying the realizable potential of geothermal energy. Therefore, the economic potentials of solar thermal (0.9–3.8 EJ) and geothermal (1.7–1.9 EJ) technologies are also the realizable economic potentials, in view of regional resource availability estimates. Likewise, the realizable economic potential for heat pumps is 1.2–1.9 EJ, with additional electricity demand to run heat pumps (3–6 EJ) being met from a mix of renewable energy sources.

In comparison to solar thermal and geothermal technologies, the potential of biomass supply depends on a number of factors, such as increases in food demand and developments in the agriculture sector, availability of resources for agriculture (water, land), and development in biomass conversion technologies. Therefore, it is a complex task to provide ranges for the technical potential of biomass supply. IRENA prepared its own estimates in a bioenergy working paper for the year 2030 based on the analysis of more than 100 countries and seven types of biomass feedstocks [115]. According to the IRENA bioenergy working paper, there is a global biomass supply potential of 95–145 EJ/y. About 40–65 EJ of this total is from agricultural residues and waste and about 25–40 EJ is from forestry products, including residues. The remainder 30–40 EJ is related to energy crops.

IRENA's [12] REmap shows that nearly 20% of the total biomass demand would be related to the industry sector if the global share of renewable energy is to double by 2030. This global average ranges from 0% in Middle East to more than half of the total final industrial energy demand in India. A quarter of all biomass is assumed to be available for the industry sector (total of 18–30 EJ, 8–18 EJ residues, and 10–12 EJ energy crops). The realizable economic potential of biomass for the industry sector, in view of resource availability, is equivalent to 19 EJ and 14 EJ for low- and high-price scenarios, respectively, according to the AmbD scenario. About 85%–95% of the total potentials are related to process heat generation. The remaining 5%–15% is use of biomass as feedstock that is estimated to have approximately 1–2 EJ in both scenarios, assuming that biomass residues and energy crops are first used for process heat generation and the remaining quantities are only available as feedstock. For electricity, the realizable potentials are 0.6 EJ and 1.1 EJ according to the AccD and AmbD scenarios, respectively (0.1 EJ solar cooling, and 0.5–1.0 EJ relocation of primary aluminum smelters).

In allocating realizable economic potential, price and temperature are important. Biomass is estimated to play a key role for high-temperature heat applications where other technologies do not provide alternatives for. In both existing and new capacity, biomass demand is estimated at 6.1 EJ (AccD scenario) and 6.7 EJ (AmbD scenario) by 2030. Demand from iron and steel (0.5–1.4 EJ) and nonmetallic minerals (4–4.4 EJ) sectors accounts for most demand for high-temperature applications with rest being in the chemical and petrochemical sector (1.2–1.4 EJ). The total potential of biomass can substitute up to 18% of the total high-temperature heat demand by 2030. The remainder of the economic potential of biomass can be used for medium-temperature heat applications in new and existing capacity (total potential of 5.3 and 7.2 EJ). The largest potential is seen in the chemical and petrochemical (1.3–1.7 EJ) and food and tobacco (0.8–1.0 EJ) sectors and in other less energy intensive sectors (3.1–4.5 EJ). The total substitution potential of biomass use for medium temperature heat applications reaches as high as 35%. At low temperatures, biomass can face competition from solar, geothermal, and heat pump resources. Biomass can, however, be used to substitute the remaining fossil fuel use in low-temperature heat applications (after accounting for the quantities substituted by other renewable energy technologies). These potentials—equivalent to 4.3 EJ (AccD scenario) and 5.7 EJ (AmbD scenario)—exist in various sectors, with the exception of energy-intensive sectors. Together with biomass, all renewable energy technologies offer a substitution potential of up half of total fossil fuel demand for low-temperature heat generation according to the AccD and AmbD scenarios, respectively. The biomass demand as feedstock for the production of materials is estimated at 1–2 EJ, which is equivalent to a substitution potential of 4%–7% according to the AccD and AmbD scenarios. Relocation of primary aluminum smelters offers 0.7 EJ (AccD scenario) and 1.0 EJ (AmbD scenario) potential related to renewable electricity use worldwide. The potential for solar cooling to substitute electricity-based equipment in the food and tobacco sector is estimated at 0.1 EJ.

The potential of renewable energy technologies could reach a total 23 and 28 EJ by 2030. Biomass is estimated to contribute to approximately 75% of this potential (17 and 22 EJ) followed by the solar thermal heat (2.5 EJ) and heat pump potential (1.5 EJ). Total potential of renewable energy technologies could raise the share of renewable energy in the fuel mix of the industry sector from 10% to up to 34%. The total potential for renewable energy is 4 EJ in the high price scenario, and 7 EJ lower according to the AccD and AmbD scenarios. More discussions on potentials for renewable energies in industrial sector are given in an excellent report by IRENA [12,115].

7.8.1 Priority Areas of Action

IRENA identified six priority areas that warrant action from both policy-makers and industrial stakeholders. These are as follows [8]:

1. **Energy-intensive Sectors**: With 75% of the total industrial energy demand and long lifetimes for these types of plants, the energy-intensive sectors need to consider renewable energy options not only as an integral part of their new build capacity but also as part of their existing capacity.
2. **Small- and medium-sized enterprises (SMEs)**: Accounting for more than 90% of all manufacturing businesses, SMEs play a crucial role in increasing the deployment rate of renewable energy technologies, providing local manufacturing opportunities and stimulating cost reductions through learning by doing.
3. **Biomass**: Among the renewable technology options, biomass has the largest substitution potential in the manufacturing industry, but immediate and internationally coordinated action is required to alleviate the serious supply constraint of sustainable sourced and low-cost biomass resources, and to deploy the most resource efficient biomass use applications.
4. **Solar thermal systems**: Solar thermal heat systems have a large technical and realizable economic potential in small-scale plants and less energy-intensive industries like the textile and food sectors, but the vicious circle of high initial capital costs and low deployment rates needs to be broken.
5. **Electrification**: With increased electrification in the industry sector, renewable energy deployment can only be achieved through technology development in both the industry and power sectors.
6. **Regional aspects**: Regional potential depends on production growth, ratio of existing and new capacity, and renewable resource availability. Energy pricing and climate policies can ensure a level-playing field and biomass resource constraints may be elevated by trade, but equally important will be specific policies to support the different industries in deploying renewable energy.

7.9   REDUCTION IN GHG EMISSION BY CLEAN
ENERGY ALTERNATIVES

Development of effective GHG mitigation strategies requires a detailed understanding of the types of industries and their energy-use patterns and associated emissions. This has recently been made possible by the U.S. Environmental Protection Agency (EPA) Greenhouse Gas Reporting Program (GHGRP). Under the Mandatory Reporting of GHGs Rule, facilities with annual direct emissions greater than or equal to 25,000 metric tons carbon dioxide-equivalent (MTCO$_2$e) are required to report to the EPA (Part98—Mandatory Greenhouse Gas Reporting 2016). Over 8,000 facilities representing nine industry sectors reported direct emissions of 3,200 million MTCO$_2$e (MMTCO$_2$e), or nearly half of U.S. total GHG emissions, for the 2014 reporting year [3–6].

NREL (National Renewable Energy Laboratory) selected fourteen key industries with the highest amount of GHG emissions and thermal heat duties. Within these industries, representative plants were selected to determine how clean heat from SMRs, SIPH, and geothermal sources could be used. The GHGRP data allowed further disaggregation of thermal energy use, enabling analysis by fuel type, combustion-unit type, and end-use for the 14 industries. The common feature of the target industries is that they convert raw materials into energy services by means of physical and chemical changes. These changes generally require thermal energy to affect solids and liquids heat-up, melting, and evaporation and to heat up reactants to initiate molecular bond-breaking and to sustain the propagation of chemical reactions. Heat demands range from low-temperature steam (50°C, 0.7 megapascal [MPa]) for steeping in corn wet-milling to high-temperature operations up to 2,200°C for electric arc furnaces. The scale of heat demand for the average facility ranges from 0.016 TJ/day (15 MMBtu/day; or 0.2 MWt) for electrochemical production of 1,330 tons per day chlorine to 26 TJ/day (25,000 MMBtu; or 300 MW$_t$) for 5,273 tons per day of potash, soda, or borate mining and processing.

The study [116] identified several technical challenges and opportunities to application of clean energy sources for industrial heat users including: (a) quality of heat required by the user (or temperature of the working fluid), (b) industry process heat-transfer modes, (c) scale of heat source versus heat user demand, which may be mitigated by selecting the appropriate source or by industrial clustering (viz., an energy park), (d) transport requirements between the heat source and industrial process-unit operations, which involves distance and the materials needed for that transport, (e) thermal energy storage needs and options, (f) hybrid heat/electricity production, (g) electrification of heating processes, and (h) hydrogen production and use as an intermediate energy source.

A summary of the findings of this study is given in Table 7.9. The study indicated that the largest end-uses of combustion energy in 2014 were CHP and/or cogeneration (37% of calculated energy use), conventional boiler use (32%), and process heating (24%). Natural gas was the most-used fuel by the target industries, accounting for 44% of calculated combustion energy use. The study also indicated that hybrid thermal/electricity generation may help balance hourly, daily, and/or seasonal electrical

TABLE 7.9

Summary of Potential Alternative Heat Supplies by Target Industry (TJ = terajoule = 10^{12}J) [116]

Target Industry	Number of GHGRP-Reporting Plants in 2014	Average Size of Plant (Production Rate)	Reported CO_2 Emissions (MMTCO2e)	Fraction of Industrial-Sector GHG Emissions (%)*	Industry Process-Heat Type/Purpose	Average Plant Heat Use in TJ/day (MMBtu/day)	Process-Heat Temperature (°C)	Potential Alternative Heat Supply**
Petroleum Refineries	141		124	8	Combustion gases/atmospheric crude fractionator and heavy naphtha reformer	8.23 (7,800)	600	SIPH, SMR (HTGR)
Gasoline		33,828 bpd						
Diesel		12,747 bpd						
Kerosene		6,755 bpd						
Iron and Steel Mills	115	603	51	3	Combustion gases/coke production	2.42 (2,290)	1,100	Hydrogen reducing agent
					Combustion gases/electricity production		1,700	***
					Electricial/steel production		2,200	***
Paper Mills	116	1,723	32	2	Steam/stock preparation	21.1 (20,000)	150	***
					Steam/drying		177	***
Paperboard Mills	73	4,427	24	1.5	Steam/stock preparation		150	***
					Steam/drying		177	***
Pulp Mills	30	474	12	0.7	Combustion gases/electricity production	0.67 (640)	800	***
					Steam/wood digesting, bleaching, evaporation, chemical preparation	1.15 (1,090)	200	***

(Continued)

TABLE 7.9 (Continued)
Summary of Potential Alternative Heat Supplies by Target Industry (TJ = terajoule = 10^{12}) [116]

Target Industry	Number of GHGRP-Reporting Plants in 2014	Average Size of Plant (Production Rate)	Reported CO_2 Emissions (MMTCO2e)	Fraction of Industrial-Sector GHG Emissions (%)*	Industry Process-Heat Type/Purpose	Average Plant Heat Use in TJ/day (MMBtu/day)	Process-Heat Temperature (°C)	Potential Alternative Heat Supply**
All Other Basic Chemical Manufacturing Industries	85	2,702	21	1.3	Steam/evaporation, chemical preparation	2.56 (2,430)	150	***
					Combustion gases/primary reformer; steam/methanol distillation	12.9 (12,200)	900	SMR, SIPH
Ethyl Alcohol Manufacturing	168	63.7	18	1.1	Combustion gases for steam/byproduct drying (corn dry mills)/pretreatment and conditioning (lignocellulosic processes)	1.76 (1,670)	266	SMR, SIPH
					Steam/distillation		233	SMR, SIPH
					Steam/electricity production		454	SMR, SIPH
Plastics Material and Resin Manufacturing	72	1,591	17	1	Steam/distillation	10.6 (10,061)	291	SMR, SIPH
Petrochemical Manufacturing	35	2,665	16	1	Combustion gases/cracking furnace	2.37 (2,250)	875	***
Alkalies and Chlorine Manufacturing	11		13	0.8	Steam/drying	4.26 (4,040)	177	SMR, SIPH
Chlorine		1,330						
Sodium Hydroxide		1,162						

TABLE 7.9 (Continued)
Summary of Potential Alternative Heat Supplies by Target Industry (TJ = terajoule = 10^{12}]) [116]

Target Industry	Number of GHGRP-Reporting Plants in 2014	Average Size of Plant (Production Rate)	Reported CO_2 Emissions (MMTCO2e)	Fraction of Industrial-Sector GHG Emissions (%)*	Industry Process-Heat Type/Purpose	Average Plant Heat Use in TJ/day (MMBtu/day)	Process-Heat Temperature (°C)
Nitrogenous Fertilizer Manufacturing	30		8	0.5	Combustion gases/primary steam reforming	7.03 (6,660)	850
Wet Corn Milling	24		18	1.1	Steam/steeping	8.06 (7,640)	50
Starch		1,461			Steam/drying		177
Corn Gluten Feed		593					
Corn Gluten Meal		137					
Corn Oil		92					
Lime and Cement	49		10	0.6	Combustion gases/heating kiln	12.45 (11,800)	1,200–1,500
Lime		507					
Cement		2,000					
Potash, Soda, and Borate Mining	11	5,273	6	0.4	Steam/calciner, crystallizer, and dryer	26 (25,000)	300

* Includes CO_2 from biomass combustion.

** SMR temperatures up to 850°C, SIPH temperatures up to 1,000°C, geothermal heat supply up to 150°C.

*** Industries with process temperatures above 1,000°C (i.e., lime and cement, iron and steel) were not addressed in the analysis estimating potential alternative heat supply, although the report discusses applicable alternatives. Likewise, industries that rely on their process byproducts for combustion fuels (i.e., pulp and paper, petrochemical manufacturing) were also excluded from the estimates of potential alternative heat supply.

cycles. Thermal energy storage concepts such as those being developed for concentrating solar systems may help coordinate grid profiles with industry heat use profiles. SMRs were identified as an option for process heat and hydrogen production for feedstock use.

HESs have been proposed as a solution to using the excess power generation capacity that exists on the electrical grid when the generation capacity exceeds demand periods. In the context of the national energy systems, the definition of a hybrid system is one that dynamically uses heat or electricity to optimize the financial efficiency of the systems by producing the highest value set of energy services and products throughout the year. These products include electricity, manufactured goods, and intermediate energy carriers that may be stored or directly used to produce the set of products. The value proposition of a "greenfield" HES concept is currently being addressed by DOE, with an emphasis on regional scenarios that include a relatively high, hypothetical penetration of renewable energy [2,116,117]. A "brownfield" HES at an industrial site would involve the addition of a thermal energy generation source that is dynamically connected to the grid. Three scenarios presented by Ruth et al. [116] provide a general view of the basic system integration possibilities for clean thermal energy and power generation.

The analysis by Ruth et al. [117] reveals that clean energy sources may economically displace with industry fossil-fired heat generation under the assumptions considered in the study; however, the value position of hybrid operations will depend on the value of electricity. The role and cost of energy storage will also drive hybrid system deployment and operation considerations. HES may connect to industry through energy storage and energy carriers that are produced using the excess power generation sources that are not tightly coupled to industry. Geographical separation, differences in SMR scales, and industrial operation cycles may be addressed with the production of an intermediate product. Potable water, hydrogen, and other intermediate chemicals such as methanol and ammonia are examples of intermediate products. Seasonal energy use patterns are an important consideration for HES. For example, agriculture residues may be processed into energy products during or following the summer-to-fall harvest season. This conveniently corresponds to the fall period when electricity demand is at the lowest level for the year. Similarly, ammonia production during the spring could take advantage of the excess electricity generation capacity while producing fertilizers needed for spring and summer agriculture demands. The trade-off of ammonia and fertilizer plant capacity factors and product storage associated with constant generation throughout the year should be taken into consideration in such cases.

7.10   CLOSING PERSPECTIVES

Manufacturing industry accounts for about one-third of total energy use worldwide. Roughly three quarters of industrial energy use are related to the production of energy-intensive commodities such as ferrous and nonferrous metals, chemicals and petrochemicals, nonmetallic mineral materials, and pulp and paper. In these sectors, energy costs constitute a large proportion of total production costs, so managers pay particular attention to driving them down. As a result, the scope to improve energy efficiency tends to be less in these most energy-intensive sectors than in those sectors where energy costs form a smaller proportion of total costs, such as the buildings and transportation sectors. This limits the overall potential for CO_2 reductions through energy efficiency measures in industry to 15%–30% on average.

Industrial production is projected to increase by a factor of four between now and 2050. In the absence of a strong contribution from energy efficiency improvements, renewable energy, CCS, and waste heat capture will need to make a significant impact if industry has substantially to reduce its consequent GHG emissions.

Although renewable energy has received a good deal of attention for power generation and for residential applications, its use in industry has attracted much less attention. Renewable energy plays only a relatively small role in industry today. Biomass currently makes by far the most significant renewable energy contribution to industry, providing around 8% of its final energy use in 2007. The present knowledge of the long-term potential for renewable energy in industrial applications suggests that up to 21% of all final energy use and feedstock in manufacturing industry in 2050 can be of renewable origin. This would constitute almost 50 exajoules a year (EJ/y), out of a total industry sector final energy use of around 230 EJ/y in the GEA Scenario M that is used as the baseline projection in this study. This includes 37 EJ/y from biomass feedstock and process energy and over 10 EJ/y of process heat from solar thermal installations and heat pumps.

The use of biomass, primarily for process heat, has the potential to increase in the pulp and paper and the wood sectors to 6.4 and 2.4 EJ/y, respectively, in 2050. This represents an almost three-fold increase in the pulp and paper sector and a more than five-fold increase in the wood sector, reaching a global average share of 54% and 67%, respectively, of the total final energy use in each sector. Other sectors, including some of the most energy-intensive industries such as chemicals, petrochemicals, and cement, also have the potential to increase their use of biomass, but they will only achieve that potential if there is a concerted effort for them to do so. For chemicals and petrochemicals, wider biomass deployment will depend mainly on investment in biorefineries that can make profits and spread risk through the production of a range of products. In the cement sector what is most needed is a proper policy framework for the management of wastes and incentives to increase their use in cement production. Interesting potential lies in the development of bio-based vehicle tyres and their subsequent use in cement kilns at the end of their useful life. This analysis suggests that, by 2050, biomass could constitute 22% (9 EJ/y) of final energy use in the chemical and petrochemical sectors and that alternative fuels could constitute up to 30% (5 EJ/y) of final energy use in the cement sector. Across all industrial sectors, biomass has the potential to contribute 37 EJ/y. But the achievement of this potential will depend on a well-functioning market and on the development of new standards and preprocessing technologies. About one-third of the potential (12 EJ/y) could be achieved through interregionally traded sustainable biomass feedstocks.

Solar thermal energy has the potential to contribute 5.6 EJ/y to industry by 2050. Almost half of this is projected to be used in the food sector, with a roughly equal regional distribution between OECD countries, China, and the rest of the world, mainly in Latin America (15%) and Other Asia (13%). Costs depend heavily on radiation intensity. They are expected to drop by more than 60%, mainly as a result of learning effects, from a range of USD 17–USD 34 per gigajoule (GJ) in 2007 to USD 6–USD 12/GJ in 2050. Heat pumps also have a part to play in low-temperature process applications and are estimated to contribute 4.9 EJ/y in 2050. Most (43%) of this will be concentrated in the food sector, mainly in OECD countries (60%), China (16%), and the Former Soviet Union (15%). Costs for useful energy supply are projected to drop by between 30% and 50%, due mainly to reduced capital costs, increased performance and more consistent, market-driven, international electricity prices, from a range of USD 9–USD 35/GJ in 2007 to USD 6–USD 18/GJ in 2050.

The competitiveness of biofuels with fossil fuels is strongly dependent on national energy policy frameworks and energy prices. In the last decade, the ratio between the highest and lowest end-use prices for natural gas for industry in different countries has at times been as high as 60. At the end of 2009, the ratio stood at 10. For coal, the ratio between different countries has been as high as 30 and, at the end of 2009, stood at 15. Renewables are not cost-competitive where fossil fuels are subsidized. They are, however, already cost-competitive in many cases and many countries with unsubsidized fossil fuels. This is even more so where CO_2 emissions carry a financial penalty that reflects their long-term economic and environmental impact. Where national energy policies subsidize fossil fuels, they strongly affect the competitiveness of renewable energy.

Overall, an increase in renewable energy in industry has the potential to contribute about 10% of all expected GHG emissions reductions in 2050. At nearly 2 gigatons (Gt) of CO_2, this represents 25% of the total expected emission reductions of the industry sector. This is equivalent to the total current CO_2 emissions of France, Germany, Italy, and Spain, or around one-third of current emissions in the United States. This potential can only be realized, however, if specific policies are developed to create a business environment conducive to private sector investment, particularly in the transition period. Current best practice shows the conditions under which the successful deployment of renewables can take place and this should guide future policy making. RD&D and cost reductions through economies of scale are the priorities. In the longer term, a price for GHG emissions of the order of USD 50/t CO_2 is needed to support the development of a market for renewable energy technologies and feedstocks in industry.

In recent years, renewable energy has increasingly attracted public and policy attention particularly for its potential to contribute to reductions in GHG emissions. Most interest has focused on the use of renewables in power generation and as biofuels. Although some attention has been paid to the potential for renewables, particularly biomass and solar thermal technologies, to contribute to heating and cooling in residential space heating applications, their use in industrial applications has received less interest. Renewable energy can be widely applied in industrial applications. The four options are (a) biomass for process heat; (b) biomass for petrochemical feedstocks; (c) solar thermal systems for process heat; and (d) heat pumps for process heat.

Several other options may also become relevant in the future but are more likely to make niche contribution. They include (a) conventional geothermal heat, which is highly location-dependent; (b) due to inability to transport heat over long distance without significant heat losses, industrial plant must be located near geothermal reservoir. This may not be possible except for a few highly specialized applications; (c) enhanced geothermal systems may make a contribution in the long run, subject to the resolution of technology issues; (d) the use of run-of-river hydro for motive power, of the kind that has been used for centuries for grinding mills; and (e) the use of wind for motive power, for example, by driving air compressors enabling the storage of energy in the form of compressed air.

In CHP plants, the waste heat from biomass electricity generation can be used very effectively in industrial applications. The full achievement of the potential of renewable energies will depend on the widespread adoption by the industry of the Best Available Technologies (BAT). The speed of the adoption of BAT is, however, subject to a number of significant barriers such as: (a) lack of information on the potential contribution of renewables and ways of achieving it; (b) cheap fossil fuels; (c) the absence of appropriate technology supply chains; (d) lack of technical capacity; (e) the high cost of capital in many developing countries; (f) a focus on upfront investment cost instead of full lifecycle cost; (g) risks associated with technology transitions and the adoption of early stage technologies; (h) restricted access to financial support to cover the extra costs of BAT; and (i) the lock-in of inefficient, polluting technologies with long lifetimes. International cooperation can help to address these barriers, especially where it is conducted in close collaboration with national governments. Unfortunately, the IEA estimates that fossil fuels still receive subsidies of around USD 550 billion a year worldwide. Similar types of subsidies for renewable sources are needed. The long-term potential for selected renewable energy sources and technologies in the industrial manufacturing sector is good.

REFERENCES

1. Innovating Clean Energy Technologies in Advanced Manufacturing, Quadrennial Technology Review an Assessment of Energy Technologies and Research Opportunities Chapter 6. (2015, September). Department of Energy, Washington, DC.
2. McMillan, C., Boardman, R., McKellar, M., Sabharwall, P., Ruth, M. and Bragg-Sitton, S. Generation and use of thermal energy in the U.S. industrial sector and opportunities to reduce its carbon emissions. Technical Report NREL/TP-6A50-66763 INL/EXT-16-39680 December 2016 Contract No. DE-AC36-08GO28308, NREL, Golden, CO.
3. U.S. Environmental Protection Agency. (2016). *Inventory of U.S. Greenhouse Gas Emissions and Sinks: 1990–2014.* U.S. Environmental Protection Agency, Washington, DC. http://www3.epa.gov/climatechange/ghgemissions/usinventoryreport.html.
4. U.S. Environmental Protection Agency. (2016). GHGRP 2014: Reported data. Data and Tools, https://www.epa.gov/ghgreporting/ghgrp-2014-reported-data, Accessed 5 April 2016.
5. U.S. Environmental Protection Agency. (2016). Greenhouse gas customized search. U.S. Environmental Protection Agency, https://www.epa.gov/enviro/greenhouse-gas-customized-search, Accessed 5 April, 2016.

6. U.S. Environmental Protection Agency. (2016). Subpart C methodologies fact sheet. U.S. Environmental Protection Agency, https://www.epa.gov/ghgreporting/subpart-c-methodologies-fact-sheet, Accessed 12 April, 2016.
7. Fischedick, M., Roy, J., Abdel-Aziz, A., Acquaye, A., Allwood, J.M. and Ceron, J.-P. (2014). 2014: Industry. In *Climate Change 2014: Mitigation of Climate Change*. Contribution of Working Group III to the Fifth Assessment Report of the Intergovernmental Panel on Climate Change. Cambridge University Press, Cambridge, NY, 739–810.
8. IRENA. (2015). A background paper to "Renewable energy in manufacturing". March 2015, IRENA, Abu Dhabi- Renewable Energy Options for the Industry Sector: Global and Regional Potential Until 2030, IRENA, Abu Dhabi.
9. Renewable Energy in Industrial Applications. (2015). An Assessment of the 2050 Potential. United Nations Industrial Development Organization, UNIDO, Vienna.
10. IEA. (2012). *Extended Energy Balances of Non-OECD Countries*. OECD/IEA, Paris.
11. EIA. (2014, April). Annual energy review. Table 2.1. http://www.eia.gov/totalenergy/data/monthly/archive/00351504.pdf. EIA. 2010 Manufacturing Energy Consumption Survey (MECS).
12. IRENA. (2014, January). *REmap 2030: A Renewable Energy Roadmap*. IRENA, Abu Dhabi, http://www.irena.org/remap/REmap_Report_June_2014.pdf.
13. IRENA. (2014, June). *Renewable Energy in Manufacturing: A Technology Roadmap for REmap 2030*. IRENA, Abu Dhabi, http://www.irena.org/remap/REmap%202030%20Renewable-Energy-in-Manufacturing.pdf.
14. Chapas, R.B. and Colwell, J.A. (2007). Industrial technologies program research plan for energy-intensive process industries. *Prepared by Pacific Northwest National Laboratory for DOE*. Energy and emissions savings correspond to 2030 projections in the report. Opportunities greater than 100 TBtu have been tabulated; see the report for additional opportunities. www.efce.info/efce_media/-p-531.pdf.
15. Alcorta, L., Bazilian, M., De Simone, G. and Pedersen, A. (2014). Return on investment from industrial energy efficiency: Evidence from developing countries. *Energy Efficiency* 7: 43–53.
16. Worrell, E., Bernstein, L., Roy, J., Price, L. and Harnisch, J. (2009). Industrial energy efficiency and climate change mitigation. *Energy Efficiency* 2: 109–123.
17. Saygin, D., Patel, M.K. and Gielen, D.J. (2010, November). Global industrial energy efficiency benchmarking: An energy policy tool. Working Paper. United Nations Industrial Development Organization (UNIDO), Vienna, http://www.unido.org/fileadmin/user_media/Services/Energy_and_Climate_Change/Energy_Efficiency/Benchmarking_%20Energy_%20Policy_Tool.pdf.
18. UNIDO. (2011). Industrial energy efficiency for sustainable wealth creation. Capturing environmental, economic and social dividends. Industrial Development Report 2011. United Nations Industrial Development Organization, Vienna, http://www.unido.org/fileadmin/user_media/Publications/IDR/2011/UNIDO%20IDR%20reprint%20for%20web%20020912.pdf.
19. Shah, Y.T. (2021). *Hybrid Power-Generation, Storage and Grids*. CRC Press, New York.
20. Taibi, E., Gielen, D. and Bazilian, M. (2012). The potential for renewable energy in industrial applications. *Renewable and Sustainable Energy Reviews* 16: 735–744.
21. IEA-SHC (IEA Solar Heating & Cooling Program). (2014). Database for applications of solar heat integration in industrial processes. SHIP database, http://ship-plants.info/.
22. IEA-SHC. (2014). Solar heat worldwide. Market and contribution to the energy supply 2012, http://www.iea-shc.org/data/sites/1/publications/Solar-Heat-Worldwide-2014.pdf.
23. Hennecke, K. (2012). Review of recent developments in solar heat for industrial processes. DLR, Cologne, http://elib.dlr.de/79849/1/SolarPACES2012_ProcessHeatOverview.pdf.

24. Renewable Energy. (2009). Directive 2009/28/EC I RED I Heat – Icax. A website report www.icax.co.uk›Renewable_Energy_Directive.

25. IEA. (2009) *Energy Technology Transitions for Industry*. IEA/OECD, Paris.

26. Bandwidth Study on Energy Use and Potential Energy Saving Opportunities in U.S. Chemical Manufacturing. Prepared by Energetics Inc. for the U.S. DOE Advanced Manufacturing Office (2015).

27. U.S. Department of Energy (DOE) and U.S. Environmental Protection Agency (EPA). (2012, August). Combined heat and power: A clean energy solution. DOE/EE-0779. http://www.epa.gov/chp/documents/clean_energy_solution.pdf. Department of Energy, Washington, DC.

28. The White House Executive Order. (2012, August 30). Accelerating investment in industrial energy efficiency. https://www. whitehouse.gov/the-press-office/2012/08/30/ executive-order-accelerating-investment-industrial-energy-efficiency. Combined heat and power: A clean energy solution. (2012, August). U.S. DOE and U.S. EPA. http:// energy.gov/sites/prod/files/2013/11/f4/chp_clean_energy_solution.pdf, a website report.

29. U.S. Department of Energy. (2015). Barriers to industrial energy efficiency. U.S. DOE Energy Efficiency and Renewable Energy, a website report by DOE, Washington, D.C. http://www.energy.gov/sites/prod/files/2015/06/f23/EXEC2014-005846_6%20 Report_signed_v2.pdf.

30. Manufacturing Energy and Carbon Footprints. (2010 MECS). U.S. DOE Office of Energy Efficiency & Renewable Energy. http:// energy.gov/eere/amo/ manufacturing-energy-and-carbon-footprints-2010-mecs.

31. Sankey Diagram of Process Energy Flow in U.S. Manufacturing Sector. (2010 MECS). The data source for the Sankey diagram is the "manufacturing energy and carbon footprints, a website report. http://energy.gov/eere/amo/sankey-diagram-process-energyflow-us-manufacturing-sector.

32. Number of Establishments by Usage of General Energy-Saving Technologies 2010. Energy Information Administration. (2013). http://www.eia.gov/consumption/ manufacturing/data/2010/pdf/Table8_2.pdf.

33. Nimbalkar, S., Thekdi, A.C., Rogers, B.M., Kafka, O.L. and Wenning, T.J. (2015). *Technologies and Materials for Recovering Waste Heat in Harsh Environment.* Oak Ridge national laboratory report, ORNL, Oak Ridge, TN.

34. Shah, Y.T. (2018). *Thermal Energy-Sources, Recovery and Applications.* CRC Press, New York.

35. Frangopoulos, C.A. (2012). A method to determine the power to heat ratio, the cogenerated electricity, and the primary energy savings of cogeneration systems after the European directive. *Energy* 45: 52–61.

36. Shah, Y.T. (2020). *Modular Systems for Energy Usage Management.* CRC Press, New York.

37. Peakman, A. and Merk B. (2019). The role of nuclear power in meeting current and future industrial process heat demands. *Energies* 12: 3664, Doi: 10.3390/en12193664www. mdpi.com/journal/energies, a MDPI publication.

38. Rankine Cycle. (2017). Wikipedia. Retrieved 28 February 2017 from https:// en.wikipedia.org/wiki/RankineCycle.

39. Vanslambrouck, B. (2010). The Organic Rankine Cycle (ORC). CHP: Technology Update. Retrieved 29 April 2015 from http://www.ibgebim.be/uploadedFiles/Contenu_ du_site/Professionnels/Formations_et_s%C3%A9minaires/S%C3%A9minaire_ URE _%28Energie%29_2010_%28Actes%29/02-ORC_ VANSLAMBROUCK.pdf, a website report.

40. Tchanche, B.F., Lambrinos, G., Frangoudakis, A. and Papadakis, G. (2011). Low-grade heat conversion into power using organic Rankine cycles—A review of various applications. *Renewable and Sustainable Energy Reviews* 15(8): 3963–3979, Doi: 10.1016/j. rser.2011.07.024.

41. Quoilin, S., Vandenbroek, M., Declaye, S., Dewaller, P. and Lemort, V. (2013). Techno-economic survey of organic Rankine cycle (ORC) systems. *Renewable and Sustainable Energy Reviews* 22: 168–186.

42. Arvay, P., Muller, M.R., Ramdeen, V. and Cunningham, G. (2011). Economic implementation of the organic Rankine cycle in industry. In: ACEEE Summer Study on Energy Efficiency in Industry, 12–22., a website report based on summer study (2011).

43. Daccord, R., Melis, J., Kientz, T., Darmedru, A., Pireyre, R., Brisseau, N., and Fonteneau, E. (2013, May 28). Exhaust heat recovery with Rankine piston expander. *Proceedings of ICE Powertrain Electrification & Energy Recovery*, Rueil-Malmaison. France, published in an open access by MDPI.

44. STOWA. (2007). Organic Rankine cycle for electricity generation, selected technologies. http://www.stowaselectedtechnologies.nl/Sheets/index.html, a website report.

45. Duffy, D. (2005). Better *Cogeneration through Chemistry: The Organic Rankine Cycle.* Distributed Energy, SOWA and Distributed Energy, November/December.

46. Heidelberg Cement. (2007). Organic Rankine cycle method. http://www.heidelbergcement.com/global/en/company/products_innovations/innovations/orc.htm.

47. Nguyen, T.-V., Knudsen, T., Larsen, U. and Haglind, F. 2014. Thermodynamic evaluation of the Kalina split-cycle concepts for waste heat recovery applications. *Energy* 71: 277–288, Doi: 10.1016/j.energy.2014.04.060.

48. Kalina cycle. (2017). Wikipedia. Retrieved 19 January 2017 from https://en.wikipedia.org/wiki/Kalinacycle.

49. Shah, Y.T. (2019). *Modular Systems for Energy and Fuel Recovery and Conversion.* CRC Press, New York.

50. Min, G., Rowe, D.M. and Kontostavlakis, K. (2004). Thermoelectric figure-of-merit under large temperature differences. *Journal of Physics D: Applied Physics* 37: 1301–1304.

51. Min, G. and Rowe, D.M. (2007) Ring-structured thermoelectric module. *Semiconductor Science and Technology* 22: 880–883.

52. Lamonica, M. (2014). A thermoelectric generator that runs on exhaust fumes. IEEE Spectrum. http://spectrum.ieee.org/energywise/green-tech/conservation/a-thermoelectric-generator-that-runs-on-exhaust-fumes.

53. Takanose, E. and Tamakoshi, H. (1993). The development of thermolectric generator for passenger car. *Proceedings of 12th IEEE International Conference on Thermoelectrics,* 9–11 November, Yokohama.

54. Ikoma, K., Munekiyo, M., Furuya, K., Kobayashi, M. and Izumi, T. (1998). Thermoelectric module and generator for gasoline engine vehicles. *Proceedings of 17th IEEE International Conference on Thermoelectrics,* 24–28 May, Nagoya.

55. LeBlanc, S. (2014). Thermoelectric generators: Linking material properties and systems engineering for waste heat recovery applications. *Sustainable Materials and Technologies* 1–2: 26–35.

56. Yodovard, P., Khedari, J. and Hirunlabh J. (2001). The potential of waste heat thermoelectric power generation from diesel cycle and gas turbine cogeneration plants. *Energy Sources* 23: 213–224.

57. Rowe, D.M. (1995). *CRC Handbook of Thermo- Electrics,* 1st ed. CRC Press. 701 s. ISBN 978-0849301469.

58. Ensescu, D. (2019). Thermoelectric energy harvesting: Basic principles and applications InTech open access paper, Doi: 10.5772/intechopen.83495, www.intech.com.

59. Riffat, S. B. and Ma, X. (2003). Thermoelectrics: A review of present and potential applications. *Applied Thermal Engineering* 23: 913–935.

60. Taguchi, T. (2007). US20070193617.

61. Dell, R. and Wei, C.-S. (2008). US20080142067.

62. Patil, D. and Arakerimath, R. (2013). A review of thermoelectric generator for waste heat recovery from engine exhaust. *Internatinal journal of research in Aeronautical and Mechanical Engineering* 1: 1–9.

63. Ismail, B.I. and Ahmed, W.H. (2009). Thermoelectric power generation using waste-heat energy as an alternative green technology. *Recent Patents on Electrical Engineering Continued as Recent Advances in Electrical & Electronic Engineering* 2(1): 27–39.

64. Hill, J.M. (2011). *Study of Low-grade Waste Heat Recovery and Energy Transportation Systems in Industrial Applications.* The University of Alabama. http://acumen.lib.ua.edu/content/u0015/0000001/0000628/u0015_0000001_0000628.pdf.

65. Hi-Z Technology. (2014). HZ-14 Thermoelectric Module. hi-z.com. Retrieved 30 April 2015 from http://www.hi-z.com/ uploads/2/3/0/9/23090410/hz-14.pdf, a website report.

66. Stevens, J.W. (2001). Optimal design of small ΔT thermoelectric generation systems. *Energy Conversion and Management* 42(6): 709–720.

67. Kaibe, H., Makino, K., Kajihara, T., Fujimoto, S. and Hachiuma, H. (2012). Thermoelectric generating system attached to a carburizing furnace at Komatsu Ltd., Awazu Plant. *9th European Conference on Thermoelectrics: ECT2011* Vol. 524, 524–527, Doi: 10.1063/1.4731609.

68. Caterpillar. (2013). Standby 1000 ekW 1250 kVA 60 Hz 1800 rpm 480 Volts. Products. Retrieved 15 January 2015 from http://www.miltoncat.com/products/NewGenerators/Documents/C32/C32 1000KW Spec Sheet EPD0157-A.pdf, a website report.

69. Kuroki, T., Kabeya, K., Makino, K. et al., (2014). Thermoelectric generation using waste heat in steel works. *Journal of Electronic Materials.* Doi: 10.1007/s11664-014-3094-5.

70. EE Times India. (2014). Industrial-size generator makes waste heat valuable. *Global Sources.* Retrieved 22 October 2014 from http://www.eetindia.co.in/ART_8800705257_1800008_NT_3a933cb6.HTM, a website report.

71. Yazawa, K. and Shakouri, A. Energy payback optimization of thermoelectric power generator systems. In: *Energy Systems Analysis, Thermodynamics and Sustainability. Nano Engineering for Energy, Engineering to Address Climate Change*, Vol. 5, 569–576, Parts A and B. Asme. Doi: 10.1115/IMECE2010-37957.

72. Hendricks, T. and Choate, W. (2006). *Engineering Scoping Study of Thermoelectric Generator Systems for Industrial Waste Heat Recovery.* U.S. Department of Energy, Washington, DC.

73. Polcyn, A. and Khaleel, M. (2009). Advanced thermoelectric materials for efficient waste heat recovery in process industries. https://www1.eere.energy.gov/ manufacturing/industries_technologies/imf/pdfs/16947_advanced_thermoelectric_materials1.pdf, a website report.

74. Thekdi, A.C. and Nimbalkar, S.U. (2014). Industrial waste heat recovery: Potential applications, available technologies and crosscutting R&D opportunities. Energy and Transportation Science Division (ETSD), Prepared for OAK ridge national laboratory Oak Ridge, TN: US Department of Energy under contract DE-AC05-00OR22725 December.

75. Melosh, N. and Shen, Z. (2012). Photon enhanced thermionic emission for solar energy harvesting. *Final Report to the Global Climate and Energy Project.* http://www-spires.slac.stanford.edu/pubs/slacpubs/15250/slacpub-15455.pdf.,a website report.

76. McCarthy, P.T., Reifenberger, R. G. and Fisher, T. S. (2014). Thermionic and photo-excited electron emission for energy conversion processes. *Frontiers in Energy Research* 2, Doi: 10.3389/fenrg.2014.00054.

77. Haddad, C., Périlhon, C., Danlos, A., François, M.-X. and Descombes, G. (2014). Some efficient solutions to recover low and medium waste heat: Competitiveness of the thermoacoustic technology. *Energy Procedia* 50: 1056–1069, Doi: 10.1016/j.egypro.2014.06.125.

78. Smoker, J., Nouh, M., Aldraihem, O. and Baz, A. (2012). Energy harvesting from a standing wave acoustic-piezoelectric resonator. *Journal of Applied Physics* 111(10): 104901, Doi: 10.1063/1.4712630.

79. Wolf, S. (2012). Industrial heat pumps in Germany – potentials, technological development and application examples. ACHEMA 2012, 13 June 2012, Frankurt am Main. http://web.ornl.gov/sci/ees/etsd/btric/usnt/03InHPsAchmaIERWolf.pdf.

80. Börjesson, M. and Ahlgren, E.O. (2010). Biomass gasification in cost-optimized district heating systems-A regional modelling analysis. *Energy Policy* 38: 168–180.

81. IEA (International Energy Agency). (2007). *Renewables for Heating and Cooling: Untapped Potential.* OECD/IEA, a report by Renewable energy technology deployment, IEA, Paris. France. https://www.iea.org/publications/freepublications/publication/ Renewable_Heating_Cooling_Fin al_WEB.pdf.

82. SCI-PAK. (2013). Rice husk boiler at textile processing mill. Sustainable and cleaner production in the manufacturing industries of Pakistan, Bremerhaven. http://www. sci-pak.org/ContactInfo/tabid/68/Default.aspx, Accessed on 15 February 2013.

83. Shivakumar, A.R., Jayaram, S.N. and Rajshekar, S.C. (2008). *Inventory of Existing Technologies on Biomass Gasification.* Karnataka State Council for Science and Technology, Bangalore, http://kscst.org.in/energy/pdf/Biomass_Gasification_ Inventory_Report_KSCST.pdf.

84. UNEP (United Nations Environmental Programme). (2006). *Energy Efficiency Guide for Industry in Asia.* UNEP, Bangkok, http://www.energyefficiencyasia.org/docs/hard-copies/EnergyGuideIndustryAsia.pdf.

85. Einstein, D., Worrell, E. and Khrushch, M. (2001). Steam systems in industry: Energy use and energy efficiency improvement potentials. *ACEEE 2001 Summer Study on Energy Efficiency in Industry Proceedings* 1, 24–27 July, 2001, Tarrytown, NY.

86. WBCSD. (2006). WBCSD Cement Sustainability Initiative (CSI), Getting the Numbers Right (GNR). A global cement database. Geneva.

87. Sampaio, R.S. (2005). *Large-Scale Charcoal Production to Reduce CO_2 Emission and Improve Quality in the Coal Based Iron-Making Industry.* Rede Nacional de Biomassa para Energia Renabio, Viçosa.

88. Dornburg, V., Hermann, B. and Patel, M.K. (2008). Scenario projections for future market potentials of biobased bulk chemicals. *Environmental Science & Technology* 42(7): 2261–2267.

89. Popovska, S. (2001). Geothermal energy direct application in industry in Europe, 283–289.

90. Rafferty, K. (2003, September). *Industrial Processes and The Potential for Geothermal Applications,* Geo-center publication GHC Bulletin. Oregon Institute of Technology, Klamath Falls, OR, 7–11.

91. IEA. (2012). *Extended Energy Balances of OECD Countries.* OECD/IEA, Paris.

92. Arason, S. (2003). The drying of fish and utilization of geothermal energy; the Icelandic experience. *International Geothermal Conference,* Session #8, September 2003, Reykjavik.

93. EGEC (European Geothermal Energy Council). (n.d.). *Key Issue 5: Innovative Applications.* Geothermal utilization for industrial processes. European Geothermal Energy Council, Brussels. http://www.erec.org/fileadmin/erec_docs/Projcet_ Documents/K4_RES-H/K4RES- H_Geothermal_ProcessHeat.pdf, a website report.

94. IEA-ETSAP (IEA Energy Technology Systems Analysis Program). (2010). *Geothermal Heat and Power.* Technology Brief E07, May 2010. IEA-ETSAP, Paris. France http:// www.iea-etsap.org/web/e- techds/pdf/e06-geoth_energy-gs-gct.pdf.

95. Brown, H.L., Hamel, B.B. and Hedman, B.A. (1985). *Energy Analysis of 108 Industrial Processes.* Faimont Press, Atlanta, GA.

96. Brady's Hot Springs, Nevada, Geo-Heat Center Quarterly Bulletin, Vol. 15(4), Geo-Heat Center, Klamath Falls, OR.

97. Rafferty, K. (1996). *Fossil Fuel-Fired Peak Heating for Geothermal Greenhouses.* Geo-Heat Center, Oregon Institute of Technology, Klamath Falls, OR.

98. Goldstein, B.A., Hiriart, G., Tester, J., Bertani, R., Bromley, C., Gutierrez-Negrin, L., Huenges, E., Ragnarsson, A., Mongillo, M.A., Muraoka, H., and Zui, V.I. Great expectations for geothermal energy to 2100, Proceedings, Thirty-Sixth Workshop on Geothermal Reservoir Engineering Stanford University, Stanford, California, January 31–February 2, 2011. SGP-TR-191.

99. UNDP (United Nations Development Programme). (2008). Market development and promotion of solar concentrator based process heat applications in India (India CSH). UNDP India, Global Environment Facility Project Document, 121. http://www.in.undp.org/content/dam/india/docs/market_development_and_promotion_of_sol ar_ concentrators_based_project_document.pdf, a website report. http://www.iea.org/publications/freepublications/publication/Aluminium_EU_ETS.pdf.

100. Hess, S. and Oliva, A. (2010). SO-PRO solar process heat. *Solar process heat generation: Guide to solar thermal system design for selected industrial processes.* Fraunhofer ISE, Freiburg. http://www.solar-process-heat.eu/fileadmin/redakteure/So- Pro/Work_Packages/WP3/Planning_Guideline/Techn_Bro_SoPro_en-fin.pdf, a website report.

101. Vannoni, C. (2007). *Solar Heat for Industrial Processes. Existing Plants and Potential for Future Applications.* University of Rome, Rome. Italy http://www.estec2007.org/2007/download/presentations/wednesday/Session%20A1/3%20van noni_presentation_small.pdf, a website slide presentations.

102. Weiss, M., Junginger, M., Patel, M.K. and Blok, K. (2010). A review of experience curve analyses for energy demand technologies. *Technological Forecasting and Social Change* 77: 411–428.

103. Weiss, W. (2010). Potential, framework conditions and build examples. *Solar Heat for Industrial Applications*, November 2010. AEE-Institute for Sustainable Technologies. Presented in Melbourne. http://www.sustainability.vic.gov.au/resources/documents/Solar_heat_for_industrial_processes.pdf, a website presentation.

104. ESTIF (European Solar Thermal Industry Federation). (2014). Solar heating and cooling technology roadmap. *European Technology Platform on Renewable Heating and Cooling*, June 2014. http://www.estif.org/fileadmin/estif/content/projects/ESTTP/Solar_H_C_Roadmap.pdf.

105. IRENA. (2015). Renewable power generation costs in 2014. January 2015, Abu Dhabi. http://costing.irena.org.

106. Solar heat for industrial processes-technology brief, IEA-ETSAP and IRENA technology brief E21. (2015, January). IRENA, Abu Dhabi.

107. Saygin, D., Patel, M.K., Worrell, E., Tam, C. and Gielen, D.J. (2011). Potential of best practice technology to improve energy efficiency in the global chemical and petrochemical sector. *Energy* 36: 5779–5790.

108. Saygin, D., Worrell, E., Patel, M.K. and Gielen, D.J. (2011). Benchmarking the energy use of energy-intensive industry in industrialized and in developing countries. *Energy* 36: 6661–6673.

109. Turton, H. (2002). The aluminium smelting industry. Structure, market power, subsidies and greenhouse gas emissions, Discussion paper number 44, January 2002. The Australia Institute, Canberra City. http://www.tai.org.au/documents/dp_fulltext/DP44.pdf, a website report.

110. UNCTAD (United Nations Conference on Trade and Development). (2000). Aluminium and the Australian Economy, A report to the Australian Aluminium Council, May 2000. UNCTAD, Geneva. http://r0.unctad.org/infocomm/francais/aluminium/Doc/australia.pdf, a website report.

111. Reinaud, J. (Oct.,2008). Climate policy and carbon leakage. Impacts of the European Emissions Trading Scheme on Aluminium, IEA Information Paper, October 2008. OECD/IEA, Paris. France. http://www.iea.org/Textbase/about/copyright.asp.
112. Concentrating Solar Power Technology Brief, IEA-ETSAP and IRENA© Technology Brief E10 – anuary 2013 www.etsap.org –www.irena.org.
113. Arvizu, D., Balaya, P., Cabeza, L., Hollands, K., Jäger-Waldau, A., Kondo, M., Konseibo, C., Stein, W., Tamaura, Y., Xu, H., Zilles, R., ABERLE, A., Athienitis, A., Cowlin, S., Gwinner, D., Heath, G., Huld, T., James, T. and Weyers, P. (2012). Direct Solar Energy. 10.1017/CBO9781139151153.007.
114. UNIDO. (2011). Industrial energy efficiency for sustainable wealth creation. Capturing environmental, economic and social dividends. Industrial Development Report 2011. United Nations Industrial Development Organization, Vienna. Austria http://www.unido.org/fileadmin/user_media/Publications/IDR/2011/UNIDO%20IDR%20reprint%20for%20web%20020912.pdf, a website report.
115. IRENA. (2014, August). *Global Bioenergy Supply and Demand Projections for the Year2030: A working paper for REmap 2030.* IRENA, Abu Dhabi.
116. Ruth, M., Cutler, D., Flores-Espino, F., Stark, G. and Jenkin, T. (2016, July). The economic potential of a nuclear-renewable hybrid energy system providing thermal energy for industry – Draft Report.
117. Bragg-Sitton, S.M. Boardman, R. and Rabiti, C., Suk Kim, J., McKellar, M., Sabharwall, P., Chen, J., Cetiner, M.S., Harrison, T.J. and Qualls, A.L. (2016, March) *Nuclear-Renewable Hybrid Energy Systems: 2016 Technology Development Program Plan.* INL/EXT-16-38165, Idaho falls, Idaho.

8 Hybrid Energy Systems for O&G Industries

8.1 INTRODUCTION

While fossil fuel industry is facing many challenges in recent years, their dominance in energy industry is likely to last for a while. There are two major applications of fossil fuels: energy and in particular power production and raw materials for chemicals we need for day-to-day life. As shown earlier, while the use of coal for power production is likely to shrink, the use of coal for chemicals will sustain and even grow. Many coal power plants are being converted to gas power plants. Petroleum (including both oil and gas) heats our homes, drives our transportation system, generates our electricity, and makes modern life possible. Continued demand from developed countries along with growth from developing economies implies demand for oil and gas will likely continue to increase [1]. Under current policies, global oil demand in 2040 is expected to be roughly 26 million barrels per day greater than in 2016 [1]. This represents a 27% increase from the average demand in 2016 of 96.2 million barrels per day.

As demand is increasing, conventional oil and gas reserves are decreasing, leading to a shift in production to unconventional sources and a growing use of enhanced oil recovery (EOR) techniques. Production from unconventional reserves and the use of EOR raise the energy intensity of an already energy-intensive industry. Nearly 10% of oil is used in the production, transportation, and refining process [2]. This number is even higher for many unconventional sources—it takes a quarter of a barrel of oil to produce a barrel of heavy oil [3]. Energy used to produce, transport, and refine oil represents major operations costs. Furthermore, many oil and gas companies have set goals or made corporate level commitments to reduce their greenhouse gas emissions, in part to address issues raised in national and global studies (USGCRP 2017) [4]. The petroleum industry faces the difficult task of meeting growing demand and growing operational energy needs while reducing operations emissions.

There are six channels driving the energy transition across the broader energy, resources, and industrials sectors including oil and gas companies, power utilities, chemical companies, and manufacturers. The six channels include decarbonizing energy sources, increasing operational energy efficiency, identifying new investment priorities, deploying new technologies, adjusting to new policy mandates, and managing consumer and shareholder expectations. Oil and gas companies will likely need to leverage all six channels to prepare for a lower-carbon future. Two which are most relevant to the subject of the present book are decarbonizing energy sources and increasing operational efficiency. These impact the following opportunities for oil and gas companies [5]:

1. Increased renewable power generation may decrease demand for natural gas and coal
2. Deep electrification will likely require extensive energy demand management services
3. Power transmission and battery storage will likely need to be expanded to significantly increase reliability
4. Renewables deployment in oil and gas operations may reduce field consumption of natural gas
5. Investment in manufacturing energy efficiency could lead to fossil fuel demand reduction
6. Reducing fugitive methane emissions could boost sales gas volumes

There are, however, several issues related to the use of oil and gas for energy production. Natural sources for oil and gas are nonrenewable. The conversion of oil and gas for power production or industrial and residential heating and cooling is highly inefficient leading to significant waste in the form of waste heat. Most importantly, the use of oil and gas leads to the emission of CO_2 in the environment which is a source of global warming concern. These issues have led to the adoption of strategies which use less oil and gas and more renewable sources for power and heating/cooling needs. However, unlike coal, oil and gas can be synthetically generated making them more sustainable than coal. The main conclusion from these issues is that going forward we need to adopt all strategies that will allow less use of oil and gas and replace whenever possible with renewable sources through hybridization processes. The improvement of efficiencies in various upstream, midstream, and downstream processes would help reduction of carbon emission.

Increasing efficiency of operations can also mitigate the rising costs associated with increasing energy requirements. So can incorporating renewable energy sources that maximize product revenue while simultaneously addressing local environmental concerns as well as meeting emissions goals. This second factor has prompted a growing interest in integrating renewable energy technologies in oil and gas operations. Furthermore, due to dramatic declines in the cost of energy generated from renewable sources, integrating renewable energy technologies can, in many cases, reduce operations costs as well. The hybrid renewable energy systems can provide many benefits to the upstream, midstream, and downstream oil and gas processes [6].

The oil and gas (O&G) industry has several aspects that are conducive to integrating renewable energy technologies. Production facilities are often both in remote locations and require large amounts of electricity that could be generated with renewable sources (wind, solar). EOR and oil refining also require large amounts of heat, which may be supplied by renewable thermal technologies (solar thermal, geothermal). Use of waste heat or gas to run cogeneration facilities can, in some cases, also be economical. However, renewable energy technologies are not applicable in all cases. Renewable energy technologies must be both reliable and economically competitive to be commercially viable.

This chapter provides an overview of where renewable energy technologies can be integrated into oil and gas operations [7,8]. The chapter discusses drivers for an increase of renewable technology integration in the oil and gas operations and provides an overview of key challenges to renewable integration and discusses factors that determine if a given technology can be reliably and affordably integrated. The chapter also discusses relevant government policies and social considerations that influence the relative benefits of renewable integration and outlines opportunities for renewable technologies to be integrated into upstream production, midstream transportation, and downstream refining processes. To this end, next several sections summarizes the discussions provided in the excellent report by Ericson et al. [7] and review by Choi et al. [8]. Finally, this chapter outlines the need for hybrid natural gas and renewable energy possibilities in power and transportation industries. This combination possesses significant synergy and offers to reduce CO_2 emission in the near future.

8.2 DRIVERS FOR HYBRID RENEWABLE ENERGY SYSTEMS FOR OIL AND GAS INDUSTRY

As pointed out by Ericson et al. [7], three parallel drivers are increasing the profitability of renewable technology integration in oil and gas operations. These drivers are (a) depletion of higher quality oil reserves leading to an increased energy intensity of petroleum operations; (b) environmental concerns in the O&G industry; and (c) falling costs for renewable generation technologies. The analysis of Erickson et al. [7] for these three drivers is presented below.

8.2.1 Depletion of High-Quality Oil Reserves

The abundant supply of shallow oil and gas reserves with ample reservoir pressure is depleted in recent years. Today's supply of oil and gas is not easy to reach. Remaining reserves are marked by deeper reservoirs, lower pressures, and lower quality. As an example, Figure 8.1 [7] displays the increase in average water and well depths in the Gulf of Mexico. Deeper water and deeper wells increase both the complexity and the energy intensity of operations. The trend toward production from lower quality reserves may also be seen in terms of the increased exploitation of heavy oil and tar sands and the increased use of EOR techniques—the injection of gas, heat, or liquid to increase field recovery rate.

While sufficient petroleum resources remain to meet demand for the foreseeable future, the shift toward marginal reserves increases the energy intensity of production. A 2017 study of five large petroleum fields concludes that the net energy ratio (NER)—the ratio of energy produced to the energy used to produce it—for each field declined by 46%–88% over the last four decades [10]. An in-depth analysis of conventional natural gas production in Canada shows a similar trend, with the ratio of energy output to energy input falling by roughly half from 1993 to 2009 [11]. A combination of declining field production and increased energy expenditures on enhanced recovery methods contributed to the decline in NER [10].

As pointed out by Ericson et al. [7], power-quality reserves lead to higher energy intensity in production, transportation, and refining of petroleum. This leads to more

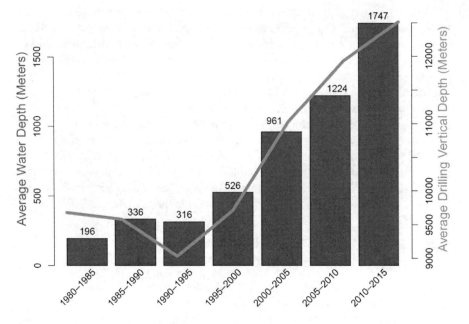

FIGURE 8.1 Average water and drilling depth in the Gulf of Mexico over time. (Data from [7,9].

demand for energy in each stage of the petroleum supply chain, which in turn leads to more possibilities for renewable technology integration to reduce energy costs. Along with cost savings, renewable technologies can reduce emissions, which is becoming an increasingly important factor.

8.2.2 Environmental Concerns in the O&G Industry

Industrial oil and gas practices have continued to advance to meet or exceed environmental regulations. Adverse incidents are rare but sometimes of high consequence. Ericson et al. [7] points out that it is well recognized that oil and gas operations emit pollutants, and continuous or episodic activities or incidents may lead to environmental degradation. Improperly drilled or completed wells can lead to oil spills, contaminated water spills, and methane leaks. Machineries such as compressors and diesel generators increase noise and air pollution. Operations can lead to traffic congestion, and drilling processes can contaminate water supplies. Finally, petroleum refining processes emit gases including carbon dioxide, carbon monoxide, methane, organic compounds, nitrogen oxides, sulfur dioxide, and hydrogen sulfide. Meeting environmental regulations can be costly, and a failure to meet requirements may lead to fines and impact a firm's social license to operate.

Operations occurring in closer proximity to urban centers and suburban developments, as well as a growing awareness of the environmental effects of operations—both local effects such as air pollution and induced seismicity and global effects such as climate change—are further increasing the importance of environmentally sound production practices. Ericson et al. [7] points out that the petroleum industry has already taken some actions. Due to industry action, greenhouse gas emissions from the major private oil companies fell by 13% between 2010 and 2015 [12] and methane emissions from natural gas wells fell by 40% between 1999 and 2012 [13]. Furthermore, many oil companies have set goals for additional emissions reductions. Because of the falling costs, renewable energy technologies could become important tools for the goals of meeting additional energy demands and stricter emission standards while reducing fuel usage and operations costs. Once again hybrid renewable energy system is the possible solution to this dilemma.

8.2.3 FALLING RENEWABLE ENERGY COSTS

Ericson et al. [7] points out price declines in the last decade have revolutionized the economics of renewable energy technologies. In 2009, the average levelized cost of electricity from solar photovoltaics (PV) was more than seven times the cost in 2017. Figure 8.2 displays the steep cost decline of electricity produced from wind and solar. Whereas power from wind and solar used to be prohibitively expensive, they are now in some cases the lowest cost source of electricity. If expected continued future cost declines for electricity from wind and solar are realized, generation from these sources will become even more competitive in the coming years. Furthermore, battery storage technology and technology for demand-side management, which can compensate for generation variability, have also seen remarkable cost declines [7].

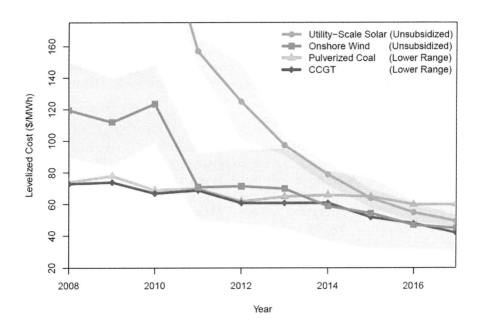

FIGURE 8.2 Levelized cost of various utility-scale generation technologies. (Data from Lazard [7,14].

Levelized Costs of Energy reports from 2008 to 2017 (bands for wind and solar represent low and high cost estimates). Because high-cost estimates for coal and natural gas incorporate the cost of carbon sequestration, the figure uses low-cost estimates instead of mid-point estimates for these technologies to better capture current production costs. Cost assumptions can be found in Ref. [14].

Ericson et al. [7] points out that the result of renewable energy cost declines is a paradigm shift in how renewable technology integration in oil and gas operations can be viewed. Whereas 10 years ago renewable technology was only applicable to cases where diesel or gas turbines could not be readily deployed, going forward renewable technology will, in a growing number of cases, be the lowest-cost solution. An understanding of where renewable generation is applicable and may be integrated can lead to both lower environmental impact and lower operation costs.

8.3 CHALLENGES TO RENEWABLE INTEGRATION

Oil and gas operations are highly intricate, requiring coordination among multiple contractors, operators, and government agencies. Operations are capital-intensive and require machines that operate reliably. The industry is also very competitive, with constant pressure to reduce costs. For renewable technologies to integrate into oil and gas operations, they must meet strict reliability and affordability metrics [7].

Cost has historically been the primary factor deterring renewable integration and is still the most important factor in determining whether a new technology can be integrated. However, even in cases where renewable generation is the lowest-cost solution, production variability and reliability concerns are still significant barriers to overcome. Operational considerations and differences in industry knowledge between oil and gas operations and renewable energy operations place an additional barrier to integration. The best solution for this concern is the use of hybrid energy system (renewable-renewable, renewable-conventional, renewable-storage).

8.3.1 VARIABILITY OF GENERATION

Ericson et al. [7] points out that the effect of variability on operations is a large determinant of whether renewable technology can be integrated. Many operations along the supply chain require continuous power. Oil rigs, pipeline compressors, and petroleum refineries operate 24/7, regardless of whether the sun is shining or the wind is blowing. In these cases, renewable generation technologies can sometimes offer operational savings by being configured into a high reliability hybrid system with backup diesel generators, gas turbines, or battery storage. However, this potentially increases system costs and complexity. In other cases, such as steam injection for EOR, variability of generation has little effect on production since steam can be produced and then stored [15]. Use cases where variability in generation does not affect production have the highest potential for renewable integration. Hybrid operation which includes combines sources of renewables with conventional sources with storage can mitigate this issue.

While use cases where variability does not affect production are ideal, there are several methods for mitigating the effects of variability. In some cases, fuel savings alone can make renewable generation viable. High diesel costs and low efficiencies of natural gas reciprocating engines and single cycle turbines can lead to renewable integration being cost effective even when backup generation capacity is required [16]. Battery storage technologies can also provide a solution in some cases. Many oil field operations, such as powering well sensors and lights and cathodic protection of pipelines from corrosion, employ solar power combined with battery storage [2].

8.3.2 SYSTEM RELIABILITY

The petroleum industry requires high levels of reliability. Gas pipelines, for example, have much higher reliability of operations than the electricity grid, which itself has a high standard of performance [17,18]. While renewable generation is well proven and shown to be reliable in a wide range of settings, systems for integrating renewable energy technologies into oil and gas operations often do not have an equally proven track record. Furthermore, reliability concerns are compounded by renewable generation variability. Systems may have to be designed to operate under a wide range of weather scenarios. Higher tolerance for extreme operating conditions and high reliability requirements (as well as space or weight constraints in some instances) raise costs, implying a renewable-generation-based system may be uneconomic even if it can provide energy at the lowest cost on average. More research and pilot projects can bridge this gap in understanding and substantiate that renewable systems can meet reliability standards. Once again hybrid operation can bring about more reliability to the operation [7].

8.3.3 OPERATIONAL CONSIDERATIONS

Oil and gas operations are often subject to harsh conditions. Machinery must be able to withstand weather and temperature extremes and must be able to operate for several years before replacement. For a technology to be scalable, it must be able to operate under a variety of settings. Ericson et al. [7] points out that the technologies that are robust and operate in a variety of settings are therefore most easily integrated. Petroleum operations are very complex. Systems that add additional complications may be unsuited for practical applications even if they offer cost savings. Turnkey solutions that do not add to operational complexities can be most readily integrated. The ideal technology for integration in oil and gas operations is a system that can be installed easily and can operate cost-effectively with little to no maintenance.

Lastly, petroleum engineering is highly specialized and often requires a different skillset than that required for operations of renewable generation technologies. Along with research to ensure reliability, work is required to bridge the gap in industry knowledge between the petroleum and renewable energy industries, such as ensuring that gas and oil operations personnel are trained in operational aspects of renewable technologies as applied to gas and oil operations. A prudent strategy for these reasons is to implement hybrid operation which can provide more flexibility in the integration process [7].

8.3.4 GOVERNMENT POLICIES

Ericson et al. [7] points out that in order to best serve stakeholders, both private and public oil companies are required to generate revenues. Thus, renewable integration must pass a test for cost effectiveness. However, the operating regulator and fiscal environment (e.g., tax codes), along with government policies and social environment, can have direct impacts on profitability. Policies such as emission regulations can greatly enhance the value of renewable technology integration, while policies such as fuel subsidies diminish their applicability. This section provides a brief overview of major policy considerations, which may affect the value of renewable integration.

Almost every country has implemented policies supporting renewable generation technologies. All but 11 countries have renewable energy targets, and most countries have renewable investment and or renewable production tax credits along with public investment or loans to support renewable energy [19]. The United States, for example, has a 30% federal investment tax credit (ITC) for qualifying renewable generation technologies and a production tax credit for many renewable generation technologies that do not receive the ITC. Several state and local policies further support renewable generation. Combined, these policies significantly increase the economics of renewable generation projects [7].

As renewable energy costs have fallen to the point where renewable prices are now in many cases competitive with other generation technologies, production and ITCs have been steadily phased out in many cases. Further phase-outs are planned and expected in the coming years. However, while tax credits will likely be lower in the future, the economics of zero-emissions generation may become more compelling in the coming years if carbon emission pricing and emission-reduction regulations are further adopted.

While renewable subsidies and emissions prices raise the value of renewable integration, some policies reduce the relative value of renewable technologies. Ericson et al. [7] notes that the most notable of such policies are subsidies on electricity, oil, or natural gas. This is important as many of the largest oil-producing countries also have the largest fossil fuel subsidies. Figure 8.3 displays countries with the highest per capita fossil fuel and electricity subsidies. There is a clear correlation between petroleum-producing countries and countries with high levels of such subsidies.

Energy subsidies for traditional fossil fuels reduce the costs of producing heat and electricity for oil and gas operations, but they can also offer opportunities for renewable technologies [20]. Fiscal costs of energy subsidies are leading to pressures for subsidy reform [21]. Energy subsidies generally reduce investment in operation technologies, meaning operations in countries with energy subsidies are often inefficient [21,22]. Renewable integration in countries attempting subsidy reform presents an opportunity to increase operational efficiency and lower costs, thereby enabling subsidy reform: subsidy reform can be politically costly and can dramatically raise production costs. The Iranian subsidy reform in 2010 raised natural gas prices by

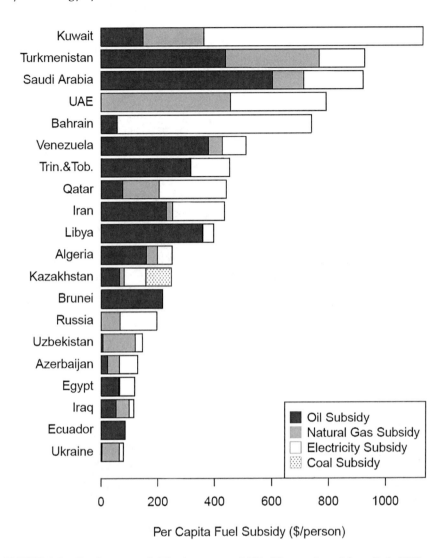

FIGURE 8.3   Total energy subsidies by country, 2016. (Figure adapted from Refs. [1,7].

over 700% and diesel prices tenfold [21]. Renewable technologies can help dampen cost shocks from subsidy reforms. Thus, while energy subsidies diminish the opportunities for renewable integration, they present opportunities for countries attempting subsidy reform [7].

The report by Ericson et al. [7] indicates that the three trends of declining reserve quality (and the resulting increase in energy requirements), growing environmental awareness, and falling renewable generation costs are leading to an uptake of renewable technologies along the supply chain. Government and industry policies are further helping to overcome integration challenges. In the following sections, we cover applications for renewable integration in upstream, midstream, and downstream oil and gas operations as outlined in excellent NREL report by Ericson et al. [7]. For each section of the supply chain, we discuss current commercial projects along with technologies that could soon reach commercialization [21,22].

8.4 HYBRID SYSTEMS

There are two types of renewable sources of energy: dispatchable like biomass, hydro, and geothermal and nondispatchable like solar and wind. Even dispatchable sources can sometimes become unpredictable. The unpredictable pattern of natural resources requires a combined utilization of sources and/or backup storage in order to provide uninterrupted and reliable power supply to its users (technical difficulties may arise due to uncontrollable weather conditions like wind speed fluctuation, day and night behavior, and summer and winter sun conditions). In other words, renewable energy is usually dependent on the weather conditions for its source of power input. For example, a hydrogenerator depends on rain to fill dams to supply flowing water, a wind turbine depends on the wind within a specific range of speed to turn the blades and run the turbine, and a solar panel depends on the position of the sun and, preferentially, no clouds to make electricity. Geothermal energy depends on the extent of geothermal energy available at any particular location which is not always predictable.

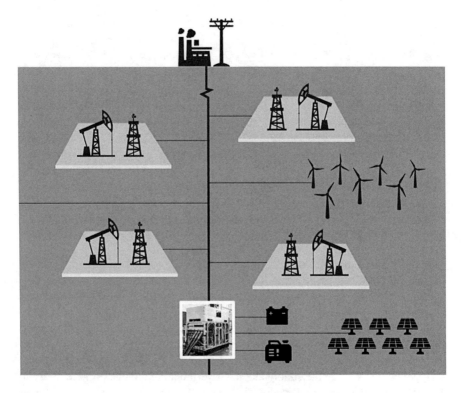

FIGURE 8.4 Schematic for a comprehensive approach to electrification of the well pad and platform via microgrids and hybrid renewable energy system [7].

A combination of two or more renewable energy sources or a renewable source with storage is more effective than the single source system in terms of cost, efficiency, and reliability (see Figure 8.4). Often it pays to retain some conventional sources as a backup. Ericson et al. [7] points out that we can easily reduce the need for fossil fuels by properly choosing a combination of renewable energy sources. The combination of two or more energy sources, working together in order to compensate for each other, is designated as a hybrid energy system. Sometimes this includes a storage system particularly if only one renewable source is used. The main advantage of a hybrid energy system is the enhancement of reliability and the cost-benefit of the system. Due to the fact that some renewable energy sources such as Solar Radiation and Wind are, most of the times, intermittent, and they are frequently combined with other power sources such as utility grid, diesel generators, or storage systems. The objective is to ensure the continuous supply of power.

Nowadays two types of hybrid systems are in operation: one combining the only production from renewable energies with storage, which are ideal for applications in isolated systems, and a second type that makes use of the production from diesel and gas generators. Solar-wind hybrid systems have as the main advantage the way they complement each other and the fact that they are exclusively renewable energy resources. The behavior of solar radiation throughout the day follows an approximately constant pattern of production, reaching its peak at noon decreasing until sunset. The wind generator can serve as a complement to the system, in the periods where there is low or nonexistent solar radiation. This characteristic gives this type of system a greater reliability in the matter of the continuity of electrical production over time.

Solar-wind-diesel/gas or simply solar-diesel/gas hybrid systems work similarly to the mentioned solar-wind system. This type of system has the advantage of reducing the consumption of fossil fuels. It is even more reliable, because diesel or gas generators work as a backup, thus ensuring the operation of the system even in periods when the remaining energy sources are not available or are not enough to guarantee the energy required. Around the world, thousands of communities and industrial sites are however not powered by the utility grid. The traditional means of electrification in such locations is diesel generation since it is a more conventional technology and more people trained in operation and maintenance. Wind turbines and solar panels are the better known renewable energy devices used in hybrid power systems. Hybrid systems usually include energy storage, so they can deliver a certain amount of energy on demand. These systems provide a high degree of energy security through the mix of generation methods and can often incorporate a storage system (battery and/or fuel cell) or to ensure maximum supply reliability and security, a small fueled generator (nonrenewable). Hybrid power systems exhibit higher reliability and lower cost of generation than those that use only one source of energy.

8.4.1 EVALUATION AND SUCCESSFUL CASE STUDIES

Some studies were conducted concerning the techno-economic assessment of an autonomous hybrid PV/diesel hybrid power system installed in a bungalow complex in Elounda, Crete. In remote areas which are far from the grids, electricity is supplied either by diesel generators or small hydroelectric plants. Under such circumstances, the supply of fuel becomes so expensive that hybrid diesel/photovoltaic generation becomes competitive with diesel-only generation. The use of hybrid power systems in industries is becoming so usual and so advanced that in some cases its users started to require to design and model stand-alone systems with software such as MATLAB/SIMULINK and to evaluate their performance. One example of such a study was considered to establish a comparison of a hybrid power system based on PV module, wind turbine, and diesel generator with a PV/wind/battery hybrid power system. From the simulation results, it was identified that a hybrid connected system with the battery can perform better in some ways than the diesel connected system.

As mentioned before, the industry is adapting and there are many cases already of implemented hybrid systems. In Jamaica, for example, it is installed as the world's biggest hybrid system. It was installed by Wind Stream Technologies in an area very close to the coast and takes advantage of winds with an average speed around 96.5 km/h. With this system, an output production of 106 MWh/y and a return of investment in less than 4 years are expected. The plant is designed to save the company approximately $2 million in energy costs over 25 years. The hybrid energy system consists of 50 Solar Mills (patented product with solar panels and vertical axis turbines) making it the largest facility in the world. EFACEC has also developed a Hybrid System Platform (called EFASOLAR3) which has the capability of integrating several energy sources such as diesel, photovoltaic, electrical grid, energy storage, wind, and hydro. Efacec is a Portuguese company that emerged, in 1948, from the union of the Belgian group ACEC (Ateliers de Construtions Électriques de Charleroi) and CUF (Companhia União Fabril), one of the largest Portuguese business groups at the time. This system can be implemented in an existing installation or in a new project, ensuring a stable operation and optimizing the energy produced by the photovoltaic plant. A hybrid system combining floating PV and hydroelectric power generation has been installed in the Alto Rabagão dam in Portugal (by EDP Group). FPV systems are producing energy on areas that are unused otherwise. The pioneer project has an installed capacity of 220 kW and is expected to produce 332 MWh.

8.5   HYBRID POWER SYSTEMS FOR OFFSHORE UNITS

DNV GL [23] is proposing a new hybrid power concept for offshore units using batteries in combination with traditional power generation equipment. The concept has already delivered multiple advantages in the maritime and automotive industries, but has not yet been applied in the O&G industry. Oil and gas operators are under constant pressure to reduce both cost and emissions—two requirements that are conflicting. An important means of achieving these reductions is by improving energy efficiency, which DNV GL suggests can be done by using a hybrid system.

The concept that DNV GL is proposing is similar to that used in the automotive industry, for example, in the Toyota Prius. The maritime industry started to adopt this technology five years ago not just because of the environmental benefits, but also because it adds financial value and performance benefits. In the automotive industry, there is typically about a 20% fuel saving with the use of a hybrid system, but in some maritime applications, there is up to a 40% saving and DNV GL expects to see the same for the O&G industry. Currently there are 17 hybrid ships either in operation or being built, plus one electric ferry. DNV GL studies show that the payback period is typically only about two years for the maritime application [23].

The reduced size of the power generator is a major advantage with a hybrid approach. Today, a conventional offshore facility is typically powered by three gas turbine generator sets, plus a spare one for redundancy. Here the total installed power is dimensioned according to the peak power requirement, but is normally operated at a much lower power level. In addition, the gas turbines are operated at low partial-load in order to have reserve capacity to be ready to react to sudden power demands. This result in low power generation efficiency, and thus, the fuel consumption and CO_2 emissions are higher per unit of power. However, with a hybrid system, the power generation becomes optimized for an average load rather than a peak load, and hence, the power generation capacity can be reduced. A battery, like those used in maritime vessels, will act as a buffer and supply power when needed or charge up when demand is low.

The initial studies done by DNV GL showed that hybrid technology can add value to a project in several different ways: (a) *Reduction* **in fuel consumption and emission.** (b) *Green field*: For a new build, the number of gas turbine generator sets or the size of them can potentially be reduced, since the power generation system can be designed based more on average power loads rather than peak loads. This will reduce CAPEX for the generators. If the number of generators is reduced then the maintenance costs will drop proportionally. Maintenance cost for an offshore gas turbine is about 18% of the turbine CAPEX per year. (c) *Brown-field*: For a brown-field development where more power is required, then it is easier to install batteries below deck rather than free up deck space for additional generators. Top side modification is very expensive and results in extended production shut-down. The battery can be modularized for easier handling and installation below deck. (d) *Deck-space*: If fewer generators are installed, then this will free up valuable deck-space that can be used for other purposes or potentially reduce the size of the facility. On a typical presalt Floating Production Storage and Offloading (FPSO) with 75 MW of installed power, the power generation equipment takes up 10%–15% of the available deck space. (e) *Performance*: Batteries can deliver instant power compared to a gas generator which needs time to spin up when increasing the load. This is very important in dynamic applications such as dynamic positioning systems or heave compensated lifting systems. (f) *Energy harvesting*: The batteries allow energy to be stored which can be used to recover energy from, for example, electric active heave drawworks used for drilling or cranes. Today, this energy is burned off as heat. (g) *Reliability*: Since the power generators will run at much steadier and ideal loads, this has a positive effect on the reliability of this equipment, thus reducing unplanned maintenance [23].

8.6 UPSTREAM: RENEWABLE INTEGRATION IN OIL AND GAS PRODUCTION

The O&G industry has a long history of integrating renewable energy in operations. One of the earliest commercial applications of solar PV panels was their use in warning lights for offshore oil installations starting in the early 1970s [2]. Often remotely located and off-grid, oil and gas operations primarily generate power by gas turbines or diesel generators, which can be expensive to operate and maintain. As production operations can have significant electricity requirements, power generation can be a substantial cost. For example, typical power requirements for an offshore platform are between 20 and 35 megawatts (MW) [16].

NREL report by Ericsion et al. [7] indicate that hybridization with renewable energy can be used to reduce fuel use and maintenance costs in upstream operations. Furthermore, renewable generation can reduce noise, reduce emissions, and increase safety. For each stage in the life of an oil field, different renewable technologies can be deployed. The three production stages are:

1. **Primary recovery**: During primary recovery, there is sufficient well pressure for oil and gas extraction. Artificial lift methods such as electric submersible pumps and rod pumps are often used to increase output. About 5%–15% of reserves are extracted during primary recovery [24].
2. **Secondary recovery**: As well pressures fall, production moves into the secondary recovery phase. Injection fluids or gases are used to increase reservoir pressure. An additional 10%–20% of reserves can be extracted from secondary recovery [25].
3. **Tertiary recovery**: Also known as EOR, tertiary recovery consists of a variety of methods to stimulate production. Examples include hydraulic fracturing, steam injection, in-situ combustion, and chemical injection. An additional 20% of total initial reserves may be extracted during tertiary recovery.

An individual field may not experience all three production regimes, and separate wells within a field may be in different regimes at any given period. However, the separation is a useful construct for both understanding general production stages and understanding where renewable generation technologies may be integrated.

Energy requirements vary throughout the life of the field. Operations energy intensity tends to be low during the primary recovery phase and then increase as well pressure declines. Hence, older fields, in the secondary and tertiary recovery phase, tend to have higher energy requirements than fields in the primary production phase. However, all phases in the production life cycle, along with drilling operations, offer opportunities for renewable technologies to lower costs and emissions. NREL analysis for these opportunities delineated by Ericson et al. [7] is presented below.

8.6.1 Electrification of Drilling and Primary Recovery

Drilling entails significant short-term energy requirements. A single drilling rig can require more than 1 MW of power to operate [26]. As drilling operations only last a few weeks, these high-power requirements are only needed for a relatively short duration. Primary recovery operations, on the other hand, have relatively lower energy requirements that are sustained over several years.

The energy for drilling and primary recovery operations is often provided by diesel or natural gas generators. Diesel fuel requirements for drilling and constructing a conventional oil well typically range between 18,000 and 24,000 gallons [27]. For shale gas wells, fuel consumption is generally much higher, often requiring more than 50,000 gallons of diesel fuel per well and requiring more than 80,000 gallons in some plays [27]. Renewable generation can reduce or eliminate the need for a generator, which can result in substantial fuel savings.

When operations are sufficiently close to electrical lines, the electricity grid can be used to power operations. Using electricity can reduce noise, emissions, and traffic congestion, which can be especially important when operations are close to urban centers. An electrical connection also allows for renewable energy to be integrated into various points of operations via a microgrid. Microgrid solutions empower users to integrate distributed generation into a versatile, reliable, and environmentally friendly operation [7].

Oil operations are frequently conducted in areas with available renewable energy resources. Oil field and well pad equipment could be converted to electric power and then connected via a microgrid with a controller that optimizes multiple clean power sources (see Figure 8.4). This approach has been successfully implemented at remote military bases, communities, and islands and could be adapted to such an industrial process with known energy demands and available resources. Power sources could consist of solar PV/wind systems, fuel cells, energy storage, hydrogen, field gas, or even grid power. This approach reduces leaks and emissions, provides resiliency during outages, and optimizes for least cost [7].

A notable example of where renewable technologies can be integrated into upstream operations is to power artificial lift pumps. When there is not enough well pressure for oil to flow to the surface, a rod beam artificial lift pump— often referred to as a "sucker rod pump" or a "Jack pump"—is used to assist extraction. In many areas, rod beam pumps are used extensively. For example, more than 80% of oil production wells operating in the western United States are installed with a rod beam pump [28]. Solar power, combined with a capacitor to store regenerative power during the rod down-stroke, can be used to power a rod beam pump. Test cases have shown significant potential energy savings [28], and solar powered oil pumps are now beginning to see commercial operation [29]. Combined with battery storage, solar power pumps could be applicable to off-grid locations with sufficient sunlight.

8.6.2 USE OF HYBRID RENEWABLE-ENERGY-POWERED SECONDARY RECOVERY

Renewable generation can be especially well suited to powering water injection pumps to stimulate oil recovery because production from water injection is not significantly affected by injection variability [30]. An especially important potential application is using offshore wind power-to-power water injection pumps. By one analysis, offshore wind was found in 2012 to be an economic and environmentally sound option for supplying electricity to offshore oil and gas platforms in some cases [16]. Costs for offshore wind generation have fallen significantly since then and are anticipated to continue to decrease as the technology matures. The average strike price for European offshore wind projects that come online between 2021 and 2025 is less than half the strike price for similar projects that have or will come online between 2016 and 2020 [31].

Ericson et al. [7] points out that as wind costs fall, it will become economically attractive to substitute power from diesel and gas generators with power from wind platforms. Finally, wind integration reduces carbon emissions, which can be beneficial to countries that are attempting to meet climate targets and in areas that price carbon emissions. Wind-powered water injection can provide water injection far from the platform, which reduces the need for lengthy water injection lines and can eliminate the need for costly modifications for oil platforms not initially designed for water injection. The Wind Powered Water Injection project recently completed an initial testing phase in which it was shown that wind power can provide water injection at competitive prices [30].

8.6.3 CONCENTRATING SOLAR AND GEOTHERMAL HEAT
FOR TERTIARY RECOVERY (EOR)

According to Ericson et al. [7], tertiary recovery, or EOR, presents several opportunities for renewable integration. Renewable technologies can both provide energy for and generate energy from EOR processes. Land and solar irradiance permitting, concentrated solar can be an economic means of generating steam for EOR. Geothermal cogeneration can also produce electricity from latent heat in wells.

Thermal EOR consists of injecting steam into the oil reservoir to facilitate flow by reducing the viscosity of the oil. Natural gas is used as an energy source to produce steam, which is then injected into the reservoir. Solar thermal EOR substitutes natural gas with concentrated solar power (CSP) as the energy source for producing steam [32]. Trough-shaped mirrors, sometimes housed in a protective greenhouse, concentrate sunlight to generate steam with temperatures up to 640°F [33].

CSP technologies are well suited for thermal EOR. There is significant overlap between regions with high solar radiation and large petroleum reserves [34]. Due to latent heat stored in the reservoir rock, oil recovery rates were found to be not greatly impacted by solar variability [33]. Finally, solar thermal EOR has lower capital costs than CSP for electricity generation because solar thermal EOR does not require a turbine to convert steam to electricity, which greatly reduces project cost and complexity.

While CSP capital costs are significantly higher than for a comparable natural gas system, fuel cost savings have been found to make solar thermal EOR competitive with traditional natural gas EOR in some cases [33]. The profitability of solar thermal EOR relative to alternatives varies with natural gas prices, cost of capital, and field characteristics, such as the amount of solar radiation and expected field lifetime. There is generally an alignment between oil reserves and solar potential, especially among countries in the Middle East. Estimates for solar thermal projects used for petroleum EOR could be economically viable and range from 19 to 44 gigawatts (GW) [34].

The technology for integrating CSP into oil production is now commercially viable in some locations [7]. The first major project to integrate CSP into oil production was in California through former Chevron subsidiary BrightSource [35]. The pilot project began operation in 2011 and operated for 4 years. More recently, the Miraah project—a $600 million joint venture initiated in 2015 between the government of Oman and the company GlassPoint Solar—has begun producing solar steam for the

Amal oilfield. When completed, the project will be the largest concentrated solar project in the world, with a capacity exceeding 1 GW and a production of 6,000 tons of steam per day [36]. Successful completion and operation of the Miraah project would prove the economic viability of solar thermal EOR at scale and lead to the adaption of similar projects in other locations.

Mature oil wells can produce up to 50 barrels of water per barrel of oil [37]. Water extracted from a well has naturally raised temperatures due to geothermal heating. While for most wells, the heating is not significant—80% of wells have temperatures below 176°F (80°C)—in some instances, temperatures can exceed 400°F (~200°C) [38]. Steam can be used to generate electricity, provide field heating, and can be reinjected for EOR. While near-term prospects for geothermal cogeneration are limited, in some cases, they can provide economic benefits [38]. In addition to heat and electricity generated, geothermal cogeneration has the additional benefit of extending field life. Extending a field's lifetime delays well abandonment costs, which improves field economics.

Steam injection and geothermal energy have important synergies. Steam injection leads to higher produced water temperatures than for conventional fields [39]. Additional energy requirements for steam injection also lead to greater demand for the electricity and heat produced from geothermal generation [39]. A trend toward deeper wells, which produce higher water temperatures, an increased use of steam injection EOR, and an increase in the pricing and regulation of carbon emissions indicate a growing opportunity for geothermal cogeneration in the future [7].

8.6.4 EXAMPLES OF SUCCESSFUL CASE STUDIES

In the O&G industry, renewable energy technology is being used to resolve problems of supplying electricity for offshore production and to supply the thermal energy required for the EOR technique [40]. PV systems have been applied at several oil and gas fields such as the Midway-Sunset, Kern River, and Louisiana Bayou in USA. A few applications of wind power systems were also identified at the oil fields of Suizhong 36-1 in China, Beatrice in UK, and Utsira Nord in Norway. In addition, several applications of geothermal power systems have been found at the oil fields of Rocky Mountain in USA and Fort Liard in Canada. Solar thermal systems were installed at the oil fields of Mckittrick in USA, Caolinga in USA, and Amal in Oman.

Recently, Halabi et al. [40] reported study results on solar energy (PV and solar thermal) technologies being used in the O&G industry. A study by Halabi et al. [40], however, had limitations because it only considered solar energy technology, among the many renewable energy technologies that are being used in the O&G industry. Therefore, it is also necessary to analyze cases in which other renewable energy technologies, such as wind power and geothermal power, are being applied.

8.6.4.1 Photovoltaic Hybrid Systems

8.6.4.1.1 The Midway-Sunset Oil Field in the United States
Chevron Texaco built a 500 kW PV system and is now operating the system at the Midway-Sunset oil field in California (USA). The Midway Sunset oil field is the largest oil production area and was discovered in 1894. The electricity produced from the PV

system at the Midway-Sunset oil field is used to power the oil production plant. It supplies about 5% of the total electricity demand. At this location, 4,800 modules were used in the PV system [41]. The system is accompanied by storage and conventional systems.

8.6.4.1.2   The Kern River Oil Field in the United States

Chevron Texaco operates the Kern River oil field in California (USA). From its discovery in 1899 through 2006, the oil field has had cumulative production of close to two billion barrels [42]. Currently, a 750 kW PV system is operating at the Kern River oil field. Approximately 7,700 flat PV modules including six different makes of thin-film PV, plus one type of traditional crystalline silicon PV, and several different makes of racking systems and inverters make up the PV system. The electricity produced by this system powers 40 oil extracting pump jacks. Chevron Texaco is considering building an additional PV system for powering more oil-extracting pump jacks [43]. The system is accompanied by storage and other conventional systems.

8.6.4.1.3   The Louisiana Bayou Oil Field in the United States

Kyocera Solar is operating 7.7, 3.9, and 6.25 kW PV systems at the Louisiana Bayou oil field in Louisiana (USA). The PV systems include PV modules, mounting structures, controller (with integrated solar charge regulation), storage batteries, system enclosure, installation/wiring kit, and instruction manual [44]. The electricity produced from the PV system is used to prevent the oil pipeline from corroding using the cathodic protection technique. The results from recent studies indicated that corrosion of transport pipes significantly affects the production and transportation of oil and gas. As a solution to these corrosion problems, Popoola et al. [45] proposed a cathodic protection method that incorporates a PV system. The overall system is hybrid in nature.

8.6.4.2   Wind Power Systems

8.6.4.2.1   The Suizhong 36-1 Oil Field in China

China National Offshore Oil Corporation (CNOOC), the biggest oil company in China, is operating the Suizhong 36-1 oil field, which is located in Liaodong Bay in the Bohai Sea. The Suizhong 36-1 oil field was discovered in 1987 and started producing oil in 1993. This is the largest oil field in the Bohai Sea and has over one billion tons of oil reserves. The Suizhong 36-1 oil field has many offshore steel jacket-type platforms. These platforms are approximately 20 years old and need either to be demolished or reused for a different purpose. In 2007, for the first time, CNOOC reused an abandoned 36-1 SPM jacket as a foundation on which to build a 1.5 MW wind turbine [46]. The electricity produced by this wind turbine supplies the production plant at the Suizhong 36-1 oil field. It has reduced the emission of greenhouse gases by 5300 tons annually [47]. The overall system is hybrid in nature.

8.6.4.2.2   The Beatrice Oil Field in the United Kingdom

An English energy company, Scottish and Southern Energy, and a Canadian multinational oil exploration company, Talisman Energy, made a joint investment to build a wind power system at the Beatrice oil field located in the North Sea. The Beatrice oil field is located 24 km off the northeast coast of Scotland and started commercial production in 1981. The Beatrice oil field comprises three offshore plat-forms. Beatrice Alpha is an oil production plant located at the center of the oil field [48]. Two 5 MW wind turbines were built near the Beatrice Alpha platform. The electricity produced from these wind turbines supplies about 30% of the total elec-tricity demand of the Beatrice Alpha platform [49]. The capital cost (CAPEX) and levelized cost of energy (LCOE) for three different floating foundation types are presented. The overall system is hybrid.

8.6.4.2.3   The Utsira Nord Oil Field in Norway

DNV GL of Norway is working on the Wind-Powered Water Injection (Win-Win) project that builds a wind power system at an offshore oil and gas production plant at the Utsira Nord oil field [50]. The main purpose of the Win-Win project is to build a floating-type wind turbine that produces electricity and supplies this electricity to an offshore platform. Furthermore, Nilsson [51] investigated creating an offshore wind farm with capacity of 288 MW, consisting of 48 wind turbines with capacity of 6 MW at the Utsira Nord oil field. In this study, it was reported that building a wind farm to produce 288 MW would generate 1,222 GWh of electricity per year. This is enough to supply 48% of total electricity demand in the neighboring Utsira High region. The overall system is hybrid.

8.6.4.3   Use of Geothermal Energy

8.6.4.3.1   The Rocky Mountain Oil Field Testing Center in the United States

In the United States, approximately 25 billion barrels of thermal water are discharged every year through oil boreholes [52]. A feasibility study on using discharged thermal water for geothermal power generation was done in 2008 [53]. Furthermore, Blodgett [54] reported to the Department of Energy, regarding the study results on the feasi-bility of building a 1 MW geothermal power system using thermal water from bore-holes in oil industrial zones in the United States.

Based on this, a 217 kW geothermal power system that uses thermal water from oil boreholes was built at the Rocky Mountain Oil Field Testing Center located in Wyoming (USA). In an area of 38 km, 150 boreholes ranging from 76 to 1500 m in depth were used, and the thermal water temperature was approximately 65°C. It was reported that the geothermal power facility generated 1918 MWh of electricity from 109 million barrels of thermal water [55]. The overall system is hybrid.

8.6.4.3.2   The Fort Liard Oil and Gas Field in Canada

The Fort Liard oil and gas field located in the northwest region of Canada produced 1.4 billion barrels of natural gas and 700 million barrels of oil in 2008 [56]. The cumulative production of natural gas until 2013 was approximately 1.6 billion bar-rels [57]. Currently, the Fort Liard oil and gas field has stopped producing oil and gas. Instead, a geothermal power generation project is in progress, using thermal water from boreholes deeper than 5000 m. Thompson and Dunn [58] designed a 700–1,000 kW geothermal power system and it is predicted to produce 2,900 MWh of electricity annually. The overall system is hybrid.

8.6.4.4   Solar Thermal Systems

8.6.4.4.1   The Mckittrick Oil Field in the United States

The Mckittrick oil field operated by Berry is located in western California (USA) and produces approximately 1,300 barrels of crude oil per day [59]. Berry launched the solar thermal EOR project with Glass Point Solar to apply the EOR technique to the Mckittrick oil field. The main purpose of this project was to install solar thermal collectors approximately 13,388 m, at the Mckittrick oil field, to collect sunlight. The sunlight was used to boil water and produce high-pressure steam. The high-pressure steam was used to apply the EOR technique. This project implemented the enclosed trough technique of Glass Point Solar for the first time. The enclosed trough technique allowed installation of solar thermal collectors in oil fields with a lot of sand and dust. An extremely lightweight mirror was installed in the structure and an automatic directing device was mounted on the mirror to focus the sunlight to a designated spot. The overall system is hybrid.

8.6.4.4.2   The Caolinga Oil Field in the United States

The Caolinga oil field operated by Chevron is located in western California, the United States. It is the eighth largest oil field in California. In October of 2011, Chevron started the Coalinga Solar EOR project with Bright Source Energy (a company focusing on solar power plants) to install a 29 MW centralized solar thermal system at the Caolinga Oil Field. In this project, 3,822 2×3 m heliostats (mirrors) were used. The heliostat reflected and focused light on a solar tower (99.7 m high) to heat the water and produce steam. According to the reports, using solar thermal technology instead of natural gas reduced the cost to create steam for EOR by approximately 120 MMBtu/h [60]. The overall system is hybrid.

8.6.4.4.3   The Amal Oil Field in Oman

The Amal oil field operated by Petroleum Development Oman (PDO) started producing oil and gas in 1967. In 2012, 920,000 barrels of oil and 2.8 BCF (billion cubic feet) of natural gas were being produced each day. PDO and Glass Point Solar launched the solar thermal EOR project at the Amal oil field and were the first in the Middle East to start such a project [61]. A 7 MW solar thermal EOR plant was built in the Amal oil field in 2012 and the enclosed trough technology of Glass Point Solar was implemented. Presently, the solar thermal EOR plant produces around 50 tons of steam each day, which is used for implementing the EOR technique. This is approximately 1% of the steam that used to be produced using only natural gas [62]. Glass Point Solar [39] is planning a project to build a 1 GW solar power plant at the Amal oil field by 2017. This project is anticipated to replace 5.6 trillion BTUs of natural gas and supply electricity to approximately 209,000 people using residual latent heat [63,64]. The overall system is hybrid.

8.6.5   CLOSING PERSPECTIVES

The Midway-Sunset oil field was a case where a PV system was applied to supply electricity to an oil and gas production site. A 500 kW PV system was built to supply a portion of the electricity required at a production plant. Furthermore, at the Kern River oil field in the United States, a 750 kW PV system was built to supply the electricity required to operate pump jacks. At the Louisiana Bayou oil field, also in the United States, 3.9, 6.25, and 7.7 kW PV systems were being used to supply electricity to preventing corrosion of oil pipelines. An overall analysis indicated that PV technology was generally used on small-scale PV systems to supply a portion of the electricity required at oil and gas production or transportation facilities.

The Suizhong 36-1 oil field in China was one of the cases of applying wind power technology. A 1.5 MW wine turbine was built upon elements of an abandoned oil drilling structure. The Beatrice oil field in England had two 5 MW turbines off-shore, that provided electricity to an oil production plant nearby (about 30% of total electricity consumption). Last, the Utsira Nord oil field in Norway had a floating wind turbine, which provided electricity to an offshore oil production plant. Overall analysis of these cases showed that marine structures built to produce oil were being efficiently reused as structural elements for marine wind turbines. Furthermore, the wind turbines were supplying a portion of the electricity consumption at the oil production plants.

Cases of the application of geothermal technology were found at the Rocky Mountain Oil Field Testing Center in the United States and at the Fort Liard oil field in Canada. In both cases, the aim was to use the thermal water discharged from oil boreholes to generate geothermal power. At the Rock Mountain oil field, a 217 kW geothermal system was designed and a feasibility test performed. The Fort Liard oil field is also testing for feasibility to build a 700–1000 kW geothermal system. Overall analysis on the cases showed that the geothermal technology was mainly being applied to reuse thermal water discharged from exhausted oil and gas fields. Cases of applying solar thermal technology to provide thermal energy were found in the Mckittrick oil field in the United States with a 300 MW solar thermal system, the Caolinga oil field in the United States with a 29 MW solar thermal system, and the Amal oil field in Oman with a 7 MW solar thermal system. Overall, the solar thermal technology was being used to supply the thermal energy required for applying the EOR technique to oil and gas production sites.

In all cases, renewable technologies were used to partially satisfy the total needs of each plant. This strategy for hybridizing conventional system with renewable system has proven to be very effective for reducing cost and emission of CO_2.

8.7 MIDSTREAM: INTEGRATION OF HYBRID RENEWABLE ENERGY SYSTEMS IN OIL AND GAS TRANSPORTATION

Midstream transportation uses less energy than upstream or downstream operations and correspondingly has the smallest potential for renewable integration [2]. Furthermore, oil transportation by ship, rail, and truck currently has limited opportunities for renew-able technologies. That being said, there are still several opportunities for renewable technologies to have a meaningful impact on operation costs and emissions.

A vast network of pipelines transports oil and gas from producers to refineries and end-users. In the United States alone, there are more than 300,000 miles of main-line natural gas pipelines with 1,400 compressor stations [65] and approximately 55,000 miles of crude oil trunk lines and 95,000 miles of refined products pipe-lines, which combined transport more than 40 million barrels of liquids per day [66]. Figure 8.5 displays the U.S. natural gas transportation network along with major fields. As can be seen, pipelines can transport supply significant distances from pro-duction areas to major demand hubs.

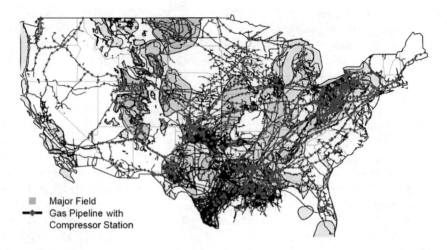

FIGURE 8.5   Natural gas transportation system. (Figure produced using data from Refs. [7,67].

Renewable technologies are already integrated into some pipeline operations, with solar energy powering sensors and providing cathodic corrosion protection (a cathodic protection system uses charged anodes attached to the pipe to prevent the pipe from corroding) [66]. However, as renewable costs decline and the technologies improve, additional opportunities will be available. Several possibilities in this regard as analyzed by Ericson et al. [7] is presented below.

8.7.1   COMPRESSOR ELECTRIFICATION, HEAT RECOVERY, AND USE OF TURBO EXPANDERS

As most oil pumping stations use electric motors for the prime mover, these systems are automatically using more renewable power as the grid greens [66]. Natural gas pipelines, however, are primarily powered by gas engines or turbines [17]. Replacing gas engines and turbines with electric motors could increase renewable integration and reduce noise, fuel use, and emissions. Electric motors have low operating costs and more efficiently accommodate a wide throughput range than gas engines or turbines [17]. As natural gas plants are ramped to account for increased variable generation, the variability of natural gas demand will increase, meaning pipelines will likely deal with larger swings in throughput. According to Ericson et al. [7], this trend supports the use of more electric motors to power natural gas compressor stations. An important barrier to using electric motors to power compressor stations is a concern of reliability. While electric motors are highly reliable, the electric grid does not always meet the operation standards of the O&G industry, especially in remote areas where compressor stations are often located [17]. Work on increasing grid reliability, reducing the price of microgrid technologies, and rewarding lower emissions from electric motors could help motivate a shift toward using more electric motors [7].

Gas turbines used to power natural gas compressor stations generate a considerable amount of heat. In some cases, this heat can be reused to generate power or used in industrial processes. There are currently 12 power generation systems, totaling 64 MW of electricity capacity, installed at natural gas compressor stations in the United States [68]. An estimated 40 MW worth of potential additional projects currently offer expected paybacks of less than 5 years [68]. However, changes in market conditions could significantly increase the economic opportunity for compression station heat recovery. Estimates of the technical potential for electricity generated from compressor station heat recovery range from 500 [69] to 1.1 GW [68]. As economies of scale are present, large systems with high utilization rates are most applicable [69]. A disadvantage of compressor cogeneration is the variability of output. Compressor stations do not always run and therefore would not always generate heat or electricity [17]. If extracted heat is used for industrial processes, then an associated process must be able to operate with variable heat. Similarly, if electricity generated provides off-grid power, then the consumer must be able to handle variable generation [7].

Compressors raise the gas pressure between 500 and 1,400 pounds per square inch (psi) to transport natural gas long distances. While higher pressures increase the economics of transportation, and while power plants and some industrial customers use high pressure gas, most customers require pressures well under 10 psi [70]. The potential energy created from raising pipeline pressure is lost when the pressure is stepped down at distribution hubs. A turboexpander can capture this potential energy to generate electricity. Turboexpanders are used in a variety of industry applications. Unfortunately, they are generally considered uneconomic for use in natural gas pipelines under current market conditions [17,69]. Primary challenges include the required cost of preheating the gas before expansion and pipeline operations causing variability in output [69]. Turboexpanders have the benefits of being a proven technology that emits significantly less criteria and other gasses per megawatt-hour generated compared to coal and gas power plants. Ericson et al. [7] indicates that the turboexpanders may therefore see economic opportunities in regions that price or regulate emissions.

8.8 DOWNSTREAM: INTEGRATION OF HYBRID RENEWABLE ENERGY SYSTEMS IN OIL REFINING

Ericson et al. [7] points out that oil refining is a complicated process that requires large capital investments and has significant energy requirements. The refining process requires heating oil to break carbon bonds and remove impurities. More than 90% of energy use in refining goes toward direct heating and steam generation [71]. As Figure 8.6 shows, the majority of energy used in the petroleum supply chain is from burning onsite oil to process and refine petroleum. Due to the large production volumes and high temperatures required, petroleum refining is the largest consumer of fuel and the largest generator of onsite greenhouse gas emissions in the U.S. manufacturing sector [2,71,72].

Energy usage in the refining process is a major operations cost. In 2010, the U.S. petroleum refining industry spent $9 billion on energy purchases, equating to roughly 50% of total noncapital costs [73,74]. Furthermore, due to shifts toward increasingly heavy and sour oil, refining energy requirements are rising [75]. Emissions from refineries also increase as oil quality decreases, which increases the difficulty of meeting environmental regulations [75].

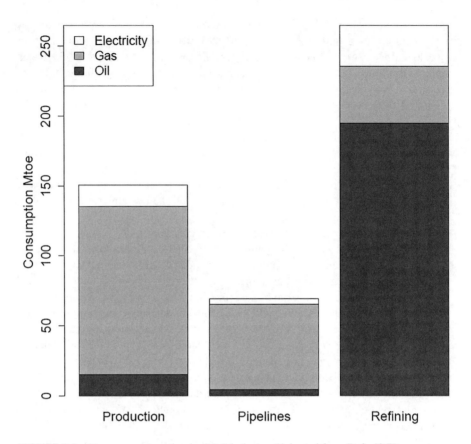

FIGURE 8.6 Energy consumption in O&G industry. (Adapted from Refs. [2,7].

Ericson et al. [7] points out that oil refineries can incorporate renewable energy technology into operations by using renewable sources to generate heat and electricity, and to produce hydrogen that is an input in the refining process. Oil refineries can also use excess heat to generate electricity. One of the opportunities is natural gas pipeline compressor electrification, compressor heat recovery, and electricity production from expansion turbines.

Unconventional reserves, such as heavy crude oils, are expected to have more contribution in achieving world demand in 2030 [1]. Heavy crude oils are more viscous and have higher boiling ranges and densities. In addition, they are usually rich in aromatics and tend to have more residual material, e.g., sulfur, nitrogen, asphaltene [2]. Normally, heavy oils cannot be processed directly in petroleum refineries, thus, oil industries will usually set up an upgrader close to the oil fields that generally consist of distillation unit, residual oil processor, and hydrotreater. Synthetic crude oils are obtained from the upgrader that satisfies the requirements for processing in downstream. The heat required for upgrading can be obtained from renewable energy. Among various renewable energies, solar energy is the most abundant resource and can be harvested and utilized in different ways, such as electricity generation via photovoltaic cells or thermal energy by concentrating panels [3].

Moreover, the locations of substantial heavy oil reserves are in the area with high solar irradiance, which indicates that using solar energy in these regions can be efficient and cost-effective [1]. NREL [7] and IEA [1] analysis on the use of cogeneration in this regard is presented below.

8.8.1 COGENERATION (HEAT AND POWER) AND USE OF HYBRID RENEWABLE ENERGY SYSTEMS

Oil refining runs at high temperatures and elevated pressures. This offers numerous opportunities for heat recovery and power recovery. Cogeneration is already common within the refining industry, representing almost 13% of all industrial cogenerated electricity [74]. However, the potential for cogeneration is still far from being fully exploited. Power recovery systems can reduce a refinery's energy intensity by 7%–10% [73].

The primary application for power recovery is with the fluid catalytic cracker (FCC) [73]. The hydrocracking process presents another opportunity for power recovery. Power is generated from the pressure difference between the reactor and fractionation stages [73]. Integration of a turboexpander on the hydrocracker at the Zeeland Refinery in the Netherlands produces 73,000 MWh/y, resulting in a payback period of less than 3 years [73]. Power recovery offers a means of reducing the energy and carbon intensity of the refining process. For many refineries, cogeneration can already offer positive economic returns. According to Ericson et al. [7], if the industry seeks to further reduce air emissions, cogeneration may become increasingly beneficial as well.

Besides power recovery, each step along the refinery process requires different temperatures and pressures. Oil is first heated to 200°F–300°F (90°C–150°C) and washed with water to remove salt and other suspended solids [73]. Atmospheric distillation then separates products based on their boiling point in the crude distillation unit (CDU), where crude oil is heated to temperatures around 750°F (390°C) [73]. Heavy fuel oils are further treated in the vacuum distillation unit (VDU) at temperatures between 730 and 850°F (390°C and 450°C) [73]. The low-pressure environment of the VDU lowers the heavy oil boiling point to facilitate separation. Additional processing occurs in the FCC, the hydrocracker, and coking unit. Each of these processes separates heavier oil products into lighter and more valuable products.

Hydrotreating mixes hydrogen with the feed stream at temperatures between 500°F and 800°F (260°C and 460°C) to remove sulfur [73]. A catalytic reformer uses a catalytic reactor to produce high-octane gasoline and hydrogen. Additional processes, such as alkylation, which uses steam, power, and various acids to produce alkylates, are also common. Figure 8.7 displays yearly estimated energy use by refining process for U.S. refineries in 2012. Hydrotreatment and the CDU account for nearly 50% of total energy requirements [7].

Ericson et al. [7] indicates that when land and sufficient sunlight are available, CSP can be used to produce heat and steam. While CSP may be unsuited for some high-end temperature requirements, steam production and heating up to 750°F (400°C), and in some cases up to 1,020°F (550°C), can be provided by current CSP technologies [76]. While most refineries do not meet the criteria of having both available land and abundant sunlight, the potential for CSP integration is still large due to the significant energy intensity of oil refining. Estimated market potential for solar

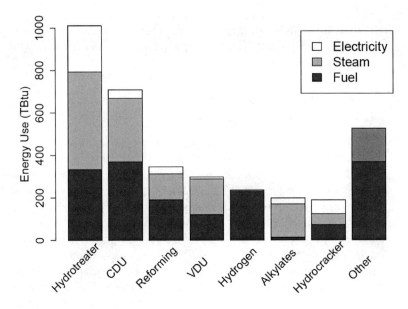

FIGURE 8.7   Energy use by refinery process. (Figure adapted from Refs. [7,75].

thermal used in oil refining ranges between 21 and 95 GW [34]. Along with high heat requirements, oil refineries have large electricity loads. Increasing the renewable energy used to run machinery, such as compressors, pumps, circulators, and controllers, can reduce operation costs and carbon emissions.

8.8.2   HYDROGEN PRODUCTION

Refineries use hydrogen in several production processes. Most importantly, hydrogen is used in the hydrotreatment process to lower the sulfur content of diesel fuel. Refinery demand for hydrogen has increased as demand for diesel fuel has risen and sulfur-content regulations have become more stringent [77]. Hydrogen used in the refining process increased by more than 50% between 2008 and 2014 [77]. While hydrogen is produced as a coproduct of the catalytic reformation of gasoline, demand for hydrogen exceeds supply produced from the refining process itself [78].

Over the recent decades, there have been many applications of solar energy in petroleum industry as presented in the review by Absi Halabi et al. [79]. One notable example was using solar concentrators to concentrated solar energy; this solar thermal energy was then used to produce steam required in the oil production process. The authors also mentioned that there were potentials of using solar energy in the petroleum industry since most of oil reserves were located at isolated area, where lands were available for solar energy plants. Therefore, it is possible to use solar energy to produce hydrogen for the crude oil upgrader. For the method of producing hydrogen using solar energy, Ngoh and Njomo [80] suggested that currently there were three possible processes, including photochemical, thermochemical, and electrochemical process. Among these processes, the photochemical process was under a stage of investigation, leaving the possible candidates to be two other processes.

In the thermochemical process, solar energy was concentrated and used to provide heat required in endothermic hydrocarbon cracking reactions, such as steam reforming of natural gas. Recent works on this thermochemical process are available; for example, Möller et al. [81] proposed process configurations of hydrogen production by solar reforming of natural gas. In their process, conventional reformer was replaced by the new innovative volumetric reactor receiver (VRR). This VRR can be heated with concentrated solar energy to more than 900°C which was adequate for steam methane reforming reaction. However, the process could only operate during the full load hours of solar energy, restricting the operating hours to 2,038 h/y. Detailed economic evaluation was carried out; the authors claimed that the hydrogen production cost of this process is only 20% more expensive compared to the conventional steam reforming of natural gas process.

More examples of thermochemical process were demonstrated by Giaconia et al. [82]. In their work, solar energy was transferred to hydrogen production process by molten salt which had been long tested as solar heat carrier and heat storage medium. The advantage of this process over the previous process by Möller et al. [81] is that it can operate during the absence of sunlight. However, due to temperature limit of molten salt that could only provide temperature up to 565°C, resulting in a low conversion of methane, the authors proposed a steam reforming process coupled with membrane reactors. The use of membrane reactor could drive the steam reforming reactions forward at low temperature (<565°C), increasing methane conversion thereby. The authors also carried out an economic analysis by estimating hydrogen production cost from each proposed process; the results showed that the cost was 40%–80% higher than conventional steam reforming process. These were because of additional equipment installed, increasing capital cost. However, the cost could be reduced in the future as solar equipment becomes less expensive, and natural gas price increases.

For electrochemical process, the most well-known process is electrolysis of water where electricity is used to decompose water in hydrogen and oxygen. It is the most developed and useful method in industries for hydrogen production [1]. It is not cost-effective on the economic viewpoint unless the electricity is generated from renewable energy, such as solar energy. In addition, utilizing renewable energy for electricity generation will allow hydrogen to be produced without CO_2 emissions. According to the literature, there had been numerous efforts in exploiting solar energy for hydrogen production. Particularly, steam reforming was the most interested subject of development. However, utilization of solar energy to produce hydrogen had not been attempted in industrial scale. Consequently, this work's aim is to use solar energy to produce hydrogen required in the crude oil upgrader process.

Hydrogen is produced primarily through steam-reforming natural gas [83]. Concentrated solar energy can complement or replace conventional fuel as the heating source for this process [83]. Hydrogen produced by electrolysis powered by renewable generation can also be used to meet oil refinery demand [78]. Costs of hydrogen production from renewables are currently higher than conventional production methods [83]. However, if renewable generation costs continue to fall, and if regulations on greenhouse gas emissions increase, hydrogen production may become an effective means of using renewable technologies to reduce costs and reduce emissions from oil and gas operations.

In the study by Likkasit et al. [83], methods of integrating solar energy into O&G industries are proposed. There are numerous applications of solar energy in O&G industries; however, this study particularly aims to use solar energy for producing hydrogen that required in hydrotreater unit of crude oil upgrader process. Each proposed method will be designed to have a hydrogen capacity that satisfies a demand in the crude oil upgrader process. Aspen Plus® and System Advisor Model [4] are used to aid in the process designs. After proper equipment size and process requirements are obtained, each method is compared in both process efficiency and economic aspects in order to determine the most applicable way of using solar energy to produce hydrogen for the crude oil upgrader process.

Thus, in the study of Likkasit et al. [83], three hydrogen production processes using solar energy are proposed, and these process are compared with the conventional steam reforming of natural gas process. Process designs were carried out, and according to the simulation results obtained from Aspen Plus, all proposed processes consumed less natural gas feeds, as well as had lower CO_2 emission rates compared to the conventional process. However, the economic analysis using the levelized cost of hydrogen production (LCHP) showed that all proposed processes had higher LCHP compared to the conventional process. The lowest LCHP is 2.514 $ per kg H_2 which belongs to the process 2: steam reforming of natural gas process using a volumetric receiver reactor, and it is 46% higher than the conventional process. For these findings, it is concluded that using solar energy to aid in the hydrogen production could lower the environmental impact; however, in terms of economic point of view, these methods are currently not economic since the solar-related equipment are still expensive. In the future, the cost of this equipment could be less expensive, making these methods become applicable and promising. The roles of biomass, electrolysis, and nuclear energy in the hydrogen production are described in detail in Chapter 11.

8.8.3   OTHER EFFORTS TO HYBRIDIZED OIL AND GAS WITH RENEWABLE ENERGY

Although NOCs (National Oil Companies) are employing various strategies to navigate the energy transition, renewable energy and energy efficiency are the main paths. Many energy efficiency improvements involve process hybridization. More than 90% of the required reduction in energy-related CO_2 emissions can be achieved through increased energy efficiency and renewable energy utilization [84]. NOCs have begun utilizing renewable energy in their facilities or, in some cases, providing renewable energy to other sectors. The latter, however, is not the main focus of most of the NOCs. Their actions are mostly policy-driven, and renewable energy production is mainly the purview of other companies in their countries. For example, ADNOC (Abu Dhabi National Oil Companies) has not developed renewable energy assets, letting the Abu Dhabi Future Energy Company, Masdar, play this role while cooperating with ADNOC [85]. NOCs' actions related to renewables are highlighted as follows.

Since 2010, CNOOC has started the production of biodiesel with an annual capacity of 60,000 tons for use in vehicles [86]. CNPC has also evaluated and begun using biomass (aviation biofuel), wind, solar, and geothermal; this includes pilot tests for the use of geothermal energy in the Huabei Oilfield and projects for photovoltaic power generation, wind power generation, and the development and utilization of geothermal energy in the Xinjiang and Liaohe oilfields [87].

Gazprom operates 1959 power generation units, generating from both fossil fuels and renewable energy, for auxiliary needs and to sell to third-party consumers in remote or off-grid areas where it is economically and technically feasible [88]. In 2017, the Gazprom Group generated 471,470 kWh of electricity from its power plants (excluding hydro) [89]. Gazprom's renewable energy production includes but is not limited to generating around 13 million MWh of electricity from hydroelectric power in 2016 through Gazprom Energo holding LLC and generating more than 360,000 kWh of electricity in 2016 by using other renewable energy sources such as solar panels, wind turbines, turboexpanders, and thermoelectric generators [90]. Additionally, Gazprom is building a 102 MW wind power plant in cooperation with the Serbian company Energo wind NIS, and it also has plans to build a geothermal power plant in northern Serbia in partnership with the Singapore company Betec [90].

KPC has also stepped into renewable energy and commenced several initiatives such as the Renewable Energy Dashboard with the goal of producing at least 15% of Kuwait's energy from renewable energy by 2030 [91]. It includes but is not limited to the following actions [91–93]:

- Launching the largest solar project in Kuwait, named Sidra 500, with electricity capacity of 10 MW.
- Building and maintaining solar panels in the parking lots of the oil complex (in cooperation with the Kuwait Institute for Scientific Research).
- Using solar to light the 6.5 km road in the north of Kuwait.
- Producing electricity from solar and wind to light the head office building and for irrigation purposes (Waha Al-Subaihiya project).
- Producing steam from CSP to use in injection operations at the South Ritqa field in the north of Kuwait.
- Building the Abdaliya Integrated Solar Combined Cycle project and establishing the first solar thermal power plant in Kuwait by 2020.
- Operating two fuel stations in the Al-Zahra and Al-Riqqa areas powered by solar energy.
- Planning to build an integrated CSP plant to produce power and steam for enhanced energy recovery.

PDO's utilization of renewables is mainly focused on solar energy; it involves projects such as a CSP plant with electricity capacity of 1021 MW to use in EOR in 2017 (a joint venture between Shell, Total, Partex, and the Government of Oman) and the Mina Al Fahal solar parking project commissioned in early 2018 [94,95].

National Iranian Oil Company (NIOC) has begun employing solar energy in some of its facilities, including installing four solar energy lighting systems in the southwest of Iran due to the availability of sunshine more than 300 days a year [96]. Petronas has produced around 13,628 MWh of solar energy from existing solar PV investment projects in 2017 [97]. Petrobras has been investing in renewable energies such as solar and wind. For instance, the Alto Rodrigues Photovoltaic Unit is a pilot plant with an installed capacity of 1.1 MW. In solar and wind energy segments, the company also has partnership with other oil companies such as CNPC (Comperj and Marlim cluster), Equinor (offshore wind energy), and the Total group (solar and onshore wind) [98].

Saudi Aramco is currently a member of the Carbon Sequestration Leadership Forum and Oil and Gas Climate Initiative (OGCI). It has launched investments in renewables projects and has plans to use renewable energy for remote facilities and well sites; they include installing 9.5 GW of solar and wind capacity by 2023, 3.3 GW of solar PV power, and 800 MW of wind in 2018 [99–101]. The company currently owns a wind power project, called the Turaif oil storage depot, with one turbine and a generation capacity of 2,750 kW [102].

8.9 PERSPECTIVES ON USE OF RENEWABLE ENERGIES FOR OIL

As oil and gas production shifts toward lower quality and unconventional reserves, energy use and emissions from operations are likely to grow in the future. Furthermore, oil and gas operations are energy-intensive which can have negative environmental impacts. Integrating renewable energy technologies into oil and gas operations offers a means of reducing fossil fuel use in the production of oil and gas, which can both lower operation costs and reduce emissions as well as conserve petroleum production for higher value uses. In some cases, renewable integration can currently provide a cost-effective and environmentally beneficial way of meeting operations energy requirements. If costs of renewable technology continue to fall, the benefits of renewable integration would continue to increase.

Renewable technologies can be integrated in all links of the oil and gas supply chain. For upstream oil and gas production, the primary applications identified are solar heating for EOR and other heating requirements, offshore wind to power offshore operations, well-pad electrification from solar and wind, and geothermal cogeneration from oil fields. Solar thermal generation for EOR and well-pad electrification is already seeing commercial operation. Wind power for offshore water injection and geothermal energy production are close to commercial operation as well. Combined, these technologies will be useful tools to reduce operation costs and emissions.

While midstream transportation offers the fewest opportunities by size of demand for renewable integration, renewable technologies can still be beneficially integrated. The use of electric motors to power natural gas compressors reduces emissions and reduces operation and maintenance costs. Heat recovery is already economic for large compressor stations, and turboexpanders may be an economic option as technology continues to advance and operation cost environments evolve. Oil refineries are the largest consumer of fuel in U.S. manufacturing and the largest generator of onsite greenhouse gas emissions in the manufacturing sector. Fuel costs also constitute half of all operations costs. Renewable technologies can be integrated to reduce fuel expenditures. Where land and sunlight are available, solar heating may be used. Renewable energy may also be used to produce hydrogen by electrolysis. Finally, further use of power recovery can reduce emissions and fuel costs.

While significant challenges exist to integrating renewable energy technologies into oil and gas operations, there are also significant economic opportunities for renewable technologies in the petroleum industry. Falling costs and growing energy intensity and environmental concerns are all leading to more compelling reasons for integrating renewable energy technologies. Oil and gas operators would be well served by further analysis of where renewable technologies can be beneficially integrated and by working toward further integrating renewable technologies into current and future operations.

8.10 NATURAL GAS-RENEWABLE SOURCES HYBRID SYSTEMS

Natural gas and renewable energy have been touted as key elements of a transition to a cleaner and more secure energy future. Still, the specific roles, values, and merits of natural gas and renewable energy in relation to long-term goals of energy security and climate change mitigation have been, and continue to be, debated. To some extent, however, they complement each other [103–105].

In the energy security arena, both natural gas and renewable energy are building blocks for a robust domestic energy economy. However, there are currently large, but not insurmountable, barriers to harnessing natural gas and renewable energy to meaningfully reduce our national reliance on imported oil for transportation. These include the current state of the petroleum dependent transportation sector and a lack of clarity and consensus on how many alternative pathways to pursue and which ones are best. Early adopters are testing alternatives and will provide valuable experience, but full deployment of one or more alternative fuels will require broader structural shifts.

Regarding climate change, the Intergovernmental Panel on Climate Change reports that to avoid the largest negative impacts, global greenhouse gas emissions would need to decline by 50%–85% from 1990 levels by 2050 [106]. Some have argued that, given the difficulty in meeting this goal, an exclusive focus on natural gas would distract from and impede progress toward the ultimate goal of large-scale deployment of a suite of low-carbon technologies, including renewable energy, energy efficiency, nuclear energy, and carbon capture and storage [107]. Others have offered roadmaps for how cost-effective deployment of natural gas and low-carbon technologies to meet emissions targets might occur [104,108] while many more have provided insightful analyses and framed relevant issues [109].

Much of the current discourse is narrowly focused on either natural gas or renewable energy as distinctly separate components or concentrates on the competitive impacts of one over the other. Another perspective is that instead to build upon embryonic efforts [6] to more closely examine the nexus of natural gas and renewable energy and explore untapped complementarities and potential synergies on a number of levels. This will lead to examination of hybrid systems of natural gas and renewable technologies.

Use of natural gas and renewable energy has grown significantly in recent years. The two forms of energy appear complementary in many respects: natural gas electricity generation enjoys low capital costs and variable fuel costs, while renewable energy generators have higher capital costs but generally zero fuel costs, excluding bioenergy (see Table 8.1 for selected examples). Natural gas is a key input for corn starch-based ethanol fuel production, and new transportation infrastructure and technology experiences could enable use of both natural gas and renewable fuels in vehicles. Both forms of energy support a future orientation toward a built environment that utilizes local energy supply and use, including distributed generation and home vehicle fueling.

Despite the complementarities and potential for greater coordinated use, the natural gas and renewable energy industries have at times viewed each other as direct competitors, especially in the power sector. As of mid-2012, the primary competitive impact of inexpensive natural gas has been over 300 terawatt-hours (TWh) of fuel switching from coal- to natural gas-fired electricity since 2008. If natural gas prices remain below roughly $5/million British thermal units (MMBtu), many developers of renewable electricity projects might be hard pressed to offer competitive power purchase prices, thus limiting the number of projects deployed. Similarly, natural gas producers and biofuel producers might compete over water, especially during drought conditions [110].

It is important to identify how the natural gas and renewable energy communities might (a) promote a new systems approach to natural gas and renewable energy technologies, (b) jointly research mutually beneficial policy and market structure options, and (c) communicate with each other, and jointly to the public, to clarify misconceptions.

8.10.1 TEMPORAL FRAMEWORK

Because the occurrence of risks along an investment's lifetime varies significantly by issue and technology, it is also important to consider these risks within a temporal framework. As reported by Lee et al. [104], Figure 8.8 provides a generalized illustrative assessment of the magnitude and timescale of the major issues for natural gas and renewable energies. A more rigorous quantitative application of this framework in the analysis of specific power project options could help inform stakeholder investment decisions [104].

8.10.2 COLLABORATIVE MARKET REDESIGN

Shale gas has brought unprecedented changes to the gas industry, prompting increased efforts by regulators and stakeholders to foster efficient market-based gas operations. In recent years, changing paradigms in energy industry and the development of new renewable technologies have forced redesign of market needs. In 2012, Federal Energy Regulatory Commission [111] hosted five regional conferences on gas-electric coordination to better understand the issues faced by gas and electricity stakeholders and how they might help address these issues through new regulations or recommendations. Based on these discussions, Cochran et al. [112,113] evaluated business case for natural gas-renewable hybrid energy system.

The growth of variable renewable energy generation in regional electricity networks has led to a flurry of novel industry and regulatory approaches to handling higher levels of variable generation. The experience of these approaches, some successful and some not, and the subsequent analyses of their results have brought to light areas of operational and market inefficiencies. They have also produced a variety of recommended technical, structural, and economic solutions to improve the assimilation of variable generation at lowest cost with market efficiency. Milligan

TABLE 8.1

Issue Matrix of Selected Natural Gas and Renewable Energy Characteristics [104]

Issues	Natural Gas Power	Wind	Solar	Bioenergy
Resource distribution	Relatively diverse for unconventional supplies; less so for conventional	Diverse but often far from load centers	Diverse but best in Southwest	Diverse but best in Midwest and Southeast
Capital cost	Low, stable	Moderate, some fluctuation	Relatively high, declining	Moderate-high, stable (early generation biofuels)
Fuel cost	Variable but currently low	None	None	Moderate
Output	Dispatchable; Variable	Variable and mostly	Dispatchable power flexible	Somewhat predictable and fuel predictable
Carbon impact	Most recent life cycle assessments conclude that both conventional and unconventional less than half that of coal	Very low	Very low	Depends; corn starch-based may be slightly less than gasoline
Environmental and social concerns	Some opposition to hydraulic fracturing; relatively clean-burning fossil fuel	Some opposition to siting; no combustion emissions; low water use	Some opposition to siting of large projects for ecosystem reasons; no combustion emissions or water use for PV	Concern over ecosystem impacts for many biofuels, water use

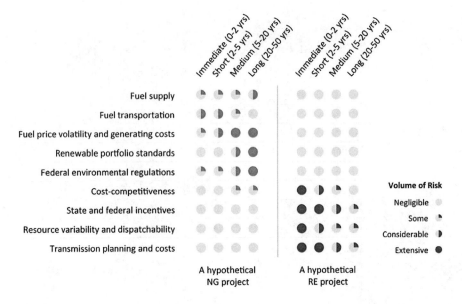

FIGURE 8.8 Illustrative framework for evaluating investment options by risk source, magnitude, and timescale [104].

et al. [114,115] has presented an in-depth discussion of operational and institutional challenges of integrating high levels of renewable energy in natural gas.

There are six major market design issues facing natural gas and renewable energy electricity generators for hybridization of natural gas and renewable sources [104]. These are (a) harmonization of day-ahead natural gas and electricity scheduling logistics, (b) flexible natural gas pipeline service options, (c) implementation of natural gas pipeline expansion and local gas storage facilities, (d) valuation of flexibility, (e) electricity dispatch timing, and (f) balancing electricity generation and load over larger geographic areas. These issues and their implications are discussed in detail by Lee et al. [104].

The pursuit of future energy systems that can meet electricity demands while supporting the attainment of societal environment goals, including mitigating climate change and reducing pollution in the air, has led to questions regarding the viability of continued use of natural gas. Natural gas use, particularly for electricity generation, has increased in recent years due to enhanced resource availability from nontraditional reserves and pressure to reduce greenhouse gasses (GHG) from higher emitting sources, including coal generation. While lower than coal emissions, current natural gas power generation strategies primarily utilize combustion with higher emissions of GHG and criteria pollutants than other low-carbon generation options, including renewable resources. Furthermore, emissions from life cycle stages of natural gas production and distribution can have additional detrimental GHG and air quality (AQ) impacts. On the other hand, natural gas power generation can play an important role in supporting renewable resource integration by (a) providing essential load balancing services and (b) supporting the use of gaseous renewable fuels through the existing infrastructure of the natural gas system. Additionally, advanced technologies and strategies including fuel cells and combined cooling heating and power (CCHP) systems can facilitate natural gas generation with low emissions and high efficiencies. Thus, the role of natural gas generation in the context of GHG mitigation and AQ improvement is complex and multifaceted, requiring consideration of more than simple quantification of total or net emissions. If appropriately constructed and managed, natural gas generation could support and advance sustainable and renewable energy.

Natural gas is unique among fossil fuels with regard to the benefits of complementing renewable resource integration. These benefits include (a) the dynamic ramping ability of modern natural gas power generation methods; (b) low criteria pollutant and GHG emissions relative to other fossil fuel generation methods, particularly during dynamic operation; (c) technical ease with which renewable fuel alternatives such as biogases can be substituted; and (d) potential for transition to 100% renewable fuel (e.g., biogas, renewable hydrogen) injection, storage, and delivery in the future.

8.10.3 PERSPECTIVES ON LOW- AND ZERO-EMISSION HYBRID GENERATION

Increasing the capacity of renewable resources will require overcoming challenges associated with electricity system incorporation. The existing structure of the U.S. and regional electrical grids and the nature of renewable resources (e.g., many with characteristic intermittencies, uncontrollability, etc.) cause difficulties for managing load balancing and other key operational parameters that can result in undesirable outcomes, e.g., enhanced power curtailment, increased costs, and large required installed capacities [116]. This is particularly true regarding the integration of high levels of wind and solar generation in terms of systems-level operation [116,117]. Wind and solar resources are characteristically intermittent at multiple timescales (e.g., hourly, daily, seasonally), and complementary strategies are needed to ensure satisfactory systems-level operation most notably balancing generation with demand. Grid operation must accommodate all load demand conditions irrespective of the availability of renewable power and, due to the rapidity at which intermittent resources come online and/or dropout, additional reserve capacity and ramping capabilities must be constantly available [118]. As a result, high penetration of renewable power dictates rapid responses by controllable generators (mostly fossil fueled today) and energy storage or other complementary technologies. Regardless of the amount of each complementary technology deployed, it is likely that high penetration of wind and solar power will lead to greater dynamic operation of existing fossil generators.

In California, natural gas power plants are most often used to provide the complementary generation needed to balance renewables and load including those with load-following or peaking capabilities [119]. In other regions of the US, such capabilities may be required from generators with higher emission rates than natural gas plants, including coal power plants. Natural gas load following and peaker plants are often simple cycle steam or combustion gas turbines with lower efficiencies and higher emissions than baseload plants [120]. Natural gas baseload generation is often provided by higher efficiency and lower emitting natural gas combined cycle (NGCC). Consequences of wind and solar integration can include increased load following and peaking generation accompanied by a decrease in generation from base-load generators visible in the modeled generator-level impacts from increasing integration of wind and solar resources in the California grid. Additionally, increasing levels of curtailed generation are observed that could have economic and energy consequences.

The system-wide impacts related to intermittencies of renewable integration can have unwanted and unforeseen GHG and pollutant emission impacts, in addition to the positive benefits of offsetting direct emissions from fossil generators [121]. Emission rates of GHG and pollutants per unit energy of electricity from natural gas power plants increase during dynamic operation events including cycling, ramping, and start/stop conditions [122,123]. Ramping has deleterious impacts on plant efficiency (and thus emissions) by two mechanisms: (a) heat rates are higher for ramping resources at full load relative to units designed to be operated at fixed levels of output and (b) the operation of gas plants at partial load results in a higher heat rate relative to full load operation [124]. All of the described stages of dynamic operation (i.e., start/stop, cycling, ramping) result in pollutant and GHG emissions that otherwise would not have been generated. The large-scale electrification of energy systems could potentially exacerbate the issues previously mentioned by increasing the size and altering the shape of the load. Therefore, it is not accurate to assume that increasing the levels of intermittent renewables will have no detrimental AQ impacts as utility grid network integration of these resources requires balancing power that includes emissions from fossil generators that could potentially reduce expected emissions benefits or even yield localized increases in emissions from certain power plants [121,125,126]. Such plants are typically associated with marginal generation or the last plants to come online in the loading order. Marginal generation is typically impacted by renewable resource integration.

Future hybrid generation can be provided by advanced technologies and strategies in place of current fossil complementary generation to improve efficiencies and avoid undesirable emission impacts. Fuel cells operating on natural gas that represent potential technologies for load balancing intermittent renewables in place of natural gas load following and peaking units in California. Emissions of NO_x from NGCC are much lower than both particularly for new state-of-the-art NGCC relative to existing load following and peaking generation [127]. Load following and peaking plants may also require up to several hours to start or stop and therefore often must be run (i.e., cycled) to be ready for any renewable generation drop off. Contrastingly, if available, renewable generation increases then the available output from gas generators may be unutilized even though they continue to operate. Therefore, increasing the amount of total generation from load following and peaking plants at the expense of baseload plants can result in additional GHG and pollutant emissions [128]. In contrast, fuel cell generation results in negligible emissions relative to other devices including combustion turbines, microturbines, reciprocating engines, and NGCC. Technologies utilizing combustion (i.e., reciprocating engines, combustion turbines, microturbines) can also operate with lower emissions than load-following and peaking generation but may require pollutant control technologies to achieve minimum emission limits. For example, microturbines must be integrated with CCHP to achieve reductions in GHG and must be integrated with selective catalytic reduction (SCR) technology to achieve low criteria pollutant emissions [129–131].

Fuel cells can deliver peaking or intermediate load-following service, which can prevent the need for new transmission and distribution infrastructure and provide peaking capacity in emissions and/or electricity infrastructure constrained areas. Fuel cells have been evaluated for providing grid support in energy systems with very high levels of intermittent resources [132] including a 100% renewable energy system in support of climate mitigation [133]. Future fuel cell or fuel cell hybrid systems can provide load following capabilities in tandem with low emissions with proper system and control configurations [134]. Clusters of 10–100 MW scale fuel cells installed at distribution substations have been proposed as a method of supporting high penetrations of intermittent renewables through the provision of baseload and load following services with very low NO_x and CO_2 [119]. Fuel cells can also be integrated in the residential sector to support and complement solar PV deployment [135].

An additional distinction of fuel cell systems is the ability to provide hydrogen fuel as an output when operating on hydrocarbon fuels. This allows fuel cells to operate as trigeneration systems producing electricity, heat, and hydrogen [136]. Incorporating hydrogen production further increases the energy efficiency of the system and provides additional energy benefits including the potential production of fuel for zero-emission hydrogen fuel cell electric vehicles. An important pathway for GHG and AQ cobenefits includes the operation of trigeneration fuel cell systems on biogas as this allows for a means of coupling very low or even net negative emission strategies in the power generation sector to the transportation sector [137].

Hybrid fuel cell-heat engine plants integrate a high-temperature fuel cell (solid oxide fuel cell [SOFC] or molten carbonate fuel cell [MCFC]) with a heat engine (e.g., gas turbine, reciprocating engine) to achieve even higher efficiency than a fuel cell alone (converting fuel cell heat to useful work) [138]. By utilizing energy synergies between the fuel cell and heat engine (e.g., fuel cell waste heat is turned into useful electrical work and compression power, and higher pressure operation increases fuel cell electricity production efficiency), enhanced performance is achieved [139]. These emerging power plants are being developed by several manufacturers and have been shown to achieve very high electrical efficiencies [140,141] with ultralow emissions even at distributed power sizes [121] and with dynamic dispatch characteristics [142,143].

Power conditioning inverters in fuel cell systems that are needed to transform DC electricity into AC can be used for system power factor correction and voltage support. The load-following capabilities of fuel cells are particularly important in regions with projected increases in intermittent wind and solar power generation that will necessitate increasing amounts of clean, efficient, load-following power generation. This would maximize the GHG reductions of renewable resources while limiting any negative effects on regional AQ that the renewable intermittencies would otherwise introduce. Therefore, emissions must be considered at the systems-level rather than simply comparing life cycle or direct emissions at the point-of-generation. For example, while it is certain that renewables must play a central role in a sustainable energy system, large-scale integration in regional energy systems is not a simple matter. This is particularly true for wind and solar power, which are likely to comprise the largest portion of future renewable capacity in most regions around the world with potential unforeseen electrical grid impacts including sites of emissions increases leading to AQ worsening. For example, emissions of NO_x and VOC from conventional load following and peaking generation could contribute to increases in ground-level ozone concentrations in regions adjacent to power plants. Similarly, direct and secondary emission impacts could drive increases in ground-level concentrations of PM. Thus, in the absence of enhanced flexibility and intelligence of the grid and the presence of low or zero-emitting and controllable complementary strategies, high levels of renewable integration could yield some localized worsening in regional AQ. This must be considered in the context that overall AQ impacts of high renewable power use will be largely beneficial as a result of net emission reductions compared to fossil generation. Therefore, the worsening described herein is directly referencing localized areas around impacted fossil generators that would be forced to operate in a highly dynamic fashion to complement renewable intermittencies. This scenario could have particular importance in many urban areas due to high population exposure and for regions currently experiencing challenges meeting ambient AQ standards. Similarly, the GHG impacts of renewables will almost certainly result in net reductions in emissions but the total reduction could be increased and AQ benefits preserved if emissions from complementary generation could be avoided or minimized.

8.10.4 ROLE OF STORAGE IN HYBRID GENERATION

In terms of providing hybrid generation, energy storage technologies are a key strategy in support of renewable hybridization. Energy storage, which also represents a complementary option for balancing intermittent resources, also can potentially reduce GHG emissions from natural gas by 21%–98% with life cycle emissions ranging from 6 to 292 gCO_2/kWh [144]. A wide range of technologies are available or under consideration for the support of renewable energy, with a comprehensive overview provided in Ref. [145]. It is well understood that energy storage has the potential to support renewable energy and provide environmental benefits [146]. Energy storage devices are more responsive than conventional generators and costs are becoming more competitive with other sources of generation [147]. Further, they can provide emission benefits as many have very low to no emissions during operation, e.g., batteries, flywheels. Energy storage can then have GHG and AQ benefits by reducing emissions associated with load balancing and spinning reserve via replacement of required fossil fuel backup generators [148,149]. Energy storage can also reduce emissions by time-shifting loads away from peak demand, reducing generation from peaking and load following plants in place of base-load generation [150]. The following pathways are recommended for providing dispatchable power to support renewables [151] and would likely provide very low to zero-emission complementary generation with AQ and GHG benefits.

1. Expand the use of high-capacity batteries including the development and commercialization of advanced technologies such as solid-state and flow batteries, supercapacitors, and super conducting magnetic energy storage.
2. Develop high-capacity compressed air storage resources including underground caverns where geologically and technically feasible.
3. Expand the use of thermal energy storage mechanisms including molten salts, solids, and other high-heat capacity media, or via phase change.

4. Construct high-capacity pumped-hydro storage facilities where feasible.
5. Develop large-scale flywheels for direct storage and recovery of kinetic energy.

However, large-scale deployment of energy storage faces barriers including, among others, economics, regulatory and utility structures, and siting of facilities [152,153]. These barriers complicate an understanding of when and how the very large capacities needed to match increasing levels of renewables will become available. Therefore, while providing complementary generation from advanced strategies, including energy storage, represents an important and valuable long-term goal, natural gas generation is currently the most feasible method of balancing fluctuating renewable power with load and may continue to be so in the near- to mid-term.

Further, energy storage may not always be the environmentally preferred option for providing reserves if not managed properly, even if it is assumed to have no operational emissions [154]. For example, if energy storage enables system operators to use lower cost generation with higher emissions that would otherwise not be possible due to reserve requirements [155]. Analyzation of adding energy storage to a power system concluded emission impacts were highly case-dependent [156]. In systems with high renewable penetration levels and significant curtailment, energy storage did provide emission reductions. However, in other systems, emission impacts were reported to be positive, neutral, or negative. It should be noted that the most appropriate comparison for the discussion here is the high renewable integration scenario and the use of otherwise curtailed power. In such a case, energy storage is likely to provide emission reduction benefits by making available renewable electricity to meet demand otherwise satisfied by dispatchable generation.

Additional options for low environmental impact hybrid generation includes Smart Grid technologies, demand response, clean dispatchable power generation, and other strategies including transportation sector integration through vehicle-to-grid services. Smart Grid technologies offer benefits relative to the current grid, including improving power quality and reliability, reducing costs, improving efficiency and conservation, facilitation of increased renewable resources, and advanced technology penetrations, and enabling enhanced supply- and demand-side energy management [157]. Implementation of Smart Grid technologies can have important direct GHG and pollutant emissions benefits by improving systems-level efficiency, thus reducing levels of necessitated generation [158,159]. Vehicle-to-grid strategies also can provide emission benefits, essentially functioning as a form of battery energy storage [160]. As with other forms of energy storage, these strategies will require further advancement, both techno-economically and in social acceptance, but represent an optimal method of providing complementary generation in place of fossil fuels.

8.10.5 Low-Carbon Renewable Fuel Storage and Transmission

An additional avenue of support for renewable resources in pursuit of GHG and AQ reductions includes the use of existing natural gas infrastructure to transport, store, and distribute renewable gaseous fuels with two prominent examples including biogas and hydrogen sourced from renewable resources. Within the US, a robust and established infrastructure network exists to support the natural gas industry, including greater than 300,000 miles of transmission pipelines, gathering systems, storage sites, processing plants, and distribution pipelines [161]. Gaseous fuels can be upgraded and injected into existing natural gas infrastructure to provide a source of flexible fuel which can be utilized in various end-use applications [162,163]. The use of renewable gaseous fuels as energy resources is desirable due to potentially net-zero GHG intensities, reduced pollutant emissions, and additional energy and environmental benefits [164]. In addition to directly providing fuels for energy services, use of low-carbon pipeline fuels can provide additional emission reductions, e.g., avoidance of emissions associated with the construction or expansion of transmission infrastructure for hydrogen or electricity. The use of renewable gaseous fuels can further support and enhance other forms of renewable energy including renewable power generation from intermittent resources. Relative to renewable resources and electrification alone, the addition of low-carbon gaseous fuels distributed via existing natural gas pipelines is complementary in meeting future GHG mitigation goals by [165]:

1. Assisting in reducing emissions from sources that experience greater challenges to electrifying (e.g., industrial process heating, heavy duty vehicles, aircraft).
2. Supporting intermittent renewable power integration by providing a form of energy storage amenable to large-scale energy magnitude and long-duration storage to balance load and demand.
3. Offsetting the need for new infrastructure (e.g., electric transmission wires) construction by enabling continued use of exiting gas pipelines.
4. Enabling widespread use of the renewable low-emissions fuel in all sectors of the economy due to the pervasiveness of the natural gas grid.
5. Helping reduce technological risk and improve flexibility for decision-makers.

Therefore, the natural gas system inherently possesses features that are, and will be, valuable to ultimate sustainability, perhaps offering the only technically feasible option (and certainly one of the most cost-effective options) for achieving massive and long-term storage of renewable electricity, and achieving 100% emissions-free energy conversion in all sectors of the economy and especially the challenging sectors (e.g., heavy duty transport and industry).

8.10.6 RENEWABLE FUEL INJECTION IN THE GRID

The potential for increased biogas resources is significant, e.g., biogas generation could potentially be expanded to provide 3%–5% of the total domestic natural gas market [166]. Following clean-up for compounds including siloxanes and hydrogen sulfide, biogas is notably similar to fossil natural gas in chemical composition (i.e., a renewable biological source of methane) enabling direct interchangeability in the natural gas system for sources of demand. Upgraded biogas can readily be injected into the existing natural gas system, facilitating the storage, transport, and distribution of a renewable gaseous fuel [167]. This further allows for biogas to be transported to numerous possible end-uses with potential for higher value than that which is available only local to the biogas resource sites [168]. For example, pipeline biogas could supply a portion of resources utilized by a centralized NGCC with benefits of scale including reduced emissions [169] and potentially lower cost relative to generation on-site at resource locations, although economics and emission impacts of biogas utilization strategies are complex and depend on many factors [164,166,170]. However, the expanded use of biogas in the U.S. would benefit from nuanced policies and programs supporting site-specific evaluations and actions, as opposed to generic approaches [164].

Biogas is generally 55%–65% methane with the remainder composed of CO_2, fractions of water vapor, traces of hydrogen sulfide and hydrogen, and other contaminants including siloxanes. An important step in facilitating long distance transport from an economic and energy sensibility standpoint is increasing the energy content [171]. Additionally, attaining pipeline quality gas for injection represents a techno-economic hurdle for biogas at present. Bio-gas composition and quality differ across sources of origin (e.g., landfills, manure digesters, gasifiers, etc.) with implications for various end-uses [172]. This requires site- and case-specific selection of upgrading technologies dependent on factors including product purity and impurities, methane recovery and loss, upgrading efficiency, and investment and operating costs [173]. Regulatory and technical standards associated with the acceptance of biogas into pipelines differ by state and country. For example, California gas contaminant standards for investor-owned utility gas are comparable to other States and have been found to be reasonably achievable using conventional gas clean-up technologies [174]. Contrastingly, minimum energy content standards are higher than other locations, and the majority of conventional and emerging biogas upgrading technologies may not provide acceptable gas [174]. Additionally bio-gas cleaning and upgrading costs can be high, resulting in the economic unfeasibility of pipeline injection for low-quantity biogas producers operating small anaerobic digestion systems. To address this issue, the following recommendations were put forth in a report to the California Energy Commission as an example of potential research and regulatory steps in support of biogas utilization and gas grid injection [174]:

1. Reduce the energy content requirements for pipeline biomethane from 990 to 960–980 Btu/scf on a HHV basis.
2. Collect data on concentrations of contaminants of concern in current California natural gas supply including instate and imported sources.
3. Address costs and provide financial support and incentives for biogas upgrading and pipeline interconnection and for small-scale DG systems.
4. Develop a streamlined application process (e.g., standardized forms, agreements) to minimize time and resources spent by all parties.

Hydrogen can be produced with very low emissions from a diverse selection of renewable pathways including by electrolysis from electricity generated by wind and solar technologies, the processing of biogas resources, and via direct solar water splitting in a process called artificial photosynthesis [175]. Although not renewable, hydrogen could also be produced from additional low-carbon pathways including cogeneration of electricity and hydrogen via high-temperature nuclear reactors and fossil pathways with the inclusion of carbon capture and sequestration (CCS), e.g., coal gasification and SMR of natural gas [176]. Even the most common current method for producing hydrogen, SMR of natural gas, provides a relatively low GHG and AQ emissions pathway to produce a fuel that has no GHG and no AQ emissions in its end-use in a fuel cell, for example [177]. While current hydrogen production methods generally rely upon fossil pathways, progressive shifts toward renewable and other low-carbon strategies can allow for the production of hydrogen in increasingly sustainable methods.

A notable hydrogen strategy for both the production of very low-carbon fuels and the support of intermittent renewable integration is the electrolysis of water via otherwise curtailed renewable electricity (most often wind and solar power), a concept often referred to as power-to-gas (P2G) [178–180]. Generally, the P2G strategy is a means of linking the electrical grid and the natural gas grid by converting electricity (typically surplus renewable electricity) into a grid compatible gaseous fuel [181].

Excess available wind and solar power can be fed to battery energy storage for short-term storage, or sent to an electrolyzer to produce renewable hydrogen, which better enables long-term and massive energy storage. Produced hydrogen can then be stored in dedicated facilities and used on-site when needed in stationary and mobile applications achieving emission reductions. For example, hydrogen can be used to power fuel cell electric vehicles with negligible pollutant emissions in place of petroleum-fueled vehicles or utilized in stationary fuel cells to efficiently generate on-site power and heat. Hydrogen may also be injected into the natural gas system directly, or reacted with an external carbon source in a methanation reaction to produce renewable methane. Similarly to hydrogen, this renewably produced methane can be injected into existing natural gas storage and distribution grids, used on-site as a stationary generation or vehicle fuel, or utilized in other established natural gas end-uses.

An important ability of P2G is the potential to support the deployment of electrical systems with very high levels of renewable electricity (e.g., 85%) by providing a means of long-term energy storage potentially from the sub-megawatt to gigawatt power scale and comprising up to 10 seconds of terawatt hours of energy [182,183]. Additionally, electrolyzers utilized in P2G systems can provide grid functions including rapid demand or supply response, spinning reserve, and frequency and voltage regulation. Typically, the most commonly considered renewable energy source for P2G systems is wind and solar power that become difficult to otherwise manage at high-use levels [178]. P2G can facilitate hydrogen production during periods of excess renewable generation that would otherwise be curtailed, which assists in addressing the spatial and temporal challenges discussed for renewable energy [183]. Additional benefits of P2G in energy storage include siting flexibility, subsecond response times, and minimal adverse environmental impact on both the production and end-use of the renewable fuel, especially if the hydrogen is converted ultimately to electricity in a fuel cell for either transportation or stationary power applications [184].

The injection of hydrogen directly (i.e., without methanation) into the existing natural gas system is being considered as a means of facilitating the production, storage, and transport of large quantities of a renewable fuel that can be used as an energy source by a range of zero-emitting end-use devices [185,186]. Use of the existing natural gas system also represents a solution to the key barrier of current lack of hydrogen infrastructure [187]. However, due to the differences in chemical properties between hydrogen and natural gas, concerns exist including safety, leakage rate and dispersion, and materials degradation including embrittlement of pipeline steel [188]. The amount of hydrogen allowable in the natural gas system varies by region and typically ranges from 0% to 12% by volume [181]. However, the majority of these concerns are being actively addressed through various research and development projects around the world and it is possible that in the future hydrogen may be injected at increasingly higher concentrations as mixed with natural gas.

Methanation of hydrogen to produce synthetic methane has some advantages over simply utilizing hydrogen, namely, the ability to serve as a drop-in fuel with regard to current infrastructure and use in existing natural gas consuming technologies. Essentially, synthetic methane is completely usable in the existing, technically mature natural gas infrastructure, e.g., can be feasibly injected into existing natural gas grids without many of the challenges associated with hydrogen [189]. This results in economic benefits as no new investments are required for transport, storage, and utilization, but also is beneficial in avoiding issues of obtaining permission from authorities and general public acceptance [190]. Different methods for methanation include both biological and catalytic rectors with implications for achievable gas quality, reactor volume, and the complexity of process technology required [181].

Major technical and economic barriers exist for P2G that must be overcome, however. For example, though water electrolysis has existed commercially for several decades, only approximately 4% of global hydrogen supply is derived from such pathways [191]. Considering electrolysis, improvements in efficiency during transient operation and reductions in capital cost are needed [181]. Currently, alkaline electrolysis is the lowest cost and most reliable technology, but has limited efficiency improvement and cost reduction potential. proton-exchange membrane (PEM) electrolysis can offer improved transient performance and have a simple modular design, but cost reduction and technical improvement including membrane lifetime of PEM electrolyzers is needed. Similarly, solid oxide electrolysis can offer the benefits of very high efficiencies but will require further technical development prior to commercialization including increased lifetime [192]. If hydrogen storage is utilized, this can also contribute to high costs [193]. Methanation as a viable commercial step would benefit from techno-economic advancement including an improved understanding of dynamic reactor operation, advancement in cool fixed-bed reactors, and others [181,194].

An additional issue associated with injection of both biogas and hydrogen into the existing natural gas supply is the potential impact on end-use combustion performance [194]. Current end-use devices (e.g., turbines, engines, industrial burners, residential and commercial appliances) are optimized for use on pure natural gas. Altering the composition of the supplied fuel could result in alterations to key parameters including heating value, Wobbe-index, knock phenomena, flame stability, blow off limits, and flashback [194]. Given the scope of this review, perhaps the most important concern is associated with potential changes in emissions. Impacts on criteria pollutants will vary for systems depending on many factors including gas composition, end-use device geometry, operational parameters, presence of clean-up technologies, etc. One concern is increases in NO_x from the addition of hydrogen, although this could also reduce hydrocarbon and CO_2 emissions [195]. However, if device operation is controlled properly, NOx emissions could be maintained at equivalent levels [196,197] or even reduced due to leaner combustion [198]. Impacts may also be multifaceted with regard to different pollutants. For example, adding hydrogen to natural gas had a range of impacts on NO_2, NO, and N_2O emissions from an industrial boiler, including inverse relationships between species for some operating conditions [199]. These results highlight the complicated nature and lack of current understanding of emission impacts from end-use devices due to the blending of renewable fuels. Further studies are needed for stationary power generation technologies under a range of operating conditions and different renewable fuel blends. Studies should also be undertaken elucidating the AQ impacts from economy-wide scenarios of renewable gas blending in future years using atmospheric modeling. Results from such assessments can provide insights into end-use device management to minimize AQ degradation, potentially through policy mechanisms and other programs.

In summary, natural gas generation can support transitions to renewable resources by (a) use in advanced conversion devices to provide complementary (hybrid) grid services efficiently and with very low emissions to maximize the benefits of intermittent renewable resources, and (b) natural gas generation and the existing natural gas system can support the use of renewable gaseous fuels with high energy and environmental benefits. This is because advanced conversion devices (including fuel cells [200]) can operate on natural gas and renewable gaseous fuels, including blends, with very low direct pollutant emissions and the existing natural gas system can support the production, storage, and distribution of renewable gaseous fuels. The ability of hydrogen to be produced via a range of pathways, including from multiple renewable feedstocks, and to be used by high-efficiency and zero-emitting fuel cell devices in a diverse a range of applications across all economic sectors (e.g., transportation, power generation, industry, the built environment) raises the possibility of energy systems with essentially produce negligible emissions of both GHGM and pollutants [201]. Advanced conversion technologies can thus provide renewable energy pathways that achieve very high GHG and AQ co-benefits [202].

Hybrid generation for renewable balancing should focus on utilizing advanced NGCC in the near-term, as these are commercially available, have favorable economics, and reasonably low emissions relative to other fossil options. Additionally, advanced low-emitting peaker plants exist that can significantly reduce emissions from current peaking generation. In the mid- to long-term, hybrid generation for renewable resources should be accomplished via advanced energy storage to the maximum possible extent, with fuel cells and fuel cell hybrid systems providing the remainder. Additionally, smart grid and related strategies should again be pursued with high priority due to the synergistic and beneficial interactions with integrating higher levels of renewable generation, particularly those with intermittencies. A research focus is needed for developing advanced complementary strategies with low to zero emissions including fuel cell systems such as those described in Reference [119], P2G systems, advanced energy storage that can technically and economically provide terawatt-hour-scale storage, and advancement of other load management services including vehicle-to-grid, demand response, and others. Due to the inherent energy and environmental benefits that are possible, use of biogas resources for energy (rather than no collection or flaring) should be targeted with priority [203]. In the near-term, gas clean-up and injection into the existing gas grid should be pursued as a means of immediately reducing the carbon intensity of the pseudo-hybrid gas system while avoiding permitting challenges.

REFERENCES

1. International Energy Agency. (2017). *World Energy Outlook 2017*. IEA.
2. Halabi, M., Al-Qattan, A. and Al-Otaibi, A. (2015). Applications of solar energy in the oil industry: Current status and future prospects. *Renewable and Sustainable Energy Reviews* 43: 296–314.
3. Wesoff, E. (2015). GlassPoint is building the world's largest solar project in an Omani oil field. A website publication of *Greentech Media*. (July, 8, 2015).
4. IPCC. (2018). Summary for policymakers. In *Global Warming of 1.5°C: An IPCC Special Report on the Impacts of Global Warming of 1.5°C Above Pre-Industrial Levels*, Cambridge University Press, 1–35.
5. Porter, S., Dickson, D., Hardin, K. and Shattuck, T. (2020). Oil, gas, and the energy transition -How the oil and gas industry can prepare for a lower-carbon future. A website report by Deloitte insight, Deloitte Co. UK.
6. Kitasei, S. (2010). *Powering the Low-Carbon Economy: The Once and Future Roles of Renewable Energy and Natural Gas*. Worldwatch Institute, Washington, DC; ICF International. (2011). *Firming Renewable Electric Power Generators: Opportunities and Challenges for Natural Gas Pipelines*. Submitted to the INGAA Foundation. ICF International.
7. Ericson, S., Engel-Cox, J. and Arent, D. (2019, January) Approaches for integrating renewable energy technologies in oil and gas operations. The Joint Institute for Strategic Energy Analysis Golden, CO 80401 www.jisea.org, Technical Report NREL/TP-6A50-72842 Contract No. DE-AC36-08GO28308, NREL, Golden, CO.
8. Choi, Y., Lee, C. and Song, J. (2017). Review of renewable energy technologies utilized in the oil and gas industry. *International Journal of Renewable Energy Research* 7: 592–598.
9. BOEM. (2017). Bureau of Energy and Ocean Management. Accessed 2017. https://www.boem.gov/.
10. Tripathi, V. and Brandt, A. (2017). Estimating decades-long trends in petroleum field energy return on investment (EROI) with an engineering-based model. *Plos One* 12: e0171083.
11. Freise, J. (2011). The EROI of conventional Canadian natural gas production. *Sustainability* 3(11): 2080–2104.
12. Hirtenstein, A. (September 2017). Big oil becomes greener with progress in cutting pollution. A website report by Bloomberg. New York, NY.
13. Bluestein, J., Mallya, H. Yandoli, L., Polchert, M. and Amarin, N. (2015). Methane emissions from the oil and gas industry: "Making Sense of the Noise". A website report by ICF International, Atlanta, Georgia.

14. Lazard. (2017). Lazard's Levelized Cost of Energy Analysis—Version 11.0. Lazard.
15. Sandler, J., Fowler, G., Cheng, K. and Kovscek, A. (2014). Solar-generated steam for oil recovery: Reservoir simulation, economic analysis, and life cycle assessment. *Energy Conversion and Management* 77: 721–732.
16. Korpas, M., Warland, L., He, W. and Tande, JO. (2012). A case-study on offshore wind power supply to oil and gas rigs. *Energy Procedia* 24: 18–26.
17. Greenblatt, J. (2015). *Opportunities for Efficiency Improvements in the U.S. Natural Gas Transmission, Storage and Distribution System*. Ernest Orlando Lawrence Berkeley National Laboratory, Berkeley, CA.
18. Liss, W. and Rowley, P. (2018). *Assessment of Natural Gas and Electric Distribution Service Reliability*. Gas Technology Institute, Des Plaines, IL.
19. REN21. (2017). *Renewables 2017 Global Status Report*. REN21, Paris.
20. El-Katiri, L. and Fattouh, B. (2017). A brief political economy of energy subsidies in the middle east and North Africa. *International Development Policy* 7: 57–87. Doi: 10.4000/poldev.2267.
21. Fattouh, B. and El-Katiri, L. (2012). *Energy Subsidies in the Arab World*. United Nations Development Program, New York, NY.
22. Sovacool, B. (2017). Reviewing, reforming, and rethinking global energy subsidies: Towards a political economy research agenda. *Ecological Economics* 135(135): 150163.
23. Markussen, C. (2014, May 8) DNV GL oil & gas, introducing hybrid power systems for offshore units. A website report Christian.Markussen@dnvgl.com.
24. Abramova, A., Abramov, V., Kuleshov, S. and Timashev, E. (January 2014). Analysis of the modern methods for enhanced oil recovery. *Energy Science and Technology* 3: 118–148. Doi: 10.13140/2.1.2709.4726.
25. Veld, K. and Phillips, O. (2010). The economics of enhanced oil recovery: Estimating incremental oil supply and CO_2 demand in the powder river basin. *Energy Journal* 31 (3): 31–55.
26. Quinlan, E., van Kuilenburg, R., Williams, T. and Thonhauser, G. (2011). The impact of rig design and drilling methods on the environmental impact of drilling operations. Presented at the AADE National Technical Conference and Exhibition, Houston, Texas, 12-14 April. AADE-11-NTCE-61.
27. Clark, C.E., Han, J., Burnham, A., Dunn, J.B. and Wang, M. (2011). *Life-Cycle Analysis of Shale Gas and Natural Gas*. Argonne National Laboratory, Lemont, IL.
28. Endurthy, A., Kialashaki, A. and Gupta, Y. (2016). *Solar Jack Emerging Technologies Technical Assessment*. Pacific Gas and Electric Company, San Francisco, CA.
29. Healing, D. (2015). Producer builds solar array to power nodding prairie Pumpjack. A website report by Calgary Herald. (Oct. 7, 2015) Canada.
30. Feller, F. (2017). WIN-WIN—wind-powered water injection. *Offshore Mediterranean Conference and Exhibition*. Ravenna, Italy.
31. Musial, W., Beiter, P., Schwabe, P., Tian, T. Stehly, T. and Spitsen, P. (2017). *2016 Offshore Wind Technologies Market Report*. U.S. Department of Energy, Washington, DC.
32. Zhong, M. and Bazilian, M. (2018). Countours of the energy transition: Investment by international oil and gas companies in renewable energy. *The Electricity Journal* 31: 82–91.
33. Sandler, J., Fowler, G., Cheng, K. and Kovscek, A. (2012). *Solar-Generated Steam for Oil Recovery: Reservoir Simulation, Economic Analysis, and Life Cycle Assessment*. Society of Petroleum Engineers.
34. Wang, J., O'Donnell, J. and Brandt, A. (2017). Potential solar energy use in the global petroleum sector. *Energy* 118: 884–892.
35. Moritis, G. (2011). Chevron starts California demo of solar-to-steam enhanced recovery. *Oil & Gas Journal* 109: 86–88.
36. Wesoff, E. (2015). *GlassPoint is Building the World's Largest Solar Project in an Omani Oil Field*. Greentech Media.

37. Xin, S., Liang, H., Hu, B. and Li, K. (2012). Electrical power generation from low temperature co-produced geothermal resources at Huabei Oilfield. *Thirty-Seventh Workshop on Geothermal Reservoir Engineering.* Stanford, CA.

38. Augustine, C. and Falkenstern, D. (2012). An estimate of the near-term electricity generation potential of co-produced water from active oil and gas wells. *GRC Transactions* 36: 187–200.

39. Ziabakhsh-Ganji, Z., Nick, H.M., Donselaar, M.E. and Bruhn, D.F. 2018. Synergy potential for oil and geothermal energy exploitation. *Applied Energy* 212: 1433–1447.

40. Halabi, M.A. Qattan, A.A and Otaibi, A.A. (2015). Application of solar energy in the oil industry-Current Status and future prospects. *Renewable and Sustainable Energy Reviews* 3: 296–314.

41. DOGGR. (2008). Annual report of the state oil and gas supervisor. California Department of Conservation Division of Oil, Gas, and Geothermal Resources, 2009. Available at: ftp://ftp.consrv.ca.gov/pub/oil/annual_reports/2008/PR0 6_Annual_2008. pdf, Accessed 17 December 2016, a website report.

42. Waldner, E. Powered by sunshine. Available at: http://www.bakersfield.com/news/ business/2006/02/10/powered-by-sunshine.html, Accessed 17 December 2016, a website report.

43. Nelder, C. Heavy oil of the kern river oil field. Available at: http://www.getreallist.com/ heavy-oil-of-the-kern-river-oil-field.html, Accessed 17 December 2016.

44. KYOCERA Solar. Oil and Gas remote solar power systems. Available at: https:// www.emarineinc.com/pdf/Kyocera/kyocera_oil_gas_remote_solar_power_systems. pdf,Accessed 17 December 2016.

45. Popoola, L.T., Grema, A.S., Latinwo, G.K., Gutti, B. and Balogun, A.S. (2013). Corrosion problems during oil and gas production and its mitigation. *International Journal of Industrial Chemistry* 4: 1–15.

46. Wang, Y., Duan, M. and Shang, J. (2009). Application of an abandoned jacket for an offshore structure base of wind turbine in Bohai heavy ice conditions. *Proceedings of the International Offshore and Polar Engineering Conference*, Japan, 384.

47. Hua, Z. (2009). Applied research on offshore oil field wind energy. Northeast Asia Petroleum Forum 2009, Japan. Available at: http://eneken.ieej.or.jp/seminar/ hokuto/2009/7-33.pdf, Accessed 17 December 2016.

48. Barrel Full, A. Beatrice oil field. Available at: http://abarrelfull.wikidot.com/beatrice-oil-field, Accessed 17 December 2016.

49. Hi Energy, Talisman Beatrice project. Available at: http://www.hi-energy.org.uk/ HI-energy-Explore/talisman-beatrice-project.htm, Accessed 17 December 2016, a website report.

50. Addison, V. Wind-powered pumps could lower offshore costs. Available at: http:// www.epmag.com/wind-powered-pumps-could-lower-offshore-costs-824466#p=full, Accessed 17 December 2016.

51. Nilsson, D. and Westin, A. (2014). *Floating Wind Power in Norway Analysis of Future Opportunities and Challenges* MS Thesis, Lund University, Lund, 1–153.

52. U.S. DOE. Geothermal power/ oil and gas coproduction opportunity. Available at: http://energy.gov/sites/prod/files/2014/02/f7/gtp_coproduction_factsheet.pdf, Accessed 17 December 2016.

53. U.S. DOE. Geothermal energy production with co-produced and geopressured resources. Available at: http://www.nrel.gov/docs/fy10osti/47523.pdf, Accessed 17 December 2016.

54. Blodgett, L. Oil and gas coproduction expands geothermal power possibilities. Available at: http://www.renewableenergyworld.com/articles/2010/07/oil-and-gas-coproduction-expands-geothermal-power-possibilities.html, Accessed 17 December 2016.

55. Williams, T., Johnson, A.L., Popovich, N. and Reinhardt, T. (2012). Operational results for geothermal co-production of electricity from oilfield operations. *AAPG Annual Convention and Exhibition*, California, 1–19.
56. Quenneville, G. Oil and gas production declines in NWT. Available at http://www.nnsl.com/business/pdfs/oil-gas.pdf, Accessed 17 December 2016, a website report.
57. AANDC. Northern oil and gas annual report 2014. Available at: https://www.aadnc-aandc.gc.ca/DAM/DAM-INTER-HQ-NOG/STAGING/texte-text/pubs_ann_ann2014_1431442627961_eng.pdf, Accessed 17 December 2016.
58. Thompson T. and Dunn, C. Ft. Liard geothermal energy project. Available at: http://ecologynorth.ca/wp-content/uploads/2011/12/Ft.-Liard-Geothermal-Project.pdf, Accessed 17 December 2016, a website report.
59. Hussain, E. World's first commercial solar EOR project begins. Available at: http://www.arabianoilandgas.com/article-8545-worlds-first-commercial-solar-eor-project-begins/, Accessed 17 December 2016.
60. Bloomberg News. Chevron's solar-powered oil extraction begins in California. Available at: http://www.mercurynews.com/2011/10/03/chevrons-solar-powered-oil-extraction-begins-in-california/, Accessed 06 September 2016.
61. EY. Solar enhanced oil recovery: An in-country value assessment for Oman. Available at: http://www.ey.com/Publication/vwLUAssets/EY-Solar-enhanced-oil-recovery-in-Oman-January-2014/$FILE/EY-Solar-enhanced-oil-recovery-in-Oman-January-2014.pdf, Accessed 17 December 2016, a website report.
62. IPICA. Solar thermal: Downstream upstream. Available at: http://www.ipieca.org/resources/energy-efficiency-solutions/power-and-heat-generation/solar-thermal/, Accessed 17 December 2016.
63. Glass Point Solar. Glasspoint unveils first commercial solar enhanced oil recovery project. Available at: https://www.glasspoint.com/glasspoint-unveils-first-commercial-solar-enhanced-oil-recovery-project/, Accessed 17 December 2016, a website report.
64. Gastli, A., Charabi, Y. and Al-Maamari, R. Potential of solar energy applications in Oman's oil industry. *The 20th Joint GCC-Japan Environment Symposium*, 22–24 November 2011, UAE, 1–24.
65. Federal Energy Regulatory Commission. (2015). *Energy Primer: A Handbook of Energy Market Basics.* FERC, Washington, DC.
66. Pharris, T.C. and Kolpa, R.L (2007). *Overview of the Design, Construction, and Operation of Interstate Liquid Petroleum Pipelines.* Argonne National Laboratory, Lemont, IL.
67. EIA. 2018. *Layer Information for Interactive State Maps.* Energy Information Administration, https://www.eia.gov/maps/layer_info-m.php, Accessed 15 April 2018
68. Elson, A., Tidball, R. and Hampson, A. (2015). *Waste Heat to Power Market Assessment.* ICF International, Atlanta.
69. Hedman, B. (2008). *Waste Energy Recovery Opportunities for Interstate Natural Gas Pipelines.* ICF International, Atlanta.
70. CPS Energy. (2011). *Summary of Gas Service Standards for Engineers, Architects, Contractors and Developers.* CPS Energy San Antonio, TX.
71. Brueske, S., Sabouni, R. and Zach, C. (2012). *U.S. Manufacturing Energy Use and Greenhouse Gas Emissions Analysis.* Energetics Incorporated, Department of Energy, Washington, DC.
72. EPA. (2018). *Inventory of U.S. Greenhouse Gas Emissions and Sinks: 1990–2016.* Environmental Protection Agency, Washington, DC.

73. Worrell, E., Corsten, M. and Galisky, C. (2015). *Energy Efficiency Improvement and Cost Saving Opportunities for Petroleum Refineries: An ENERGY STAR Guide for Energy and Plant Managers.* United States Environmental Protection Agency, Washington, DC.

74. Worrell, E. and Galisky, C. (2005). *Energy Efficiency Improvements in the Petroleum Refining Industry.* Ernest Orlando Lawrence Berkeley National Laboraory, Berkeley, CA.

75. Karras, G. (2010). Combustion emissions from refining lower quality oil: What is the global warming potential. *Environmental Science* 44: 9584–9589.

76. Kurup, P. and Turchi, C. (2015). *Initial Investigation into the Potential of CSP Industrial Process Heat for the Southwest United States.* NREL, Golden, CO.

77. Hicks, S. and Gross, P. (2016). *Hydrogen for Refineries is Increasingly Provided by Industrial Suppliers.* A website report, U.S. Energy Information Administration, Washington, D.C.

78. Philibert, C. (2017). *Renewable Energy for Industry: From Green Energy to Green Materials and Fuels.* International Energy Agency. Paris.

79. Absi Halabi, M., Al-Qattan, A. and Al-Otaibi, A. (2015). Application of solar energy in the oil industry-Current status and future prospects. *Renewable and Sustainable Energy Reviews* 43: 296–314.

80. Ngoh, S.K. and Njomo, D. (2012). An overview of hydrogen gas production from solar energy. *Renewable and Sustainable Energy Reviews* 16: 6782–6792.

81. Möller, S., Kaucic, D. and Sattler, C. (2006). Hydrogen production by solar reforming of natural gas: A comparison study of two possible process configurations. *Journal of Solar Energy Engineering* 128: 16–23, Doi: 10.1115/1.2164447.

82. Giaconia, A., Monteleone, G., Morico, B. et al., (2015). Multi-fuelled solar steam reforming for pure hydrogen production using solar salts as heat transfer fluid. *Energy Procedia* 69: 1750–1758.

83. Likkasit, C., Maroufmashat, A., Elkamel, A. and Ku, H.-m. (2016). Integration of renewable energy into oil & gas industries: Solar-aided hydrogen production. *International Conference on Industrial Engineering and Operations Management*, Detroit.

84. IRENA. (2018). *Global Energy Transformation: A Roadmap to 2050 Int.* Renewable Energy Agency, Abu Dhabi.

85. Graves, L. (2016). Oil and gas still the main focus for adnoc press release. Retrieved from www.thenational.ae/business/oil-and-gas-still-the-main-focus-for-adnoc-1.140288.

86. CNOOC. (2012). Sustainability report China national offshore oil corporation. Retrieved from www.cnooc.com.cn/data/upload/nb2012en.pdf, a website report.

87. CNPC. (2018a). Climate change. Retrieved from www.cnpc.com.cn/en/climate/common_index.shtml.

88. Gazprom. (2018). Energy saving. Retrieved from www.gazprom.com/nature/energy-conservation/, a website report.

89. Gazprom. (2017). Pjsc gazprom environmental report 2017. Retrieved from www.gazprom.com/f/posts/60/709300/gazprom_ee_2017_2.pdf.

90. Gazprom. (2016). Gazprom group's sustainability report 2016. Retrieved from www.gazprom.com/f/posts/44/307258/sustainability-report-2016-en.pdf.

91. KPC World and Bouresly, A.M. (2016) KNPC draws up detailed plans to meet the needs of local market for petroleum products. KPC World 77 15.

92. K-Pulse. (2018). KOC launches renewable energy dashboard K-Pulse 5 28–31. Retrieved from www.kpc.com.kw/press/KPCPublications/KPulse/KPulse-en-issue-5.pdf.

93. KPC. (2017). Brightness in integration. Annual report 2017 Kuwait Petroleum Corporation. Retrieved from www.kpc.com.kw/press/KPCPublications/AnnualReports/AnnualRep2017-eng.pdf.

94. PDO. (2017). Sustainability report 2017 petroleum development of Oman. Retrieved from www.pdo.co.om/en/news/publications/Publications%20Doc%20Library/_Annual-report-2018-04-BothSN.pdf.

95. Power Technology. (2018). Miraah solar thermal project. Retrieved from www.power-technology.com/projects/miraah-solar-thermal-project/.

96. Makvandi M. (2018). Using solar energy lighting system at the national drilling company. National Iranian Oil Company. Retrieved from http://nioc.ir/portal/Home/ShowPage.aspx?Object=NEWSID=2d001dd3-7240-4909-b58e-a3db8994809cLayoutID=caffb005-7022-4586-b08e-b6dbd9df0e2e CategoryID=40295507-1e70-49ec-af82-e5eeba6870bc, a website report

97. Petronas. (2017). Annual report 2017. Retrieved from www.petronas.com/ws/sites/default/files/2018-08/petronas-annual-report-2017_0.pdf.

98. Petrobras. (2018). Annual report Petrobras (Brazilian Petroleum Corporation). Retrieved from www.investidorpetrobras.com.br/en/annual-reports/integrated-report/annual-report.

99. Bloomberg. (2017). Saudi Aramco said to weigh up to $5 billion of renewable deals Gulf News. Retrieved from https://gulfnews.com/business/sectors/energy/saudi-aramco-said-to-weigh-up-to-5-billion-of-renewable-deals-1.1969857, a website report.

100. Dhahran. (2018). Renewable best practice Saudi Aramco. Retrieved from www.saudiaramco.com/en/news-media/news/2018/sharing-renewable-best-practices-joint-ventures.

101. Habboush, M. (2018). Saudi Arabia plans up to $7 billion of renewable energy projects this year Bloomberg. Retrieved from www.bloomberg.com/news/articles/2018-01-16/saudi-arabia-plans-up-to-7-billion-of-renewables-this-year.

102. Wind Power. (2018). Wind energy market actors wind energy market intelligence. Retrieved from www.thewindpower.net/players_en.php.

103. Shojaeddini, E., Naimoli, S., Ladislaw, S. and Bazilian, M. (2019). Oil and gas company strategies regarding the energy transition. *Topical Review Progress in Energy* 1(1) Published 16 July 2019 © 2019 IOP Publishing Ltd, Doi: 10.1088/2516-1083/ab2503.

104. Lee, A., Zinaman, O. and Logan, J. (2012, December). Opportunities for synergy between natural gas and renewable energy in the electric power and transportation sectors. Technical Report NREL/TP-6A50-56324 Contract No. DE-AC36-08GO28308, National Renewable Energy Laboratory, Golden, CO.

105. Deutsche Bank Climate Advisors. (2010, November). *Natural Gas and Renewables: A Secure Low Carbon Future Energy Plan for the United States*, New York; International Energy Agency. (2012, May). The golden age of natural gas, Paris; Wesoff, E. (2011, January). Natural gas and renewables: A perfect match? Greentech Media.

106. IPCC (2007). Fourth assessment report of the intergovernmental panel on climate change: Summary for policymakers, Working Group III (Table SPM 5); Intergovernmental Panel on Climate Change. Cambridge University Press, Cambridge, UK.

107. Schrag, D.P. (2012). Is shale gas good for climate change? *Dædalus, the Journal of the American Academy of Arts & Sciences* 141(2): 72–80.

108. Fulton, M. and Melquist, N. (2010). *Natural Gas and Renewables: A Secure Low Carbon Future Energy Plan for the United States*. DB Climate Change Advisors, New York; Fulton, M. and Melquist, N. (2011). *Natural Gas and Renewables: The Coal to Gas and Renewables Switch is on!* DB Climate Change Advisors, New York.

109. Moniz, E.J., Jacoby, H.D. and Meggs, A.J.M. (2011). *The Future of Natural Gas: An Interdisciplinary MIT Study*. Massachusetts Institute of Technology, Cambridge, MA; Verrastro, F. and Branch, C. (2010). *Developing America's Unconventional Gas Resources: Benefits and Challenges*. Center for Strategic and International Studies, Washington, DC.

110. Hargreaves, S. (2012, May 31). Drought Strains U.S. Oil Production: More than 60% of Nation is in Some Form of Drought. KJCT8.com, Grand Junction. A website report by CNN Money (May, 31, 2012).

111. Federal Energy Regulatory Commission (FERC) *Technical Conference on Coordination between Natural Gas and Electricity Markets for the Central Region*, 6 August, 2012, St. Louis, Missouri.

112. Cochran, J., Bird, L., Heeter, J. and Arent, D.A. (2012). *Integrating Variable Renewable Energy in Electric Power Markets: Best Practices from International Experience, Summary for Policymakers.* NREL/TP-6A20-53730. National Renewable Energy Laboratory, Golden, CO.

113. Cochran, J., Zinaman, O., Logan, J. and Arent, D. (2014, February) Exploring the potential business case for synergies between natural gas and renewable energy. Technical Report NREL/TP-6A50-60052. Contract No. DE-AC36-08GO28308, National Renewable Energy Laboratory (NREL), Golden, CO.

114. Milligan, M., Kirby, B. and Beuning, S. (2010). *Potential Reductions in Variability with Alternative Approaches to Balancing Area Cooperation with High Penetrations of Variable Generation.* NREL/MP-550-48427. National Renewable Energy Laboratory, Golden, CO, 78 p; Milligan, M., Ela, E., Hein, J., Schneider, T., Brinkman, G. and Denholm, P. (2012). *Exploration of High-Penetration Renewable Electricity Futures. Vol. 4 of Renewable Electricity Futures Study.* NREL/TP-6A20-52409-4. National Renewable Energy Laboratory, Golden, CO.

115. Milligan, M., Ela, E., Hein, J., Schneider, T., Brinkman, G. and Denholm, P. (2012). *Exploration of High-Penetration Renewable Electricity Futures. Vol. 4 of Renewable Electricity Futures Study.* NREL/TP-6A20-52409-4. National Renewable Energy Laboratory, Golden, CO, http://www.nrel.gov/analysis/re_futures/

116. Xie, L., Carvalho, P., Ferreira, L.A. et al., (2011). Wind integration inpower systems: Operational challenges and possible solutions. *Proc IEEE* 99: 214–232. Available at: http://ieeexplore.ieee.org/stamp/stamp.jsp?arnumber=5607275.

117. Verbruggen, A., Fischedick, M., Moomaw, W. et al., (2010). Renewable energy costs, potentials, barriers: Conceptual issues. *Energy Policy* 38: 850–861.

118. GE Energy Consulting. (2010). *Western Wind and Solar Integration Study.* NREL/SR-550-47434. National Renewable Energy Laboratory (NREL), Golden, CO. Available at: http://www.nrel.gov/grid/wwsis.html.

119. Shaffer, B., Tarroja, B. and Samuelsen S. (2015). Dispatch of fuel cells as transmission inte-grated grid energy resources to support renewables and reduce emissions. *Applied Energy* 148: 178–186.

120. U.S EPA. (2015). The emissions and generation resource integrated database (eGRID)2012. Available at: http://www.epa.gov/energy/egrid-2012-summary-tables.

121. Katzenstein, W. and Apt, J. (2008). Air emissions due to wind and solar power. *Environmental Science and Technology* 43: 253–258.

122. Liik, O., Oidram, R. and Keel, M. (2003). Estimation of real emissions reduction caused by wind generators. *Proceedings of the International Energy Workshop*, Laxenburg, 24–26. Available at: https://docs.wind-watch.org/liik-emis-sionsreduction.pdf, a website report.

123. Valentino, L., Valenzuela, V., Botterud, A., Zhou, Z. and Conzelmann G. (2012). System-wide emissions implications of increased wind power penetration. *Environmental Science & Technology* 46: 4200–4206.

124. Nyberg, M. (2014). *Thermal Efficiency of Gas-Fired Generation in California: 2014 Update.* California Energy Commission, Sacramento, CA. CEC-200-2014-005. Available at: http://www.energy.ca.gov/2014publications/CEC-200-2014-005/CEC-200-2014-005.pdf2014, a website report.

125. Mills, A., Wiser, R., Milligan, M. and Malley, M. (2009). Comment on air emissions due to wind and solar power. *Environmental Science & Technology* 43: 6106–6107.

126. Novan, K.M. (2011). Shifting wind: The economics of moving subsidies from power produced to emissions avoided. In: *Proceedings of the 30th Anniversary Meeting of the International Energy Workshop*, Stanford University, Palo Alto, CA. Available at: https://web.stanford.edu/group/emf-research/new-emf.stanford.edu/files/docs/273/Novan_IEW_Presentation.pdf, a website report.

127. California green lights. 570-MW Palmdale CSP hybrid plant. A website report, 28 August 2015 www.energy.ca.gov/.

128. Inhaber, H. (2011). Why wind power does not deliver the expected emissions reductions. *Renewable and Sustainable Energy Reviews* 15: 2557–2562.

129. Darrow, K., Tidball, R., Wang, J. and Hampson A. (2015). Catalog of CHP technologies. Available at: http://www.epa.gov/sites/production/files/2015-07/documents/catalog_of_chp_technologies.pdf, a website report.

130. Moore, M.J. (2002). *Micro-Turbine Generators*. John Wiley & Sons, New Yok, NY.

131. Barsali, S., De Marco, A. Giglioli, R., Ludovici, G. and Possenti A. (2015). Dynamic modelling of biomass power plant using micro gas turbine. *Renewable Energy* 80: 806–818.

132. Lund, H. (2005). Large-scale integration of wind power into different energy systems. *Energy* 30: 2402–2412.

133. Mathiesen, B.V., Lund, H. and Karlsson, K. (2011). 100% Renewable energy systems, climate mitigation and economic growth. *Applied Energy* 88: 488–501.

134. Mueller, F., Jabbari, F., Gaynor, R. and Brouwer, J. (2007). Novel solid oxide fuel cell system controller for rapid load following. *Journal of Power Sources* 172: 308–323.

135. Maclay, J.D., Brouwer, J. and Samuelsen, G.S. (2006). Dynamic analyses of regenerative fuel cellpower for potential use in renewable residential applications. *International Journal of Hydrogen Energy* 31: 994–1009.

136. Margalef, P., Brown, T.M., Brouwer, J. and Samuelsen, S. (2012). Efficiency comparison of tri-generating HTFC to conventional hydrogen production technologies. *International Journal of Hydrogen Energy* 37: 9853–9862.

137. Kast, J.F. (2014). *Dynamic Modeling, Design, and Performance Evaluation of Large Scalehigh Temperature Fuel Cell Tri-Generation Systems Thesis*. University of California, Irvine, CA.

138. Roberts, R. and Brouwer, J. (2006). Dynamic simulation of a pressurized 220 kW solid oxidefuel-cell-gas-turbine hybrid system: Modeled performance compared to measured results. *Journal of Fuel Cell Science and Technology* 3: 18–25.

139. Spiegel, R.J., Trocciola, J. and Preston, J. (1997). Test results for fuel-cell operation on landfill gas. *Energy* 22: 777–786.

140. Samuelsen, S. and Brouwer, J. (2009). Fuel cell/gas turbine hybrid. In: *Encyclopedia of Electrochemical Power Sources*, first ed., J. Garche (Ed.), Elsevier Science, Netherland, pp. 124–134.

141. Chan, S. and Ho, H. (2002). Modelling of simple hybrid solid oxide fuel cell and gas turbine power plant. *Journal of Power Sources* 109: 111–120.

142. Costamagna, P., Magistri, L. and Massardo, A. (2001). Design and part-load performance of ahybrid system based on a solid oxide fuel cell reactor and a micro gas turbine. *Journal of Power Sources* 96: 352–368.

143. Mueller, F., Jabbari, F. and Brouwer, J. (2009). On the intrinsic transient capability and limita-tions of solid oxide fuel cell systems. *Journal of Power Sources* 187: 452–460.

144. Denholm, P. and Kulcinski, G.L. (2004). Life cycle energy requirements and greenhouse gasemissions from large scale energy storage systems. *Energy Conversion and Management* 45: 2153–2172.

145. Lefebvre, D. and Tezel, F.H. (2017). A review of energy storage technologies with a focus on adsorption thermal energy storage processes for heating applications. *Renewable & Sustainable Energy Reviews* 67: 116–125.

146. Sørensen, B.E. (2017). *Renewable Energy: Physics, Engineering, Environmental Impacts, Economics and Planning.* Academic Press, New York.

147. Luo, X., Wang, J., Dooner, M. and Clarke J. (2015). Overview of current development in electrical energy storage technologies and the application potential in power system operation. *Applied Energy* 137: 511–536.

148. CESA. (2010). Energy storage—a Cheaper and cleaner alternative to natural gas-fired peaker plants. California Energy Storage Alliance. Available at https://www.ice-energy.com/wp-content/uploads/2016/04/cesa_peaker_vs_stor-age_2010_06_16.pdf, a website report.

149. Sioshansi, R. and Denholm, P. (2009). Emissions impacts and benefits of plug-in hybrid electric vehicles and vehicle-to-grid services. *Environmental Science & Technology* 43: 1199–**11204.**

150. Lin, J. (2010). *Imperative of Energy Storage for Meeting California's Clean Energy Needs. Senate Energy, Utilities, and Communications Committee.* California EnergyStorage Alliance (CESA), Berkley, CA, a website report.

151. Ghoniem, A.F. (2011). Needs, resources and climate change: Clean and efficient conversion technologies. *Progress in Energy and Combustion Science* 37:15–51.

152. Elkind, E.N., Weissman, S. and Hecht, S. (2010). *The Power of Energy Storage: How to Increase Deployment in California to Reduce Greenhouse Gas Emissions.* University of California, Berkeley School of Law, Berkeley, CA, White paper: Available at: https://www.law.berkeley.edu/files/Power_of_Energy_Storage_July_2010.pdf.M.A. Mac Kinnon et al. / Progress in Energy and Combustion Science 64 (2018) 62.9291.

153. Rastler, D. (2010). *Electricity Energy Storage Technology Options: A White Paper Primer on Applications, Costs and Benefits.* Electric Power Research Institute, Palo Alto, CA. 1020676. Available at: http://large.stanford.edu/courses/2012/ph240/doshay1/docs/EPRI.pdf.

154. Fisher, M.J. and Apt, J. (2017). Emissions and economics of behind-the-meter electricity storage. *Environmental Science & Technology* 51: 1094–10101.

155. Lin, Y., Mathieu, J.L. and Johnson, J.X. (2016). Stochastic optimal powerflow formulation to achieve emissions objectives with energy storage. *Proceedings of the Power Systems Computation Conference (PSCC)*, 2016, IEEE, 1–7, Genova, Italy.

156. Lin, Y., Johnson, J.X. and Mathieu, J.L. (2016). Emissions impacts of using energy storage for power system reserves. *Applied Energy* 168: 444–456.

157. Harris, C. and Meyers, J.P. (2010). Working smarter, not harder: An introduction to the "smart grid. *Electrochemical Society Interface* 19(3):45–48.

158. DOE. (2008). The smart Grid: An introduction. U.S. Department of Energy. Available at: http://energy.gov/sites/prod/files/oeprod/DocumentsandMedia/DOE_SG_Book_Single_Pages%281%29.pdf, a website report.

159. EPRI. (2008). The green grid: Energy savings and carbon emissions reductions enabled by a smart grid. Available at: https://www.smartgrid.gov/files/The_Green_Grid_Energy_Savings_Carbon_Emission_Reduction_En_200812.pdf, a website report.

160. Tarroja, B., Zhang, L., Wifvat, V., Shaffer, B. and Samuelsen, S. (2016). Assessing the stationary energy storage equivalency of vehicle-to-grid charging battery electric vehicles. *Energy* 106: 673–690.

161. Moniz, E.J., Jacoby. H.D., Meggs. A. et al., (2011). *The Future of Natural Gas.* Massachusetts Institute of Technology, Cambridge, MA.

162. Han, J., Mintz, M. and Wang, M. (2011). Waste-to-wheel analysis of anaerobic-digestion-based renewable natural gas pathways with the GREET model. ArgonneNational Laboratory (ANL). ANL/ESD/11-6. Available at: http://www.osti.gov/bridge/purl.cover.jsp?purl=/1036091/.

163. Jury, C., Benetto, E., Koster, D., Schmitt, B. and Welfring, J. (2010). Life cycle assessment of bio-gas production by monofermentation of energy crops and injection into the natural gas grid. *Biomass Bioenergy* 34: 54–66.

164. Chai, X., Tonjes, D.J. and Mahajan, D. (2016). Methane emissions as energy reservoir: Context, scope, causes and mitigation strategies. *Progress in Energy and Combustion Science* 56: 33–70.

165. E3. (2014). *Decarbonizing Pipeline Gas to Help Meet California's 2050 Greenhouse Gas Reduction Goal.* Energy and Environmental Economics, Inc., San Francisco, CA. Available at: http://origin-qps.onstreammedia.com/origin/multivu_arch-ive/ENR/1241844-Decarbonizing-Pipeline-Gas.pdf, a website report.

166. Murray, B.C., Galik, C.S. and Vegh, T. (2014). *Biogas in the United States: An Assessment of Market Potential in a Carbon-Constrained Future.* Duke University, Durham, NC. NI R 14-02. Available at: https://nicholasinstitute.duke.edu/sites/default/files/publications/ni_r_14-02_full_pdf.pdf, a website report.

167. Renewable Natural Gas Systems. (2015). Utility scale RNG gas. A website report by AMERESCO.

168. Persson, M., J€onsson, O. and Wellinger, A. (2007). Biogas upgrading to vehicle fuel standards and grid injection. IEA Bioenergy Task 372006. Available at: https://www.iea-biogas.net/files/daten-redaktion/download/publi-task37/upgrading_rz_low_final.pdf, a website report.

169. Brzozowski, C. (2014). Getting the gas out. *MSW Management* 1: 28–34.

170. Lombardi, L., Carnevale, E. and Corti, A. (2006). Greenhouse effect reduction and energy recovery from waste landfill. *Energy* 31: 3208–3219.

171. Appels, L., Baeyens, J., Degreve, J. and Dewil, R. (2008). Principles and potential of the anaerobic digestion of waste-activated sludge. *Progress in Energy and Combustion Science* 34: 755–781.

172. Rasi, S., L€antel€a, J. and Rintala, J. (2011). Trace compounds affecting biogas energy utilization: A review. *Energy Conversion and Management* 52: 3369–3375.

173. Sun, Q., Li, H., Yan, J., Liu, L., Yu. Z. and Yu, X. (2015). Selection of appropriate biogas upgrading technology-a review of biogas cleaning, upgrading and utilization. *Renewable and Sustainable Energy Reviews* 51: 521–532.

174. Ong, M.D., Williams, R.B. and Kaffka, S.R. (2015). *Comparative Assessment of Technology Options for Biogas Clean-up.* California Energy Commission. CEC-500-11-020. Available at: http://biomass.ucdavis.edu/files/2015/10/Biogas-Cleanup-Report_FinalDraftv3_12Nov2014-2.pdf.

175. Tachibana, Y., Vayssieres, L. and Durrant, J.R. (2012). Artificial photosynthesis for solar water-splitting. *Nature Photonics* 6: 511–518.

176. Barreto, L., Makihira, A. and Riahi, K. (2003). The hydrogen economy in the 21st century: A sustainable development scenario. *International Journal of Hydrogen Energy* 28: 267–284.

177. Tarroja, B., Shaffer, B. and Samuelsen, S. (2015). The importance of grid integration for achievable greenhouse gas emissions reductions from alternative vehicle technologies. *Energy* 87: 504–519.

178. Gahleitner, G. (2013). Hydrogen from renewable electricity: An international review ofpower-to-gas pilot plants for stationary applications. *International Journal of Hydrogen Energy* 38: 2039–2061.

179. Power-to-Gas. (2020). Wikipedia, The free encyclopedia, last visited 27 August 2020.

180. Patel, S. Why power-to-gas may flourish in a renewables-heavy world, senior associate editor (@sonalcpatel, @POWERmagazine), 1 December 2019.

181. G€otz, M., Lefebvre, J., M€ors, F. et al., (2016). Renewable power-to-gas: A technological and economic review. *Renewable Energy* 85: 1371–1390.

182. Carmo, M., Fritz, D.L., Mergel, J. and Stolten, D. (2013). A comprehensive review on PEM water electrolysis. *International Journal of Hydrogen Energy* 38: 4901–4934.

183. Jentsch, M., Trost, T. and Sterner, M. (2014). Optimal use of power-to-gas energy storage systems in an 85% renewable energy scenario. *Energy Procedia* 46: 254–261.

184. CHBC. (2015). Power-to-Gas: The case for hydrogen white paper. California Hydrogen Business Council. Available at: https://californiahydrogen.org/sites/default/files/CHBC%20 Hydro-gen%20Energy%20Storage%20White%20Paper%20FINAL.pdf, a website report.

185. Brouwer J. (2010). On the role of fuel cells and hydrogen in a more sustainable and renewable energy future. *Current Applied Physics* 10: S9–S17.

186. Dickinson, R.R., Battye, D.L., Linton, V.M. and Ashman, P.J. (2010). Alternative carriers for remote renewable energy sources using existing CNG infrastructure. *International Journal of Hydrogen Energy* 35: 1321–1329.

187. Yang, C. and Ogden, J. (2007). Determining the lowest-cost hydrogen delivery mode. *International Journal of Hydrogen Energy* 32: 268–286.

188. Iabidine Messaoudani, Z., Rigas, F., Hamid, M.D.B. and Hassan, C.R.C. (2016). Hazards, safety and knowledge gaps on hydrogen transmission via natural gas grid: A critical review. *International Journal of Hydrogen Energy* 41: 17511–1725.

189. Schaaf, T., Gr€unig, J., Schuster, M.R., Rothenfluh, T. and Orth, A. (2014). Methanation of CO_2-storage of renewable energy in a gas distribution system. *Energy, Sustainability and Society* 4: 2.

190. Lehner, M., Tichler, R., Steinm€uller, H. and Koppe, M. (2014). *Power-to-Gas: Technology and Business Models*. Springer. Available at: https://link.springer.com/content/pdf/10.1007/978-3-319-03995-4.pdf, a website report.

191. Basile, A. and Iulianelli. A. (2014). *Advances in Hydrogen Production, Storage, and Distribution*. Woodhead Publishing, New York.

192. Zheng, Y., Wang, J., Yu, B. et al., (2017). A review of high temperature co-electrolysis of H_2O and CO_2 to produce sustainable fuels using solidoxide electrolysis cells (SOECs): Advanced materials and technology. *Chemical Society Reviews* 46: 1427–1463.

193. Klumpp, F. (2016). Comparison of pumped hydro, hydrogen storage and compressed air energy storage for integrating high shares of renewable energies—potential, cost-comparison and ranking. *Journal of Energy Storage* 8: 119–128.

194. Bekkering, J., Broekhuis, A. and Van Gemert, W. (2010). Optimisation of a green gas supply chain–A review. *Bioresource Technology* 101: 450–456.

195. Wang, J., Huang, Z., Fang, Y., Liu, B., Zeng, K., Miao, H. and Jiang D. (2007). Combustion behaviors of a direct-injection engine operating on various fractions of natural gas-hydrogen blends. *International Journal of Hydrogen Energy* 32: 3555–3564.

196. Ma, F., Wang, Y., Liu, H., Li, Y., Wang, J. and Zhao, S. (2007). Experimental study on thermal efficiency and emission characteristics of a lean burn hydrogen enriched natural gas engine. *International Journal of Hydrogen Energy* 32: 5067–5075.

197. Ma, F., Wang, Y., Liu, H., Li, Y., Wang, J. and Ding, S. (2008). Effects of hydrogen addition on cycle-by-cycle variations in a lean burn natural gas spark-ignition engine. *International Journal of Hydrogen Energy* 33: 823–831.

198. Park, C., Kim, C., Choi, Y., Won, S. and Moriyoshi, Y. (2011). The influences of hydrogen on the performance and emission characteristics of a heavy duty natural gas engine. *International Journal of Hydrogen Energy* 36: 3739–3745.

199. Colorado, A., McDonell, V. and Samuelsen, S. (2017). Direct emissions of nitrous oxide from combustion of gaseous fuels. *International Journal of Hydrogen Energy* 42: 711–719.

200. Poeschl, M., Ward, S. and Owende, P. (2012). Environmental impacts of biogas deployment -Part I: Life cycle inventory for evaluation of production process emissions to air. *Journal of Cleaner Production* 24: 168–183.

201. Ogden, J.M. (1999). Prospects for building a hydrogen energy infrastructure. *Annual Review of Energy and the Environment* 24: 227–279. Doi: 10.1146/annurev. energy.24.1.227.
202. Alves, H.J., Bley Junior, C., Niklevicz, R.R., Frigo, E.P., Frigo, M.S. and Coimbra-Araujo C.H. (2013). Overview of hydrogen production technologies from biogas and the applications in fuel cells. *International Journal of Hydrogen Energy* 38: 5215–5225.
203. B€orjesson, P., Prade, T., Lantz, M. and Bj€ornsson, L. (2015). Energy crop-based biogas as vehicle fuel—the impact of crop selection on energy efficiency and greenhouse gas performance. *Energies* 8: 6033–6058.

9 Hybrid Energy Systems for Computing and Electronic Industries

9.1 INTRODUCTION

It is not possible to think about computing and networking today without considering data centers. They are where computing and networking equipment is concentrated for the purposes of collecting, storing, processing, distributing, or allowing access to large amounts of data. Their construction costs about \$20B a year worldwide [1], and they cause approximately as much CO_2 emissions as the airline industry [2]. According to a report by Lawrence Berkeley National Laboratory, data centers in the United States consumed about 70 billion kWh of energy in 2016 [3]. This corresponded to 1.8% of all the energy consumed in the United States that year. The United States Department of Energy currently states up to 3% of all electricity may be consumed by data centers today [4]. Globally, electricity demand by data centers in 2018 was an estimated 198 TWh, or almost 1% of demand for electricity in the world [5]. It is estimated that the electrical energy consumption by the information technology industry across the globe is less than only the United States and China and is more than the third country on the list [1]. On the other hand, global Internet traffic has tripled between 2015 and 2019 and is expected to double by 2022 [6]. Increased use of artificial intelligence and the large number of sensors expected to be deployed in the Internet-of-Things era bring up the question how much more energy use in data centers will increase. The crucial question is whether it will increase exponentially as Internet data traffic is growing and what this growth will mean for the release of CO_2 worldwide. The latter question becomes important when one considers that 63.5% of electricity generation in the United States during 2018 was from fossil fuels [7].

The Internet search engine Google disclosed in 2011 that its data centers consume 260 MW and made the estimation that this is sufficient to power 200,000 homes [8]. Google estimated that one Google search is responsible for emitting 0.2 g of CO_2 per year [9]. It also estimated its carbon footprint as being 1.5 million metric tons in 2011 [10]. Considering the increase in Internet traffic during this period, the number may be even higher today. On the other hand, in 2019, the company announced that it has made purchases of renewable energy equivalent to 1,600 MW, increasing its share of renewable energy sources to 5,500 MW, making it 40% of its total energy consumption.

Today, streaming video is the most significant part of global data traffic. The Internet company Cisco predicts that streaming video will make up 82% of all Internet traffic by 2021, up from 73% in 2012. Currently, one-third of Internet traffic in North America is already dedicated to streaming Netflix services [1]. The global nongovernmental environment organization Greenpeace shows a strong interest in energy consumption by Internet companies. To that end, Greenpeace began benchmarking the energy performance of the Information Technology sector in 2009, challenging the largest Internet companies to substantially increase their use of renewable energy. The advocacy group uses an annual report to rate big Internet and cloud companies on their use of renewable power. In the report in 2017, Greenpeace denounced Netflix for substantial energy inefficiency [11]. Netflix does not own data centers; instead it uses contractors such as Amazon Web Services (AWS). Greenpeace also denounced AWS for being completely nontransparent about the energy footprint of its massive operations. AWS has some of its largest operations in Northern Virginia, which receives less than 3% of its energy from renewable sources [1]. Decisions by Internet and cloud companies can be surprising. Northern Virginia, with such a bad record for renewable energy, has the largest concentration of data centers in the world [1]. The location where energy is generated is actually very important in terms of its efficiency. For example, to generate 1 kWh of energy, it takes 3 g of CO_2 in Norway, 100 g in France, 600 g in Virginia, and 800 g in New Mexico. Most of the energy is used to keep the processors cool. Very surprisingly, most of the world's largest data centers are located where the temperature is hot [1]. However, through various efforts, Internet and cloud companies seem to be listening. Netflix declared in 2017 that they now purchase renewable energy certificates to match their nonrenewable energy use and fund renewable energy production from sources such as wind and solar [12]. AWS has declared that it is committed to achieving 100% renewable energy use for their global infrastructure [13]. The significance of energy efficiency in data centers is well articulated by Ayanoglu [14].

Besides data centers, the energy use in power electronics is also rapidly expanding. Here while energy use in individual device is low, the number of devices is rapidly expanding. What makes low power electronic devices different from data centers is that for these devices, energy can be harvested from natural source through innovative hybrid processes and energy systems. While in data centers measures to reduce energy consumption will come from energy efficiency measures and external use of hybrid renewable energy to reduce use of fossil fuels, in small power electronics, multisource harvesting measures, and self-powering of micro- and nanodevices through these measures will dominate the industry.

In this chapter, we consider the use of hybrid energy for these two widely different industries that are exploding in their growth and energy consumption. In this digital information and communication age, the need and use of data are becoming more and more important. In recent years, significant efforts are made by tech giants to introduce renewable energy sources in the energy need of data centers to (a) reduce costs, (b) reduce use of fossil energy, and thereby (c) reduce carbon emission. This has led to the use of hybrid energy system and processes. Portable electronic devices, although use very small amount of energy individually, are increasing in numbers across the globe in a giant way. These devices can be designed to harvest its own energy need through hybrid energy and processes. As shown in this chapter, the best method appears to be the harvesting of hybrid energy and implementing hybrid processes where more than one source or a source and storage can best handle the energy needs of the devices.

9.2 THE CASE OF HYBRID APPROACH FOR DATA CENTERS

Energy consumption of the computing infrastructure has become a major concern for industry and society. Today's data centers, the backbone of the computing infrastructure, are limited in scale by the costs associated with power (distribution, cooling, density). Studies estimate that power-related costs represent already almost 50% of the operating cost of a data center and they are growing faster than compute-related costs (i.e., server and network equipment). Energy efficiency is now a first-class design concern at all levels—computation and data processing, power distribution at the rack and server level, power generation and transmission, etc. Companies such as Microsoft and Google are deploying new data centers near cheap power sources to mitigate energy costs. Processor manufacturers are pursuing their roadmap of multicore architectures [15] and low-power designs [16]. Several research proposals deal with power efficient designs and protocols for specific workloads [17] office environments [18,15] and high speed networks [19].

One important specific aspect is energy-efficient clusters for large data centers. As a first step, one can consider the current trends in server designs and try to exploit them to one's advantage. Traditionally, power efficient designs attempt to find the right balance between two distinct, and often conflicting, requirements: (a) deliver high performance at peak power (i.e., maximize compute capacity for a given power budget) and (b) scale power consumption with load (i.e., energy proportionality and very low power operations). A fundamental challenge in finding a good balance is that, when it comes to processor design, the mechanisms that satisfy the two requirements above are significantly different. High performance requires mechanisms to mask memory and input/output (I/O) latencies using large multilevel caches (today's server processors use three cache levels with the last-level cache projected to soon reach 24 MB [20]), large translation look-aside buffers, out-of-order execution, high-speed buses, and support for a large number of pending memory requests. These mechanisms result in large transistor counts leading to high leakage power and overall high power consumption. In a modern processor, less than 20% of the transistor count is dedicated to the actual cores [21,22].

Low power designs, on the other hand, focus on those processor features with low-power operations. For example, the Atom processor [14] includes an in-order pipeline that can execute two instructions per cycle, a small L2 cache, and power-efficient clock distribution. This results in a strongly reduced transistor count with low leakage power and limited power consumption at low load. Further, Atom design is focused on allowing quick and frequent transitions to a very low power state (e.g., 80 mW with less than 100 µs exit latency [16]). Proposals like fast array of wimpy nodes (FAWN) [17] and Marlowe [23] explore these features to build arrays of low power servers that operate efficiently for specific I/O bound workloads.

These observations lead to a dichotomy between low power and high performance system designs. Choosing the most appropriate design for an energy efficient datacenter is far from straightforward. First, data center workloads are diverse—some (e.g., I/O-bound map/reduce like) lend themselves rather easily to low-power designs while others (e.g., transactions, encryption) depend on high performance and fast response times to satisfy stringent service-level agreements (SLAs). Second, the workload dynamics—including job arrival patterns and completion times—may reverse the conclusion of static workload analysis. Finally, the processor is just one contributor to the overall power consumption. Other system components such as the motherboard (e.g., I/O and memory controllers), DRAM banks, and power supplies contribute to a large fraction of the overall power consumption and tend not to be optimized for low-power operation. Given these challenges, a hybrid data center architecture that mixes low-power systems with high-performance ones makes sense.

The study by Chun et al. [24] makes the case for hybrid approaches. The study compares the performance of different systems under data center-like workloads using a quad-core, dual-socket Xeon system and two low-power Atom-based personal computers (PCs). These systems were chosen because they are representative of high-performance systems currently common in data centers and of low-power platforms that are used today to build energy-efficient netbooks. Workloads that are representative for large data centers can be classified into three broad categories:

Web services: this is the classical web workload to serve pages to users. The data requested is usually a small object in a large dataset (e.g., an item on sale in an e-commerce site such as Amazon). The first request may lead to a database query but subsequent requests are cached in memory for fast retrieval—mem cached [25] is an example of this approach for large clusters and currently used by LiveJournal, Facebook, and others. The study uses a simple Apache/PHP benchmark to emulate this class of applications.

Data mining: this second class is representative of large-scale data analysis workloads that process a data set in a distributed fashion. This is typically done to populate the index used in search engines or for machine learning operations, e.g., to drive recommendation engines. To emulate this workload, we use Hadoop [26], an open-source MapReduce [27] implementation, with a pseudocluster configuration on a single server. The maximum number of mappers and reducers running on a node is set to twice the number of cores in the servers: this set-up showed the best performance. The study considers two applications that make large use of disk I/O and are available in the Hadoop distribution: "word count" over 10 GB of data and "sort" over 1 GB.

Compute-intensive: the third type of workloads model central processing unit (CPU)-intensive applications such as image processing or video encoding. They may operate on a smaller data set but require a significant amount of computation for each data object. To emulate this class of workloads on a data center environment, the study uses the Hadoop "pi" application that estimates the value of π using the Monte Carlo method and use FFm peg to convert a file from Windows Media (.wmv) to Flash Video (.flv).

Chun et al. [24] compared the performance/watt and energy proportionality of the three platforms over the above workloads. Except Web/PHP, the performance was measured as the rate of execution (i.e., one over total execution time). For Web/PHP, the study measured the number of concurrent users supported under a certain latency SLA (99% of requests are served within 100 ms). The performance was compared to the power consumption of the system measured at the wall socket. To easily compare the workloads in one graph, we normalized the results to the performance/watt of the dual-core Atom 330 system.

The comparison showed that there was no clear winner in terms of performance per watt. Depending on the workload, different platforms showed best performance per watt (i.e., power efficiency). For data mining workloads (I/O bound), both low power architectures showed a clear advantage (Atom 330 is 3-4x better than Xeon), while for more traditional web or compute-intensive workloads, the Xeon server was still the platform of choice. This suggests that mixing different platforms can be better in terms of power efficiency with diverse workloads. Further, the least power-hungry architectures, the Atom N270, exhibited the best performance/watt for Word Count compared to the other servers, but very little gain in performance/watt compared to Xeon processors for other workloads. This is due to the specific mini-PCI (peripheral component interconnect) solid-state drive used in the system that provides good read throughput but very low write throughput. Word Count benefits from this characteristic as it is mainly reads.

In order to understand how power consumption scales with load, power consumption of a Xeon server and multiple Atom servers as a function of the number of transactions per second were considered by Chun et al. [24]. To compare the performance over the entire range of load levels, they considered the case of adding more Atom-based PCs to handle the additional load. Based on this analysis, two observations were made. First, from the experiments, the study showed how a set of Atom-based platforms could be used to mimic an energy-proportional system. As load increases, the aggregate consumes power proportional to load at a macro level. Second, at a micro level, both platforms in isolation showed a quite narrow range of power consumption across a wide range of load levels. This is in line with prior work [28] that indicates that other system components, not the CPU, are responsible for the poor scaling of power consumption.

Chun et al. [24] also considered temporal framework of workload. Reports from data center operators indicate that servers run between 10% and 50% of their maximum utilization levels [28]. Servers process a continuous stream of task requests and operators try to distribute evenly across the data center to avoid high loads and meet latency SLAs. To explore the potential of hybrid solutions, Chun et al. [24] examined two possible hybrid solutions using one Xeon and one Atom platform by comparing them with Xeon-only or Atom-only solutions with goal not to design a specific strategy but rather to explore the performance and feasibility of such solutions for simple scenarios. The hybrid solutions that work well with multiple Atom and multiple Xeon platforms can be more challenging. To model the computing capabilities of the low power and high-performance platforms, Chun et al. [24] use two parameters: number of threads (T) that corresponds to the number of task requests that can be served in parallel and the computing capacity (C) normalized to Xeon performance. The actual ratio between Xeon and Atom was derived from the SPEC (standard performance evaluation corporation) power experimental results. The study only considered Atom 330 in modeling because Atom 330 and Atom N270 showed similar execution times. For simplicity, in this synthetic workload, the study assumed identical tasks that require a constant processing time. Finally, the study computed the power consumption which grows linearly between the minimum idle power (P_{idle}) and the power at full utilization ($P_{100\%}$) as observed experimentally.

With this analysis, Chun et al. [24] showed that very simple hybrid solutions (even with just one Atom and one Xeon platform) achieve good energy proportionality. Never migrating tasks appear to be a feasible strategy that leads to believe that simple software solutions could be within reach. The analysis showed that if running a task on an Atom platform satisfies data center SLAs, then a hybrid solution can preserve that guarantee. In short, this analysis showed that (a) low-power and high-performance platforms exhibit different power performance based on the workload and clearly a single solution cannot satisfy the wide range of applications seen in today's data centers; (b) many components contribute to the overall power consumption and servers have a narrow dynamic range; (c) the use of (simple) hybrid solutions may help in designing a data center architecture that gives low latency, good performance/watt, and energy proportionality in a wide range of workloads. Chun et al. [24] also outlined design options for hybrid approach and defined the design questions to be addressed and the challenges involved in exploiting the full potential of hybrid data centers in practice. Chun et al. [24] considered capacity planning and resource scheduling as well as hardware and software architecture in the design options. They indicated that while FAWN [17] and work of Lim et al. [23] are reported in the literature, their approach is the first to argue for the case for hybrid approaches and to explore different hybrid data center design options. Chun et al. [24] showed how hybrid data centers have the potential to provide energy efficient operations without sacrificing the performance levels that today's data center provides. They, however, admit that there exists a wide spectrum of possible solutions—some reachable in the short term (discrete solutions) and others that require large investments (heterogeneous cores). They all come with a different set of trade-offs and design challenges and further work is required to carefully evaluate each solution. Chun et al. [24] believed that hybrid data centers represent a good opportunity for future green-computing infrastructure.

9.3 HYBRID PROCESSES TO IMPROVE ENERGY EFFICIENCY OF DATA CENTERS

A measure used to gauge the energy efficiency of data centers is called the Power Use Effectiveness (PUE). PUE is the ratio of the amount of energy used in the center by the amount to run the processors. It is a number larger than or equal to 1 and it is desirable to make it close to 1. In other words, a smaller PUE is better because it shows that smaller energy is used for operations other than running the processors. An industry analysis a decade ago found an average PUE of 2.5. An organization, known as the Uptime Institute, publishes average PUE figures for the industry. In 2009, this number was declared to be 2.5 [29]. It is encouraging that this number is actually dropping very fast. In 2011, the Uptime Institute declared the PUE for the industry to be 1.8 [30]. An Uptime Institute study in 2014 studied the PUE of cloud data centers from Google and Facebook public disclosures plus AWS internal data, all of which show PUEs under 1.2 [31]. These numbers appear to be very good, since in 2008, the Uptime Institute declared that the typical data center has an average PUE of about 2.5, but that number could be reduced to about 1.6 employing best practices [32].

Some simple measures help in improving the PUE in data centers. These are decommissioning or repurposing servers which are no longer in use, powering down servers when not in use, replacing inefficient servers, and virtualizing or consolidating servers. Technology helps as well, by making use of intelligent power management, energy monitoring software, and efficient cooling systems. There is a data center described in Reference [4] that uses air conditioning only 33 h/y although the data center is always functioning. This is achieved by using an intelligent and hybrid cooling system.

Some drastic measures have been taken by the information technology industry to combat the problem of energy consumption. Considering the fact that cooling is a very important part of the overall PUE, the computing company Microsoft introduced an underwater data center [33,34]. Of course, it is not clear that warming the world's rivers and oceans is a good practice. But at the end of the day, the demand for significantly more computations will inevitably be with us, and it is important to keep the energy efficiency of the Internet and the cloud at a high level. It is good to lower the PUE of a data center as close to 1 as possible, but it is actually necessary to go a step further and make sure that the energy consumption by the search itself results in a minimal level of energy consumption.

Document [35] lists 12 methods to reduce energy consumption in data centers. These 12 methods can be grouped into changes in the information technology infrastructure, airflow management, and managing air conditioning. In terms of information technology, virtualizing servers, decommissioning inactive servers, consolidating lightly used servers, removing redundant data, and investing in technologies that use energy more efficiently can provide substantial improvements. One of the improvements in terms of managing air flow is a "hot aisle/cold aisle" layout where the backs of servers face each other so that the mixing of hot and cold air is avoided. In order to further reduce mixing hot and cold air, containing or enclosing servers is recommended. Improving air flow by means of simple measures such as using structured cabling to avoid restricting air flow is recommended. Finally, adjusting the temperature and humidity, employing air conditioning with variable speed fan drives, bringing in outside cooling air, and using the evaporative cooling capacity of a cooling tower to produce chilled water are recommended to potentially make significant changes.

There are a number of technological achievements to improve data center energy efficiency. Reference [36] proposes a resource management system by consolidating virtual machines according to current utilization of resources, virtual network topologies established between virtual machines, and thermal state of computing nodes. Virtualization is an important tool in data center energy efficiency. To that end, Reference [37] provides a survey of existing virtualization techniques. Reference [38] introduces an optimization based on scheduling tasks according to their thermal potential and with the goal of keeping the temperature low. This chapter introduces gains in temperature and cost reduction compared to other techniques. Traffic engineering is employed in Ref. [39] to assign virtual machines. This paper states, based on experimental results, 50% energy savings was achieved. Reference [40] provides an analysis of how increased ambient temperature will affect each component in a data center and concludes that there is an optimum temperature for data center operation that will depend on each data center's individual characteristics. In terms of shutting down inactive servers, Reference [41] introduces a technique that predicts the number of virtual machine requests, together with the amount of CPU and memory resources of these requests, provides accurate estimations of the number of physical machines that will be needed, and reduces energy consumption of cloud data centers by putting to sleep unneeded PMs. The study shows the technique achieves substantial savings in energy consumption [42].

Ebrahimi et al. [43] reviewed data center cooling technology, operating conditions, and corresponding waste heat recovery opportunities. The study indicated that a major source of waste energy is being created by data centers through the increasing demand for cloud-based connectivity and performance. In fact, recent figures show that data centers are responsible for more than 2% of the U.S. total electricity usage. Almost half of this power is used for cooling the electronics, creating a significant stream of waste heat. The difficulty associated with recovering and reusing this stream of waste heat is that the heat is of low quality. In this study, the most promising methods and technologies for recovering data center low-grade waste heat in an effective and economically reasonable way are identified and discussed. A number of currently available and developmental low-grade waste heat recovery techniques including district/plant/water heating, absorption cooling, direct power generation (piezoelectric and thermoelectric (TE)), indirect power generation (steam and organic Rankine cycle), biomass colocation, and desalination/clean water are reviewed along with their operational requirements in order to assess the suitability and effectiveness of each technology for data center applications. Based on a comparison between data centers operational thermodynamic conditions and the operational requirements of the discussed waste heat recovery techniques, hybrid processes of absorption cooling and organic Rankine cycle are found to be among the most promising technologies for data center waste heat reuse.

Ayanoglu [14] and the report by the Lawrence Berkeley National Laboratory [3] have an encouraging conclusion. It states that improved energy is almost canceling out growing capacity. In 2014, data centers in the United States consumed 70 billion kWh. If energy efficiency levels remained as they were in 2010, the energy consumption by data centers today would be 160 billion kWh. The surprising reality is that the estimate for 2020 is only 73 billion kWh [3]. However, although short-term predictions appear to be good, there are still concerns for the long-term future, such as ten years in the future [42].

9.4 ROLE OF HYBRID RENEWABLE ENERGY FOR DATA CENTERS

Interest in renewables is driven by two key requirements: environmental sustainability and energy savings. Power consumption within a data center environment is substantial, accounting for almost half of total operating expenses. When one factors in escalating and unpredictable energy costs, along with fees associated with pending carbon emission legislation, it is understandable that operators are increasingly focused on the role renewables can play in reducing and/or stabilizing energy costs.

Hyperscale data centers like Google's data center have stood at the forefront in renewable energy initiatives, as pioneers exploring their inherent business, environmental, and social value. Google sites its data centers based on a number of factors related to reliable service delivery, in locations that may not offer the best potential for renewable power. As a result, it is using renewable energy in the form of wind and solar by engaging in power purchase agreements to power over 35% of its operations, an approach that encourages the development of renewables on the part of utility providers. Facebook is building facilities in Iowa that will be 100% powered by a local wind project the social network helped to develop as well as a data center in the cool Swedish climate to take advantage of the local hydropower. Microsoft, for its part, has built in Quincy, Washington, using 100% hydropower, and is experimenting with powering a 200 kW data center in Wyoming with biogas from a local Cheyenne wastewater treatment facility. A third approach is demonstrated by companies like Apple, which has built into its Maiden, NC, facility a 100 acre, 20 MW plant that enables total reliance on renewables, or Verizon Communications, which announced plans in 2013 to invest $100 million in the installation of solar panels and fuel cells at more than a dozen facilities that will use and generate the clean resource.

Renewables of choice for data centers include rooftop solar, wind, geothermal, and waste heat reclamation. Solar has limitations in light of the high cost of photovoltaic (PV) solar arrays, climatic requirements, and space restrictions. However, it has become one of the more widely used approaches in the data center environments since rooftop real estate, when available, is virtually free, reducing the cost of implementation. Wind turbines are less widely used, largely because of real estate needs and cost, although interest in this resource is growing. Projects are also underway for geothermal (particularly in the U.S. mid-west) and in waste heat recovery. Battery storage is also in top of mind in renewable management discussions as storage can mitigate issues with intermittency. While the technology is viable, large-scale implementation is out of reach for many enterprises from financial and space perspectives. Once pricing comes down, battery storage promises to play an important role in overcoming some of the reliability concerns that are inherent in renewable power. Renewables plus battery storage or multiple renewables (like solar and wind) or a mix of renewable with conventional diesel source makes all systems hybrid in nature.

When considering a renewables strategy, facilities operators should take into account two points of convergence: the need to better manage energy consumption onsite and the potential to generate loads that can build value through the creation of renewable energy credits or the sale of surplus energy back to the main utility grid. A key driver behind the adoption of renewables is the implementation of time-of-use rates by energy providers. Data center operators are continuously exploring alternatives to manage demand and/or mitigate reliance on the grid during peak rate times (i.e., periods of highest electricity demand on utilities when additional fees are charged); use of onsite renewables (or sale into the grid) during peak rate times could help manage costs. While renewables have presented challenges from a capital cost perspective, the landscape is changing. Cost for renewables such as solar has recently come closer to achieving parity with the cost of energy from utility grid, due in part to incentive programs introduced by governments in many jurisdictions to offset the higher implementation costs associated with renewable generation. In addition, once in place, solar, wind, and geothermal provide a free and limitless energy source, ultimately translating into lower cost of ownership for the facility.

9.4.1 TECHNOLOGY CAPABILITIES

In decision-making around renewables, there are a number of dynamics that come into play. The first question to ask is how the data center consumes electricity and to what extent consumption is controllable. In other words, can renewables play a role in delivering controllable, interruptible load and can they be integrated in a way that enables them to provide added value to the data center? And can control systems be installed that would allow interface with the utility for sale back of excess energy? Recognizing the existing limitations of renewables, data center operators can still realize their benefits by introducing microgrid platforms that integrate energy from multiple resources, including power from the grid, diesel generation, and renewables. This hybrid energy approach is proving especially useful in regions where grid-delivered power is costly or unreliable.

Power management technologies are key to the successful integration of renewables, and many generation solutions aimed initially at the plant or utility provider may be adapted to suit the data center environment. Leading suppliers like ABB provide (solar) inverters, low-voltage products, monitoring and control systems, grid connections, as well as stabilization and integration products. ABB's product line provides an example of the kinds of technologies needed to create a renewables platform for the data center. The ABB solar inverters enable data centers to convert direct current (DC) electricity generated by solar modules into alternating current (AC) with 98%+ efficiency, and the company's hybrid power plant will integrate one or several fuel-based generators and one renewable energy generator to enable combined generation. ABB adds engineered solutions for integrating renewables into an existing fuel-based generation grid (integrated wind or PV power plant) to increase stability and optimize energy flow. In order to manage the erratic fluctuations in frequency and voltage characteristic of renewable energy, ABB offers a flywheel-based generator and software which controls power in and out of the flywheel, acting as what ABB calls a "high inertia shock absorber that can instantly smooth out power fluctuations." For data center facilities interested in building their own plant and load, these power management technologies can help solve for issues with renewables [44].

9.4.2 IMPLEMENTATION CHALLENGES

The reality is that renewable options typically struggle to address 100 percent of the enterprise data center's power needs. There are few scenarios in which onsite or even rooftop solar PV would fully supply a data center's energy needs simply because there is not enough space available to house all the solar arrays that would be needed. In addition, weather conditions can have a dramatic impact on generation capacity: ultimately solar is not dispatchable power—i.e., capable of reliably delivering the required level of power when it is needed. Energy production can be extremely variable, depending on time of day and weather conditions: a cloud passing over the sun could drop power generation from 5 MW to 100 kW in a matter of seconds. And while rooftop solar may have a natural "home" that does not add to operational costs, wind generation would require substantial real estate to generate energy at any meaningful levels. An underground source of energy, geothermal also comes with significant and costly logistics and infrastructure requirements. All this means that insertions of renewable sources are generally carried out in hybrid manner.

The challenge for many data center facilities interested in onsite generation is that renewable technologies as they stand today are rarely a whole solution for the supply of energy 24/7 due to intermittency (in the case of wind and solar), real estate limitations, and implementation costs or all of the above. But combining various resources such as rooftop solar, biofuel, battery, and fuel cells and backing them up with diesel generation that is onsite for redundancy purposes can help to leverage and extend the role of renewables in the data center. The choice of resources depends on a number of variables, including climate, local utility provider fees, and available access to renewable resources. As Google and others have demonstrated, data centers can also sidestep the real estate and infrastructure cost challenges through partnerships with local power providers whose stock in trade is renewable energy production.

9.4.3 BENEFITS

There is no technical reason why most data center facilities cannot consider renewable energy alternatives (though facilities located within high-rise building complexes in highly congested urban environments may encounter special challenge). However, deployment decisions do not depend solely on questions of technology viability, rather the decision becomes one of the basic economics: does a renewables strategy equate to real savings? Renewables, for example, would represent an interesting discussion point for a data center operation in New York City where space is limited but per kWh costs for grid energy, which is distributed by cable, are approximately double that in New Jersey, which uses an overhead distribution model. When making the case for renewables, there are a number of questions that need to be addressed in planning, ranging from reliability and resiliency requirements to the role renewables can play as a primary, complementary, or backup energy provider. The organization may also wish to weigh the importance of renewables in corporate social responsibility strategy, an intangible benefit that Facebook attempted to ignore to its peril. Since renewables possess several challenges, the safest approach is to hybridize renewables first with conventional sources or storage and the slowly increasing contribution of renewables as business and technological maturity implies.

The argument for renewables in data centers is increasingly compelling. Not only are renewables a central component in reducing an environmental footprint, they are also now viewed as a reliable backup source of power that can reduce reliance on what in some regions are overloaded utility grid systems. While hyperscale data centers account today for the lion's share of investment in renewable strategies, enterprise-level operations are showing increasing interest as the economic value proposition of alternative energy becomes clearer. Some operations are already well entrenched in renewables strategies; others are still in the planning stages and exploring issues such as what aspects of energy use are controllable, what are the existing energy contracts, capacity needs, incentive program opportunities, and their limitations (e.g., runtime, renewal qualification). When these questions are answered, operators will be in a better position to make informed decisions around what percentage of supply can be sources through renewables, and what enabling technologies make the most economic and environmental sense. It is, however, clear that insertion of hybrid renewables will decarbonize the data center operations.

9.4.4 Hybrid Renewable Energy Green-Works Framework for Data Centers

In order to address the above-described issues, Li et al. [45] proposed management of green data centers managed by hybrid renewable energy systems. The global server power demand reaches approximately 30 GW in total [46], which account for over 250 million metric tons of CO_2 emissions per year [47]. Faced with a growing concern about the projected rise in both server power demand and carbon emissions, academia and industry alike are now focusing more attention than ever on nonconventional power provisioning solutions. For instance, recently, there have been vigorous discussions on renewable energy-driven computer system design with respect to carbon-aware scheduling [48–50], renewable power control [51–54], and cost optimization strategies [55,56]. In addition, Microsoft, eBay, HP, and Apple have announced projects that use green energy sources like solar/wind power, fuel cells, and biogas turbines to minimize their reliance on conventional utility power [57–60]. It has been estimated that these eco-friendly IT solutions could reduce almost 15% global CO_2 emissions by 2020, leading to around $900 billion of cost savings [63].

As mentioned above, the expected growth in renewable power generation poses new challenges for data center operational resilience. A number of the renewable energy sources are *intermittent power supply*, such as wind turbine and solar array. They are free sources of energy but incur power variability problems. Several emerging green power supplies, such as fuel cells and biofuel-based generators, are typically used as *baseload power supply*. They are stable and controllable power sources, but not fast enough to respond instantaneously to quick changes in server power demand. In case the intermittent power supply drops suddenly or the baseload power supply cannot follow an unexpected power demand surge, *backup power supply* (e.g., batteries, supercapacitors) must be used to handle the power shortfall. As we move toward a smarter grid, data centers are expected to be powered by hybrid renewable energy systems that combine multiple power generation mechanisms [62]. With an integrated mix of complementary power provisioning methods, one can overcome the limitations of each single type of power supply, thereby achieving better energy reliability and efficiency.

The existing schemes can be classified into three broad categories: (a) **load shedding**, which focuses on utilizing intermittent power [51,54], typically reduces load when renewable power drops; (b) **load boosting**, which uses both intermittent and backup power [53,63], takes advantage of the stored energy to maintain desired performance when the current green power generation is inadequate; and (c) **load following**, which assumes both baseload and backup power [64], leverages tunable generators to track data center load demand. Since prior proposals lack the capability of managing renewable energy mix, they can hardly gain the maximum benefits from hybrid renewable energy systems and, consequently, yield suboptimal design trade-offs.

The study by Li et al. [45] explores diversified multisource power provisioning for green high-performance data centers today and in the future. The study proposes Green Works, a framework for managing data center power across several layers from data center server to onsite renewable energy mix. Green Works comprises two key elements: the *green workers*, which are multiple platform-specific power optimization modules that use different supply/load control strategies for different types of renewable energy systems; and *green manager*, a hierarchical coordination scheme for green workers. Green Works tackles the challenges of integrating and coordinating heterogeneous power supplies with a three-tiered hierarchical coordination scheme. Each layer of the hierarchy is tailored to the specific timing and utilization requirements of the associated energy sources. In addition, power management modules in different layers of the hierarchy can also interact with each other within the framework. This allows one to further improve the power management effectiveness of hybrid renewable energy systems.

Green Works emphasize a multiobjective power management. It jointly manages green energy utilization, backup energy availability, and workload performance. Specifically, three types of green workers can be defined: (a) **baseload laborer**, which adjusts the output of the baseload power to track the coarse-grained changes in load power demand; (b) **energy keeper**, which regulates the use of the stored renewable energy to achieve satisfactory workload performance while maintaining desired battery life; and (c) **load broker**, which could opportunistically increase the server processing speed to take advantage of the excess energy generation of the intermittent power supply. All the three modules are able to distill crucial runtime power profiling data and identify appropriate control strategies for different types of renewable generation.

The study by Li et al. [45] makes three main contributions:

1. **It proposes Green Works**, a hierarchical power management framework for green data centers powered by renewable energy mix. It enables cross-source power management coordination, thereby greatly facilitating supply-load power matching.
2. **It proposes a multisource-driven multi-objective power management** that takes advantage of a hierarchical power management framework. The proposed technique enables Green Works to maximize the benefits of the hybrid renewable energy systems without heavily relying on any single type of power supply.
3. **It evaluates Green Works using real-world work-load traces and green energy data**. The study shows that Green Works could achieve less than 3% job runtime increase, extend battery lifetime by 23%, increase uninterruptible power supply (UPS) backup time by 12%, and maintain the same energy efficiency as the state-of-the-art design.

9.4.4.1 Hybrid Renewable Energy Systems

Green Works proposes that there are three types of renewable power supplies that one can leverage to power a data center. Some green power supplies, such as solar panels and wind turbines, are affected by the availability of ambient natural resources (i.e., solar irradiance or wind speed). They are referred to as *intermittent power supply* since their outputs are time-varying. Several emerging green power supplies—including fuel cells, biofuel-based gas turbines, and biomass generators—can offer controllable green energy by burning various green fuels. They are referred as *baseload power supply* since they can be used to provide stable renewable power to meet the basic data center power demand (e.g., idle power). In addition, energy storage devices (ESDs) such as batteries and supercapacitors are also critical components that provide *backup power supply*. They can be used to temporarily store green energy or improve power quality.

Looking ahead, data centers in the smart grid era are expected to be powered by *hybrid renewable energy systems* that combine all three types of power supplies. Different power supplies are typically implemented as small, modular electric generators (called microsources) near the point of use. To manage such an integrated renewable energy mix, microgrid is proposed as a coordinated cluster/network of supply and load [65]. Although the microgrid allows its customer to import power from the utility, Green Works focuses attention on minimizing the reliance on utility power due to sustainability and cost concerns. In previous studies, energy source management and data center load management are largely decoupled. Furthermore, existing microgrid control strategies often focus on power supply scheduling [66] and recent proposals on power-aware data center mainly emphasize demand response control [67,68]. In contrast, Li et al. [45] propose load/supply cooperative power management across several layers from servers to hybrid renewable energy systems.

9.4.4.2 Energy Balance Challenge

Many system-level events can cause power demand fluctuations, such as dynamic power tuning via dynamic voltage and frequency scaling (DVFS), on/off server power cycles, and random user request. Unexpected variations in intermittent power supplies, unfavorably combined with data center workload fluctuation, could make the power mismatch problem even worse. Therefore, matching data center load to the variable power budget is often the crux of eliminating power disruptions in a green data center. Managing multisource powered system can be a great undertaking. Microsources like batteries, flywheels, fuel cells, and gas turbine have different characteristics and operating timeframes. Most baseload green power systems cannot meet the needs of fast supply-load power matching. For example, both fuel cells and gas turbines need time to be committed and dispatched to a desired output level. They provide a slow energy balance service called *load following*, which typically occurs every tens of minutes to a few hours [69]. Li et al. [45] showed that the load matching can be achieved by using real-world data center traces and renewable energy data sets. The example used was intermittent wind supply. Load following alone, however, cannot eliminate fine-grained power mismatch. When the wind power is stable, fluctuating load can be the main cause of power mismatch; when wind power output varies, it can significantly increase mismatch events. Although increasing the baseload power output can reduce the chance of brownout, it will significantly increase the operational expenditure.

One also cannot heavily rely on utility power grid and energy backup to manage the demand-supply power mismatches. First, it requires additional standby power capacity, which is economically unfavorable. Energy backup services are typically much more costly than the load following services [70]. Second, grid-inverter and battery incur round-trip energy loss, which degrade overall system efficiency. Third, heavy reliance on backup power supply can be risky. As recent survey indicates, data centers in the US experience 3.5 times of utility power loss per year with an average duration over 1.5 hours [71]. It also shows that UPS battery failure and capacity exceeded are the top root causes of unplanned outages. Without appropriate coordination, the demand–supply power mismatch can cause frequent battery discharging activities, which not only decrease the battery lifetime but also frequently deplete the stored energy that is crucial for handling emergencies. Li et al. [45] explored a holistic approach for eliminating the supply-load mismatch problems in green data centers. Specifically, the study looked at how cross-source power management and coordination will help to improve energy balance and data center resilience.

9.4.4.3   The Green Works Framework

Green Works is a hierarchical power management scheme that is tailored to the specific timing and utilization requirements of different energy sources. It provides coordinated power management across intermittent renewable power supplies, controllable baseload generators, onsite batteries, and data center servers. The intention of the study was to provide an initial power management framework for data centers powered by renewable energy mix. In the smart grid era, data centers must increase their awareness of the attributes of power supplies to achieve the best design trade-offs.

The study adopts typical microgrid power distribution scheme for managing various renewable energy resources. Various renewable energy systems are connected to the power feeder through circuit breakers and appropriate interfaces. Green Works is a middleware that resides between front-end computing facilities and back-end distributed generators. It manages various onsite energy sources through a microgrid central controller, which is a typical power management module in the microgrid system. The controller is able to adjust onsite power generation through communication with the dedicated power interface connected to each distributed generators. Green Works also communicates with the UPS battery rack, the cluster-lever power meters, and the server-level power control module. It cooperatively adjusts power supplies and workload performance levels, and thereby eliminates demand–supply power mismatch.

Green Works comprises two key elements: the **green workers and the green manager**. The former are platform-specific power management modules for managing different types of microsources and the later coordinates these modules. In this study, we define three types of green workers: **baseload laborer** (B), **energy keeper (E), and the load broker (L)**. The **baseload laborer** controls the output of distributed generators such as fuel cells and biofuel generators. It is responsible for providing a specific amount of baseload power to satisfy the basic power needs (i.e., data center idle power). It can also provide load-following services [69] at each coarse-grained time interval. The **energy keeper** is able to provide necessary power support if intermittent power supply drops suddenly or load surge happens. It also monitors the capacity utilization and the health status of the battery packs. Green Works uses distributed battery architecture (at server cluster level) since it has better energy efficiency, reliability, and scalability [72]. The **load broker** is responsible for managing the fine-grained power mismatch between the fluctuating data center load and the intermittent power supply. Green Works leverage the performance scaling capability (via CPU frequency scaling) of server system to match load power demand to time-varying green energy budget.

Although the hybrid renewable energy systems are often centrally installed at the data center facility level, improving the overall efficiency requires a multilevel, cooperative power management strategy. Green Works uses a three-tier control hierarchy for power management coordination. It organizes different types of green workers in the power management hierarchy based on their design goals. In the top tier of the hierarchy is the baseload laborer. Green Works put the load laborer at the data center facility level since it is where the baseload power generator is integrated. Managing baseload power budget at data center level facilitates load-following control, thereby minimizing over-/under-generation of the baseload renewable energy. Green Works manages the intermittent renewable power supply at the cluster level, or power distribution unit level. At this level, DVFS shows impressive peak power management capabilities [73] and could be leveraged to manage the supply–load power mismatch. During runtime, the load broker calculates the total renewable power generation based on the baseload power budget and the assigned renewable power. When the total renewable power generation is not enough, the load broker will decrease server processing speed evenly or request stored energy (from the energy keeper), depending on whichever yields the best design trade-off.

The energy keeper resides in the third tier of the hierarchy. This allows Green Works to provide backup power directly to server racks if local demand surge happens or power budget drops. Such distributed battery architecture [72] has many advantages such as high efficiency and reliability. In this study, we leverage it for managing fine-grained supply-load power mismatches. The main advantage of multilevel cooperative power management scheme is that it facilitates cross-source power optimization. For example, Green Works allows data centers to schedule additional baseload generation to release the burden of the energy backup when the capacity utilization of onsite batteries is high. It also allows them to request additional stored renewable energy to boost server performance if necessary.

The study by Li et al. [45] also proposes multisource-driven multiobjective power management for Green Works. The basic idea is to take advantages of the cross-source coordination capability of Green Works to balance the usage of different types of energy sources. To achieve this goal, the study develops a novel three-stage coordination scheme that synergistically combines battery-aware power management, workload-aware power management, and variability-aware power management to achieve the best design trade-offs. The details of these three stages are described by Li et al. [45]. The study also develops a simulation framework for data centers powered by renewable generation mix and HOMER software.

The study evaluates the benefits of applying Green Works to data centers powered by hybrid onsite green power supplies. It compares Green Works to two state-of-the-art baselines: **Shedding and Boosting**. *Shedding* is a widely used load management schemes for emerging renewable energy powered data centers [74,75]; *Boosting* represents recent data center power management approaches that emphasis the role of ESDs [76,77]. Both baselines use UPS and server load scaling to manage fine-grained power short-fall and adjust baseload output level at each end of the control period. The only difference between the two is that *Shedding* gives priority to load scaling, while *Boosting* gives priority to UPS stored energy.

9.4.4.3.1 Execution Time

The study evaluates data center performance in terms of average job turnaround time increase compared to an oracle (which always ensures full processing speed with zero service downtime). On average, the job execution time increases of *Shedding*, *Boosting*, and Green Works are 5.4%, 2.1%, and 2.4%, respectively. Compared to *Shedding*, *Boosting* shows less execution time increase since it trades off UPS capacity for performance. As Green Works seeks a balanced power management across different power supplies, it yields slightly higher energy transitions initiative (ETI) compared to *Boosting*. The performance of the worst 5% jobs could significantly affect the SLAs of data centers. The worst-case result of *Shedding* is 28%. Surprisingly, Green Works (12%) reduces the maximum job execution time increase by 33%, compared to *Boosting* (18%).

9.4.4.3.2 Energy Efficiency

The main sources of inefficiency in green data centers are the battery round-trip power loss and the power conversion loss in the grid-tied inverter. The study assumes a typical battery system of 80% round-trip energy efficiency and a power inverter of 92% energy efficiency. Green Works could maintain the same energy efficiency as *Shedding* and *Boosting*. ESDs should be always taken care of. A lower autonomy time can pose significant risk as the backup generator may not be ready to pick up the load. Without appropriate power management and coordination, data centers have to increase their installed UPS capacity, which is both costly and not sustainable.

9.4.4.3.3 Control Sensitivity

The results were obtained for the impact of the x% shedding mechanism on various performance matrix of Green works and the impact of the control intervals on the performance of multisource-driven multiobjective control. Both results were favorable.

9.4.4.3.4 Battery Lifetime

Typically the rated lifetime of a valve-regulated lead–acid battery is 3–10 years [45]. Green Works showed a near-threshold battery life (8.3 years). It means that multisource multiobjective power management can maximally leverage batteries without degrading their life significantly. In contrast, *Boosting* shows a mean lifetime of 6.7 years, and *Shedding* shows a mean lifetime of 19.7 years. Typically, the battery lifetime is not likely to exceed 10 years [78]. The reason *Shedding* overestimates battery life is that the system underutilizes batteries. Since batteries may fail due to various aging problems and self-discharging issues, it is better to fully utilize it.

9.4.4.3.5 UPS Back-up Time

Another advantage of Green Works is that it can optimize the mean UPS autonomy time. The autonomy time is also known as backup time. It is a measure of the time for which the UPS system will support the critical load during an unexpected power failure. On average, the mean autonomy time is *Shedding* (88%), *Boosting* (70%), and Green Works (78%). Green Works could ensure rated backup time (the discharge time of a fully charged UPS) for 20% of the time. *Shedding* maintains its rated backup time for 50% of the time and the number for *Boosting* is only 10%. This is because *Boosting* uses UPS battery much more aggressively than *Shedding*.

The design of Li et al. [45] Green Works mainly focus on certain specific type of green energy sources (i.e., intermittent power or baseload generators) and overlook the benefits of cross-source coordination. In summary, Green Works facilitates multisource-based green data center design and enables data centers to make informed power management decisions based on the available baseload power output, renewable power variability, battery capacity, and job performance. Green Works could achieve less than 3% job runtime increase, extend battery life by 23%, increase UPS backup time by 12%, and still maintain desired energy utilization efficiency [45–98].

9.5 HYBRID STORAGE DEVICES FOR DATA CENTERS

In order to improve operation of data center operated by hybrid energy systems, Sun et al. [99] devised hierarchical and hybrid storage devices for data centers. Modern data center investments comprise one-time infrastructure costs that are amortized over the lifetime of data center (capex) and monthly recurring operating expenses (opex) [100]. The capex is directly impacted by data center's peak power requirement, which determines the provisioned capacity of power infrastructure and is estimated at $10–20 per Watt [101]. The opex is charged by utility company based on high power tariff scheme and dynamic energy pricing policy [102], and has been steadily increasing [103]. Reducing both capex and opex of a data center has become a key enabler to ensure its economic success. Because (a) the capex depends on the largest provisioning power and (b) up to 40% opex is caused by the peak power tariff [104], *power capping* is widely studied for modern data centers to reduce the peak power, thereby simultaneously reducing the capex and opex. Majority of power capping techniques focus on (a) throttling computing devices [105,106], (b) shifting the workload peak draw temporally/spatially [107,108], and (c) improving nonpeak/idle power efficiency of servers [109,110]. These solutions can have adverse performance consequences depending on the workload behavior. This will become a problem for any workload that has performance constraints or SLAs.

Recently, a new approach has been introduced that leverages and overprovisions ESDs in data centers for facilitating capex/opex reductions [101,102,104,111,112], without performance overhead. After overprovisioning, ESDs can be leveraged for power capping [101] and peak power shaving [104,113]. Reference works [101,104,113] have demonstrated that the benefits from capex and opex reductions outweigh the extra costs associated with ESD overprovisioning. ESDs in data center are commonly made of lead-acid batteries and utilized as centralized UPS [114,115]. Potential ESDs also include Li-ion batteries, supercapacitors, flywheels, and compressed air energy storage (CAES). Lead-acid and Li-ion are the most widely adopted ESDs, due to their good reliability and high energy density [116–118]. Particularly, Li-ion batteries have significantly high energy density, high efficiency, long cycle life, and environmental friendliness, and therefore is one of the most promising technologies in electrical energy storage [115]. Supercapacitors have much higher power density due to the electrochemical double-layer structure [117,118]. Flywheels depend on the momentum of rotating wheel/ cylinder to provide temporary power [118], while air in CAES is compressed to store electrical energy and decompressed to discharge energy [102,118]. Some key character- istics for various types of ESDs are described in great detail in my previous book [119]. Most applications employ a single energy storage technology, while for certain applica- tions with more advanced requirements, it is desirable to use two or more ESDs with complementary characteristics, either by combining different batteries or by integrating a battery with a supercapacitor, flywheel, etc. [115]. The power delivery architecture of a data center with the centralized ESD structure is illustrated in Figure 9.1.

Various research has studied the issue of charging strategies for ESDs. A model- predictive control-based charging strategy for Li-ion batteries is formulated in Ref. [120]. Various types of models for ultracapacitors are also examined [121,122]. Traditional power delivery facilities in data centers adopt centralized ESD struc- ture developed by Intel [123]. In this architecture, the 480V AC power from the utility grid or alternatively the diesel generator must first go through AC-DC-AC double conversion with the centralized ESD connected in-between (because the ESD is essentially DC). This AC-DC-AC double conversion can guarantee seam- less transition from the utility grid to ESDs and then to diesel generator, but it is the primary source of inefficiency in this design. This power architecture may result in 20%–30% power loss [123]. Some state-of-the-art data centers by Google [124], Microsoft [125], and Facebook [126] employ the distributed single-level ESD struc- ture, where ESDs are integrated into rack or server level and directly connected to the corresponding DC power buses. In the distributed rack-level ESD structure from Microsoft [15], ESDs are directly connected (without converters) to the rack- level DC bus. On the other hand, in the distributed server-level ESD structure from Google [124], ESDs are directly connected to the server-level DC bus. Compared with the centralized counterpart, the distributed single-level ESD structure achieves less transmission power loss and thereby higher efficiency. However, the distributed single-level ESD structure may encounter serious volume/real-estate constraints since the space inside each rack is precious and limited, thereby restricting the ESD size and capability in performing power capping.

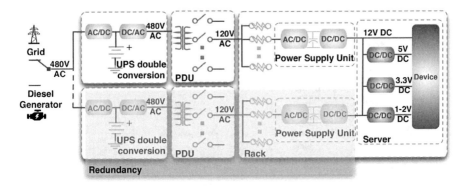

FIGURE 9.1 The power delivery architecture of a data center with the centralized ESD structure [99].

A hierarchical ESD structure could address this shortcoming by placing ESDs to data center, rack, and server levels, with the potential of taking advantages of both centralized and distributed ESD structures [102]. Moreover, proper deployment and control of multiple types of ESDs might achieve high capability in power capping and high energy capacity simultaneously [102,117,127,128]. To fully realize the potential benefits of the hierarchical ESD structure, the study proposes a comprehensive design, control, and provisioning framework, including (a) designing the *power delivery architecture* (i.e., detailed and feasible connections of ESDs, power buses, power converters), supporting the hierarchical ESD structure, and potentially hybrid ESDs (with more than one ESD type) for some levels, which is lacking in literature; (b) control and provisioning of the hierarchical ESD structure including runtime ESD charging/discharging control and design-time determination of ESD type, homogeneous/hybrid options, ESD provisioning at each level. The proposed hierarchical ESD structures borrow the best features of centralized ESD structure from Intel [123] and distributed single-level ESD structures from Google [124] and Microsoft [125], therefore simultaneously mitigate efficiency (and redundancy) shortcomings of the centralized structure and the space limitation of distributed single-level structures.The power delivery architectures of a data center with distributed single level ESDs by Microsoft and Google versions are illustrated in Figure 9.2

The proposed *Hier-Homo* and *Hier-Hybrid* architectures have respective advantages and applicable conditions. The *Hier-Homo* architecture is relatively straightforward to implement and control, whereas the *Hier-Hybrid* is a more general architecture that can be reduced to *Hier-Homo* if only the primary ESDs are employed. The *Hier-Hybrid* architecture is more advanced to further reduce the cost, more complicated in control and provisioning, and therefore more suitable for large-scale data center systems.

The *Hier-Homo* architecture combines the advantages of both centralized and distributed single-level ESD structures while hiding their weaknesses. It avoids

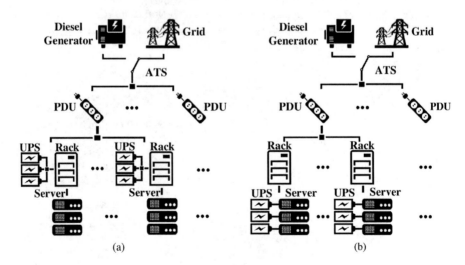

FIGURE 9.2 The power delivery architectures of a data center with the distributed single-level ESDs [99]. (a) Microsoft version, (b) Google version.

AC-DC-AC double conversion if the data center-level ESD/UPS is not in use (by switching to the high efficiency mode) and directly connects rack-level and server-level ESDs to corresponding DC buses, thereby significantly reducing the power losses. When power outage happens, the ESDs at each level can immediately provide backup power during the grid to diesel generator transition. High power supply reliability can be achieved without redundancies.

Since different types of ESDs exhibit distinct characteristics such as energy density, power density, etc., it may be beneficial to incorporate hybrid ESDs to accommodate different power demands, e.g., utilizing supercapacitors for short-term high-peak power demands while deploying batteries to accommodate long-term and relatively low-peak power demands. The hybrid ESD at each level comprises two parts: (a) an ESD bank with relatively stable terminal voltage, termed the *primary* ESD, which is directly connected to the DC bus (similar to the homogeneous ESDs in the *Hier-Home* architecture), (b) an additional ESD bank, termed the *secondary* ESD, which is connected to the DC power bus through a bidirectional converter (or other power conversion and auxiliary circuitry if flywheels or CAES are adopted). The "bi-directional" property is necessary due to the requirement of charging and discharging of secondary ESD. Please note that the hybrid ESD is a general concept, in that it can possibly comprise only the primary ESD without secondary ESD (similar to the ESD in *Hier-Homo* architecture), or only the secondary ESD.

The *Hier-Hybrid* architecture also avoids AC-DC-AC double conversion if the data center-level ESD/UPS is not in use. The primary ESDs in the *Hier-Hybrid* architecture are also directly connected to the corresponding DC buses, similar to ESDs in the *Hier-Homo* architecture. However, the secondary ESDs are connected to the corresponding DC buses through a DC/DC converter, which will inevitably incur power loss. Therefore, the actual benefit for applying hybrid ESDs needs to be scrutinized when accounting for the more sophisticated power delivery architecture and control mechanisms [99–132]. Illustrations of proposed Hier-Homo and Hier-Hybrid architectures are described in Figures 9.3a and 9.3b respectively.

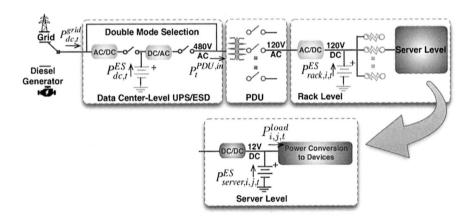

FIGURE 9.3a   Illustration of the proposed *Hier-Homo* architecture [99].

(*Continued*)

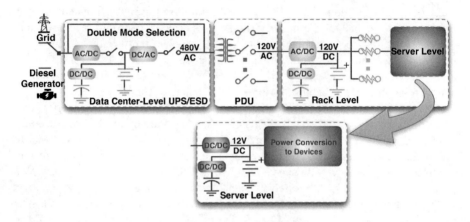

FIGURE 9.3b   Illustration of the proposed *Hier-Hybrid* architecture [99].

9.6   FORMS OF HYBRID ENERGY IN DATA CENTERS

There are numerous ways hybrid energy can be used in the data centers. Here we examine few options.

9.6.1   HEAT INTEGRATION

Practically all electricity consumed in a data center is eventually transformed into heat that must be removed in order to keep the data center running. Even with energy-efficient cooling, large amounts of waste heat are released that could have been utilized. Wahlroos et al. [133] suggest focusing on energy reuse efficiency instead of PUE as an efficiency metric to capture the use of waste heat. The waste heat from data centers is of low grade, and the temperature depends on the technology used for cooling. For example, air cooling typically results in outlet temperatures of 25°C–35°C, while outlet temperatures of up to 60°C are possible with liquid cooling, and novel two-phase systems can provide outlet temperatures up to 90°C [43]. There are numerous ways in which waste heat can be utilized, such as the most common being direct use of waste heat for nearby offices or apartment blocks. It has also been suggested that small-scale distributed servers could be used as heat sources in individual apartments [134], where temperatures of 25°C–35°C would be sufficient to meet heating demands. On a larger scale, utilizing waste heat in district heating systems is becoming common in Northern Europe [133–136]. For use in district heating, low-grade heat must be upgraded with heat pumps in order to reach temperatures of 75°C–120°C for distribution in district heating networks, although fourth-generation district heating can be run with supply temperatures of 45°C–55°C [135,137].

Several data centers are currently developing this concept in Scandinavia [135]. Stockholm Data Parks [138] is an example where several data centers are colocated around a central heat pump station that increases the temperature of the waste heat and supplies the district heating network. This combined siting reduces the risk of the district heating company being reliant on a single data center operator. A third way of utilizing waste heat is for drying or heating in industrial processes. This is a development of the industrial symbiosis concept, where geographical proximity enables waste heat recovery/use between different plants. Drying of biomass is one option, as well as preheating in thermal power plants, thus increasing the efficiency [43]. Waste heat can also be used for low-temperature applications such as greenhouses and aquaculture.

9.6.2 Demand Response

Integrating renewables in their electricity mix has become interesting for some data center operators in recent years as a means of reducing their carbon footprint and for branding. Numerous workload management strategies that align data center activity with the availability of renewables have been suggested to this end. Although the objective so far has often been to increase the utilization of onsite renewables or to take advantage of low grid prices, the same strategies could be used to engage the data center in power flexibility markets. Various strategies can be used to achieve a load shift. Scheduling delay-tolerant workloads within a data center can lead to a temporal load shift [139,140]. For operators with geographically distributed data centers, workloads can be shifted between data centers in the network, leading to a spatial load shift [141,142,143]. By implementing constraints in the algorithms, workloads, and thus power demand, can be shifted without any negative impact on the quality of service.

In an electricity system dominated by solar and wind power, a major challenge will be to develop a reliable power system where flexible load management will be a key feature [144]. Data center load management strategies have considerable potential in this respect. As data centers are large electricity consumers, their participation in demand response is meaningful for system stability. They are also highly connected and automated, with components that are already monitored and responsive to control signals. These characteristics make them suitable candidates for demand response, with minimal infrastructure investments [145].

An U.S. field study by Ghatikar et al. [146] demonstrated that data centers can participate in demand response events with a duration of up to several hours and with a response time of 2–22 minutes. The study focuses on current possibilities and on automated response markets, and includes both geographic and temporal flexibilities. The amount of workloads that could be shifted was estimated to correspond to 35% of the average power demand in this case study. Clausen et al. [147] applied these/their results to the current situation in Denmark, where data centers consuming about 500 GWh/y and 60 MW, were estimated to be able to provide 22 MW of demand response. In a bottom-up calculation, Koronen et al. [148] estimated the maximum theoretical potential for demand response for data centers to be in the range of 38%–80% of the installed power demand in 2030. The size of the potential demand response as a share of the data center's average or peak power will, however, vary from case to case, depending on the data center. Furthermore, the duration of the applicable demand response events is typically in the range of minutes to hours, but depends on the type of workload handled by the data center. The possibility of shifting the load geographically depends on the delay sensitivity of the data management.

9.6.3 INNOVATIVE USE OF BACKUP POWER

Data centers are normally equipped with UPS units to ensure continuous and high-power quality, and for handling grid power failures. The UPS unit is part of the emergency power or backup power system which typically includes diesel generators. Typical UPS battery units have the capacity to power the data center at its maximum power requirement for 5–30 minutes [149], which allows the diesel generators to be started. Interest is growing in the more active use of UPS units for peak shaving and frequency regulation in power systems with increasing shares of variable renewable power generation (see, e.g., Ref. 150).

Backup diesel generators are intended to be used only in the case of a power outage. From a power system perspective, this represents a large and underutilized potential for grid balancing in the future. With increasing needs for grid balancing and energy storage, as well as power technology development, considerable economic benefits may be obtained by designing microgrid systems for grid services and balancing, rather than for emergency backup power only [151,152]. New clean technologies also allow for operation in areas where diesel generator emissions would have posed a problem. Fuel cell systems appear to be particularly promising in this regard, but the potential and various technological options for data centers seem to be relatively unexplored [151,153]. It is quite conceivable that future data centers will be highly integrated into power systems by routinely utilizing capacity that was formerly perceived as backup emergency power. Several data center operators are aiming at 100% renewable electricity on an annual basis. It would be a logical follow-on from that to use 100% renewable electricity at every stage by using emerging electricity storage options such as power-to-gas storage and fuel cells, or flow batteries, in a microgrid [153]. The viability of such a system will depend partly on the regulations governing power production, and on markets, grid access, and permits.

9.7 HYBRID ENERGY HARVESTING FOR PORTABLE ELECTRONICS

Energy harvesting technology covers the conversion of solar/light, vibration/kinetic, wind/fluidic, magnetic, and thermal energies into electricity, via various mechanisms such as the PV, piezoelectric, electromagnetic, electrostatic, triboelectric, magnetostrictive, TE, and pyroelectric effects. Recently, a comprehensive review of energy harvesting research has been published by Yang et al. [154]. It summarizes single-source energy harvesters and focuses on hybrid and multisource energy harvesters with novel materials and structures as well as those integrated with energy storage and/or sensors. We briefly summarize here important findings of this review and other reported results.

Single-source harvesting for portable electronics includes kinetic energy harvesters, piezoelectric harvesters, triboelectric harvesters, harvesters based on electrostatic and electrostrictive effects as well as those based on magntostrictive effects. Because of energy losses, each of the above-mentioned effects has a limit on the extractable electricity and hence a limit on the energy conversion efficiency. Therefore, the hybridization of several kinetic energy harvesting effects into the same harvester not only complements each other's disadvantages but also helps to increase the overall efficiency. Hybrid harvesters for kinetic energy are the most common single-source harvesters. A bidirectional, piezoelectric-electromagnetic hybrid harvester built around a tube-shaped frame, combining two piezoelectric cantilevers with cuboidal magnets as tip masses, a suspended cylindrical magnet, and a set of coils, has been reported by Fan et al. [155]. Another piezoelectric-electromagnetic hybrid harvester designed for higher operating frequencies (60–120 Hz) has been reported by Li et al. [156]. Biomechanical energy (e.g., human body movement) is another important kinetic energy source which is attractive in the application of wearable electronics. An electromagnetic-triboelectric hybrid harvester has been designed and fabricated for the energy generated from the swinging behavior of human arms during locomotion [157]. A electromagnetic-triboelectric hybrid harvester has been created for oceanic energy harvesting by Feng et al. [158]. In this harvester, seven Al electrodes covered by polytetrafluoroethylene (PTFE) were fixed on an acrylic substrate. In a capacitor charging test, the hybrid harvester achieved 54% and 150% efficiency increases compared to those of its electromagnetic and triboelectric individual counterparts [158].

9.7.1 HYBRID, MULTISOURCE ENERGY HARVESTERS

Rationally designed materials and technologies have been developed in the past decades for the conversion of various types of energy, such as solar, thermal, mechanical, and chemical energy, into electricity. These existing approaches, however, were investigated and developed on the basis of drastically different physical principles and diverse engineering approaches to specifically harvest a certain type of energy, while the other types of energy were wasted. Innovative approaches had to be developed for the conjunctional harvesting of multiple types of energy through the use of integrated structures/materials, so that all available energy resources can be effectively and complementarily utilized [159]. On a smaller scale, the temporal/spatial distribution and availability of energy sources for driving micro and nano sensors (MNSs) vary drastically. The concurrent harvesting of multiple energy types from the ambient environment by a single integral device has therefore emerged as a promising approach toward the sustainable and maintenance-free operation of MNSs. Ever since the first demonstration of a nanotechnology-enabled hybrid cell (HC) by the Wang research group for the simultaneous harvesting of multiple types of energy with a single device [160], this technology has been advancing at an increasing pace.

9.7.1.1 Magnetic and Kinetic Energy

In practice, a single energy source may not always be powerful and stable enough for a harvester to generate sufficient electricity for the end usage. Multisource energy harvesting through structural hybridization or multifunctional materials addresses this issue. By employing magnetostrictive and piezoelectric materials at the same time, ambient magnetic waves dissipated from electric devices and power transmission lines can be <1 mT magnetic input energy. This power density was an order of magnitude higher than that using piezoelectric cantilevers to harvest kinetic energy [161]. A wireless sensor network (WSN) for the IoT (Internet of Things) could be powered by such harvested energy for timing and sensing (pressure and temperature) functions. Although simultaneously harvesting kinetic energy via the piezoelectric effect was not realized in this work, it is definitely an option for such a hybrid structure. A specially designed piezoelectric cantilever with magnets as tip masses has also been reported to simultaneously harvest magnetic and vibrational energy [161]. An output power increase of up to 300% was achieved when simultaneously harvesting both energy sources compared to that of harvesting either single source [162].

The handheld-electronics market has grown remarkably, with realization of diversified products including mobile phones, personal digital assistants, cameras, and healthcare devices. This market drives significant progress in the development of low-power electronics and wireless-communication technology. Harvesting energy from ambient vibrations to recharge the batteries in such devices is a potential route to realizing handheld electronics with infinite lifespans, i.e., self-powered electronics. The human body itself is a potential source of energy for harvesting. Other sources of vibrations include equipment and engines used in household appliances. State-of-the-art microelectromechanical systems (MEMS) energy harvesters [1,2] offer intriguing possibilities to realize such self-renewing power sources. Energy extracted from vibrations is stored in chip-compatible, rechargeable batteries such as thin-film lithium-ion types, which further power the loading application (e.g., a wireless-sensor node) by means of a regulator circuit. Vibration-based MEMS energy harvesters use one of three energy-transduction mechanisms: electrostatic, electromagnetic, or piezo-electric. Electrostatic energy harvesters collect energy from capacitance changes during the vibration cycle, while electromagnetic energy harvesters collect energy from the current generated in coils by variations in magnetic flux induced by the movement of a permanent magnet. On the other hand, piezo-electric materials are perfect candidates for harvesting power from ambient vibration sources, because they can efficiently convert mechanical strain to an electrical charge without any additional power [163,164].

Most reported approaches work by employing one of these three mechanisms. The study by Lee [164] explores a hybrid approach that combines two mechanisms. To harvest energy from vibrations, an inertial mass is needed to collect kinetic energy. Magnets that provide this inertial mass are placed at the end of a piezo-electric cantilever. The study used cofired multilayer piezo-electric cantilever elements from Piezo Systems Inc. These cantilevers are built up from a number of lead zirconate titanate (PZT) layers, each $30\,\mu m$ thick, and screen-printed electrodes a few micrometers thick between each pair of layers in a piezo-electric ceramic pattern. The cantilever's length, width, and total thickness are 22, 9.6, and 0.65 mm, respectively. When the device is excited by external vibrations, their kinetic energy can be harvested through both electromagnetic and piezo-electric transduction mechanisms. Previously reported data indicate that piezo-electric energy harvesters present a high power density and are more suitable for microsystem applications, while electromagnetic energy harvesters are good at relatively large applications. The major advantages of the device proposed by Lee et al. [164] are low cost and the capacity to harvest energy by both piezo-electric and electromagnetic mechanisms. The position and weight of the magnets can be used to optimize the output power and modify the resonant frequency of the PZT cantilever. In the future, this hybrid energy harvester could be an alternative power source for wireless-sensor nodes. More details on the device are given by Lee et al. [164].

9.7.1.2 Kinetic and Solar Energy

Kinetic and solar energy can be considered the most pervasive and commonly coexisting ambient energy sources, where harvesting of both simultaneously can be significantly beneficial. A self-powered lantern based on a triboelectric-PV hybrid structure has been built for harvesting wind and light energy [165]. In this structure, a transparent polylactic acid (PLA) tube was connected to a central axis rod via bearings and transparent, interdigitated indium tin oxide (ITO) electrodes were coated on the inner wall of the tube. A rod coated with a layer of fluorinated ethylene propylene (FEP) film was placed on top of the ITO electrodes. A dye-sensitized solar cell (DSSC) was also mounted on the tube. In addition, a soft lithium battery (SLB) and 10 light-emitting diodes (LEDs) were connected. The PLA tube rotated when driven by the wind, and the freely moving FEP rod generated friction with the ITO electrodes, thus harvesting the wind energy through the triboelectric effect. Meanwhile, solar energy was harvested by the DSSC. The harvested energy was stored in the SLB and used for lighting the LEDs. Being charged for 0.88 hour, the SLB could be discharged for 4.12 hour with 10 µA current [165]. This performance was equivalent to 100% and 33% efficiency increases compared to those of the individual triboelectric generator and solar cell, respectively. Apart from structural hybridization as introduced above, multisource energy harvesters can also be realized with multifunctional materials. Details have been given in the reference [166]. Recently, a polyvinylidene fluoride (PVDF)-ZnO composite has been reported [167]. The composite consisting of 33 wt% ZnO nanowires and PVDF polymer helped to increase the output voltage by over 300% compared to that using PVDF only.

Lee et al. [168] fabricated a tandem device which integrates a PVDF nanogenerator and silicon (Si) nanopillar solar cell. The SiNP solar cell was fabricated using a plasma etching technique and doping process by rapid thermal annealing and furnace annealing. The PVDF nanogenerator was stacked on top of the Si nanopillar solar cell using a spinning method. The optical properties and the device performance of nanowire solar cells were characterized, and the dependence of device performance versus annealing time or method has been investigated. Furthermore, the PVDF nanogenerator was operated with a 100 dB sound wave and a 0.8 V peak-to-peak output voltage was generated. This tandem device can successfully harvest energy from both sound vibration and solar light, demonstrating its strong potential as a future ubiquitous energy harvester.

9.7.1.3 Wind and Thermal Energy

The pyroelectric effect is known for the energy harvesting from temperature fluctuations. A pyroelectric harvester has been reported to utilize a vortex to generate the temperature fluctuation and thus harvest wind energy via the pyroelectric effect [169]. The vortex generator was a right-angled substrate on which a PVDF film was deposited on the horizontal part. The wind flowed along the horizontal direction and when it collided with the vertical part of the substrate, energy was harvested. The main advantage of this harvester was that it was able to capture energy even with a very weak air flow (down to 1 m/s velocity) [169]. Although the piezoelectric effect was not mentioned, PVDF is also ferroelectric and well known for its coexhibition of piezoelectric and pyroelectric effects. Therefore, this structure could naturally become a multisource harvester by employing the piezoelectric effect at the same time. Stronger ferroelectric materials, e.g., [170,171] could also replace the PVDF in order to improve the piezoelectric and pyroelectric performance.

9.7.1.4 Solar, Kinetic, and Radio Frequency Energy

A recently published patent has released the design of an umbrella apparatus which managed to incorporate PV, piezoelectric, electromagnetic, and radio frequency (RF) energy harvesters into different parts of an umbrella [172]. The canopy of the umbrella was replaced by a multilayer lamination combining PV and piezoelectric energy harvesters. The lamination consisted of an inverted polymer solar cell layer [173], a PVDF piezoelectric layer, electrode layers, and other supportive layers. When the energy harvesting canopy was open, solar and kinetic (i.e., wind and raindrops) energy could be harvested. Meanwhile, the open canopy could spin when subjected to wind force, thus driving a miniature windmill (electromagnetic energy harvester) embedded in the shaft of the umbrella. In addition, the shaft of the umbrella could act as a monopole antenna and an RF energy harvesting circuit could be installed in the shaft in order to harvest RF energy. Furthermore, the shaft, in conjunction with a movable ferrule, contained a magnet and coils. When the canopy was closed, the umbrella could be used as a cane. By striking the ferrule on the ground while walking, the magnet would be forced to move through the coils, forming another electromagnetic harvester. Apart from these energy harvesting components, the necessary DC-DC converters, switches, conductive leads, and a rechargeable battery were also installed in the umbrella [172]. Personal electronic devices could be charged through a port of the umbrella so that the entire umbrella became a self-powered, portable charging station.

Since the amount of power available by energy harvesting is quite limited, there has been interest in utilizing multiple forms of external energy simultaneously—such as light and heat, or light and vibrations—in order to collect a sufficient amount for practical use. In the past, this has been achieved by combining different kinds of devices, which leads to higher costs. Fujitsu Laboratories has developed a new hybrid harvesting device that captures energy from either light or heat, which are the most typical forms of ambient energy available for wide-scope application. This makes it possible for a single device to capture energy from either heat or light without combining two harvesting devices. In addition, as it can be manufactured from inexpensive organic materials, device production costs can remain low. This technology includes (a) new structure for hybrid generating devices, (b) development of an organic material for hybrid generating device. Until now, PV cells—which generate electricity from light, and TE devices—which generate electricity from temperature differentials, have only been available as separate devices. This new technology from Fujitsu Laboratories doubles the energy-capture potential through the use of both ambient heat and light in a single device. If either the ambient light or heat is not sufficient to power the sensor, this technology can supply power with both sources, by augmenting one source with the other. In addition, the technology can also be used for environmental sensing in remote areas for weather forecasting, where it would be problematic to replace batteries or run electric lines.

9.7.1.5   HCs for the Harvesting of Solar and Mechanical Energy

The first nanotechnology-enabled HC was developed by the Wang research group in 2009 for harvesting solar and mechanical energy with a single energy harvester. The device essentially integrates a DSSC and a piezoelectric nanogenerator (NG), both of which are based on an array of ZnO nanowires (NWs), on a common substrate [160]. The cathode of the NG and the anode of the DSSC were integrated on the same silicon substrate to form a serial connection between the DSSC and the NG. The DSSC and NG units in the HC can work independently when a source of either solar or mechanical energy is available. It has also been demonstrated that the HC can harvest both the solar and the mechanical energy simultaneously and synergistically. However, reliability issues imposed by solvent leakage and evaporation as a well as the low power output hinder the practical application of this HC. A prototype of a compact HC in which an NG based on a ZnO-NW array was integrated with a solid-state DSSC showed enhanced performance and durability owing to the introduction of a solid-state electrolyte and convolute structures formed between the NG and the DSSC [174].

Choi et al. later described a flexible HC based on a ZnO-NW array that overcomes the disadvantages of the above HC prototypes of cross-talk and additional assembly processes [175]. The ZnO-NW array in this flexible HC not only serves as the NG but also acts simultaneously as the solar cell part of the device by integrating with an infiltrated organic polymer. One significant characteristic of this flexible HC is that the output signals from the solar-cell part take the form of a DC, whereas the output signals from the NG originally occur as an AC. By controlling the mechanical straining process, the AC signals can be converted into DC-like signals. Owing to the controllability of the output behavior, the performance of the HC can be synergistically enhanced by the contribution of the NG part. Lee et al. demonstrated a conceptually similar HC prototype based on the integration of a ZnO-NW NG with infiltrated quantum dots, which surrounded the NWs. This HC was specifically developed for harvesting sound and solar energy simultaneously [176]. A further demonstration of the integration of multiple energy harvesters together with a storage device along a single fiber involved the use of ZnO NWs and graphene [177]. This approach allows simultaneous harvesting of solar and mechanical energy and *in situ* storage of this harvested energy for potential applications in flexible and wearable electronics.

9.7.1.6   HCs for the Harvesting of Biomechanical and Biochemical Energy

There has been an increasing need for sustainably powered implantable wireless micro/nanodevices for in vivo biomedical applications, preferably without the incorporation of batteries. One viable approach is to concurrently harvest energy from multiple energy sources within the biological entity. Inherently, mechanical and biochemical energy due to body motion, muscle stretching, and metabolic processes are abound in the biological entity. A prototype hybrid energy-scavenging device was developed to address the above application needs through the direct harvesting of mechanical and biochemical energies in a biofluid environment [178]. This hybrid energy scavenger consists of a piezoelectric PVDF-nanofiber NG for harvesting mechanical energy, such as from respiration and blood flow in the vessels, integrated with a flexible enzymatic biological fuel cell (BFC) for harvesting the biochemical energy from the chemical processes between glucose and O_2 in biofluid. These two energy harvesting approaches, integrated within one single device, can work either individually or synergistically. This HC for harvesting biomechanical and biochemical energy has similar disadvantages to previous HCs: the separate arrangement of the two components on the substrate without sophisticated integration leads to engineering problems, such as cross-talk, and hence deteriorated overall performance. To solve this problem, a compact structure was developed by the integration of a ZnO-NW NG and a BFC on single carbon fiber [179]. The NG for harvesting mechanical energy is based on a textured ZnO-NW film and is grown radially on the carbon fiber, which serves as both the core electrode and the substrate for ZnO growth. The BFC for converting chemical energy from the ambient biofluid is fabricated at the other end of the same carbon fiber. Elimination of the separating membrane and mediator significantly reduced the size of the BFC relative to that of conventional BFCs. The integrated structure improves the performance as well as the adaptability of the HC for harvesting biomechanical and biochemical energy.

9.7.1.7 HCs for the Harvesting of Solar and Thermal Energy

During the PV conversion process in solar cells, a big proportion of the wasted energy is converted into heat, which leads to a temperature rise in the solar cells. Furthermore, incident photons with longer wavelengths, which cannot participate in PV conversion, may also be converted into heat. To improve the conversion efficiency and fully utilize the solar spectrum, Guo et al. designed an HC to harvest solar energy as well as the concurrently generated heat [180]. The two-compartment hybrid tandem cell consists of a DSSC and a thermoelectric cell (TC). Solar energy is first converted into electricity in the DSSC, and the heat induced during this process is then transmitted to the TC for subsequent TE conversion. This HC is more efficient than a single harvester and fully utilizes the energy from the solar spectrum. Recently, a novel PV–TE hybrid device composed of a series-connected DSSC, a solar-selective absorber, and a TE generator was reported with a significantly enhanced efficiency of 13% [181].

Although the concept of hybrid energy harvesting and the proposed approaches described above are promising, several practical issues need to be addressed before real applications of these prototypes are possible. One of the biggest issues is network matching between different energy harvesters. The power output from different harvesters differs significantly. Strategic approaches for matching and reconciling the different outputs should therefore be implemented. On the other hand, solutions to the current problems might also impart increased cost as well as difficulties in manufacturing. Overall, it can be anticipated that the concept of hybrid energy harvesting will play a critical role in the implementation of novel sustainable micro-/nanotechnology with more flexibility and adaptability. It can also be expected that more sophisticated hybrid energy-scavenging devices that are capable of concurrently harvesting even more types of energy may be developed. Japanese company Fijitsu [182] has developed a device that harvest energy from both heat and light.

9.7.1.8 Hybrid Energy via Microscale Waste Heat Applications

The thermoelectric generator (TEG) devices are especially suitable for waste heat harvesting for low-power generation to supply electric energy for microelectronic applications. Wearable TEGs harvest heat generated by the body to generate electricity. For this reason, it is possible to use waste human body heat to power a TEG watch device. In this case, the wristwatch can capture the TE energy. Now, body-attached TEGs are commercially available products including watches operated by body temperature and thin film devices. Some manufacturers produce and commercialize wristwatches with an efficiency about 0.1% at 300 mV open circuit voltage from 1.5 K temperature drop and 22 µW of electric output power under TEG normal operation. A thermoclock wristwatch produces a voltage of 640 mV and gives a power of 13.8 µW for each °C of temperature difference. A wristwatch with 1,040 thermoelements generates in the same conditions at about 200 mV [183]. The wearable TEG performance is affected by the utilization of the free air convection cooling on the cold side of TEG, the low operating temperature difference between the body and environment, as well as the demand for systems that are thin and lightweight, being practical for long-term usage [184].

Furthermore, various microelectronic devices, such as WSNs, mobile devices (e.g., mp3 player, smartphones and iPod), and biomedical devices, are developed. The TE energy harvesters are microelectronic devices such as cardiac pacemakers, pulse oximeter, wireless communication, electrocardiography, electromyography, etc., made of inorganic TE materials, at different dimensions, with a lifetime of about 5 years [184]. For these microelectronic devices, standard batteries are used. These batteries are made of various inorganic materials (such as nickel, zinc, lithium, lead, mercury, sulfuric acid, and cadmium) that are not friendly for the human body. In this case, the body-attached TEGs could be an alternative solution because the materials used are nontoxic [184].

As pointed out by Ensescu [184], a TEG can also be applied in a network of body sensors. In this case, the device has been fixed in a body zone, where the maximum body heat has been obtained and also maximum energy. This equipment is capable of storing about 100 µW on the battery, leading to an output voltage of 2.4 V. A TEG can also be designed to be used on the wrist. The output voltage of the device was 150 mV under normal conditions and an electric output power of 0.3 nW. Waste human body heat can also be used to power a TE "watch battery." In this application, thermocouples can be prepared by depositing germanium and indium antimonide on either side of a 1 mm thick insulator which served as a simulated watch strap. It was estimated that 2,875 thermoelements connected in series would be required to obtain the 2 V required to operate the watch [183–189].

Growing applications like autonomous microsystems or wearable electronics urgently look for microscale power generators. One possibility is to convert waste heat into electrical power with a micro-TE power generator. Micro-TE power generators can be fabricated using integrated circuit technology. In some cases, alternate n- and p-type thermoelements are ion implanted into an undoped silicon substrate. Metalization of thermoelement-connecting strips and output contacts enable several hundred thermocouples to be connected electrically in series and occupy an area approximately 25 mm². The miniature generator can be designed [183–189] specifically to provide sufficient electrical power to operate an electronic chip in a domestic gas-monitoring system. In excess of 1.5 volts could be produced when a temperature difference of a few tens of degrees was established across the module. In this case, any available waste heat source, such as the surface of a hot water pipe, would provide sufficient heat flux to thermoelectrically generate the required chip-voltage. More recently, Glatz et al. [185] presented a novel polymer-based wafer-level fabrication process for microthermoelectric power generators for the application on nonplanar surfaces. Various microelectronic devices, such as WSNs, mobile devices (e.g., mp3 player, smartphones, and iPod), and biomedical devices which include cardiac pacemakers, pulse oximeter, hearing aid, etc., are also developed. For these microelectronic devices, standard batteries are used. Significant literature on microscale applications of hybrid waste heat by thermoelectricity has been reviewed [183–189].

9.7.1.9   Harvester-Sensor Integrations

The IoT is undoubtedly where energy harvesting technology is aiming to have a significant impact. As the IoT requires an extensive number of sensors and WSNs, energy harvesters need to show their compatibility when integrated or utilized in different cases rather than only giving consideration to their output performance. As energy harvesters as individual components are becoming mature, there is an increasing number of publications demonstrating the feasibility of integrated energy harvesters with sensors or electricity consuming devices.

Piezoelectric energy harvesters have been integrated with structural health monitoring sensors for various constructions and bioactivity monitoring circuits used in implantable biomedical devices. In the former case, piezoelectric patches were attached on a pipe which was undergoing forced vibrations, for the purposes of damage detection and indicators for control [190]. In the latter case, a self-powered neural activity monitor consisting of an impact-based piezoelectric harvester, a power transfer circuit, and a neural signal monitoring circuit was proposed [191]. Both of them successfully demonstrated enough energy generated by the harvesters to power all the devices, as well as the feasibility and reliability of these methods. Numerous other devices using thermomagneto-piezoelectric hybrid harvester-sensor and electromagnetic-triboelectric hybrid harvester-sensor integrations are reported by Chun et al. [192], Shen et al. [193], and Yang et al. [194].

9.7.2   Self-Powered Hybrid Micro-/Nanosystems

The current rapid advancement of micro-/nanotechnology will gradually shift its focus from the development of discrete devices to the development of more complex integrated systems that are capable of performing multiple functions, such as sensing, actuating/responding, communicating, and controlling, by the integration of individual devices through state-of-the-art microfabrication technologies. Furthermore, it is highly desired for these multifunctional MNSs to operate wirelessly and self-sufficiently without the use of a battery, especially in applications such as remote sensing and implanted electronics. As the dimensions of individual devices shrink, the power consumption decreases accordingly to a reasonably low level, so that energy scavenged directly from the ambient is sufficient to drive the devices.

Wang research group demonstrated the first integrated self-powered nanosystem on a large scale without an external power supply. Within this system, the NW pH sensor and the UV sensor were driven by an integrated ZnO-NW NG [195]. Self-powered nanoelectronic systems based on discrete core/shell silicon-NW PV devices have also been demonstrated [196], and a DSSC based on high-density arrays of vertically aligned ZnO nanotubes has been investigated as the power source to drive a humidity sensor [197]. Intensive research effort has since been invested in the development of self-powered MNSs, and various prototypes have been reported. These are well reviewed by Wang et al. [198]. It can be anticipated that self-powered MNSs will play a critical role in the implementation of implantable electronics, remote and mobile environmental sensors, nanorobotics, intelligent MEMSs/NEMSs, and portable/wearable personal electronics. Self-powered MNSs are also key components of large-scale fault-tolerant sensor networks. The design/fabrication flow for the development of future self-powered MNSs should be amenable to scale up and, critically, be compatible with the microfabrication technology. More work in this area is needed.

9.8 ENERGY HARVESTERS INTEGRATED WITH ENERGY STORAGE AND/OR END USERS

9.8.1 HARVESTER-STORAGE INTEGRATIONS

In practice, it is common to connect a solar cell with a Li- ion battery for simultaneous energy harvesting and storage. However, a monolithic photo-battery has been invented to substitute this two-component solar cell-battery combination [199]. Apart from their electrochemical properties, Li-ion intercalation materials are also mechanically active. When a stress is applied, they can exhibit a mechanical-electrochemical coupling that increases the voltage of the battery [200]. Therefore, a study into using Li-ion batteries to simultaneously harvest energy has been carried out [200]. The efficiency of this harvester-storage integration was only 0.012%, much smaller than that of conventional kinetic energy harvesters such as piezoelectric, electromagnetic, and triboelectric systems [201]. The theoretical efficiency was predicted to be 2.9%. A supercapacitor has also been reported for the same purpose of simultaneously harvesting and storing energy [201]. The electrolyte started to flow when subjected to pressure, causing the electrokinetic supercapacitor to operate in both harvesting and storage modes. The efficiency could reach 0.03%–0.1% under one bar pressure or regulated external load.

9.8.2 HYBRID NANOGENERATORS

The harvesting, storage, and utilization of energy are becoming a worldwide issue due to the energy crisis, environmental pollution, and the fast development of electronics for our daily life [202–204]. Recent developments in hybrid devices are based on nanogenerators and energy storage systems through integration, hybridization, and all-in-one designs for self-charging energy systems for future electronics. Basically, the nanogenerator can be divided into representative two devices depending on piezoelectric and triboelectric properties, called the piezoelectric nanogenerator (PENG) and triboelectric nanogenerator (TENG), respectively. The first demonstrated nanogenerator was the PENG using a ZnO NW [205], which was operated by the piezoelectric property. The formation of a piezoelectric potential or piezo-potential was arisen from the breakage of central symmetry in the ZnO crystal structure by external force. The ZnO-based PENGs have been developed as various types [206, 207–209]. This basic principle and the model of power generation apply to other PENGs based on various piezoelectric materials, such as PZT [210–212]. Material selection and structural design are the key factors for the development of PENGs, which are based on the coupling of piezoelectric materials and flexible substrates. The all-polymer-based flexible TENG based on triboelectrification and electrostatic induction was invented in 2012 [213], which could convert mechanical energy into electricity. Interestingly, triboelectrification can be found everywhere in the surrounding environment and in most common materials used every day in our daily lives. The detailed mechanisms for PENG and TENG are described in the literature [213,214,215]. Both PENG and TENG generate AC pulse output. In order to provide stable power using nanogenerators, energy storage systems (ESSs) such as supercapacitors and rechargeable batteries are essential for future electronic devices. Thus, the selection of the materials, structural design, and circuit connections in designs for hybridizing ESSs and nanogenerators should be properly considered and designed because most nanogenerators have thin, lightweight, flexible substrates, and require bendable, and stretchable device formation [214].

In addition, wired and textile ESSs and other new functions are required according to the developments of advanced nanogenerators with higher output power and new application in future.

PENG technology is highly reliable, has stable operations, smaller device area, and diverse application fields. To date, extensive fabrication methods, growth of various one-dimensional (1D)/two-dimensional (2D) inorganic piezoelectric nanostructures (NSs) on plastic substrates [215], flexible piezoelectric polymer films, and device designs (planar, stretchable, cylindrical, or fiber) were developed to improve the PNG (literature uses PNG and PENG interchangeably) technology as a prominent energy-harvesting approach for creating the sustainable independent power source to drive the low-power consumed electronic devices/sensors. Moreover, the device compatibility, electrical output performance (nW/cm^2 to $\mu W/cm^2$) under various harsh environments, and flexibility issues were optimized to think about the real-time commercialized PNG or PENG (piezo electric nanogenerator) product. On the other side, few PNGs have dual functionality such as wearable/portable independent power source to drive the commercial electronic devices and can also work as a self-powered sensor (or a battery-free sensor) to measure/monitor the various physical, chemical, biological, and optical stimuli [215].

The study by Nagamalleswara et al. [215] suggests that PNG (or PENG) technology based on the type of materials is classified into three categories such as (a) inorganic NSs-based PNG, (b) polymer matrix-based PNG, and (c) composite (polymer + inorganic NPs) PNG as shown in Figure 9.4 [215]. Following the development of PENG by ZnO nanowire mentioned above, many other researchers across the world extend the growth of other NSs and its device designs for inorganic PNG, but it has few ample drawbacks such as typical growth process of piezoelectric NSs, brittleness, lower force limits, failure instability, and leakage current issues [215]. Further, PNG was implemented with the flexible PVDF and its copolymers due to its high flexibility, easy process to prepare flexible films, low electrical output performance, and accepting large mechanical force. However, it has one major disadvantage such

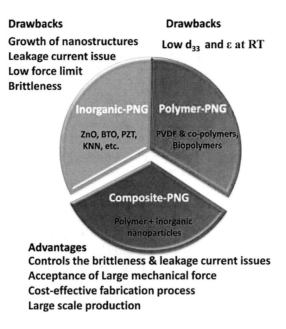

Drawbacks
Growth of nanostructures
Leakage current issue
Low force limit
Brittleness

Drawbacks
Low d_{33} and ε at RT

Advantages
Controls the brittleness & leakage current issues
Acceptance of Large mechanical force
Cost-effective fabrication process
Large scale production

FIGURE 9.4   The schematic represents the overview of PNGs (or PENGs): classification, drawbacks, and advantages [215].

as low piezoelectric coefficient and relative permittivity at room temperature than the inorganic piezoelectric NSs. Nagamalleswara et al. [215] overcome the issues of the inorganic and polymer PNGs by developing the composite PNG technology. The major key factor in designing the efficient, flexible composite PNG device is the development of multifunctional hybrid or composite piezoelectric structures. It indirectly depends on the selection of individual high-performance inorganic and polymer materials and cost-effective fabrications processes. These kinds of composite PNGs are highly suitable to work as sustainable independent power sources as well as self-powered sensors to measure various physical parameters such as physical, optical, biological, and chemical stimuli. Zhao et al. [216] developed hybrid piezo-/TENG for highly efficient and stable rotation energy harvesting.

The all-polymer-based flexible TENG based on triboelectrification and electrostatic induction was invented in 2012 [213], which could convert mechanical energy into electricity. Interestingly, triboelectrification can be found everywhere in the surrounding environment and in most common materials used every day in our daily lives. Both PENG and TENG generate AC pulse output. In order to provide stable power using nanogenerators, ESSs such as supercapacitors and rechargeable batteries are essential for future electronic devices. Thus, the selection of the materials, structural design, and circuit connections in designs for hybridizing ESSs and nanogenerators should be properly considered and designed because most nanogenerators have thin, lightweight, flexible substrates, and require bendable, and stretchable device formation [213]. In addition, wired and textile ESSs and other new functions are required according to the developments of advanced nanogenerators with higher output power and new application in future.

To test the efficient charging of the battery for hybrid devices, recently, researchers have studied how to integrate the TENGs with lithium-ion batteries (LIBs) using a rectifier. In 2016, Pu et al. demonstrated efficient charging of LIBs by a rotating TENG with pulsed output current [217]. Fast Li-ion extraction from the typical electrode materials $LiFePO_4$ and $Li_4Ti_5O_{12}$ was achieved by the TENG at a rotation speed of 250 rpm. The estimated coulombic efficiency of the TENG charging and the following 0.5 C discharging can be higher than 90%, comparable with constant current charging. Interestingly, improvement of the power utilization efficiency (up to 72.4%) in transferring power from the TENG to the $LiFePO_4$-$Li_4Ti_5O_{12}$ full battery was achieved by optimizing the coil ratio of a transformer. High efficiency was achieved when the impedance of the TENG was reduced to close to that of the battery cell. In addition, they showed that a 1 h charging of a commercial LIB by the rotating TENG (600 rpm, 36.7 transformer coil ratio) can exhibit a discharge capacity of 130 mAh. In fact, the energy conversion efficiency is very important when the generated output power is stored in ESSs. Nan et al. prepared the cathodic material $Li_3V2(PO_4)3/C$ and compared the storage efficiency with most popular cathodic materials: $LiCoO_2$, $LiFePO_4$, and $LiMn_2O_4$ [218]. They showed that the selection of electrode materials is important for efficient charging in hybrid devices.

The term "all-in-one hybrid device" means that a single device has both an energy harvesting nanogenerator and an ESS in the device without complicated connections. The all-in-one hybrid device design has many advantages for compact, simple, and portable devices, but more efforts are needed to fabricate the devices because of the problem of matching device and material characteristics, such as flexibility, coatings, compositions, electrochemical properties, etc., between the nanogenerator and the ESS. Very recently, researchers have reported and suggested new concepts for all-in-one devices.

In 2016, Wang et al. reported a new nanoenergy cell (NEC) that uses high-density piezoelectric nanowires to harvest mechanical energy and has a large electrolyte (phosphoric acid/polyvinylalcohol (H_3PO_4/PVA) gel electrolyte)-nanowire interface to store electricity in the all-in-one system consisting of a PENG and an electric double-layer supercapacitor (EDLC) [219]. The device achieved a continuous output current for over 90 s, and the mechanical-electric energy conversion efficiency of the NEC was over 10 times higher than that of the PENG without increasing the device volume or reducing the efficiency. Interestingly, Ramadoss et al. made an all-in-one device from a PENG and a pseudo-capacitor based on PVDF-ZnO and MnO_2, respectively [220]. The device exhibited self-charging capability under palm impact (aluminum-foil-based device to 110 mV over 300 s; fabric-based device to 45 mV over 300 s). Most recently, Song et al. demonstrated an integrated sandwich-shaped, self-charging power unit (SCPU) with a wrinkled poly(dimethyl siloxane)-based TENG and a carbon nanotube (CNT)/paper-based solid-state EDLC [221]. During vibrations, the device can be utilized to simultaneously harvest and store the mechanical energy as electrochemical energy, and it could be charged to 900 mV in 3 h under the compressive stress at 8 Hz. This study showed that their developed novel all-in-one device is a promising candidate for flexible electronics and wearable devices.

Guo et al. reported the concept of an all-in-one shape-adaptive self-charging power package based on a TENG and EDLC that has been simultaneously demonstrated for harvesting body motion energy to sustainably drive wearable/portable electronics. Subsequent research made advances on washability, flexibility, water proof characteristics, and stretchability of the all-in-one devices [222].

9.8.3   WEARABLE DEVICES OF ESSs AND NANOGENERATORS

Recently, wearable electronics, such as smart phones, smart watches, healthcare sensors, smart glasses, etc., have been developed and become an important part of our lives [223]. To utilize these electronics, wearable energy devices are consequently required. Conventional ESSs (batteries and supercapacitors) require frequent and inconvenient charging, however. Therefore, wearable self-charging power devices that combine energy-harvesting and energy-storage technologies could be potential solutions. Pu et al. reported wearable hybrid devices combining nanogenerators with a supercapacitor [224] and a flexible battery [225], respectively. They reported the facile and scalable fabrication of an all solid-state flexible yarn supercapacitor and its integration with a TENG cloth for a self-charging power textile [224]. To fabricate wearable, a TENG-battery self-charging power textile and a flexible battery are essential [225].

Very recently, Wen et al. [226] reported a multifunctional hybridized self-charging power textile system designed to simultaneously collect outdoor sunshine and random body motion energy by a DSSC and a TENG, respectively, and then storing the energy in a fiber-shaped supercapacitor [226]. A single-fiber-DSSC unit showed an overall power conversion efficiency of 5.64% and fiber-TENG can take advantage of human motions, such as jogging, to deliver an output current of up to 0.91 mA. The fiber-supercapacitor showed excellent pseudo-capacitance of 1.9 mF cm^{-1} from $RuO_2 \cdot x\ H_2O$. Due to the properties of all-fiber-shaped devices, this textile system can be easily woven into electronic textiles to fabricate smart clothes that can operate wearable electronic devices. These multifunctional hybrid self-powered devices will result in practical human benefits.

To develop and improve the new systems and performances of hybridized energy devices, further research efforts are needed on the design of devices, materials, integration, and better understanding of various energy harvesting and storage devices. Through this creative hybridization technology, it is expected that other energy harvesting devices using clean environmental energy sources will improve the harvesting capability of nanogenerators. In future, superior energy-harvestable nanogenerators with excellent charging efficiency can be applied for integration with next-generation ESSs with high energy storage capability, such as the metal-air battery, Li-sulfur battery, Li-silicon battery, etc.

9.8.4   CMOS TECHNOLOGY-BASED HARVESTERS AND SYSTEMS

As sensors (especially large sensors for the IoT) are made with CMOS technology focusing on the full integration of energy harvesters (and/or storage), interface circuits, and sensors. CMOS technology provides an approach to integrating all microscale components on one chip. Therefore, a number of CMOS-based power conditioning circuits (e.g., converters, rectifiers, etc.) for energy harvesting have been reported, including those allowing multiple energy sources as the input [227–233]. A low-power batteryless energy harvesting system implemented in a CMOS–130 nm technology for IoT applications used a dual-mode DC-DC converter to harvest solar and kinetic energy via PV and piezoelectric transducers, respectively [234]. A supercapacitor was used as the storage. On-chip, there were also a programmable switch to optimize the efficiency of the system and a maximum power point tracking circuit (MPPT). This circuit realized self-starting and a peak efficiency of 90.5% (the ratio of rectified energy to harvested energy) [234]. Meanwhile, a triple-source hybrid circuit implemented in a CMOS–180 nm technology managed simultaneously to extract thermal energy via a thermoelectric harvester and vibration via electromagnetic and piezoelectric harvesters, and then deliver a single DC output [235]. The system included an on-chip cross-coupled charge pump to boost the thermoelectric voltage, a low drop-out AC-DC doubler to rectify the electromagnetic output, and a combination of negative voltage converter, synchronous power extraction, DC-DC converter, and external inductor for the piezoelectric signal. Tested with an electromagnetic harvester worn on the wrist of a jogger, a commercial low-volume PZT-based harvester, and a thermoelectric generator, the system provided up to 110 μW output power in the simultaneous multimode operation, equivalent to 460% of the power delivered by a stand-alone circuit [235].

In several other studies [236–238], RF energy harvesters were integrated with a power management circuit, temperature sensor, or image sensor on the same chip. The RF-DC converter using internal threshold voltage cancellation with an auxiliary transistor block was implemented in a CMOS–180 nm technology together with an MPPT [237]. This RF harvesting system delivered nearly 11 μW output power and 39.3% power conversion efficiency with −15 dBm, 900 MHz input [237]. Park et al. [238] reported a multifunctional CMOS active pixel which could simultaneously complete the tasks of imaging and energy harvesting. In contrast to conventional CMOS electron-based imaging pixels, this work adopted a hole-based imaging method, being able to self-power the image capturing function at 15 fps with the energy harvested from >60 klux (The lux (symbol: lx) is the SI derived unit of illuminance, measuring luminous flux per unit area) (2018). Other relevant circuit designs for ultralow power management used in energy harvesting powered IoT systems, with a special focus on multiple energy sources, have recently been reviewed [239].

9.9 HYBRID ENERGY STORAGE FOR LOW-POWER EMBEDDED SYSTEMS APPLICATIONS

As mentioned earlier, ESS may also be categorized by their scale, which is directly related to the total capacity, portability, and target applications. Small-scale and large-scale ESSs have distinctive performance requirements. Portable ESSs are used for powering portable electronics, and so have strict constraints on their volume and weight. High energy density per unit volume or unit weight is thus the key criterion for portable ESS. LIB is a most promising type of energy storage element for today's portable applications. Electrical energy is increasingly becoming a major concern for today's foot soldiers [240] who must carry more and more advanced power-hungry electronics. Almad et al. [199] have developed photo-rechargeable organo-halide perovskite batteries. Some portable applications such as military radios also require high-power capability with small and light form factor. Battery-supercapacitor HESS may deal with the high-power demand of such applications by using supercapacitors as an energy buffer [241,242].

Low-power sensor nodes [243,244] employ a battery-supercapacitor. Due to very limited capability to produce power from energy harvesting devices such as PV cells, reducing the power loss during charge/discharge cycles is important. They take advantage of the high cycle efficiency of the supercapacitor while using the battery for low-leakage long-term energy storage. Other types of HESS, a CAES may utilize supercapacitors to keep track of the maximum efficiency operating point. A HESS can also take advantage of high-power power density of superconducting magnetic energy storage (SMES) and high-energy density of batteries. A fuzzy control logic determines the power split between the SMES and batteries [245–251].

Energy storage for portable electronics has unique requirements compared with ESS for residential (grid connected) or HEV (hybrid electric vehicle) usages. First, it has strict constraints on the size and weight, and requirements for high-energy density per unit volume or unit weight (which is the key criteria for portable ESS). Hence, LIB is the most promising type for man portable applications nowadays due to its high-energy density. Electrical energy is increasingly becoming a major concern for today's foot soldiers who must carry more and more advanced power-hungry electronics. Moreover, some portable applications such as military radios and bio-sensors require small and light form factor, high-energy capacity, and high power capability for a short period of time. Battery-supercapacitor HESS may deal with the high-power demand of such applications by using supercapacitors as an energy buffer [241,242]. Finally, relatively simple structure and control policy are desirable for ESS in portable electronics due to the limited computation capability.

A battery-supercapacitor HESS is a promising candidate to address the above-mentioned requirements. It has a simple architecture and high-energy density due to the usage of Li-ion battery, and can deal with the high-power demand by using supercapacitors as an intermittent energy buffer [241,242]. One can enhance the total service time of the battery-supercapacitor HESS by use of a constant-current charger circuit compared with a conventional hybrid architecture that simply connects the battery and supercapacitor in parallel. Earlier dual-battery as the hybrid power source for a portable electronic system was used in order to exploit both the rate- capacity characteristic and the relaxation-induced recovery effect. The study by Kim et al. [252] maximizes the utilization of the battery capacity under a given performance constraint using continuous-time Markov decision process, by modeling the workload arrival times, device service times, and battery selection times as stationary stochastic processes known in prior. However, this assumption may not be realistic.

As an even more interesting application, battery-supercapacitor HESS can be employed in wireless sensor nodes with energy harvesting in order to significantly extend the lifetime of the wireless sensors (which is typically restricted by the battery's cycle life) to even achieve near-perpetual operation. Lifetime extension of the battery is achieved by relying mostly on the supercapacitor as an energy buffer and reducing the frequency in charging/discharging the battery, because the supercapacitor has a nearly infinite cycle life. This is a desirable feature of wireless sensors since frequent replacements of battery may not be an easy task after deployment. The analysis [252] predicts that the sensor nodes will operate for 43 years under 1% load, 4 years under 10% load, and 1 year under 100% load, even under a simple control algorithm far from optimal. Similar potential applications of HESS include medical devices with energy harvesting from heat or vibration in order to significantly prolong the device's total lifetime. Besides achieving near-perpetual operation, the HESS can reduce the power loss during charge/discharge cycles, which is also critical due to the very limited capability and intermittent nature to produce power from energy harvesting devices such as PV cells. Effective control algorithms should be investigated to take advantage of the high cycle efficiency of the supercapacitor while using the Li-ion battery as a low-leakage long-term energy storage.

The sensor nodes of WSNs are mostly inactive to achieve longer runtimes. Conventionally, power being mainly provided by BESS, which is either primary or secondary energy source. However, considering high resistance of BESS, a significant loss occurs due to sleep and leakage current, moreover due to internal BESS impedance a sizable voltage drop is also experienced at the activation instant of the node, limiting the required power to be extracted. Considering these anomalies in, a parallel combination of BESS-SCSS (battery energy storage system-supercapacitor energy storage system) and HESS (hybrid energy storage system) has been proposed. Here, with a selection of three BESS, namely, silver oxide batteries, lithium batteries, and NiMH (nickel-metal hydride) batteries, it has been observed that hybridization with SCSS is feasible with only the first two as the resistance of NiMH is already optimally low enough. However, for silver oxide and lithium batteries, the combination with a low resistance SCSS significantly increases the runtime of the sensor nodes with a feasible decrement in leakage current losses.

The properties of HESS have been exploited to enhance the feasibility of wireless power transfer (WPT) applications in electric vehicles (EVs). Here a three-mode energy management strategy based on the State of Charge (SoC) conditions of HESS components has been proposed [252]. Hence, in addition, to regulating the energy input from the WPT and manage energy distribution between the BESS-SCSS, this study proposes the implementation of a short period of negative current periods in the charging state of the DC-DC converters. This depolarization pulse significantly reduces the stress on the internal resistance of BESS. Accordingly, the study in Ref. [242] presents the architecture of HESS in PV-based WSN. Proposing a sizing formulation for the HESS using a statistical approach on the historical power data, this study briefly addresses the improvement of HESS lifetime and endurance under fluctuating current conditions.

A wireless power supply with BESS-SCSS HESS has been developed for powering implantable devices, such as intracranial pressure (ICP) monitors. The shunt used to drain excess fluid from the brain in hydrocephalus patients fails regularly. Here, the most valuable diagnostic tool is a pressure sensor implanted in the brain. In this respect, WPT based on inductive power transfer technology is implemented on HESS where the HESS-powered ICP can meet both the short- and long-term operational requirements, and further, the study recorded that SCSS provides a quick charge back-up for collection of data.

Recent work [253–255] focuses on joint optimization of the embedded electronic system and its hybrid power supply. Reference [253] presents HypoEnergy, a framework for extending the lifetime of the hybrid battery-supercapacitor power supply. HypoEnergy studies the hybrid supply lifetime optimization for a preemptively known workload (a given set of tasks). The optimization problem is formulated and mapped to a multiple-choice knapsack problem and solved using the dynamic programming method. The authors evaluate the efficiency and applicability of the HypoEnergy framework using iPhone load measurements. The authors further extend their work to the set-up of multiple supercapacitors and workload that is not given a priori [254], and they use the machine learning technique to derive a near-optimal adaptive management policy for the hybrid power supply. Recently, Kim et al. [252] propose to use a model-free reinforcement learning (RL) technique for an adaptive dynamic power management (DPM) framework in embedded systems with bursty workloads, using a hybrid power supply comprised of LIBs and supercapacitors [241,242,250]. The study proposes a hierarchical power management framework with two dedicated power managers: one Supply PM for the hybrid power supply and one Device PM for the embedded device, in order to reduce the online computation overhead. The study also uses continuous-time Q-learning for the device PM to deal with bursty workloads and use discrete-time Q-learning for the supply PM. The supply PM makes decision at a much lower frequency than the device PM, and they exchange information with each other about embedded system states (such as busy, idle, sleep), number of waiting requests, SoC's of the battery and supercapacitor, and so on. The proposed RL-based DPM approach enhances the power efficiency by up to 9% compared to a battery-only power supply.

9.9.1 Integrating Faradaic and Capacitive Storage Mechanisms

Nonplanar electrode architectures may play an important role in future hybrid electric energy storage (EES) systems. Designed to be heterogeneous at the nanoscale, this approach enables devices to benefit fully from multielectron faradaic processes with minimum diffusion limitations and without phase transformations that hinder kinetics. The prospect of designing architectures with minimal expansion offers a route toward significantly decreasing the volume/mass required for nonactive components, since packaging can be simplified and bulky current collectors eliminated. In developing these architectures, not only conventional materials such as carbon and metal oxides and their combination should be considered, but also new materials in reduced dimensionalities that intrinsically possess metallic conductivity and redox-capable chemistries, such as Mxenes and 1-T dichalcogenides, should be explored. Concurrently, there is a need to develop new electrolytes that are compatible (no parasitic reactions or corrosion) with all components of the device and can withstand many thousands of cycles with stable performance over a wide voltage window.

A concept for future energy storage outlined by Lukatskaya et al. [256] is depicted in the schematic shown in Figure 9.4. In the opinion of the authors, future energy storage systems will be hybrid devices combining the best features of metal-ion batteries

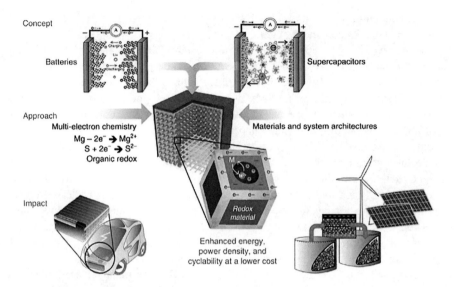

FIGURE 9.5 Redefining electrical energy storage [256]—Conceptual presentation of development of fully integrated rechargeable hybrid battery-supercapacitor (supercapbattery) electrical ESDs [256].

and Ecs. Such devices are based on hierarchical electrode architectures consisting of interconnected thin scaffolds of an active solid-electrode material containing designed micro-/mesopores that provide a large surface area in contact with a liquid or gel electrolyte, enabling fast charge–discharge with minimal strain generation. Electrons should ideally be provided from a conductive scaffold forming the electrode backbone, thus all but eliminating current collectors. Utilization of multielectron chemistries of both electrolyte and electrode materials will increase the amount of energy stored. The impact of these developments may start with portable electronics but will go well beyond, impacting large-scale EES by facilitating storage of electrical energy from renewable sources such as solar and wind. The concept, approach and impact of redefining electrical energy storage are illustrated in Figure 9.5.

Conceptual presentation of development of fully integrated rechargeable hybrid battery-supercapacitor (supercapbattery) electrical ESDs [256].

9.10 POWER ELECTRONICS FOR RENEWABLE ENERGY SYSTEMS

Hybrid renewable energy power systems are positioned to become the long-term power solution for portable, transportation, and stationary system applications [257]. Hybrid power systems are virtually limitless in possible set-ups and configurations to produce the desired power for a particular system. A hybrid system can consist of solar panels, wind power, fuel cells, electrolyzers, batteries, capacitors, and other types of power devices. Hybrid systems can be set up with power electronics to handle low, high, and variable power requirements. For example, solar panels can be used to convert solar energy into electrical energy when sunlight is directly hitting the PV panels for maximum efficiency, and then power from wind turbines can be used when wind speed and direction is ideal. The energy from these devices can be stored in batteries and used for electrolysis to produce hydrogen. The hydrogen can then be fed to fuel cells to provide power for long periods of time or portable or transportation applications. Power electronics provides a key element in stabilizing, boosting, and managing the power when necessary [257].

The electrical output of a specific power system may not provide the input needed for a certain device. Many applications, such as grid or residential power, require AC power. Other devices such as cell phones require DC power. The output of fuel cells and batteries, however, is DC voltage with an intensity that depends on the number of cells stacked in series. An inverter can be used to change the output from DC to AC power when needed. Also, many renewable energy systems can have slow start-up times and can be slow to respond to higher power needs. Therefore, systems usually have to be designed to compensate for high or intermittent power requirements. Power converters can be used to regulate the amount of power flowing through a circuit.

Most renewable energy technologies only provide a certain voltage and current density (depending upon the load) to the power converter. The power converter must then adjust the voltage available from the fuel cell to a voltage high enough to operate the load. A DC-DC boost converter is required to boost the voltage level for the inverter. This boost converter, in addition to boosting the fuel cell voltage, also regulates the inverter input voltage and isolates the low and high voltage circuits. A hybrid fuel cell/LIB charger system includes the following major components: the fuel cell, the LIB, a constant voltage regulation system, and a smart battery charger. A rechargeable LIB can be located inside the fuel cell unit to maintain the microcontroller in a low-power standby or programmed-timer sleep state for several days. The battery will also enable immediate system start-up and power during system shutdown. The battery will be automatically charged whenever the fuel cell is running. The internal battery charging circuit will stop charging the Li-ion battery once it has reached a certain voltage or has been charged for a specific amount of time [257].

The two basic power electronics areas that need to be addressed in renewable energy applications are power regulation and inverters. The electrical power output of fuel cells, solar cells, and wind turbines are not constant. The fuel cell voltage is typically controlled by voltage regulators, DC/DC converters, and other circuits at a constant value that can be higher or lower than the fuel cell operating voltage. Multilevel converters are of interest in the distributed energy resources area because several batteries, fuel cells, solar cells, and wind turbines can be connected through a multilevel converter to feed a load or grid without voltage-balancing issues. The general function of the multilevel inverter is to create a desired AC voltage from several levels of DC voltages. For this reason, multilevel inverters are ideal for connecting an AC grid either in series or parallel with renewable energy sources such as PVs or fuel cells or with ESDs such as capacitors or batteries. Multilevel converters also have lower switching frequencies than traditional converters, which results in reduced switching losses and increased efficiency.

Advances in fuel cell technology require similar advances in power converter technology. By considering power conversion design parameters early in the overall system design, a small, inexpensive converter can be built to accompany a reasonably sized solar panel, wind turbine, or fuel cell for high system power and energy density [257].

9.10.1 DC-TO-DC CONVERTERS

A DC-to-DC converter is used to regulate the voltage because the output of a renewable energy system varies with the load current. Many fuel cell and solar cell systems are designed for a lower voltage; therefore, a DC-DC boost converter is often used to increase the voltage to higher levels. A converter is required for these renewable energy systems because the voltage varies with the power that is required. A typical fuel cell drops from 1.23 V DC (no-load) to below 0.5 V DC at full load. Consequently, a converter will have to work with a wide range of input voltages [257].

DC-to-DC converters are important in portable electronic devices such as cellular phones and laptop computers where batteries are used. These types of electronic devices often contain several subcircuits, that each has its voltage level requirement that is different than supplied by the battery or an external supply. As the battery's stored power is drained, a DC-to-DC converter offers a method to increase voltage from a partially lowered battery voltage which saves space instead of using multiple batteries to accomplish the same task [257].

9.10.2 INVERTERS

Renewable energy can be used in both homes and businesses as the main power source. These energy systems will have to connect to the AC grid. The renewable energy system output will also need to be converted to AC in some grid-independent systems. An inverter can be used to accomplish this. The resulting AC current can be at the required voltage and frequency for use with the appropriate transformers and control circuits. Inverters are used in many applications from switching power supplies in computers to high voltage DC applications that supply bulk power. Inverters are commonly used to apply AC power from DC sources such as fuel cells, solar panels, and batteries [257].

Electronics are an important part of the devices that are used every day and a critical part of hybrid energy systems. These components help to transform DC into AC, help to increase the voltage of an energy system, regulate the power that a system provides, and/or creates the proper waveforms and timing that a motor requires. Without integrating these electronics into the system, the voltage and power produced by an energy system would not be very useful. Therefore, power electronics is an essential part of every hybrid energy system [257].

REFERENCES

1. Pearce, F. (2018, April). Energy hogs: Can the world's huge data centers be made more efficient? Yale Environment 360 [online]. Available at: https://e360.yale.edu/features/energy-hogs-can-huge-data-centers-be-made-more-efficient, a website report
2. Vaughan, A. (2015, September 25). How viral cat videos are warming the planet. *The Guardian*.
3. Shehabi, A., Smith, S., Sartor, D., Brown, R., Herrlin, M., Koomey, J., Horner, E,M.N., Azevedo, I. and Lintner, W. (2016, June). United States Data Center Energy Usage Report. This work was supported by the Federal Energy Management Program of the U.S. Department of Energy under Lawrence Berkeley National Laboratory Contract No.

DE-AC02-05CH1131, LBNL-1005775 [online]. Available at: http://eta-publications.lbl. gov/sites/default/files/lbnl-1005775 v2.pdf, LBNL, Berkley, CA.

4. United States Department of Energy. Energy 101: Energy efficient data centers [online]. Available at: https://www.energy.gov/eere/videos/energy-101-energy-efficient-data-centers.

5. Sverdlik, Y. (2020, February 27). Study: Data centers responsible for 1 percent of all electricity consumed worldwide, a website report by Data Center Knowledge.

6. International Energy Agency. (2019, May). Data centres and data transmission networks: Tracking clean energy progress [online], Amsterdam. Available at: https://www. iea.org/tcep/buildings/datacentres/.

7. U.S. Energy Information Administration. (2019, March). What is U.S. electricity generation by energy source? [online]. Available at: https://www.eia.gov/tools/faqs/faq. php?id=427&t=3.

8. Glanz, J. (2011, September 8). Google details, and defends, its use of electricity. *The New York Times.*

9. Holzle, U. (2009, January). Powering a google search. [online]. Available at: https:// googleblog.blogspot.com/2009/01/powering-google-search.html.

10. Clark, D. (2011, September 8). Google discloses carbon footprint for the first time. *The Guardian.*

11. Rodriguez, A. (2017, January). Greenpeace says binge-watching all those TV shows is bad for the environment. [online]. Available at: https://qz.com/882078/greenpeace-says-that-binge-watching-netflix-nflx-and-amazon-prime-amzn-is-bad-for- the-environment/.

12. Hunt, N. (June, 2017). Renewable energy at Netflix: An update. [online]. Available at: https://media.netflix.com/en/company-blog/renewable-energy-at-netflix-an-update.

13. Amazon Web Services. (2019). AWS & sustainability. [online]. Available at: https://aws. amazon.com/about-aws/sustainability/.

14. Ayanoglu, E. (2019, November 13). Energy Efficiency in Data Centers, IEEE TCN. A website publication by IEEE communication society.

15. Nedevschi, S., Chandrashekar, J., Liu, J., Nordman, B., Ratnasamy, S. and Taft, N. (2009, April). Skilled in the art of being idle: Reducing energy waste in networked systems. *NSDI* 9: 381–394.

16. Gerosa, G., Curtis, S., D'Addeo, M., Jiang, B., Kuttanna, B., Merchant, F., Patel, B., Taufique, M.H. and Samarchi, H. (2009). A sub-2 w low power ia processor for mobile internet devices in 45 nm high-k metal gate CMOS. *IEEE Journal of Solid-State Circuits* 44: 73–82.

17. Andersen, D.G., Franklin, J., Kaminsky, M., Phanishayee, A., Tan, L. and Vasudevan, V. (2009, October 11–14). FAWN: A fast array of wimpy nodes. In *SOSP*, 20. *SOSP'09*, Big Sky, MT. Copyright 2009 ACM 978-1-60558-752-3/09/10.

18. Agarwal, Y., Hodges, S., Chandra, R., Scott, J., Bahl, V. and Gupta, R. (2009, April). Somniloquy: Augmenting network interfaces to reduce PC energy usage. *NSDI.* USENIX Association NSDI '09: 6th USENIX Symposium on Networked Systems Design and Implementation 365–380.

19. Nedevschi, S., Popa, L., Iannaccone, G., Ratnasamy, S. and Wetherall, D. (2008, April). Reducing network energy consumption via rate-adaptation and sleeping. *NSDI* 8: 323–336.

20. Kumar, R., Farkas, K.I., Jouppi, N.P., Ranganathan, P. and Tullsen, D.M. (2003). Single-ISA heterogeneous multi-core architectures: The potential for processor power reduction. *IEEE/ACM International Symposium on Microarchitecture*, San Diego, CA

21. George, V. Jahagirdar, S., Tong, C. (2007). Penryn: 45-nm next generation Intel Core 2 processor. *IEEE Asian Solid State Circuits Conference*, Jeju Island, Korea.

22. Rusu, S., Tam, S., Muljono, H., Ayers, D., Chang, J., Cherkauer, B., Stinson, J., Benoit, J., Varada, R., Leung, J., Limaye, R.D. and Vora, S. (2007, January). A 65-nm Dual-Core Multithreaded Xeon® Processor with 16-MB L3 Cache. *IEEE J. of Solid State Circuits*, 42 (1): 17–25.

23. Lim, K., Ranganathan, P., Chang, J., Patel, C., Mudge, T. and Reinhardt, S. (2008) Understanding and designing new server architectures for emerging ware house-computing environments. *ACM SIGARCH Computer Architecture News* 36(3): 315–326.

24. Chun, B.-G., Iannaccone, G., Iannaccone, G., Katz, R., Lee, G. and Niccolini L. (2009, October 10). *An Energy Case for Hybrid Datacenters, HotPower'09,* Big Sky. MT.

25. Fitzpatrick, B. (2004, August). Distributed caching with memcached. *Linux Journal*. A website report (2004).

26. Hadoop. hadoop.apache.org.

27. Dean, J. and Ghemawat, S. (2004). MapReduce: Simplified data processing on large clusters. *OSDI*. 1-13.

28. Barroso, L. and Holzle, U. (2007, December). The case for energy-proportional computing. *Computer* 40: 33–37.

29. Fontecchio, M. and Rouse, M. (2009, April). Power usage effectiveness (PUE). [Online]. Available at: https://searchdatacenter.techtarget.com/definition/power-usage-effectiveness-PUE.

31. Amazon Web Services. (2019). AWS & sustainability. [online]. Available at: https://aws.amazon.com/about-aws/sustainability/.

30. Miller, R. (2011, May). Uptime Institute: The average PUE is 1.8. [Online]. Available at: https://www.datacenterknowledge.com/archives/2011/05/10/uptime-institute-the-average-pue-is-1-8.

32. Szalkus, M. (2008, December). What is Power Usage Effectiveness? [Online]. Available at: https://www.ecmweb.com/design/what-power-usage-effectiveness.

33. CBCI News. (2018, August). Microsoft's underwater datacenter now has live video feeds for your viewing pleasure.- a website report (2018).

34. Paul, I. (2016, February). Microsoft's audacious Project Natick wants to submerge your data in the oceans. [Online]. Available at: https://www.pcworld.com/article/3027934/microsofts-project-natick-wants-to-submerge-your-data-in-the-oceans.html.

35. Energy Star. 12 ways to save energy in data centers and server rooms. [Online]. Available at: https://www.energystar.gov/sites/default/files/asset/document/DataCenter-Top12-Brochure-Final.pdf.

36. Beloglazov, A. and Buyya, R. Energy efficient resource management in virtualized cloud data centers. *2010 10th IEEE/ACM International Conference on Cluster, Cloud and Grid Computing*, 826–831.

37. Bari, M.F., Boutaba, R., Esteves, R., Granville, L.Z., Podlesny, M., Rabbani, M.G., Zhang, Q. and Zhani, M.F. (2013). Data center network virtualization: A survey *IEEE Communications Surveys Tutorials* 15(2): 909–928, Second Quarter.

38. Tang, Q., Gupta, S.K.S. and Varsamopoulos, G. (2008, November). Energy-efficient thermal-aware task scheduling for homogeneous high-performance computing data centers: A cyber-physical approach. *IEEE Transactions on Parallel and Distributed Systems* 19(11): 1458–1472.

39. Wang, L., Zhang, F., Aroca, J.A., Vasilakos, A.V., Zheng, K., Hou, C., Li, D. and Liu, Z. (2014, January). GreenDCN: A general framework for achieving energy efficiency in data center networks. *IEEE Journal on Selected Areas in Communications* 32(1): 4–15.

40. Patterson, M.K. (2008, May). The effect of data center temperature on energy efficiency. *2008 11th Intersociety Conference on Thermal and Thermomechanical Phenomena in Electronic Systems*, 1167–1174.

41. Dabbagh, M., Hamdaoui, B., Guizani, M. and Rayes, A. (2015, September). Energy-efficient re- source allocation and provisioning framework for cloud data centers. *IEEE Transactions on Network and Service Management* 12(3): 377–391.

42. Jones, N. (2018, September). How to stop data centres from gobbling up the world's electricity. *Nature* 561: 163–166.

43. Ebrahimi, K., Jones, G. and Fleischer, A. (2014). A review of data center cooling technology, operating conditions and the corresponding low-grade waste heat recovery opportunities. *Renewable and Sustainable Energy Reviews* 31: 622–638, Doi: 10.1016/j.rser.2013.12.007.

44. Adapting renewable energy to the data center. A website article by ABB (2015).

45. Li, C., Wang, R., Li, T., Qian, D. and Yuan, J. Managing green datacenters powered by hybrid renewable energy systems. *Proceedings of the 11th International Conference on Autonomic Computing (ICAC '14)*, June 18–20, 2014, Philadelphia, PA ISBN 978-1-931971-11-9 Open access to the Proceedings of the 11th International Conference on Autonomic Computing (ICAC '14) by USENIX. https://www.usenix.org/conference/icac14/technical-sessions/presentation/li_chao

46. DCD Industry Census (2011). 2011: Forecasting energy demand. *USENIX Association 11th International Conference on Autonomic Computing* 271. www.dcd-intelligence.com.

47. Data Center Carbon Calculator (2020). http://www.apcmedia.com/salestools/WTOL-7DJLN9_R0_EN.swf, a website report.

48. Ren, C., Wang, D., Urgaonkar, B. and Si-vasubramaniam, A. (2012). Carbon-aware energy capacity planning for datacenters. *IEEE International Symposium on Modeling, Analysis & Simulation of Computer and Telecommunication Systems*, Washington, DC.

49. Deng, N., Stewart, C., Gmach, D., Arlitt, M. and Kelley, J. (2012). Adaptive green hosting. *International Conference on Autonomic Computing*, San Jose, CA.

50. Haque, M., Le, K., Goiri, I., Bianchini, R. and Nguyen, T. (2013). Providing green SLAs in high performance computing clouds. *International Green Computing Conference*, Arlington, VA.

51. Li, C. Zhang, W., Cho, C. and Li, T. (2011). SolarCore: Solar energy driven multi-core architecture power management. *IEEE International Symposium on High- Performance Computer Architecture*, Washington, DC.

52. Deng, N., Stewart, C., Kelley, J., Gmach, D. and Arlitt, M. (2012). Adaptive green hosting. *International Conference on Autonomic Computing*, San Jose, CA.

53. Li, C., Qouneh, A. and Li, T. (2012). iSwitch: Coordinating and optimizing renewable energy powered server clusters. *International Symposium on Computer Architecture*, Portland, OR.

54. Goiri, I., Le, K., Nguyen, T., Guitart, J., Torres, J. and Bianchini, R. (2012). GreenHadoop: Leveraging green energy in data-processing frameworks. *ACM EuroSys*, Bern.

55. Le, K., Bianchini, R., Martonosi, M. and Nguyen, T.D. (2010). Capping the brown energy consumption of internet services at low cost. *International Conference on Green Computing*, Hangzhou.

56. Liu, Z., Lin, M., Wierman, A., Low, S. and Andrew, L. (2011). Greening geographical load balancing. *ACM International Conference on Modeling and Measurement of Computer Systems*.

57. http://biomassmagazine.com/articles/8351/microsoft-data-center-to-install-biogas-fuel-cell-power-plant.

58. http://www.datacenterknowledge.com/archives/2012/05/30/hp-developing-net-zero-data-center- concept/., a website report.

59. http://green.ebay.com/greenteam/ebay/blog/Building-a-Greener-Company/26, a website report.

60. http://www.apple.com/environment/renewable-energy/, a website report.
61. Enabling the Low Carbon Economy in the Information Age. http://www.smart2020.org.
62. Burch, G. (2001). Hybrid renewable energy systems. *Natural Gas/Renewable Energy Workshops*, U.S. Department of Energy.
63. Goiri, I., Katsak, W., Le, K., Nguyen, T.D. and Bianchini, R. (2013). Parasol and GreenSwitch: Managing datacenters powered by renewable energy. *International Conference on Architectural Support for Programming Languages and Operating Systems*, Houston, TX.
64. Li, C., Zhou, R. and Li, T. (2013). Enabling distributed generation powered sustainable high-performance data center. IEEE 19th International Symposium on High Performance Computer Architecture (HPCA), Shenzhen, pp. 35–46. doi:10.1109/HPCA.2013.6522305.
65. Lasseter, R. and Piagi, P. (2004). Microgrid: A conceptual solution. *IEEE Annual Power Electronics Special- ISTS Conference*, Aachen.
66. Salomonsson, D., Soder, L. and Sannino, A. An adaptive control system for a DC Microgrid for data centers. *IEEE Transactions on Industry Applications* 44(6): 1910–1917.
67. Wang, R., Kandasamy, N., Nwankpa, C. and Kaeli, D. (2013). Datacenters as controllable load resources in the electricity market. *International Conference on Distributed Computing Systems*, Philadelphia, PA.
68. Deng, W., Liu, F., Jin, H. and Wu, C. (2013). SmartDPSS: Cost-minimizing multi-source power supply for datacenters with arbitrary demand. *International Conference on Distributed Computing Systems*.
69. Zareipour, H., Bhattacharya, K. and Canizares, C. (2004). Distributed generation: current status and challenges. *The 36th Annual North American Power Symposium*.
70. The importance of flexible electricity supply. (2011). Solar Integration Series. Technical Report, U.S. Department of Energy.
71. National Survey on Data Center Outages. (2010). Ponemon Institute, White Paper.
72. Kontorinis, V., Zhang, L., Aksanli, B., Sampson, J., Homayoun, H., Pettis, E., Rosing, T. and Tullsen, D. (2012). Managing distributed UPS energy for effective power capping in data centers. *International Symposium on Computer Architecture*, Portland, Oregon.
73. Fan, X., Weber, W. and Barroso, L. (2007). Power provisioning for a warehouse-sized computer. *International Symposium on Computer Architecture*, San Diego, CA.
74. Ghatikar, G., Ganti, V., Matson, N. and Piette, M. (2012). Demand response opportunities and enabling technologies for data centers: Findings from field studies. Technical Report, Lawrence Berkeley National Laboratory.
75. Xu, H., Topcu, U., Low, S., Clarke, C. and Chandy, K. (2010). Load-shedding probabilities with hybrid renewable power generation and energy storage. *The 48th Annual Allerton Conference on Communication, Control, and Computing., Monticello, Illinois*
76. Wang, D., Ren, C., Sivasubramaniam, A., Urgaonkar, B. and Fathy, H. (2012). Energy Storage in Datacenters: What, Where, and How much? ACM *International Conference on Modeling and Measurement of Computer Systems*, New York.
77. Govindan, S. Sivasubramaniam, A. and Urgaonkar, B. (2011). Benefits and limitations of tapping into stored energy for datacenters. *International Symposium on Computer Architecture*, New York.
78. McCluer, S. (2003). Battery technology for data centers and network rooms: Lead-acid battery options. APC White Paper #30. By SchneiderElectric–DataCenterScienceCenter, http://www.concordebattery.com/products/techinical_info/thermal_runaway.htm, 09/2001, DCSC@Schneider-Electric.com.
79. Kirby, B. and Hirst, E. (2000). Customer-specific metrics for the regulation and load-following ancillary services. Technical report, ORNL.
80. The role of distributed generation and Combined Heat and Power (CHP) systems in data centers. (2007). Technical Report, US EPA.

81. Chowdhury, S. and Crossley, P. (2009). *Microgrid and Active Distribution Networks.* The Institute of Engineering and Technology. Volume 6 of IET renewable energy series, ISBN: 1849190143, 9781849190145, London.
82. Fuel Cell Technologies Program Multi-year Research. Development and demonstration plan. Technical Report, US Department of Energy.
83. Bindner, H., Cronin, T., Lundsager, P., Manwell, J., Abdulwahid, U. and Gould, I. (2005). Lifetime modelling of lead acid batteries. Technical Report, Risø National Laboratory.
84. Jongerden, M. and Haverkort, B. (2008). Which battery model to use? *The 24th UK Performance Engineering Workshop,* Imperial College, London.
85. Getting started guide for HOMER version 2.1. (2005). National Renewable Energy Laboratory.
86. National Wind Technology Center (NWTC). http://www.nrel.gov/wind/.
87. Pelley, S., Meisner, D., Zandevakili, P., Wenisch, T. and Underwood, J. (2010). Power Routing: Dynamic power provisioning in the data center. *International Conference on Architectural Support for Programming Languages and Operating Systems,* Pittsburgh, PA.
88. Ahmad, F. and Vijaykumar, T. (2010). Joint optimization of idle and cooling power in data centers while maintaining response time. *International Conference on Architectural Support for Programming Languages and Operating Systems,* Pittsburgh, PA.
89. Logs of Real Parallel Workloads. http://www.cs.huji.ac.il/labs/parallel/workload.
90. Host Power Management in VMware vSphere 5. (2010). Technical Report, VMware.
91. Ranganathan, P. Leech, P., Irwin, D. and Chase, J. (2006). Ensemble-level power management for dense blade servers. *International Symposium on Computer Architecture,* New York.
92. SPEC power_ssj (2008). http://www.spec.org/power_ssj2008/.
93. Li, C., Wang, R., Goswami, N., Li, X., Li, T. and Qian, D. (2013). Chameleon: Adapting throughput server to time-varying green power budget using online learning. *International Symposium on Low Power Electronics and Design,* Beijing.
94. Li, C., Hu, Y., Zhou, R., Liu, M., Liu, L., Yuan, J. and Li, T. (2013). Enabling datacenter servers to scale out economically and sustainably. *International Symposium on Microarchitecture,* Davis, CA.
95. Goiri, I., Beauchea, R., Le, K., Nguyen, T.D., Haque, M., Guitart, J., Torres, J. and Bianchini, R. (2011). GreenSlot: Scheduling energy consumption in green datacenters. *International Conference on for High Performance Computing, Networking, Storage and Analysis,* New York.
96. Xu, J. and Fortes, J. (2011). A multi-objective approach to virtual machine management in datacenters. *International Conference on Autonomic Computing,* Karlsruhe.
97. Banerjee, P., Patel, C., Bash, C. and Ranganathan P. (2009). Sustainable data centers: Enabled by supply and demand side management. *Design Automation Conference,* San Francisco, CA.
98. Govindan, S., Wang, D., Sivasubramaniam, A. and Urgaonkar, B. (2012). Leveraging stored energy for handling power emergencies in aggressively provisioned datacenters, battery emergency. *International Conference on Architectural Support for Programming Languages and Operating Systems,* London.
99. Sun, M., Xue, Y., Bogdan, P., Tang, J, Wang, Y. and Lin, X. (2018). Hierarchical and hybrid energy storage devices in data centers: Architecture, control and provisioning. *PLoS One* 13(1): e0191450, Doi: 10.1371/journal.pone.0191450.
100. Barroso, L.A., Clidaras, J. and Hölzle U. (2013). The datacenter as a computer: An introduction to the design of warehouse-scale machines. *Synthesis Lectures on Computer Architecture* 8(3): 1–154, Doi: 10.2200/S00516ED2V01Y201306CAC024.

101. Kontorinis, V., Zhang, L.E., Aksanli, B., Homayoun, J.S.H., Pettis, E., Tullsen, D.M. and Rosing, T.S. (2012). Managing distributed UPS energy for effective power capping in data centers. *Proceedings of the ACM International Symposium on Computer Architecture (ISCA)*, Portland, OR, 488–499.

102. Wang, D., Ren, C., Sivasubramaniam, A., Urgaonkar, B. and Fathy, H. (2012). Energy storage in datacenters: What, where, and how much? *Proceedings of the ACM International Conference on Measurement and Modeling of Computer Systems (SIGMETRICS)*, London, 187–198.

103. Fan, X., Weber, W.D. and Barroso, L.A. (2007). Power provisioning for a warehouse-sized computer. *Proceedings of the ACM International Symposium on Computer Architecture (ISCA)*, San Diego, CA, 13–23.

104. Govindan, S., Wang, D., Sivasubramaniam, A. and Urgaonkar, B. (2012). Leveraging stored energy for handling power emergencies in aggressively provisioned datacenters. *Proceedings of the ACM Architectural Support for Programming Languages and Operating Systems (ASPLOS)*, London, 75–86.

105. Cochran, R., Hankendi, C., Coskun, A.K. and Reda, S. (2011). Pack & cap: Adaptive DVFS and thread packing under power caps. *Proceedings of the IEEE/ACM International Symposium on Microarchitecture (MICRO)*, Porto Alegre, 175–185.

106. Gandhi, A., Harchol-Balter, M., Das, R. and Lefurgy, C. (2009). Optimal power allocation in server farms. *ACM SIGMETRICS Performance Evaluation Review* 37: 157–168. [Google Scholar].

107. Gandhi, A., Gupta, V., Harchol-Balter, M. and Kozuch, M.A. (2010). Optimality analysis of energy-performance trade-off for server farm management. *Performance Evaluation* 67: 1155–1171, Doi: 10.1016/j.peva.2010.08.009 [Google Scholar].

108. Ranganathan, P., Leech, P., Irwin, D. and Chase, J. (2006). Ensemble-level power management for dense blade servers. *Proceedings of the ACM International Symposium on Computer Architecture (ISCA)*, Boston, MA, 66–77.

109. Barroso, L.A. and Hölzle, U. (2007). The case for energy-proportional computing. *IEEE Computer* 40: 33–37, Doi: 10.1109/MC.2007.443 [Google Scholar].

110. Li, C., Zhang, W., Cho, C.B. and Li, T. (2011). Solarcore: Solar energy driven multi-core architecture power management. *2011 IEEE 17th International Symposium on High Performance Computer Architecture*, IEEE, 205–216.

111. Aksanli, B., Rosing, T. and Pettis, E. (2013). Distributed battery control for peak power shaving in datacenters. *Green Computing Conference (IGCC), 2013 International*, IEEE, 1–8.

112. Aksanli, B., Pettis, E. and Rosing, T. (2013). Architecting efficient peak power shaving using batteries in data centers. *2013 IEEE 21st International Symposium on Modelling, Analysis and Simulation of Computer and Telecommunication Systems*, IEEE, 242–253.

113. Govindan, S., Sivasubramaniam, A. and Urgaonkar, B. (2011). Benefits and limitations of tapping into stored energy for datacenters. *Proceedings of the ACM International Symposium on Computer Architecture (ISCA)*, San Jose, CA, 341–352.

114. Ton, M. and Fortenbury, B. (Dec., 2005). High performance buildings: Data centers uninterruptible power supplies (UPS).pp. 44, a website report, work performed for "High-Performance High-Tech Buildings" project of LBNL, Berkley, CA. Available at: http://hightech.lbl.gov/documents/ups/final_ups_report.pdf.

115. Hu, X., Zou, C., Zhang, C. and Li, Y. (2017). Technological developments in batteries: A survey of principal roles, types, and management needs. *IEEE Power and Energy Magazine* 15(5): 20–31, Doi: 10.1109/MPE.2017.2708812 [Google Scholar].

116. Linden, D. and Reddy, T.B. (2002). *Handbook of Batteries*. McGraw-Hill. [Google Scholar], New York.

117. Pedram, M., Chang, N., Kim, Y. and Wang, Y. (2010). Hybrid electrical energy storage systems. Low-Power Electronics and Design (ISLPED). *2010 ACM/IEEE International Symposium*, IEEE, 363–368.

118. Chen, H., Cong, T.N., Yang, W., Tan, C., Li, Y. and Ding, Y. (2009). Progress in electrical energy storage system: A critical review. *Progress in Natural Science* 19: 291–312, Doi: 10.1016/j.pnsc.2008.07.014[Google Scholar]. Open Access. Elsevier, Netherland.

119. Shah, YT. (2021). *Hybrid Power- Generation, Storage and Grids*. CRC Press, New York.

120. Zou, C., Hu, X., Wei, Z. and Tang, X. (2017). Electrothermal dynamics-conscious lithium-ion battery cell-level charging management via state-monitored predictive control. *Energy* 141(Supplement C): 250–259, Doi: 10.1016/j.energy.2017.09.048 [Google Scholar].

121. Zhang, L., Wang, Z., Hu, X., Sun, F. and Dorrell, D.G. (2015). A comparative study of equivalent circuit models of ultracapacitors for electric vehicles. *Journal of Power Sources* 274(Supplement C): 899–906, Doi: 10.1016/j.jpowsour.2014.10.170 [Google Scholar].

122. Zhang, L., Hu, X., Wang, Z., Sun, F. and Dorrell, D.G. (2016). Fractional-order modeling and state-of-charge estimation for ultracapacitors. *Journal of Power Sources* 314(Supplement C): 28–34, Doi: 10.1016/j.jpowsour.2016.01.066 [Google Scholar].

123. Fortenbery, B., EPRI, E.C. and Tschudi, W. (2008). DC power for improved data center efficiency. Available at: http://hightech.lbl.gov/documents/data_centers/dcdemofinalreport.pdf.

124. Google Inc. (2009). Google datacenter video tour. Available at: http://www.google.com/about/datacenters/efficiency/external/2009-summit.html.

125. Miller, R. (2011). Microsoft reveals its specialty servers. Racks. Available at: http://www.datacenterknowledge.com/archives/2011/04/25/microsoft-reveals-its-specialty-servers-racks/.

126. Facebook Inc. (2011). Open compute project. Available at: http://www.opencompute.org/.

127. Liu, L., Li, C., Sun, H., Hu, Y., Gu, J., Li. T., Xin, J. and Zheng, N. (2015). Heb: deploying and managing hybrid energy buffers for improving datacenter efficiency and economy. *ACM SIGARCH Computer Architecture News ACM* 43: 463–475.

128. Koushanfar, F. (2010). Hierarchical hybrid power supply networks. *Proceedings of the 47th Design Automation Conference*. ACM, New York, 629–630.

129. Meisner, D., Gold, B.T. and Wenisch, T.F. (2009). PowerNap: Eliminating server idle power. *Proceedings of the ACM Architectural Support for Programming Languages and Operating Systems (ASPLOS)*, Washington, DC, 205–216.

130. Meisner, D., Sadler, C.M., Barroso, L.A., Weber, W.D. and Wenisch, T.F. (2011). Power management of online data-intensive services. *Proceeding of the ACM International Symposium on Computer Architecture (ISCA)*, San Jose, CA, 319–330.

131. Li, C., Zhou, R. and Li, T. (2013). Enabling distributed generation powered sustainable high-performance data center. *High Performance Computer Architecture (HPCA2013), 2013 IEEE 19th International Symposium*, IEEE, 35–46.

132. Hu, X., Martinez, C.M. and Yang, Y. (2017). Charging, power management, and battery degradation mitigation in plug-in hybrid electric vehicles: A unified cost-optimal approach. *Mechanical Systems and Signal Processing* 87(Part B): 4–16, Doi: 10.1016/j.ymssp.2016.03.004 [Google Scholar].

133. Wahlroos, M., Pärssinen, M., Manner, J. and Syri, S. (2017). Utilizing data center heat in district heating – impacts on energy efficiency and prospects for low-temperature district heating networks. *Energy* 140: 1228–1238.

134. Woodruff, J.Z., Brenner P., Buccellato A.P.C. and Go D.B. (2013). Environmentally opportunistic computing: A distributed waste heat reutilization approach to energy-efficient buildings and data centers. *Energy and Buildings*, Doi: 10.1016/j.enbuild.2013.09.036.

135. Wahlroos, M., Pärssinen, M., Rinne, S., Syri, S. and Manner, J. (2018). Future views on waste heat utilization–Case of data centers in Northern Europe. *Renewable and Sustainable Energy Reviews* 82: 1749–1764.

136. Davies, G.F., Maidment, G.G. and Tozer, R.M. (2016). Using data centres for combined heating and cooling: an investigation for London. *Applied Thermal Engineering* 94: 296–304.

137. Lund, H., Werner, S., Wiltshire, R., Svendsen, S., Thorsen, E., Hvelplund, F. and Vad, M.B. (2014). 4th Generation District Heating (4GDH): integrating smart thermal grids into future sustainable energy systems. *Energy* 68: 1–11.

138. Stockholm Data Parks sets new standards for sustainable data … stockholmdataparks. com, a website report (2019).

139. Liu, Z., Chen, Y., Bash, C., Wierman, A., Gmach, D., Wang, Z., Marwah, M. and Hyser, C. (2012). Renewable and cooling aware workload management for sustainable data centers. In *Proceedings of the 12th ACM SIGMETRICS/PERFORMANCE Joint International Conference on Measurement and Modeling of Computer Systems*, New York.

140. Goiri, Í., Haque, M.E., Le, K., Beauchea, R., Nguyen, T.D., Guitart, J., Torres, J. and Bianchini, R. (2014). Matching renewable energy supply and demand in green data centers. *Ad Hoc Networks* 25(February): 520–534.

141. Liu, Z., Lin, M., Wierman, A., Low, S., Andrew, L.L.H. and Member, S. (2015). Greening geographical load balancing. *IEEE/ACM Transactions on Networking* 23(2): 657671.

142. Chen, C., He, B. and Tang, X. (2012). Green-aware workload scheduling in geographically distributed data centers. *4th IEEE International Conference on Cloud Computing Technology and Science Proceedings*, Taipei, IEEE, 82–89.

143. Toosi, A.N., Qu, C., de Assunção, M.D. and Buyya, R. (2017). Renewable-aware geographical load balancing of web applications for sustainable data centers. *Journal of Network and Computer Applications* 83(January): 155–168.

144. Haas, R., Lettner, G., Auer, H. and Duic, N. (2013). The looming revolution: How photovoltaics will change electricity markets in Europe fundamentally. *Energy* 57: 38–43.

145. Krioukov, A., Goebely, C., Alspaugh, S., Chen, Y., Culler, D. and Katz, R. (2011). Integrating renewable energy using data analytics systems: Challenges and opportunities. *IEEE Data Engineering Bulletin* 34(1): 3–11.

146. Ghatikar, R., Ganti, V., Matson, N. and Piette, M. (2012). Demand response opportunities and enabling technologies for data centers: findings from field studies. Report number: LBNL-5763E, Affiliation, Lawrence Berkeley National Laboratory.

147. Clausen, A., Ghatikar, G. and Nørregaard Jørgensen, B. (2014). Load management of data centres as regulation capacity in Denmark. *Proceedings International Green Computing Conference,* 3–5 November 2014, Dallas, TX. Doi: 10.1109/IGCC.2014.7039161.

148. Koronen, C., Åhman, M. and Nilsson, L. (2020). Data centers in future European energy systems-energy efficiency, integration and policy. Project Low Carbon Industry.

149. Guo, Y. and Fang, Y. (2013). Electricity cost saving strategy in data centers by using energy storage. *IEEE Transactions on Parallel and Distributed Systems* 24(6): 1149–1160.

150. Shi, Y., Xu, B., Wang, D. and Zhang, B. (2018). Using battery storage for peak shaving and frequency regulation: joint optimization for superlinear gains. *IEEE Transactions on Power Systems* 33(3): 2882–2894.

151. Ma, Z., Eichman, J. and Kurtz, J. (2018). *Fuel Cell Backup Power System for Grid Service and Micro-Grid in Telecommunication Applications: Preprint.* NREL/CP–5500–70990. National Renewable Energy Laboratory, Golden, CO.

152. Luo, X., Wang, J., Dooner, M. and Clarke, J. (2015). Overview of current development in electrical energy storage technologies and the application potential in power system operation. *Applied Energy* 137: 511–536.

153. Navigant. (2017). Data centers and advanced microgrids: Meeting resiliency, efficiency, and sustainability goals through smart and cleaner power infrastructure, Navigant report by Peter Asmus, commissioned by Schneider Electric.

154. Bai, Y., Jantunen, H. and Juuti, J. (2018, November). Hybrid, multi-source, and integrated energy harvesters. *Frontiers in Materials* 5(65), Doi: 10.3389/fmats.2018.00065.

155. Fan, K., Liu, S., Liu, H., Zhu, Y., Wang, W. and Zhang, D. (2018). Scavenging energy from ultra–low frequency mechanical excitations through a bi–directional hybrid energy harvester. *Applied Energy* 216: 8–20, Doi: 10.1016/j.apenergy.2018.02.086.

156. Li, P., Gao, S. and Cong, B. (2018). Theoretical modeling, simulation and experimental study of hybrid piezoelectric and electromagnetic energy harvester. AIP Advances 8: 035017, Doi: 10.1063/1.5018836.

157. Maharjan, P., Toyabur, R.M. and Park, J.Y. (2018). A human locomotion inspired hybrid nanogenerator for wrist–wearable electronic device and sensor applications. *Nano Energy* 46: 383–395, Doi: 10.1016/j.nanoen.2018.02.033.

158. Feng, L., Liu, G., Guo, H., Tang, Q., Pu, X., Chen, J., Wang, X., Xi, Y. and Hu, C. (2018). Hybridized nanogenerator based on honeycomb–like three electrodes for efficient ocean wave energy harvesting. *Nano Energy* 47: 217–223, Doi: 10.1016/j.nanoen.2018.02.042.

159. Wang, Z.L. (2011). *Nanogenerators for Self-Powered Devices and Systems.* Georgia Institute of Technology, SMARTech digital repository, Atlanta.

160. Xu, C., Wang, X.D. and Wang, Z.L. (2009). Nanowire Structured Hybrid Cell for Concurrently Scavenging Solar and Mechanical Energies. *Journal of the American Chemical Society* 131: 5866–5872.

161. Annapureddy, V., Na, S., Hwang, G. et al., (2018). Exceeding milli–watt powering magneto–mechano–electric generator for standalone–powered electronics. *Energy & Environmental Science* 11: 818–829, Doi: 10.1039/C7EE03429F.

162. Hu, Z., Qiu, J., Wang, X., Gao, Y., Liu, X., Chang, Q., Long, Y. and He, X. (2018). An integrated multi-source energy harvester based on vibration and magnetic field energy. *AIP Advances* 8: 056623, Doi: 10.1063/1.5006614.

163. Wang, Z.L. and Wu, W. (2012). Nanotechnology-enabled energy harvesting for self-powered micro-/nanosystems. *Angewandte Chemie International Editi*on 51: 11700–11721 Wiley-VCH Verlag GmbH & Co. KGaA, Weinheim.

164. Lee, C. (2010, October 18). Hybrid energy harvesters could power handheld electronics, SPIE. A website report by International society of optics and photonics.

165. Cao, R., Wang, J., Xing, Y., Song, W., Li, N., Zhao, S., Zhang, C. and Li, C. (2018). A self– powered lantern based on a triboelectric–photovoltaic hybrid nanogenerator. *Advanced Materials Technologies* 3: 1700371, Doi: 10.1002/admt.201700371.

166. Yang, B., Jantunen, H. and Juuti, J. (2018). Energy harvesting research: the road from single source to multisource. *Advance Material* 30: 1707271, Doi: 10.1002/adma.201707271.

167. Ma, J., Zhang, Q., Lin, K., Zhou, L. and Ni, Z. (2018). Piezoelectric and optoelectronic properties of electrospinning hybrid PVDF and ZnO nanofibers. *Materials Research Express* 5, Doi: 10.1088/2053-1591/aab747.

168. Lee, D.-Y., Kim, H., Li, H., Jang, A.-R., Lim, Y.-D., Cha, S., Park, Y., Kang, D.J. and Yoo, W. (2013). Hybrid energy harvester based on nanopillar solar cells and PVDF nanogenerator. *Nanotechnology* 24: 175402, Doi: 10.1088/0957-4484/24/17/175402.

169. Raouadi, M.H. and Touayar, O. (2018). Harvesting wind energy with pyroelectric nanogenerator PNG using the vortex generator mechanism. *Sensors and Actuators A: Physical* 273: 42–48, Doi: 10.1016/j.sna.2018.02.009.

170. Bowen, C.R., Kim, H.A., Weaver, P.M. and Dunn, S. (2014). Piezoelectric and ferroelectric materials and structures for energy harvesting applications. *Energy & Environmental Science* 7: 25–44, Doi: 10.1039/C3EE42454E.

171. Bowen, C.R., Taylor, J., LeBoulbar, E., Zabek, D., Chauhan, A. and Vaish, R. (2014). Pyroelectric materials and devices for energy harvesting applications. *Energy & Environmental Science* 7: 3836–3856, Doi: 10.1039/C4EE01759E.

172. Lu, R.P., Ramirez, A.D. and Pascoguin, B.M.L. (2018). Multi-source energy harvesting device. Patent No US20180069405A1: United States of America as represented by Secretary of the Navy, San Diego, CA.

173. He, Z., Zhong, C., Su, S., Xu, M., Wu, H. and Cao, Y. (2012). Enhanced power–conversion efficiency in polymer solar cells using an inverted device structure. *Nature Photonics* 6: 591–595, Doi: 10.1038/nphoton.20 12.190.

174. Xu, C. and Wang, Z.L. (2011). Compact hybrid cell based on a convoluted nanowire structure for harvesting solar and mechanical energy. *Advanced Material* 23: 873–877.

175. Choi, D., Lee, K.Y., Jin, M.J. et al. (2011). Control of naturally coupled piezoelectric and photovoltaic properties for multi-type energy scavengers. *Energy & Environmental Science* 4: 4607–4613.

176. Lee, M., Yang, R., Li, C. and Wang, Z.L. (2010). Nanowire–Quantum Dot Hybridized Cell for Harvesting Sound and Solar Energies. *The Journal of Physical Chemistry Letters* 1: 2929–2935.

177. Bae, J., Park, Y.J., Lee, M., Cha, S.N., Choi, Y.J., Lee, C.S., Kim, J.M. and Wang, Z.L. (2011). Single-Fiber-Based Hybridization of Energy Converters and Storage Units Using Graphene as Electrodes. *Advanced Material* 23: 3446–3449.

178. Hansen, B.J., Liu, Y., Yang, R.S. and Wang, Z.L. (2010). Hybrid nanogenerator for concurrently harvesting biomechanical and biochemical energy. *ACS Nano* 4: 3647–3652.

179. Pan, C.F., Li, Z.T., Guo, W.X., Zhu, J. and Wang, Z.L. (2011). Fiber-Based Hybrid Nanogenerators for/as Self-Powered Systems in Biological Liquid. *Angewandte Chemie* 123: 11388–11392; *Angewandte Chemie International Edition* (2011) 50: 11192–11196.

180. Guo, X.Z., Zhang, Y.D., Qin, D., Luo, Y.H., Li, D.M., Pang, Y.T. and Meng, Q.B. (2010). Hybrid tandem solar cell for concurrently converting light and heat energy with utilization of full solar spectrum. *Journal of Power Sources* 195: 7684–7690.

181. Wang, N., Han, L., He, H.C., Park, N.H. and Koumoto, K. (2011). A novel high-performance photovoltaic–thermoelectric hybrid device. *Energy & Environmental Science* 4: 3676–3679.

182. Fujitsu hybrid energy harvesting device generates electricity from heat or light. (2010, December). A website report by Fijitsu lab, Japan.

183. Watkins, C.B., Shen, B. and Venkatasubramanian, R. (2005). Low-grade-heat energy harvesting using superlattice thermoelectrics for applications in implantable medical devices and sensors. ICT 2005. *24th International Conference on Thermoelectrics*, Doi:10.1109/ICT.2005.1519934. Corpus ID: 6085987 (2005).

184. Ensescu, D. (2019). Thermoelectric energy harvesting: Basic principles and applications. InTech open access paper, DOI: 10.5772/intechopen.83495. www.intech.com.

185. Glatz W., Schwyter E., Durrer L. and Hierold C. (2009). Bi2Te3-based flexible micro thermoelectric generator with optimized design. *Journal of Micrelectromechanical Systems* 18(3): 763–772.

186. Jaziri, N., Boughamoura, A., Müller, J., Mezghani, B., Tounsi, F. and Ismail, M. (2019). A comprehensive review of thermoelectric generators: Technologies and common applications. Energy Reports ISSN 2352-4847, https://doi.org/10.1016/j.egyr.2019.12.011. Vol. 5, pp. 2139 (Dec., 2019). Elsevier, Netherland.

187. Rowe, D.M. (1995). *CRC Handbook of Thermo- electrics*, 1st ed. CRC Press, 701 s. ISBN 978-0849301469.

188. Riffat, S.B. and Ma, X. (2003). Thermoelectrics: a review of present and potential applications. *Applied Thermal Engineering* 23: 913–935.

189. Ismail, B.I., Ahmed, W.H. (2009). Thermoelectric power generation using waste-heat energy as an alternative green technology. *Recent Patents on Electrical Engineering Continued as Recent Advances in Electrical & Electronic Engineering* 2(1): 27–39.

190. Cahill, P., Pakrashi, V., Sun, P., Mathewson, A. and Nagarajaiah, S. (2018). Energy harvesting techniques for health monitoring and indicators for control of a damaged pipe structure. *Smart Structures and Systems* 21: 287–303, Doi: 10.12989/sss.2018.21.3.000.

191. Kim, S., Ju, S. and Ji, C. (2018). Impact–based piezoelectric energy harvester as a power source for a neural activity monitoring circuit. *International Journal of Grid and Distributed Computing* 11: 51–62, Doi: 10.14257/ijgdc.2018.11.3.05.

192. Chun, J., Kishore, R. A., Kumar, P., Kang, M., Kang, H. B., Sanghadasa, M. and Priya, S. (2018). Self–powered temperature–mapping sensors based on thermo–magneto–electric generator. *ACS Applied Materials & Interfaces* 10: 10796–10803, Doi: 10.1021/acsami.7b17686.

193. Shen, Q., Xie, X., Peng, M., Sun, N., Shao, H., Zheng, H., Wen, Z. and Sun, X. (2018). Self–powered vehicle emission testing system based on coupling of triboelectric and chemoresistive effects. *Advanced Functional Materials* 28: 1703420, Doi: 10.1002/adfm.201703420.

194. Yang, H., Liu, W., Xi, Y. et al., (2018). Rolling friction contact–separation mode hybrid triboelectric nanogenerator for mechanical energy harvesting and self–powered multifunctional sensors. *Nano Energy* 47: 539–546, Doi: 10.1016/j.nanoen.2018.03.028.

195. Xu, S., Qin, Y., Xu, C., Wei, Y.G., Yang, R.S. and Wang, Z. L. (2010). Self-powered nanowire devices. *Nature Nanotechnology* 5: 366–373.

196. Pan, C.F., Fang, Y., Wu, H. et al. (2010). Generating electricity from biofluid with a nanowire-based biofuel cell for self-powered nanodevices. *Advanced Material* 22: 5388–5392.

197. Han, J., Fan, F., Xu, C., Lin, S., Wei, M., Duan, X. and Wang, Z.L. (2010). ZnO nanotube-based dye-sensitized solar cell and its application in self-powered devices. *Nanotechnology* 21: 405203.

198. Wang, S., Ke, Y., Huang, P. and Hsieh, P. (2018). Electromagnetic energy harvester interface design for wearable applications. *IEEE Transactions on Circuits and Systems II: Express Briefs* 65: 667–671, Doi: 10.1109/TCSII.2018.2820158.

199. Ahmad, S., George, C., Beesley, D.J., Baumberg, J.J. and De Volder, M. (2018). Photo–rechargeable organo–halide perovskite batteries. *Nano Letters* 18: 1856–1862, Doi: 10.1021/acs.nanolett.7b05153.

200. Schiffer, Z.J. and Arnold, C.B. (2018). Characterization and model of piezoelectrochemical energy harvesting using lithium ion batteries. *Experimental Mechanics* 58: 605–611, Doi: 10.1007/s11340-017-0291-1.

201. Yang, P., Qu, X., Liu, K. et al., (2018). Electrokinetic supercapacitor for simultaneous harvesting and storage of mechanical energy. *ACS Applied Materials & Interfaces* 10: 8010–8015, Doi: 10.1021/acsami.7b18640.

202. Liu, C., Li, F., Ma, L. and Cheng, H. (2010). Advanced Materials for Energy Storage. *Advanced Material* 22: 28, Doi: 10.1002/adma.200903328, Google ScholarCrossref.

203. Hochbaum, A.I., Chen, R., Delgado, R., Liang, D., Garnett, W.E., Najarian, C., Majumdar, M.A. and Yang, P. (2008). Enhanced thermoelectric performance of rough silicon nanowires *Nature* 451: 163, Doi: 10.1038/nature06381, Google ScholarCrossref.

204. Hu, L., Choi, J.W., Yang, Y., Jeong, S., Mantia, F.L., Cui, L.-F. and Cui, Y. (2009). Highly conductive paper for energy-storage devices *Proceedings of the National Academy of Sciences of the United States of America* 106: 21490, Doi: 10.1073/pnas.0908858106, Google ScholarCrossref.

205. Wang, Z.L. and Song, J.H. (2006). Piezoelectric Nanogenerators Based on Zinc Oxide Nanowire Arrays. *Science* 312: 242, Doi: 10.1126/science.1124005, Google ScholarCrossref.

206. Yang, R.S., Qin, Y., Dai, L.M. and Wang, Z.L. (2009). Power generation with laterally packaged piezoelectric fine wires. *Nature Nanotechnology* 4: 34, Doi: 10.1038/nnano.2008.314, Google ScholarCrossref.

207. Xu, S., Qin, Y., Xu, C., Wei, Y., Yang, R. and Wang, Z.L. (2010). Self-powered nanowire devices. *Nature Nanotechnology* 5: 366, Doi: 10.1038/nnano.2010.46, Google ScholarCrossref.

208. Lee, M., Chen, C.Y., Wang, S., Cha, S.N., Park, Y.J., Kim, J.M., Chou, L.J. and Wang, Z.L. (2012). A hybrid piezoelectric structure for wearable nanogenerators. *Advanced Material* 24: 1759, Doi: 10.1002/adma.201200150, Google ScholarCrossref.

209. Choi, D., Choi, M.-Y., Choi, W.M. et al. (2010). Fully rollable transparent nanogenerators based on graphene electrodes. *Advanced Material* 22: 2187, Doi: 10.1002/adma.200903815, Google ScholarCrossref.

210. Qi, Y., Jafferis, N.T., Lyons, K., Lee, C.M., Ahmad, H. and McAlpine, M.C. (2010). Piezoelectric ribbons printed onto rubber for flexible energy conversion. *Nano Letters* 10: 524, Doi: 10.1021/nl903377u, Google ScholarCrossref.

211. Qi, Y., Kim, J., Nguyen, T.D., Lisko, B., Purohit, P.K. and McAlpine, M.C. (2011). Enhanced piezoelectricity and stretchability in energy harvesting devices fabricated from buckled PZT ribbons. *Nano Letters* 11: 1331, Doi: 10.1021/nl104412b, Google ScholarCrossref.

212. Park, K.I., Son, J.H., Hwang, G.T. et al., (2014). Highly-Efficient, Flexible Piezoelectric PZT Thin Film Nanogenerator on Plastic Substrates. *Advanced Material* 26: 2514, Doi: 10.1002/adma.201305659, Google ScholarCrossref.

213. Fan, F.R., Tian, Z.Q. and Wang, Z.L. (2012). Flexible triboelectric generator. *Nano Energy* 1: 328, Doi: 10.1016/j.nanoen.2012.01.004, Google ScholarCrossref.

214. Wang, S., Lin, Z., Niu, S., Lin, L., Xie, Y., Pradel, K.C. and Wang, Z.L. (2013). Motion Charged Battery as Sustainable Flexible-Power-Unit. *ACS Nano* 7: 11263, Doi: 10.1021/nn4050408, Google ScholarCrossref.

215. Alluri, N.R., Chanderashkear, A. and Kim, S.-J. (2018, June 27). Hybrid structures for piezoelectric nanogenerators: Fabrication methods, energy generation, and self-powered applications. Energy Harvesting, Reccab Manyala, IntechOpen, DOI: 10.5772/intechopen.74770. Available at: https://www.intechopen.com/books/energy-harvesting/hybrid-structures-for-piezoelectric-nanogenerators-fabrication-methods-energy-generation-and-self-po.

216. Zhao, C., Zhang, Q., Zhang, W., Du, X., Zhang, Y., Gong, S., Ren, K., Sun, Q. and Wang, Z.L. (2019, March). Hybrid piezo/triboelectric nanogenerator for highly efficient and stable rotation energy harvesting. *Nano Energy* 57: 440–449.

217. Pu, X., Liu, M., Li, L., Zhang, C., Pang, Y., Jiang, C., Shao, L., Hu, W. and Wang, Z.L. (2016). Efficient Charging of Li-Ion Batteries with Pulsed Output Current of Triboelectric Nanogenerators. *Advanced Science* 3: 1500255. Doi: 10.1002/advs.201500255, Google ScholarCrossref.

218. Nan, X., Zhang, C., Liu, C., Liu, M., Wang, Z.L. and Cao, G. (2016). Highly Efficient Storage of Pulse Energy Produced by Triboelectric Nanogenerator in Li3V2(PO4)3/C Cathode Li-Ion Batteries. *ACS Applied Materials & Interfaces* 8: 862. Doi: 10.1021/acsami.5b10262, Google ScholarCrossref.

219. Wang, F., Jiang, C., Tang, C., Bi, S., Wang, Q., Du, D. and Song, J. (2016). High output nano-energy cell with piezoelectric nanogenerator and porous supercapacitor dual functions- A technique to provide sustaining power by harvesting intermittent mechanical energy from surroundings. *Nano Energy* 21: 209. Doi: 10.1016/j.nanoen.2016.01.018. Google ScholarCrossref.

220. Ramadoss, A., Saravanakumar, B., Lee, S.W., Kim, Y.-S., Kim, S.J. and Wang, Z.L. Piezoelectric-Driven Self-Charging Supercapacitor Power Cell. *ACS Nano* 9: 4337 (2015). Doi: 10.1021/acsnano.5b00759. Google ScholarCrossref.
221. Song, Y., Cheng, X., Chen, H. et al., (2016). Integrated self-charging power unit with flexible supercapacitor and triboelectric nanogenerator. *Journal of Materials Chemistry* 4: 14298. Doi: 10.1039/c6ta05816g. Google ScholarCrossref.
222. Guo, H., Yeh, M.-H., Lai, Y.-C., Zi, Y., Wu, C., Wen, Z., Hu, C. and Wang, Z.L. (2016). All-in-one shape-adaptive self-charging power package for wearable electronics. *ACS Nano* 10(11): 10580–10588. Publication Date: 7 November, 2016, Doi: 10.1021/acsnano.6b06621.
223. Wang, X., Lu, X., Liu, B., Chen, D., Tong, Y. and Shen, G. (2014). Flexible energy-storage devices: Design consideration and recent progress. *Advanced Material* 26: 4763, Doi: 10.1002/adma.201400910, Google ScholarCrossref.
224. Pu, X., Li, L., Liu, M., Jiang, C., Du, C., Zhao, Z., Hu, W. and Wang, Z.L. (2016). Wearable self-charging power textile based on flexible yarn supercapacitors and fabric nanogenerators. *Advanced Material* 28: 98, Doi: 10.1002/adma.201504403, Google ScholarCrossref.
225. Pu, X., Li, L., Song, H., Du, C., Zhao, Z., Jiang, C., Cao, G., Hu, W. and Wang, Z.L. (2015). A self-charging power unit by integration of a textile triboelectric nanogenerator and a flexible lithium-ion battery for wearable electronics. *Advanced Material* 27: 2472, Doi: 10.1002/adma.201500311, Google ScholarCrossref.
226. Wen, Z., Yeh, M.-H., Guo, H. et al., (2016). Self-powered textile for wearable electronics by hybridizing fiber-shaped nanogenerators, solar cells, and supercapacitors. *Science Advances* 2: e1600097, Doi: 10.1126/sciadv.1600097, Google ScholarCrossref.
227. Camarda, A., Tartagni, M. and Romani, A. (2018). A−8 mV/+15mV double polarity piezoelectric transformer–based step–up oscillator for energy harvesting applications. *IEEE Transactions on Circuits and Systems I: Regular Papers* 65: 1454–1467, Doi: 10.1109/TCSI.2017.2741779.
228. Katic, J., Rodriguez, S. and Rusu, A. (2018). A high–efficiency energy harvesting interface for implanted biofuel cell and thermal harvesters. *IEEE Transactions on Power Electronics* 33: 4125–4134, Doi: 10.1109/TPEL.2017.27 12668.
229. McCullagh, J. (2018). An active diode full–wave charge pump for low acceleration infrastructure–based non–periodic vibration energy harvesting. *IEEE Transactions on Circuits and Systems I: Regular Papers* 65: 1758–1770, Doi: 10.1109/TCSI.2017.27 64878.
230. Taghadosi, M., Albasha, L., Quadir, N.A., Rahama, Y.A. and Qaddoumi, N. (2018). High efficiency energy harvesters in 65nm CMOS process for autonomous IoT sensor applications. *IEEE Access* 6: 2397–2409, Doi: 10.1109/ACCESS.2017.2783045.
231. Yi, H., Yin, J., Mak, P. and Martins, R.P. (2018). A 0.032–mm 0.15–V three–stage charge–pump scheme using a differential bootstrapped ring–VCO for energy–harvesting applications. *IEEE Transactions on Circuits and Systems II: Express Briefs* 65: 146–150, Doi: 10.1109/TCSII.2017.2676159.
232. Yi, H., Yu, W., Mak, P., Yin, J. and Martins, R.P. (2018). A 0.18–V 382–mu W bluetooth low–energy receiver front–end with 1.33–nW sleep power for energy–harvesting applications in 28–nm CMOS. *IEEE Journal of Solid-State Circuits* 53: 1618–1627, Doi: 10.1109/JSSC.2018.2815987.
233. Yoon, E., Park, J. and Yu, C. (2018). Thermal energy harvesting circuit with maximum power point tracking control for self–powered sensor node applications. *Frontiers of Information Technology & Electronic Engineering* 19: 285–296, Doi: 10.1631/FITEE.1601181.
234. Elhebeary, M.R., Ibrahim, M.A.A., Aboudina, M.M. and Mohieldin, A.N. (2018). Dual–source self–start high–efficiency microscale smart energy harvesting system for IoT. *IEEE Transactions on Industrial Electronics* 65: 342–351, Doi: 10.1109/TIE.2017.2714119.

235. Ulusan, H., Chamanian, S., Pathirana, W.P.M.R., Zorlu, O., Muhtaroglu, A. and Kulah, H. (2018). A triple hybrid micropower generator with simultaneous multi–mode energy harvesting. *Smart Materials and Structures* 27: 014002, Doi: 10.1088/1361–665X/aa8a09.

236. Saffari, P., Basaligheh, A., Sieben, V.J. and Moez, K. (2018). An RF–powered wireless temperature sensor for harsh environment monitoring with non– intermittent operation. *IEEE Transactions on Circuits and Systems I: Regular Papers* 655: 1529–1542, Doi: 10.1109/TCSI.2017.2758327.

237. Khan, D., Abbasizadeh, H., Kim, S. et al., (2018). A design of ambient RF energy harvester with sensitivity of–21 dBm and power efficiency of a 39.3% using internal threshold voltage compensation. *Energies* 11: 1258, Doi: 10.3390/en11051258.

238. Park, S., Lee, K., Song, H. and Yoon, E. (2018). Simultaneous imaging and energy harvesting in CMOS image sensor pixels. *IEEE Electron Device Letters* 39: 532–535, Doi: 10.1109/LED.2018.2811342.

239. Estrada–Lopez, J.J., Abuellil, A., Zeng, Z. and Sanchez–Sinencio, E. (2018). Multiple input energy harvesting systems for autonomous IoT end–nodes. *Journal of Low Power Electronics and Applications* 8: 6, Doi: 10.3390/jlpea8010006.

240. Rodrigues, C., Gomes, A., Ghosh, A., Pereira, A. and Ventura, J. (2019). Power-generating footwear based on a triboelectric-electromagnetic-piezoelectric hybrid nanogenerator. *Nano Energy*: 62, Doi: 10.1016/j.nanoen.2019.05.063.

241. Shin, D., Kim, Y., Seo, J., Chang, N., Wang, Y. and Pedram, M. (2011). Battery supercapacitor hybrid system for high-rate pulsed load applications. *Proceedings of the Design, Automation and Test in Europe Conference and Exhibition (DATE)*, 1–4. Grenoble, France.

242. Shin, D., Kim, Y., Wang, Y., Chang, N. and Pedram, M. (2012). Constant-current regulator-based battery-supercapacitor hybrid architecture for high-rate pulsed load applications. *Journal of Power Sources* 205: 516–524.

243. Jiang, X., Polastre, J. and Culler, D. (2005). Perpetual environmentally powered sensor networks. *Proceedings of the International Symposium on Information Processing in Sensor Networks (IPSN)*, Los Angeles, CA, 463–468.

244. Park, C. and Chou, P. (2006). AmbiMax: Autonomous energy harvesting platform for multi-supply wireless sensor nodes. *Proceedings of the Communications Society Conference on Sensor, Mesh and Ad Hoc Communications and Networks*, Reston, VA, 168–177.

245. Ise, T., Kita, M. and Taguchi, A. (2005). A hybrid energy storage with a SMES and secondary battery. *IEEE Transactions on Applied Superconductivity* 15(2): 1915–1918.

246. Nie, Z., Xiao, X., Kang, Q., Aggarwal, R., Zhang, H. and Yuan, W. (2013). SMES-battery energy storage system for conditioning outputs from direct drive linear wave energy converters. *IEEE Transactions on Applied Superconductivity* 23(3): 5000705.

247. Shim, J.W., Cho, Y., Kim, S., Min, S.W. and Hur, K. (2014). Synergistic control of SMES and battery Energy storage for enabling dispatchability of renewable energy sources. *IEEE Transactions on Applied Superconductivity* 23(3): 5701205.

248. Bayram, k., Atacak, İ. and Elmas, C. (2015). Simulation of a hybrid compressed air/ Li-ion battery energy storage system for electric vehicles. *Conference Paper, The third International Symposium on Innovative Technologies in Engineering and Science (ISITES 2015)*, January 2015, Valencia.

249. Omsin, P., Sharkh, S.M. and Moshrefi-Torbati, M. (2019). A hybrid SS-CAES system with a battery. *2019 16th International Conference on Electrical Engineering/ Electronics, Computer, Telecommunications and Information Technology (ECTI-CON)*, Pattaya, Chonburi, 159–162, Doi: 10.1109/ECTI-CON47248.2019.8955397.

250. Schiffer, Z.J. and Arnold, C.B. (2018). Characterization and model of piezoelectro-chemical energy harvesting using lithium ion batteries. *Experimental Mechanics* 58: 605–611, Doi: 10.1007/s11340-017-0291-1.

251. Yang, P., Qu, X., Liu, K., Duan, J., Li, J., Chen, Q., Xue, G., Xie, W., Xu, Z. and Zhou, J. (2018). Electrokinetic supercapacitor for simultaneous harvesting and storage of mechanical energy. *ACS Applied Materials & Interfaces* 10: 8010–8015, Doi: 10.1021/acsami.7b18640.

252. Kim, Y., Wang, Y., Chang, N. and Pedram, M. (2013). Computer-aided design and optimization of, hybrid energy storage systems. *Foundations and Trends in Electronic Design Automation* 7(4): 247338, Doi: 10.1561/1000000035.

253. Mirhoseini, A. and Koushanfar, F. (2011). Hypo Energy hybrid super capacitor battery power-supply optimization for energy efficiency. *Proceedings of the Design, Automation and Test in Europe Conference and Exhibition (DATE)*, 1–4.

254. Mirhoseini, A. and Koushanfar, F. (2011). Learning to manage combined energy supply systems. *Proceedings of the International Symposium on Low-Power Electronics and Design (ISLPED)*, 229–234.

255. Yue, S., Zhu, D., Wang, Y. and Pedram, M. (2012). Reinforcement learning-based dynamic power management in mobile computing systems equipped with hybrid power supply. *Proceedings of IEEE International Conference on Computer Design (ICCD)*, 81–86. Montreal, Canada.

256. Lukatskaya, M., Dunn, B. and Gogotsi, Y. (2016). Multidimensional materials and device architectures for future hybrid energy storage. *Nature Communications* 7: 12647, 10.1038/ncomms12647.

257. Colleen Spiegel. (2018). Power electronics for renewable energy systems. A website report added to fuels cell systems by fuel cell store. (May, 2018).

258. Nehalem-EX. www.intel.com/pressroom/ archive/releases/20090526comp.htm, a website report.

10 Hybrid Energy Systems for Water Industry

10.1 INTRODUCTION

More than 70% of the Earth's surface is covered with water; however, most of it is not suitable for human consumption. The magnitude of all water resources on the Earth is approximately 1.4 billion km³, of which roughly 97.5% is placed in the oceans and only 2.5% could be found as freshwater in the atmosphere, icebergs, lakes, rivers, and groundwater and just 0.014% of the total resources are available for humans [1–11].

Nowadays, global demands for access to freshwater are increasing rapidly as its resources are decreasing due to increased use of natural resources and climate change, especially in arid areas. The increase in usage of water is also related to the population and industrial growth. Water becomes the scarcest thing in some parts of the world as the availability is becoming limited due to the increasing contamination and environmental activities around the globe [3,4]. There are 1.2 billion people living on this earth today with no access to safe drinking water; typically two million people die annually of diarrhea and about one-third of the world's population lack satisfactory sanitation [5]. The high demand of freshwater resources and growing environmental awareness give rise to the use of reclaimed wastewater as a new source of water supply [6].

The distribution of freshwater around the world is not uniform. Unbalanced distribution caused some parts of the groundwater resources to have convenient access in several specific areas with low population such as the northern parts of Russia, Scandinavia, Canada, Alaska, and southern parts of South America. Additionally, areas with a high population or areas with industrial growth are more vulnerable to water stress and areas that are in arid regions also have a degree of water stress based on the ratio of water consumption to the amount of available water. Obviously the upstream water use on downstream stress has a direct effect on water distribution [12]. The index of water stress is essentially linked to per capita water use. Based on the defined indexes, various levels of water stress have been calculated by Falkenmark. In this classification, the water stress index for lower than 20% represents no stress and if it becomes more than 70% expresses extreme stress [13]. The measure of water stress is the ratio of total water use (domestic, industrial, or agricultural) to produced renewable water, including run-off in rivers and underground sources with little depth. Based on this analysis, about 2.8 billion people on the planet will face the problem of scarcity or water stress up to the year 2025, and by 2050, this value will be reached at 4 billion. In the future, some areas including South and Central America, Eastern Europe, and Asia will face water scarcity [3]. In addition to increasing population growth, pressure on water resources comes from increasing per capita consumption. There are different maps to determine current and predicted water stress in the future. The worldwide populace has nearly quadrupled in the course of recent years (last century), and it arrived at 6.5 billion until the end of the twentieth century. Regardless of just little varieties in per capita water utilization after this period, growth in total water utilization brought about an expansion in the population under water scarcity in the 20th century.

There are two parts to the water industry. First is to create fresh and potable water from seawater or other sources containing large amount of salt by the process of desalination. Second is to treat wastewater created by residential, commercial, and industrial use of water. Both desalination process and wastewater treatment process consume significant amount of energy. This energy consumption is most likely to rise significantly as world faces shortage of freshwater in the 21st century. Furthermore, all the energy required for desalination comes from fossil fuel sources. The use of these resources faces several major problems, such as the uneven distribution of these resources on the Earth, price fluctuations of energy carriers around the world, the difficulty of transporting these resources to remote areas, and the environmental problems caused by the use of these fuels. Nowadays, most of the desalination plants are located in the Persian Gulf, which has suitable potentials with high availability and low cost of fossil fuels. It is, however, imperative that in future, the use of fossil fuel in water industry needs to be curtailed and just like all other industries, water industry needs to be decarbonized [1–11].

Wastewaters are commonly categorized as domestic wastewater or industrial wastewater. Domestic wastewater refers to wastewater generated from "nonmanufacturing activities" occurring in residential homes which includes sewage (from toilets) and gray water (from bathrooms and kitchens). There are many types of industrial wastewater based on the different industries and contaminants; each sector produces its own particular combination of pollutants. Wastewaters are typically contaminated with physical, chemical, and biological composition which has tremendous negative impact on environment, where it has the ability to destroy many animal habitats, and cause irreparable damage to many ecosystems. Wastewater treatment processes are designed to achieve improvements of the wastewater quality. The two main reasons for collecting and treating wastewater are to prevent water-borne transmission of disease and to preserve the aquatic environment [5–11]. Physical composition in wastewater such as suspended solids can lead to the development of sludge deposits and anaerobic conditions when untreated wastewater is discharged in the aquatic environment. On the other hand, constituents such as biodegradable organics can lead to depletion of natural oxygen resources and to the development of septic conditions. Nutrients such as nitrogen and phosphorus, when discharged to the aquatic environment, can lead to the growth of undesirable aquatic life and cause groundwater pollution when discharged in excess. Many compounds found in wastewater have characteristics of carcinogenic, mutagenic, tetratogenic, or have high acute of toxicity [5–11].

Energy required for desalination of seawater or treatment of wastewater is significant and largely based on fossil fuels. Along with all other industries, the use of this energy needs to be decarbonized by use of renewable fuels and nuclear energy and improving the efficiency of the processes. The insertion of renewable fuels and nuclear energy in desalination and wastewater treatment is best done by using hybrid energy and processes concepts in the beginning and slowly increasing the contributions of renewable sources and nuclear energy over time. As mentioned before, the use of renewable sources generally required storage or multiple sources to maintain sustainability of the process. Cogeneration using nuclear reactor is the most effective way of nuclear energy in the water treatment processes. In this chapter, we examine the progress made in the use of hybrid energy (with fossil-renewable, multiple renewable, and nuclear-renewable sources with storage) and hybrid processes for various types of water treatments [1–13].

10.2 DESALINATION

Desalination is a water treatment process that separates salts from saline water to produce potable water or water that is low in total dissolved solids (TDS). Globally, the total installed capacity of desalination plants was 61 million m^3/d in 2008 [14]. Seawater desalination accounts for 67% of production, followed by brackish water at 19%, river water at 8%, and wastewater at 6%. The most prolific users of desalinated water are in the Arab region, namely, Saudi Arabia, Kuwait, United Arab Emirates, Qatar, Oman, and Bahrain [15].

Desalination plants consume a lot of energy, though, and they are not as green as they could be. Modern day facilities that desalinate water use high-energy consuming process of reverse osmosis (RO). Not only does the environment suffer from the energy expenditures of desalination plants, but so does the economy. The desalination plant in San Diego costs more than $1 billion, and it only provides about 7% of drinking water to the city of San Diego. Of the world 71% of water (29% land), only 4% is drinkable. The areas that are most affected by the shortage of freshwater are arid and semiarid areas in Asia and North Africa. In 2002, according to a UNESCO report, the deficit of drinking water in the world was around 230 billion m^3/year, and this amount was expected to rise to 2,000 billion by 2025. In fact, a report from the World Economic Forum published in 2015 highlighted the problem and indicated that the lack of freshwater could be the great global threat for the next several decades.

According to this report, there are approximately 19,000 desalination plants throughout the world, providing water to municipal and industrial users. Almost half of the global installed desalination capacity is in the Middle East, followed by the European Union with 13%, the United States with 9%, and North Africa with 8.5%. The largest desalination plant—the $3.8 billion Al-Jubail 2 in Saudi Arabia—has 948,000 m^3/d multiple-effect distillation–thermal vapor compressor (MED-TVC) capacity, plus 2,745 MW power generation using gas turbines. The Saudi Saline Water Conversion Corporation (SWCC) takes about 62% of output to supply Riyadh. China is building a 1 million m^3/d RO plant to supply potable water for Beijing.

Seawater reverse osmosis (SWRO) desalination stands for saltwater RO. This high-pressure system used to desalinate saltwater requires a high amount of energy. The amount of energy consumed from a desalination plant, which supplies water to 300,000 people, is equivalent to one jumbo jet's power. Energy consumption is one of the biggest hurdles desalination faces. It consumes an average of 10–13 kWh per every thousand gallons. Desalination is also viewed as one of many factors contributing to climate change and global warming. The ocean is home to many creatures, and desalination poses a threat to ocean biodiversity and marine habitats. If fossil energy is used to provide energy for desalination process, carbon emission can be significant [14–27].

As shown later, desalination can be done by a number of different techniques besides RO. In recent years, modular desalination operation is preferred to lower the upfront costs, improve efficiency, reduce the time for plant installation, and make the process more flexible. The use of renewable energy (RE) such as solar, geothermal, wind, and water along with the use of small modular nuclear reactors also reduce energy consumption and improve its return on investment. The use of RE also makes the process more environment-friendly. Unlike fossil energy, nuclear energy is clean. Freshwater is a major priority in sustainable development. It is generally obtained from streams and aquifers, desalination of seawater, mineralized groundwater, or treatment of urban wastewater. A study in 2006 by the UN's International Atomic Energy Agency (IAEA) showed that 2.3 billion people lived in water-stressed areas, and 1.7 billion of them have access to <1,000 m^3 of potable water per year. With population growth, these figures will increase substantially. Water can be stored, while electricity at utility scale cannot. This suggests two synergies with base-load power generation for electrically driven desalination: undertaking it mainly in off-peak times of the day and week, and load shedding in unusually high peak times.

Cumulative investment in desalination plants reached about US$21.4 billion in 2015 and is expected at least to double by 2020 according to a 2016 report by market analyst, Research and Markets. The report, Seawater and Brackish Water Desalination, includes a prediction that investment by 2020 should top $48 billion, showing a compound annual growth rate of 17.6%. The report assesses the market for large industrial or municipal facilities with a capacity >1,000 m^3/d. It highlights a growing gap between freshwater resources and demand from all sectors. Most desalination today uses fossil fuels, and thus contributes to increased levels of greenhouse gases (GHGs) [14–27].

10.2.1 Desalination Process Alternatives

Desalination can be achieved by using a number of techniques. Total world capacity in 2016 was 88.6 million m^3/d of potable water, in almost 19,000 plants. Of this, 73% is membrane desalination, and 27% thermal, though in the year till July 2016, 93% of new capacity contracted was membrane based. Industrial desalination technologies use either phase change or involve semipermeable membranes to separate the solvent or some solutes. Thus, desalination techniques may be classified into two main categories [19,27]: (a) **Phase-change or thermal processes**—where base water is heated to boiling. Salts, minerals, and pollutants are too heavy to be included in the steam produced from boiling and therefore remain in the base water. The steam is cooled and condensed. The main thermal desalination processes are multistage flash (MSF) distillation, multieffect distillation (MED), and vapor compression (VC), which can be thermal (TVC) or mechanical (MVC). (b) **Membrane or single-phase processes**—where salt separation occurs without phase transition and involves lower energy consumption. The main membrane processes are RO and electrodialysis (ED) and ED reversal. RO requires electricity or shaft power to drive a pump that increases the pressure of the saline solution to the required level. ED also requires electricity to ionize water, which is desalinated by using suitable membranes located at two oppositely charged electrodes. Other processes include solar still distillation, humidification–dehumidification (HD), membrane distillation (MD), and freezing. Three other membrane processes that are not considered desalination processes, but that are relevant, are microfiltration, ultrafiltration, and nanofiltration. The ion-exchange process is also not regarded as a desalination process but is generally used to improve water quality for some specific purposes, e.g., boiler feed water [21,27].

All processes require a chemical pretreatment of raw seawater—to avoid scaling, foaming, corrosion, biological growth, and fouling—as well as a chemical posttreat-ment. The two most commonly used desalination technologies are MSF and RO systems. In the past 15 years, membrane technology has developed to a mature and reliable technique, which is usable for (drinking) water purification. Within this tech-nology, RO has been widely used. This method requires that the pressure difference between two sides of membrane is higher than osmotic pressure in order for water to permeate through the membrane. The osmotic pressure of seawater is around 26 bar [14–27]. In 1999, about 78% of global production capacity comprised MSF plants and RO accounted for a modest 10%. But by 2008, RO accounted for 53% of world-wide capacity, whereas MSF consisted of about 25%. Although MED is less common than RO or MSF, it still accounts for a significant percentage of global desalination capacity (8%). ED is only used on a limited basis (3%) [20].

RO is a common method for seawater desalination, and the U.S. market is antici-pated to reach approximately $344 million in capital expenditures and about $195 mil-lion in operational expenditures by 2020 [16]. This is a significant increase from the 2015 capital and operational expenditures, approximately $129 million and $124 mil-lion, respectively, and these trends expected to continue to rise as water demands and shortages increase. Globally, the seawater desalination market reached approximately $2.6 billion in 2015 in capital expenditures, with a similar growth rate anticipated to hit over $4.5 billion in 2020. Operational expenditures are on the same order of magnitude, approximately $3.8 billion in 2015 and projected $5.2 billion by 2020. For seawater desalination, energy consumption is the largest component of the operational expenditures, making up approximately 36% of the total operational expenditures. In the United States alone, this accounts for about $45 million per year in electricity consumption using the 2015 market size and approximately $70 million using the 2020 projections [16]. Currently, nearly all energy consumption is largely provided by fossil fuels.

10.2.2 Efficiency Improvement through Hybrid Processes

Currently, Gulf Cooperation Council (GCC) countries' cumulative desalination capacity is 26 million cubic meter (m³) per day, equivalent to 36% of total global capacity. GCC countries are set to ramp up their desalination infrastructure 6%–10% annually till 2040, extending total capacity to almost twofold. These beyond limit extensions of fossil fuel operated desalination capacities will have significant impact on regional economy. For example, Saudi Arabia consuming nearly 300,000 barrels of oil for thermal desalination to fulfill daily water demand and similar challenges are faced by other GCC countries. Traditionally, in GCC region, thermal desalination processes (MSF & MED) are preferred over membrane-based (SWRO) processes due to two main reasons: (a) extensive pretreatment requirement for SWRO processes due to high salinity in Arabian Gulf and (b) energy efficiency of thermally driven processes by utilizing residual low grade heat due to hybrid water-and-power projects [28–30].

Thermal desalination processes gained confidence by operating for over 30 years in GCC region, and they are well integrated into power generation infrastructure. In terms of sea areas, Arabian Gulf is the largest intake facility for desalination capacities and producing $12.1\,m^3/d$. United Arab Emirates is dominating with 23% desalination capacity in Gulf followed by Saudi Arabia 11% and Kuwait 6%. In the Red Sea area, desalination plants have a total production capacity of $3.6\,m^3/d$, dominating by cogeneration plants (72%). Saudi Arabia accounts for 92% of the desalinated capacities in the Red Sea region, thermal desalination dominating by 78% installations ($2.6\,m^3/d$).

In addition to less energy efficient, conventional desalination processes have enormous impact on marine life and environment in terms of volume of brine rejection and CO_2 emissions. It is estimated that the brine rejection will increase to $240\,km^3$, and emission will be approximately 400 million tons of carbon equivalents per year by 2050. In the Gulf, 23.7 metric tons (mt) of chlorine, 64.9 mt of antiscalants, and 296 kg of copper rejection are estimated from desalination installations, and in Red Sea, these rejections are 5.6 mt of chlorine, 20.7 mt of antiscalants, and 74 kg of copper [28]. Both SWRO and thermal desalination processes showed severe impact in terms of energy consumption, chemical rejection, and CO_2 emission. Impact of SWRO processes is worse than thermal processes. Therefore, energy-efficient desalination processes (innovative hybrid cycles) and transitioning toward alternate renewable energies are two key innovation pillars needed to address future sustainable desalination water supplies in the region.

Hybrid desalination projects (SWRO/MSF, MSF/MED), for energy efficiency, were proposed and implemented to few facilities such as Jeddah-II, Fujairah II, and Ras AL-Khair in the past [28]. Unfortunately, hybrid concept was unable to get much industrial implementation due to operational limitations. Today, industrial-scale desalination processes pairing with innovative cycles are well needed to pave the way for future desalination in GCC under COP21 goal. An innovative adsorption (AD) cycle hybridization with SWRO and MED are carried out to meet COP21 accord. The proposed AD cycle utilized low-grade waste heat or RE such as solar or geothermal to produce fresh water. The integration of AD cycle with conventional desalination processes helps to overcome their operational limitations. For example, in first case, RO + AD integration can boost overall recovery to over 80% as compared to 35%–40% of RO alone. This will help to protect marine pollution by reducing pretreatment chemical rejection into the sea. In second case, its hybridization with thermal processes such as MED + AD help to overcome last stage operational temperature limitations of conventional MED system by extending to as low as 7°C as compared to 40°C in conventional processes. It also helps to boost water production to almost twofold with the same energy input. In both cases, CO_2 emission reduces as AD cycle utilizes only low-grade industrial waste heat or RE. The detailed experimentation of both mentioned cases and their economic analysis shows the superiority of hybrid cycles over conventional processes in terms of energy efficiency, marine, and environmental impact.

Hybrid thermal-membrane plants have a more flexible power-to-water ratio, efficient operation even with significant seasonal and daily fluctuations of the electricity and water demand, less primary energy consumption, and an increase in plant efficiency, thus improving economics and reducing environmental impacts. MSF plus RO or MED-TVC plus RO hybrid plants exploit the best features of each technology for different quality products or a blended product. Several thermal distillation processes capable of using waste heat from power generation are in use: MSF distillation process using steam was an earlier example. It works by flashing a portion of the water into steam in multiple stages of what are essentially countercurrent heat exchangers, and it accounted for 23% of world capacity in 2012 [14–27]. It is more energy-intensive than MED, but it can cope with suspended solids and any degree of salinity. The Japan Atomic Energy Agency has designed a 600 MW HTR (high-temperature reactor) called the GTHTR300, which produces 300 MW and uses the waste heat in MSF desalination, the projected water cost being half that of using gas-fired CCGT. An increasing number of plants uses MED with 8% world capacity in 2012, or multieffect VC (MVC or VCD) distillation or a combination of these, e.g., MED-TVC with thermal VC [29].

In GCC countries, 42% desalinated water is produced by membrane processes. Conventional SWRO processes consume 7–17 kWh_{pe}/m^3 primary energy (equivalent to 3–8 kWh_{elec}/m^3) for seawater desalination and they emit 3.0 kg/m^3 CO_2 to the environment [31]. Most of the SWRO plants are operated under SWCC, and their overall recovery varies from 17% to 40% depending upon feedwater quality. Four major plants, namely, (a) Jeddah, (b) Rbigh, (c) Jubail, and (d) Shuqaiq, operational data were collected and analysis results are presented.

The present share of thermally driven desalination processes is about 58% within the GCC countries. Typically, the energy requirements for such processes are reported as 2.0 kWh_{elec}/m^3 electricity and 60–70 kWh_{th}/m^3 of thermal energy. For energy efficiency, the thermally driven desalination processes are designed as an integral part of a cogeneration plant, producing both electricity and water from the temperature-cascaded processes. The thermal energy is low-grade bleed steam extracted from the last stages of steam turbines. Based on the exergy destruction analysis of the primary energy input, the gas turbines consumed 75% ± 2% and steam turbines (via the heat recovery from turbine exhaust) extracted 21% ± 2% of input primary energy, leaving a mere 3% ± 1.5% of the total exergy input to the thermally driven desalination processes. Consequently, the overall primary energy required by thermal desalination processes is merely 6.58 kWh_{pe}/m^3 (equivalent of 4.25 kWh_{pe}/m^3 from electricity + 2.33 kWh_{pe}/m^3 from the thermal input) [28,30,32]. All major integrated thermal desalination plants in Saudi Arabia are under SWCC and their recovery is about 40%. Since thermal desalination processes are robust, they need less chemicals as compared to SWRO processes.

All reported MED plants are operating between top brine temperature (TBT) 60°C and low-brine temperature (LBT) 40°C. The MSF operational range is slightly wider, between 120°C and 40°C. The high temperature of MED/MSF is controlled by scaling and fouling chances and LBT is by ambient conditions. Limited rage of operation puts the cap on the performance of thermally driven desalination systems even they are dominating in GCC region [28,30].

Thermal desalination processes, MED/MSF, can be integrated with AD cycle to enhance their performance by extending their operational range. In these cycles, low-temperature heat is supplied to first steam generator only and their performance depends on number of recoveries. The AD cycle hybridization can extend the last stage operational range to as low as 7°C as compared to conventional operational range of 40°C. This extension of LBT helps to insert more number of recoveries and hence boosts the performance.

Padron et al. [33] modeled hybrid systems with base in the RE to compare many different design options based on their technical and economic merits. The power requirements were guaranteed to RO Autonomous Desalination Systems, with a capacity of up to $50\,m^3$ of daily production. The HOMER Hybrid Optimization Model Tool was used to create optimal designs for these RE systems. The input assumptions in the model were the electric demand of the desalination plant, the technical specifications of the equipment, as well as the potentials of solar radiation and the wind speeds. Although this study could be applied to different scenarios where an isolated system is found, in this work, it was applied to two islands. In particular, these islands have got a relatively high wind profile, along with a high solar radiation profile. Several configurations have been considered by this simulation software and different optimal results attending to the use of RE have been obtained for this particular case. In fact, real data from two particular islands (Lanzarote and Fuerteventura) were used, although the results could be extended to other similar scenarios. The study demonstrated an efficient and environment-friendly hybrid desalination process for the first time with different energy mix for future water supplies. The following advantages were observed clearly by implementation of desalination hybridizations [28]:

1. Energy consumption and chemical discharge saving up to 99% and 150%, respectively, by the SWRO hybridization with AD cycle.
2. Thermal process integration with AD cycle will save up to 38% energy and up to 80% chemical rejection to sea.
3. CO_2 emission saving by SWRO+AD up to 99% and by MEDAD (hybrid of the conventional multi-effect distillation (MED) and an adsorption cycle (AD)) up to 30%.
4. Overall recovery of up to 80% can be achieved without scaling and fouling chances due to low-temperature operation.
5. Integration will help to reduce overall impact to low or moderate level, acceptable level under COP21 goal.
6. Integration will help to secure future water demand for expected GDP growth rate with minimal impact on environment and by implementing different energy mix for higher energy efficiency.

Thus, hybridized energy processes are more sustainable to meet COP21 goals for decarbonization.

10.2.3 ROLE OF RENEWABLE ENERGY IN DESALINATION

As mentioned earlier, currently, desalination processes almost entirely use fossil fuel sources. Going forward, RE needs to be infused in these processes. The best way to achieve this is to carry out the processes using hybrid energy of fossil fuel sources and renewable sources. Nuclear energy can also play a significant role. Before we examine the role of hybrid energy, it is worthwhile to first examine how renewable sources can support desalination processes.

As a global sunbelt region, solar energy has received particular attention due to abundance availability in the region and falling cost of technology, particularly photovoltaic (PV). RE plans of GCC will result in cumulative 2.5 billion barrels of oil equivalent saving from 2015 to 2030 equivalent to USD 55–87 billion savings. Implementation of renewable energy sources (RES) and decrease of fossil fuel consumption will reduce a cumulative total of 1 gigaton (Gt) of CO_2 emission by 2030. This will result in 8% reduction in the region's per capita carbon footprint, in line with the countries' Intended Nationally Determined Contributions submissions to the Paris climate conference (COP21). In addition to CO_2 savings, energy efficiency and RE application will reduce 16% of water consumption in power generation sector. This will save 11 trillion liters of water per year that will not only have ecological benefits but will also reduce energy consumption for water desalination [28,34]. GCC countries' targets RE applications and processes energy efficiency. Saudi Arabia planned to inject 54 GW electricity from renewable sources by 2040 [22].

In GCC region, energy-intensive desalination processes are the major contributor to satisfy the increasing water demand with the development of infrastructure. The regional water demand is expected to increase to fivefold by 2050. The water is utilized during fossil fuel extraction, industrial processing, domestic purposes, cooling and power generation. The analyst predicted that the scale of water utilized for energy production only will increase from 583 billion cubic meters (bcm) in 2010 to 790 bcm of water in 2013, resulting in even higher demand for desalination in the region [35]. Desalination technologies development and alternate energy mix are required urgently to coup the BAU (business as usual) trend.

Estimates show that currently only a small percentage of all freshwater produced in the world is from renewable sources. By developing the technology, lowering equipment prices and increasing attention to the environmental problems of fossil fuels, utilizing RE is growing. By providing a wide variety of conventional desalination methods driven by various types of RE technologies in the world, water and energy legislators should choose different methods to meet the needs based on the local potentials by paying attention to the desalination processes and power systems. In some cases, concentrated solar power for thermal desalination or electricity generated by the PV plants for membrane desalination systems can be used in arid areas. Definitely, the most problem of using renewable sources is their unsteady natures, which using storage systems or combining with other renewable sources can solve this problem. This chapter provides extensive information about renewables, desalination, and performance analysis of power systems. RO technique is a practical process in desalination which 69% of desalination plants use this system [36,37]. Solar energy is an important source of energy for hybrid systems. The geothermal has a steady performance at a specified depth. Ultimately, obtained results from energy and exergy analysis would have provided a better insight.

The insertion of renewable sources in desalination is done a number of different ways. The use of biomass in desalination is not in general a promising alternative since organic residues are not normally available in arid regions and growing of biomass requires more freshwater than it could generate in a desalination plant. In some cases, RE is used alongside conventional fossil fuels to reduce the amount of fossil fuels. In other cases, renewable sources are used with storage for process sustainability. In some cases, renewable sources are used along with the grid to satisfy both power and heating needs for desalination processes. In the case of nuclear reactor, desalination is carried out as a part of cogeneration process. Finally, in some cases, true hybrid renewable sources are used to integrate with desalination. Here we consider following five cases to illustrate the role of hybrid energy and processes in decarbonization of desalination industry.

A. Solar thermal-solar PV and hybrid solar photovoltaic-thermal (PVT) with or without storage
B. Hybrid wind energy
C. Hybrid geothermal energy
D. Hybrid wave energy
E. Small modular nuclear reactors with combined heat and power (CHP)

10.3 HYBRID SOLAR ENERGY FOR DESALINATION

Solar energy can drive the desalination units by either thermal energy and electricity generated from solar thermal systems or by PV systems or both. The cost distribution of solar distillation is dramatically different from that of RO and MSF. The main cost is in the initial investment. However, once the system is operational, it is extremely inexpensive to maintain, and the energy has minimal or even no cost. Solar-assisted desalination systems are divided into two parts: solar thermal-assisted systems and solar PV-assisted systems. The most effective way is to use hybrid solar thermal-PV system as mentioned above. All solar energy-driven systems would be hybrid because they will involve storage and/or backup fossil fuel system.

10.3.1 SOLAR-THERMAL SYSTEMS

Solar thermal energy can be harnessed directly or indirectly for desalination. Collection systems that use solar energy to produce distillate directly in the solar collector are called direct-collection systems, whereas systems that combine solar energy collection devices with conventional desalination units are called indirect systems. In indirect systems, solar energy is used either to generate the heat required for desalination and/or to generate electricity used to provide the required electric power for conventional desalination plants such as MED and MSF plants. Direct solar desalination requires large land areas and has a relatively low productivity. However, it is competitive with indirect desalination plants in small-scale production due to its relatively low cost and simplicity. The combination of solar-thermal systems and desalination involves hybrid processes.

10.3.1.1 Direct Solar Thermal Desalination

Direct systems are those where the heat collection and distillation processes occur in the same equipment. Solar energy is used to produce the distillate directly in the solar still. The method of direct solar desalination is mainly suited for small production systems, such as solar stills, and it is used in regions where the freshwater demand is low. This device has low efficiency and low water productivity due to the ineffectiveness of solar collectors to convert most of the energy they capture, and to the intermittent availability of solar radiation. For this reason, direct solar thermal desalination has so far been limited to small-capacity modular units, which are appropriate in serving small communities in remote areas having scarce water. Solar still design can generally be grouped into four categories: (a) **basin still**, (b) **tilted-wick solar still**, (c) **multiple-tray tilted still**, and (d) **concentrating mirror still**. The basin still consists of a basin, support structure, transparent glazing, and distillate trough. Thermal insulation is usually provided underneath the basin to minimize heat loss. Other ancillary components include sealants, piping and valves, storage, external cover, and a reflector (mirror) to concentrate light. Single-basin stills have low efficiency, generally below 45%, and low productivity (4–6 L/m^2/d) due to high top losses. Double glazing can potentially reduce heat losses, but it also reduces the transmitted portion of the solar radiation [38]. On a much smaller scale, a solar micro desalination unit [39] may be used in remote areas and is capable of producing about 1.5 L/d.

A tilted-wick solar still uses the capillary action of fibers to distribute feedwater over the entire surface of the wick in a thin layer. This allows a higher temperature to form on this thin layer. Insulation in the back of the wick is essential. A cloth wick needs frequent cleaning to remove sediment built-up and regular replacement of wick material due to weathering and ultraviolet (UV) degradation. Uneven wetting of the wick can result in dry spots that reduce efficiency [40]. In a multiple-tray tilted still, a series of shallow horizontal black trays are enclosed in an insulated container with a transparent glazing on top. The feedwater supply tank is located above the still, and the vapor condenses and flows down to the collection channel and finally to the storage. The construction of this still is fairly complicated and involves many components that are more expensive than simple basin stills. Therefore, the slightly better efficiency it delivers may not justify its adoption [41]. The concentrating mirror solar still uses a parabolic mirror for focusing sunlight onto an evaporator vessel. The water is evaporated in this vessel exposed to extremely high temperature. This type of still entails high construction and maintenance costs [42].

10.3.1.2 Indirect Solar Thermal Desalination

Indirect solar thermal desalination methods involve two separate systems: the collection of solar energy by a solar collecting system, coupled to a conventional desalination unit. Processes include HD, MD, solar pond-assisted desalination, and solar thermal systems such as solar collectors, evacuated-tube collectors, and concentrating collector (concentrating solar power(CSP)) systems driving conventional desalination processes such as MSF and MED.

10.3.1.2.1 HD Process

These units consist of a separate evaporator and condenser to eliminate the loss of latent heat of condensation. The basic idea in HD process is to mix air with water vapor and then extract water from the humidified air by the condenser. The amount of vapor that air can hold depends on its temperature. Some advantages of HD units are the following: low-temperature operations, able to combine with RES such as solar energy, modest level of technology, and high productivity rates. Two different cycles are available for HD units: HD units based on open-water closed-air cycle and HD units based on open-air closed-water cycle.

10.3.1.2.2 Membrane Distillation

MD is a separation/distillation technique where water is transported between "hot" and a "cool" stream separated by a hydrophobic membrane, permeable only to water vapor, which excludes the transition of liquid phase and potential dissolved particles. The exchange of water vapor relies on a small temperature difference between the two streams, which results in a vapor pressure difference, leading to the transfer of the produced vapor through the membrane to the condensation surface. In the MD process, the seawater passes through the condenser usually at about 25°C and leaves at a higher temperature, and then it is heated to about 80°C by an external source such as solar, geothermal, or industrial waste [43]. The main advantages of MD lie in its simplicity and the need for only small differentials to operate. However, the temperature differential and the recovery rate determine the overall efficiency for the process. Thus, when it is run with a low-temperature differential, large amounts of water must be used, which adversely affects its overall energy efficiency. Membrane desalination is a promising process, especially for situations where low-temperature solar, geothermal, waste, or other heat is available.

10.3.1.2.3 Solar Pond-Assisted Desalination

Salinity-gradient solar ponds are a type of heat collector, as well as a mean of heat storage. Hot brine from a solar pond can be used as a heat source for MSF or MED desalination units. Solar ponds can store heat because of their unique chemically stratified nature. A solar pond has three layers: (a) upper or surface layer, called the upper convection zone; (b) middle layer, which is the non-convection zone or salinity-gradient zone; and (c) lower layer, called the storage zone or lower convection zone. Salinity increases with depth from near pure water at the surface to the bottom, where salts are at or near saturation. Salinity is relatively constant in the upper and lower convection zones, and increases with depth in the nonconvection zone. Saline water is denser than freshwater; therefore, the water at the bottom of the pond is more dense (has a higher specific gravity) than water at the surface. The solar pond system is able to store heat because circulation is suppressed by the salinity-related density differences in the stratified water. Convection of hot water to the surface is repressed by the salinity (density) gradient of the nonconvection zone. Thus, although solar energy can penetrate the entire depth of the pond, it cannot escape the storage zone [44].

10.3.1.2.4   Concentration Solar Thermal Desalination

Concentrating solar thermal power technologies are based on the concept of concentrating solar radiation to provide high-temperature heat for electricity generation within conventional power cycles using steam turbines, gas turbines, or Stirling and other types of engines. For concentration, most systems use glass mirrors that continuously track the position of the sun. The four major CSP technologies are parabolic trough, Fresnel mirror reflector, power tower, and dish/engine systems. Debate continues as to which of these is the most effective technology [45].

The primary aim of CSP plants is to generate electricity, yet a number of configurations enable CSP to be combined with various desalination methods. When compared with PVs or wind, CSP could provide a much more consistent power output when combined with either energy storage or fossil-fuel backup. There are different scenarios for using CSP technology in water desalination [46]. Parabolic trough coupled with MED or RO systems is widely used. However, an analysis presented in Reference [47] suggests that, for several locations, CSP/MED requires 4%–11% less input energy than CSP/RO. Therefore, before any decision can be made on the type of desalination technology to be used, it is recommend that a detailed analysis be conducted for each specific location, evaluating the amount of water, salinity of the input seawater, and site conditions. It appears that CSP/MED provides slightly better performance at sites with high salinity such as in closed gulfs, whereas CSP/RO appears to be more suitable for low-salinity waters in the open ocean. One additional advantage of the RO system is that the solar field might be located away from the shoreline. The only connection between the two is the production of electricity to drive the RO pumps and other necessary auxiliary loads. A hybrid solar thermal desalination process is illustrated in Figure 10.1 [2].

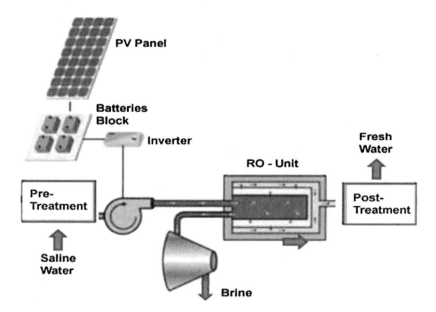

FIGURE 10.1   Hybrid solar thermal-desalination process [2].

The solar collectors are in various types such as flat plate, vacuum tube, or central concentrator that can be combined with a wide variety of thermal desalination units. High-pressure steam production technology is widely used in various industries, e.g., steam engines. In solar thermal systems, the pressure tank can be charged by the heat absorbed by the sun and then the high-pressure steam can be used to move the turbine and generate electricity. The same process can now be used to move the RO system. However, this cycle will require a steam cooling section for returning to the pressure tank. The pressure for RO desalination is 55 bar and for brackish water 10–15 bar. These pressures are much less than the pressures required driving the turbines (the pressure range of these turbines is 75–120 bar) [48]. Solar collectors are basically divided into three categories based on the temperature: (a) Low-temperature uncoated flat solar collectors which increase the temperature slightly more than the ambient temperature. These cannot operate thermal desalination systems. (b) Medium-temperature flat-panel and vacuum tube solar collectors that produce temperatures of more than 430°C. These collectors use water or air as a heat transfer and heat of these collectors can be supplied indirectly by means of a heat exchanger for desalination. (c) High-temperature solar collectors that include parabolic trough collectors and central receivers. The heat generated in these types of collectors can be used directly in thermal desalination or indirectly through the electrical energy produced in membrane desalination.

10.3.1.3 Perspectives on Pilot and Commercial Scale Operations

Although the strong potential of solar thermal energy to seawater desalination is well recognized, the process is not yet developed at the commercial level. The main reason is that the existing technology, although demonstrated as technically feasible, cannot presently compete, on the basis of produced water cost, with conventional distillation and RO technologies. However, it is also recognized that there is still potential to improve desalination systems based on solar thermal energy.

Among low-capacity production systems, solar stills and solar ponds represent the best alternative in low freshwater demands. For higher desalting capacities, one needs to choose conventional distillation plants coupled to a solar thermal system, which is known as indirect solar desalination [49]. Distillation methods used in indirect solar desalination plants are MSF and MED. MSF plants, due to factors such as cost and apparent high efficiency, displaced MED systems in the 1960s, and only small-size MED plants were built. However, in the last decade, interest in MED has been significantly renewed, and the MED process is currently competing technically and economically with MSF [50]. Recent advances in the research of low-temperature processes have resulted in an increase of the desalting capacity and a reduction in the energy consumption of MED plants providing long-term operation under remarkable steady conditions [51]. Scale formation and corrosion are minimal, leading to exceptionally high plant availabilities of 94%–96%.

Many small systems of direct solar thermal desalination systems and pilot plants of indirect solar thermal desalination systems have been implemented in different places around the world [52]. Among them are the de Almería (PSA) project in 1993 and the AQUASOL project in 2002. The study of these systems and plants will improve our understanding of the reliability and technical feasibility of solar thermal technology application to seawater desalination. It will also help to develop an optimized solar desalination system that could be more competitive against conventional desalination systems. On a commercial basis, CSP technology will take some years until it becomes economic and sufficiently mature for use in power generation and desalination [53].

The operating temperature of the MED system is lower, so it requires steam at lower temperatures and pressures, so combining CSP with MED can work much better. A 40 m³/d low-temperature solar site has been installed on the La Desirade Island in the Caribbean region. The device consisted of 14 evaporation stages, operating at temperatures between 26°C and 65°C [54]. Another multistage solar distiller is installed on Takami Island, Japan, with an average annual capacity of 10 m³/d. This unit has a minimum production capacity of 5 m³/d in winter and a maximum of 16.4 m³/d in summer. Alongside the complex, there is also a hot water storage tank of 38 m³ which automatically starts the distillation process when the temperature of the tank reaches 75°C in summer and 62°C in winter.

10.3.2 SOLAR PV-BASED APPLICATION

A PV or solar cell converts solar radiation into direct current (DC) electricity. It is the basic building block of a PV (or solar electric) system. An individual PV cell is usually quite small, typically producing about 1 or 2 W power. To boost the power output, the solar cells are connected in series and parallel to form larger units called modules. Modules, in turn, can be connected to form even larger units called arrays. Any PV system consists of a number of PV modules, or arrays. The other system equipment includes a charge controller, batteries, inverter, and other components needed to provide the output electric power suitable to operate the systems coupled with the PV system. PV systems can be classified into two general categories: flat-plate systems and concentrating systems. Concentrator-photovoltaic (CPV) system has several advantages compared with flat-plate systems: CPV systems increase the power output while reducing the size or number of cells needed; a solar cell's efficiency increases under concentrated light.

PV is a rapidly developing technology, with costs falling dramatically with time, and this will lead to its broad application in all types of systems. Today, however, it is clear that PV/RO and PV/ED will initially be most cost-competitive for small-scale systems installed in remote areas where other technologies are less competitive. RO usually uses alternating current (AC) for the pumps, which means that DC/AC inverters must be used. In contrast, ED uses DC for the electrodes at the cell stack, and hence, it can use the energy supply from the PV panels without major modifications. Energy storage is again a concern, and batteries are used for PV output power to smooth or sustain system operation when solar radiation is insufficient. Two types of PV/RO systems are available in the market: BWRO (brackish water reverse osmosis) and SWRO PV/RO systems. Different membranes are used for brackish water and much higher recovery ratios are possible, which makes energy recovery less critical [55].

Brackish water has a much lower osmotic pressure than seawater; therefore, its desalination requires much less energy and a much smaller PV array in the case of PV/RO. Also, the lower pressures found in BWRO systems permit the use of low-cost plastic components. Thus, the total cost of water from brackish water PV/RO is considerably less than that from seawater, and systems are beginning to be offered commercially [56]. Many of the early PV/RO demonstration systems were essentially a standard RO system, which might have been designed for diesel or mains power, but powered from batteries charged by PV. This approach generally requires a rather large PV array for a given flow of product because of poor efficiencies in the standard RO systems and batteries. Large PV arrays and the regular replacement of batteries typically make the cost of water from such systems rather high. The osmotic pressure of seawater is much higher than that of brackish water; therefore, its desalination requires much more energy, and, unavoidably, a somewhat larger PV array. Also, the higher pressures found in seawater RO systems require mechanically stronger components. Thus, the total cost of water from seawater PV/RO is likely to remain higher than that from brackish water, and systems have not yet passed the demonstration stage.

ED uses DC for the electrodes; therefore, the PV system does not include an inverter, which simplifies the system. Currently, there are several installations of PV/ED technology worldwide. All PV/RD applications are of a stand-alone type, and several interesting examples are outlined below. In the city of Tanote, in Rajasthan, India, a small plant was commissioned in 1986 that features a PV system capable of providing 450 peak watts (W_p) in 42 cell pairs. The ED unit includes three stages, producing 1 m^3/d water from brackish water (5,000 ppm TDS). The unit energy consumption is 1 kWh/kg of salt removed [57]. A second project is a small experimental unit in Spencer Valley, New Mexico (USA), where two separate PV arrays are used: two tracking flat-plate arrays (1,000 W_p power, 120 V) with DC/AC inverters for pumps, plus three fixed arrays (2.3 kW_p, 50 V) for ED supply. The ED design calls for 2.8 m^3/d product water from a feed of about 1,000 ppm TDS. This particular feedwater contains uranium and radon, apart from alpha particles. Hence, an ion-exchange process is required prior to ED. Unit consumption is 0.82 kWh/m^3 and the reported cost is 16 US\$/m^3 [58,59]. A third project is an unusual application in Japan, where PV technology is used to drive an ED plant fed with seawater, instead of the usual brackish water of an ED system [51]. The solar field consists of 390 PV panels with a peak power of 25 kW_p, which can drive a 10 m^3/d ED unit. The system, located on Oshima Island (Nagasaki), has been operating since 1986. Product-water quality is reported to be below 400 ppm TDS, and the ED stack is provided with 250 cell pairs.

A solar thermal desalination system built at Al-Azhar University uses PV cells to power the device. The complex contains three-stage evaporation and is suitable for the desalination of brackish water. In low-capacity desalination markets such as remote and low populated areas, wind and solar power generation methods are well underway. The most important features that make solar energy attractive to the Middle East and in Northern Africa are as follows:

1. The unlimited energy source is easily accessible.
2. It can be used in combined cycles (electricity generation, cooling, and desalination).
3. Saving solar energy or working in combination with fossil fuels ensures continued use of this renewable resource.
4. Solar concentrating plants can be produced in capacities from a few kW to several hundred MW.

Furthermore, the PV panels can be powered and used for desalination by RO or ED. In the PV method, solar energy is converted directly into electricity by silicon solar cells. Backup batteries should be used to stabilize the RO system and prevent the detrimental effect of changes in solar energy levels at different times. One of the major disadvantages of using solar collectors in desert areas is the presence of dust on collectors, which drastically reduces their heat capacity [60,61]. The most important problem with this technology is its high cost which restricts its use for desalination in small and remote areas. A PV system can produce 110–120 kWh per square meter of energy collector, so the amount of collector area needed to produce $1\,m^3$ of freshwater per day with a small desalination plant ($8\,kW\,h/m^3$) will be about $26.5–28\,m^2$ of standard technology. The use of solar energy to power the RO process is a growing technology, especially for remote areas with low desalination capacities.

10.3.3 HYBRID SOLAR THERMAL-SOLAR PV

A proposed design of PVT-integrated active solar still was tested in India by Kumar and Tiwari [62,63]. This PVT-active solar still is self-sustainable and can be used in remote areas. Compared with a passive solar still, the daily distillate yield was found 3.5 times higher, and 43% of the pumping power can be saved. Based on $0.05\,m$ water depth, the range of CPBT (cost payback time) can be shortened from 3.3–23.9 to 1.1–6.2 years (depending on the selling price of distilled water) and the energy payback time (EPBT) from 4.7 to 2.9 years. The hybrid active solar still is able to provide higher electrical and overall thermal efficiency, which is about 20% higher than the passive solar still. On the other hand, Gaur and Tiwari [64] conducted a numerical study to optimize the number of collectors for PVT/w hybrid active solar still. The number of PVT collectors connected in series has been integrated with the basin of a solar still.

A c-PVT water desalination system was also proposed by Mittelman et al. [65], in which a c-PVT collector field was to couple to a large-scale multiple-effect evaporation thermal desalination system. Small dish concentrator type was used in the numerical analysis. The vapor formed in each evaporator condenses in the next (lower temperature) effect and thus provides the heat source for further evaporation. Additional feed preheating is to be provided by vapor process bleeding from each effect. The range of TBT is from 60°C to 80°C. Through numerical analysis, this approach was found competitive relative to other solar-driven desalination systems and even relative to the conventional reverse-osmosis desalination. Because of the higher ratio of electricity to heat generation, the high concentration option with the use of advanced solar cells can be advantageous.

10.4 HYBRID WIND ENERGY FOR DESALINATION

Small desalination units can be combined with independent wind energy converters and have a high potential for converting saline and brackish water into freshwater. Wind turbines can be operated in an on-grid connection [66]. Wind energy is often used in hybrid manner involving storage or other energy sources. The Kwinana desalination unit in South Perth, Western Australia, is a successful example of the combination of RO and wind turbines. The site produces about 140 ML/d of freshwater. The electricity required for the unit is supplied by an 80-MW wind power plant [67]. The ENERCON desalination unit is specifically designed to operate with a wind source. The system, combined with a RO desalination unit, can operate at variable capacities (from 12.5% to 100%). Consequently, it can adapt well to changing weather conditions and wind fluctuations [68]. The use of wind energy for the desalination of seawater, especially for coastal areas with high wind potential, is one of the promising approaches. Energy generated by wind turbines in both electrical and mechanical energy can be used for desalination processes (especially for RO and MVC methods). Different desalination units are installed around the world based on small wind turbines. As with solar energy, fluctuations in power output over time are one of its problems. By combining this system with other RE resources and by using energy storage reservoirs such as batteries, one can observe the improvements in the operating conditions of this source [69]. A combination of wind turbine, power grid and desalination plant is illustrated in Figure 10.2 [2].

Park et al. investigated the efficiency and cost of water production by several small wind turbine models connected to the RO system in Ghana. The results showed that the 1-kW wind turbine is the most efficient and the least costly unit. Therefore, it is the best option to power the RO system [71]. Impact of various factors on the pricing of each freshwater unit such as climatic conditions, site layout, turbine power, the concentration of saltwater, operating conditions, production capacity, cost of RO, and wind turbine is reviewed by Garcia-Rodriguez et al. [72]. Peñate et al. introduced a hybrid system for the use of wind energy by RO. The design has the variable capacity with a nominal production capacity of 1,000 m³/d which the results are compared with constant production mode. The results showed that due to the variable nature of wind energy, the degassing units can gradually adapt to the energy supplied by the variable sources and maximize the amount of freshwater per year [73]. Segurado et al. have been exploring ways to increase the share of renewables on the island of Saint Vincent for combined systems. These methods have led to two scenarios for storing excessed wind energy. The main purpose of this system is to reduce costs by raising the share of wind energy in the integrated hybrid system [70]. The results demonstrated that

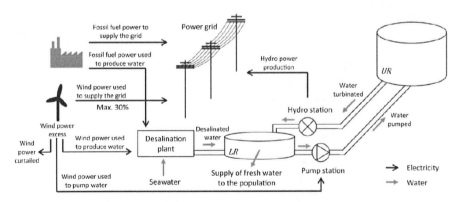

FIGURE 10.2 Combining wind turbine, power grid, and desalination plant for producing freshwater [2,70].

the island's freshwater capacity is better to be provided by small desalination units (see Figure 10.2). In addition by increasing the capacity of storage tank, the share of wind energy in the system can be increased (with the same installed wind turbine capacity), and the desalination capacity of the island can be 25% higher by 2020. Also, it used wind power which provided 56% more freshwater by wind energy and overall expected costs would be reduced by about 7%. Gökçek and his colleagues have investigated the technical and economic use of small-scale wind energy to drive a RO unit. The results confirmed that the cost of producing freshwater for a stand-alone system varies from 2962 to 6457 USD per m^3 based on the size of the turbine (6–30 kW). According to calculations using a 30-kW turbine, CO_2 reduction will be 80 t/y [74].

The solar energy is available during the day, not at night and, conversely, the wind speed is higher at night. To maximize the use of RE, a hybrid system consisting of solar and wind resources recommended. Ismail et al. have studied and modeled the performance of MED and MVC desalination methods and used renewable resources for desalination at 100 m^3/h. The results showed that the location of the desalination unit, the intensity of solar radiation, wind speed, ambient temperature, and water salinity are important factors in system performance [75]. Soshinskaya et al. [76] have investigated the techno-economic potential of using RE arrangements for industrial desalination in the Netherlands. Modeling shows that there is a great potential for electricity generation from hybrid renewable resources in the region, which can meet the electricity needs of the potable water production unit between 70% and 96% independently by using PV cells and wind turbines [76]. Mentis et al. [77] have developed a method for designing and optimizing desalination capacity with hybrid wind and solar RE units in the Greek Islands. In this work, various parameters such as the demand water and the energy required for desalination have been considered. Information on technical performance, available resources, and economic calculations has also been used. Modeling results showed that freshwater prices ranged from 1.45 EUR/m^3 for large islands to 2.6 EUR/m^3 for small islands, which is well below the current price of water in these areas (7–9 EUR/m^3) [77]. The production of freshwater by combination of solar and wind systems is illustrated in Figure 10.3 [2].

Khan et al. [79] pointed out that RO desalination based on a hybrid renewable energy system (HRES) has emerged as a cleaner alternative. They presented a comprehensive analysis of the trends and technical developments of PV-RO, Wind-RO, and hybrid PV-Wind-RO for a wide range of capacities over the past three decades. Designing and modeling HRES-RO desalination systems using different combinations of RES are thoroughly analyzed and the technical aspects of their performance are presented. The application of a range of optimization and sizing software tools available for conducting prefeasibility analysis and the comparison of the available software tools for HRES-RO desalination are also presented. The study also demonstrated that the replacement of fossil fuel with RE for desalination will significantly decrease GHG emissions. The review also highlights the effect of solar and wind profiles on the economics of desalination powered by renewables. The economic analysis indicates a significant decrease in the cost of water production by hybrid PV-wind-RO systems, implying good prospects for the technology in the near future. Finally, the study provides a flowchart depicting the steps involved in installing a hybrid PV-wind-RO system in KSA.

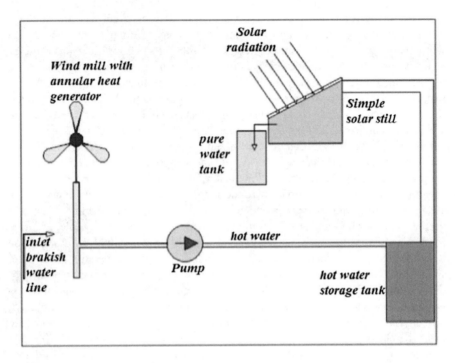

FIGURE 10.3 Producing freshwater by combining solar and wind systems [2,78].

For remote communities that have high water costs and high renewables penetration (e.g., solar or wind), there is the potential to design hybrid systems that can be used for water production, electricity production, or load balancing. This can be achieved by diverting flow from the reverse-osmosis system to an electric generator to produce electricity. An electric motor can be installed on the reverse-osmosis pump to pull excess electricity from the grid as needed for load balancing and the reverse-osmosis system. For large islands, wind turbines are preferred, while PV cells can also be added to the coupling system. In smaller islands, the computation is much more complex and the answer is not clear because the cost per kilowatt of small wind turbines is far higher than large wind turbines. In these islands, the use of PV cells is competitive and only restricts the design of the cells and their placement. Jordan is the right country for using solar and wind energy sources. The return on investment for some projects using wind energy is 6 years and for PV cells is 2.3 years [77]. The results show that the optimal use of renewable sources for electricity generation can be increased by up to 76%, leading to a significant reduction in fossil fuel consumption and CO_2 emissions. Studies demonstrated that the economic and environmental benefits of using a hybrid system for producing freshwater on an island can be up to 36%. Smaoui et al. have been looking for a way to find the optimal size of an independent hybrid system including PV, wind, and hydrogen and to provide the energy needed to desalinate seawater in the south of Tunisia. When the electricity generated by wind and solar sources exceeds the load, this power is used in the electrolysis unit to generate hydrogen. The hybrid system driven by wind, PV and hydrogen for the desalination is illustrated in Figure 10.4 [2]. In this case, as far as the integrated system faces a lack of inputted power, fuel cells (FC) would provide demanded power for the desalination process. Hydrogen storage was selected because of low maintenance costs, low noise, and the ability to operate in harsh conditions [80]. A hybrid system driven by wind, PV and hydrogen for the desalination process is illustrated in Figure 10.4. The introduced system provides freshwater for 14,400 persons in winter. The mean annual water consumption is 193.6 m³/h, and the highest water consumption is 530 m³/h. Based on the results of this paper, the combination of wind and PV systems effectively increases the reliability of the power system and reduces the need for energy storage (thus reduces installation costs).

An innovative wind energy system for the desalting process has been introduced by Moh'd A et al. (see Figure 10.3). They generated thermal energy by the wind system to preheat the water. During winter as well as at night, the device relies on the thermal energy generated. In the summer and during the days, most of the energy is supplied by the sun [78]. The ultimate aim of the project is to use a hybrid system to build an independent desalination plant in remote areas. In a wind turbine, the frictional heat energy raises the temperature of the oil, and then in heat exchanger, its thermal energy transfers to seawater. The advantage of this method is to increase the desalination capacity and efficiency of the desalination plant as well as the ability to continue operating the desalination plant at night and cloudy days. Henderson et al. have studied the possibility of using a hybrid wind and diesel system that includes a water desalination unit in the Isle of Man area—New Hampshire. The system was designed to supply power during peak hours, to produce, and to store potable water. The researchers noted that seawater desalination is technically an attractive option for long-term energy storage and power management [81]. There are numerous studies on the combination of

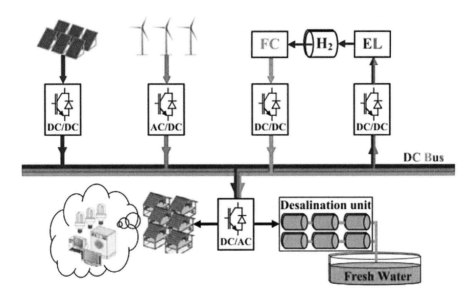

FIGURE 10.4   The hybrid system is driven by wind, PV, and hydrogen for the desalination process [2,80].

wind energy and PV panels for freshwater production, and the results indicated that by using hybrid systems, power supply guaranteed through the year [82]. A hybrid system combining wind, PV and RO systems is illustrated in Figure 10.5 [2].

Remote areas with potential wind energy resources such as islands can employ wind energy systems to power seawater desalination for freshwater production. The advantage of such systems is a reduced water production cost compared with the costs of transporting the water to the islands or to using conventional fuels as power source. Different approaches for wind desalination systems are possible. First, both the wind turbines as well as the desalination system are connected to a grid system. In this case, the optimal sizes of the wind turbine system and the desalination system as well as avoided fuel costs are of interest. The second option is based on a more or less direct coupling of the wind turbine(s) and the desalination system. In this case, the desalination system is affected by power variations and interruptions caused by the power source (wind). These power variations, however, have an adverse effect on the performance and component life of certain desalination equipment. Hence, backup systems, such as batteries, diesel generators, or flywheels might be integrated into the system. Main research in this area is related to the analysis of the wind plant and the overall system performance as well as to developing appropriate control algorithms for the wind turbine(s) as well as for the overall system. Regarding desalinations, there are different technologies options, e.g., ED or VC. However, RO is the preferred technology due to low specific energy consumption [83].

Wind turbines can be used to supply electricity or mechanical power to desalination plants. Like PV, wind turbines represent a mature, commercially available technology for power production. Wind-driven desalination has particular features due to

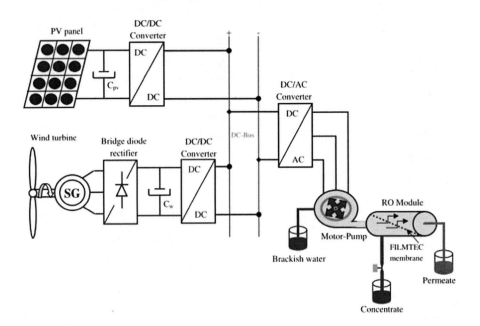

FIGURE 10.5   Producing freshwater by combining wind, PV, and RO systems [2,82].

the inherent discontinuous availability of wind power. For stand-alone systems, the desalination unit has to be able to adapt to the energy available; otherwise, energy storage or a backup system is required. Wind energy is used to drive RO, ED, and VC desalination units. A hybrid system of wind/PV is usually used in remote areas. Few applications have been implemented using wind energy to drive an MVC unit. A pilot plant was installed in 1991 at Borkum, an island in Germany, where a wind turbine with a nominal power of 45 kW was coupled to a 48 m^3/d MVC evaporator. A 36-kW compressor was required. The experience was followed in 1995 by another larger plant at the island of Rügen. Additionally, a 50 m^3/d wind MVC plant was installed in 1999 by the Instituto Tecnologico de Canarias (ITC) in Gran Canaria, Spain, within the Sea Desalination Autonomous Wind Energy System (SDAWES) project [84]. The wind farm is composed of two 230-kW wind turbines, a 1,500-rpm flywheel coupled to a 100-kVA synchronous machine, an isolation transformer located in a specific building, and a 7.5-kW uninterruptible power supply located in the control dome. One of the innovations of the SDAWES project, which differentiates it from other projects, is that the wind generation system behaves like a mini power station capable of generating a grid similar to conventional ones without the need to use diesel sets or batteries to store the energy generated.

Regarding wind energy and RO combinations, a number of units have been designed and tested. As early as 1982, a small system was set at Ile du Planier, France [55], which as a 4-kW turbine coupled to a 0.5-m^3/h RO desalination unit. The system was designed to operate via either a direct coupling or batteries. Another case where wind energy and RO were combined is that of the Island of Drenec, France, in 1990 [85]. The wind turbine, rated at 10 kW, was used to drive a seawater RO unit. A very interesting experience was gained at a test facility in Lastours, France, where a 5-kW wind turbine provides energy to a number of batteries (1,500 Ah, 24 V) and via an inverter to an RO unit with a nominal power of 1.8 kW. A 500 L/h seawater RO unit driven by a 2.5-kW wind generator (W/G) without batteries was developed and tested by the Centre for Renewable Energy Systems Technology (CREST) UK. The system operates at variable flow, enabling it to make efficient use of the naturally varying wind resource, without the need of batteries [86].

The ED process is interesting for brackish water desalination since it is able to adapt to changes of available wind power and it is most suitable for remote areas than RO. Modeling and experimental test results of one of such system installed at the ITC, Gran Canaria, Spain, are presented in Reference [87]. The capacity range of this plant is 192–72 m/d. Excellent work on wind/RO systems has been done by ITC within several projects such as AERODESA, SDAWES, and AEROGEDESA [88]. In addition in Coconut Island off the northern coast of Oahu, Hawai, a brackish water desalination wind-powered RO plant was analyzed. The system directly couples the shaft power production of a windmill with the high-pressure pump; 13 L/min can be maintained for a wind speed of 5 m/s [89,90].

Additionally, a wind/RO system without energy storage was developed and tested within the JOULE Program (OPRODES-JORCT98-0274) in 2001 by the University of Las Palmas. The RO unit has a capacity of 43–113 m^3/h, and the W/G has a nominal power of 30 kW [86]. The European Community Joule III project funded different research programs and demonstration projects of wind desalination systems on Greek and Spanish islands. In addition, an excellent job on combining wind/RO was done by ENERCON, the German wind turbine manufacturer. ENERCON provides modular and energy-efficient RO desalination systems driven by wind turbines (grid-connected or stand-alone systems) for brackish and seawater desalination. Market-available desalination units from ENERCON range from 175 to 1,400 m³/d for seawater desalination and 350–2,800 m³/d for brackish water desalination. These units in combination with other system components, such as synchronous machines, flywheels, batteries, and diesel generators, supply and store energy and water precisely according to demand [91]. Several industrial wind/RO installations for seawater include (a) Ile de Planier, France, with 500 L/h capacity (commission in 1983); (b) Fuerteventura island, PUNTAJANDIA project with a capacity of 2,333 L/h (commission in 1995); (c) Therasia island, Greece, with a capacity of 200 L/h (commission in 1997); (d) Pozo Izquierdo, Gran Canaria, AEROGEDESA project with a capacity of 800 L/h (commission in 2003); and (e) CREST, UK, with a capacity of 500 L/h (commission in 2004) [27].

Besides the ones mentioned above, an RO system is driven by a wind power plant, in the Island of the County Split and Dalmatia, as reported in Reference [92]. An RO plant in the Middle East is a 25 m³/d plant connected to a hybrid wind–diesel system [93]. In Drepanon, Achaia, near Patras (Greece), has a wind-powered RO system [94]. Finally, European Commission (1998) presents other facilities at (a) Island of Suderoog (North Sea), with 6–9 m³/d; (b) Island of Helgoland, Germany (2.480 m³/h); (c) Island of St. Nicolas, West France (hybrid wind–diesel); and (d) Island of Drenec, France (10 kW wind energy converter). More information for large stand-alone wind desalination systems [89,90], for small systems [95], and for an overview of the research activities in North America [96] are available.

10.5 HYBRID GEOTHERMAL ENERGY FOR DESALINATION

Geothermal energy is widely distributed throughout the earth and can be used for heating and power generation that could be used in both thermal and membrane desalination processes. Geothermal resources can produce steam and hot water. The use of geothermal resources for energy production ranked third among all types of renewables. In Iceland, 86% of heating and 16% of demanded power are supplied by geothermal energy [97]. By the end of 2004, geothermal energy provided 57 TWh as electricity and 76 TWh as heating. While geothermal energy is not as common in use as solar (PV or solar thermal collectors) or wind energy, it presents a mature technology that can be used to provide energy for desalination at a competitive cost. Furthermore, and comparatively to other RE technologies, the main advantage of geothermal energy is that the thermal storage may not be necessary, since it is both continuous and predictable [98,99]. A high-pressure geothermal source allows the direct use of shaft power on mechanically driven desalination, while high-temperature geothermal fluids can be used to power electricity-driven RO or ED plants. Kalogirou showed that at a specified depth from the ground, the temperature remained constant throughout the year [100].

Geothermal resources are divided into three categories (based on temperature): the low temperature for sources with a temperature below 100°C, the middle temperature sources with a temperature between 100°C and 150°C, and high-temperature sources with a temperature more than 150°C. Geothermal wells with a depth of 100 m can be reasonably used for desalination purposes. Using geothermal energy resources can supply energy all day long. This energy source can be used continuously in two ways. One method is the combination with the MED unit. Geothermal units in the range of 70°C–90°C are ideal for the MED method of desalination [101]. In the second method, geothermal energy is first used to generate electricity, and then, the electricity can be used to set up a RO desalination unit. There are other hybrid processes in which hot steam is extracted from the ground to circulate the power-generating turbine, and then, the steam released from the turbine comes into the MED desalination plant. This method can be used very efficiently to absorb geothermal energy and can be used for high water salinity areas such as the Persian Gulf and the Red Sea.

The first geothermal energy-powered desalination plants were installed in the United States in the 1970s [98,99,102,103,104], testing various potential options for the desalination technology, including MSF and ED. An analysis [105] discussing a technical and economic analysis of an MED plant, with a capacity of $80 \, m^3/d$, powered by a low-temperature geothermal source and installed in Kimolos, Greece, showed that high-temperature geothermal desalination could be a viable option. A study [106] presented results from an experimental investigation of two polypropylene-made HD plants powered by geothermal energy [107]. Recently, a study [108] discussed the performances of a hybrid system consisting of a solar still in which the feedwater is brackish underground geothermal water.

The combinations of geothermal and desalting processes have been investigated in several theoretical and experimental studies [109–112]. In particular, a detailed review of the use of geothermal energy for desalination has been carried out by Goosen et al. who examined examples of these systems in Greece, Algeria, and Mexico. They concluded that this type of energy can be very effective in some parts of the world [113]. A case study of a geothermal-desalination unit (75°C–90°C) with production capacity equaled to $600–800 \, m^3/d$ has been investigated [114]. Calise et al. have studied the solar-geothermal system for analyzing the performance of the multipurpose hybrid system (power generation, heating, cooling, and desalination). This multifunctional system included CSP, PV, geothermal wells, and MED desalination as well as a set of lithium bromide absorption chillers and other accessories such as a water reservoir and a heat exchanger [115].

Solar collectors continuously receive electrical and thermal energy at 100°C and are able to inject the surplus power into the grid. The absorbed heat energy can be used for a variety of purposes such as ambient heating as well as the supply of potable water. They also use this solar power to launch an absorption chiller in the summer. Finally, the rest of the thermal energy is combined with the low-temperature thermal energy of the geothermal well (80°C) and used in the MED unit for desalination. Geothermal energy is also used to supply hot water at 45°C. In winter, solar energy reduced noticeably, resulting in less heat and electricity production. For this reason, much of the energy required for desalination comes from geothermal energy in winter. The results of the performance analysis of this system are well accepted, and from the economic point of view, the system's adaptability is greatly enhanced by the increased use of hot water. The design requires a heat storage tank to meet

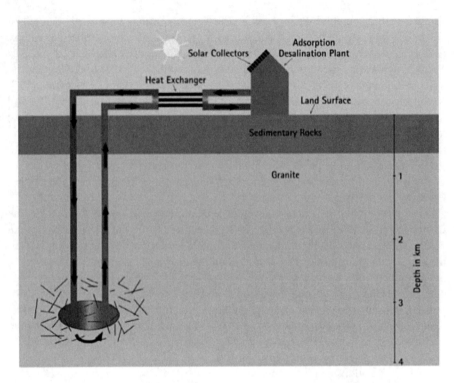

FIGURE 10.6 The combined system consisted of solar and geothermal energies for fresh-water production [2,116].

the need for desalination units at night and geothermal energy at this time [116]. A hybrid system combining solar and geothermal energies for freshwater production is illustrated in Figure 10.6 [2].

10.6 HYBRID WAVE ENERGY

Wave power use is a developing technology with high investment costs, which has challenges in its implementation. Sustainable use of this energy requires new strategies to reduce the investment cost and possible extraction of energy from small sources. With the help of this technology, produced power can be used in desalination units (especially RO) [117]. The total produced power by the waves in the world is estimated to be about 2 TW [118]. Studies by Davies in 2005 were conducted to combine wave power with desalination plants. The research showed that along the arid and sunny beaches, the desalination unit with the power of the waves can optimally supply the water needed to irrigate a strip of land with 0.8 km width by waves with a minimum height of one meter. The amount of energy available to the waves varies from region to region. The ratio of available power to water scarcity in Morocco is 16% and Somalia is 100%. It should be noted that the use of wave energy for desalination units has so far been limited to laboratory cases. The biggest problem with wave power is that it fluctuates over time (like wind energy),

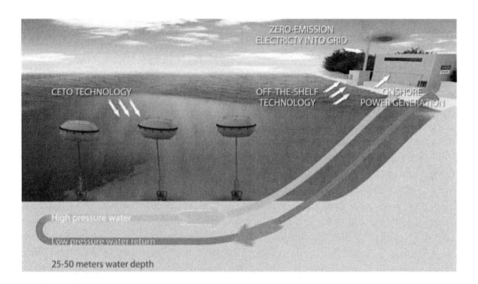

FIGURE 10.7   The combined system consisted of CETO technology and DPP [2,124].

which is a failure in the sustainability of desalination units [119–123]. In future, wave energy needs to be combined with other sources in hybrid manner.

A desalination pilot plant (DPP) based on wave power was installed on Garden Island, Australia. The utilized device is called CETO. This innovative system uses numerous buoys to exploit the wave power to pump the seawater into a RO unit for producing freshwater, in which the production capacity is $150 \, m^3/d$ [124]. The world's first commercial-scale wave energy was installed in Australia and connecting to the electricity grid uses produced power for the desalination process [125]. With the aid of DEIM converters, the production of electric power in Pantelleria Island in Italy with wave power potential equaled to 7 kW/m became achievable. In this case, the consumption of diesel fuels (1,391 tons) and CO_2 emission (4,406 t/y) was eliminated. In this way, by integrating renewable sources with desalination processes, energy dependence on fossil fuels was avoided [126]. Effective specifications for wave energy convertor (e.g., energy requirements and structure features) were carried out by Salter et al. [127]. A hybrid system consisted of CETO technology and DPP is illustrated in Figure 10.7 [2].

10.6.1   BARGE-WAVE AND BELOW WATER ENERGY CONVERSION

The deep-sea RO desalination facility to meet the needs of providing freshwater and reducing CO_2 emission has been investigated and designed. In this system, energy consumption is significantly lower than in other systems. The introduced system has a longer membrane lifetime in comparison with other strategies [128]. The U.S. Department of Interior funded the Subfloor Water Intake Structure System (SWISS), currently utilized in desalination plants in California and Japan [129–134]. The SWISS approach is to install a permanent subfloor well/intake system for the source water for the traditional shore structures. The in situ sand provides the filtration media [132].

The barge wave energy conversion was studied in the late 1970s by Haren [129,133,134] at MIT (Massachusetts Institute of Technology). He found that the optimum articulated-barge configuration was a three-barge system. In the 1980s, McCabe showed that the efficiency of the three-barge system could be substantially improved by suspending an inertial-damping plate below the center barge. McCabe, then, produced a prototype of the system, coined the Wave Pump (MWP). The MWP was primarily designed as a producer of potable water. Ocean Energy System is in the business of designing and manufacturing articulated-barge systems to produce potable water by RO desalinization of seawater. U.S. Patent Publication No. 2009/0084296 (McCormick) describes a system directed to a wave-powered device having enhanced motion making use of an amplified wave energy conversion system (AWECS). The AWECS basically comprises a forward barge, a rear barge, and an intermediate or center barge, all of which arranged to float on a body of water having waves. The barges are hingedly coupled together so that they can articulate with respect to each other in response to wave motion. The AWECS also includes high-pressure pumps that straddle and pivotably connect the barge pairs, e.g., at least one pump connects the forward barge and the intermediate barge, and at least another pump connects the rear barge and the intermediate barge. The pumps are designed to draw in the water through a prefilter, pressurize the water, and deliver the water to an onboard RO desalinization system. That system includes an RO membrane. As an incoming wave makes contact with the forward barge first, the hydraulic fluid in the pump(s) coupled between the forward barge and the center barge are driven in a first direction; as the wave continues, the hydraulic fluid in the pump(s) coupled between the rear barge and the center barge are driven in a second opposite direction. The end results are bidirectional hydraulic pumps.

In U.S. Provisional Patent Application Ser. No. 61/707,206, filed on September 28, 2012, there is disclosed an AWECS arranged for producing electrical energy from the wave energy. To that end, it makes use of an AWECS similar to that described above, except that it also makes use of a commercially available rotary-vane pump to drive a generator to produce the electricity. In particular, the invention of that Provisional Application entails a floating device having a first portion (e.g., a first barge) movably coupled (e.g., hinged) to a second portion (e.g., a second barge); at least one hydraulic or pneumatic pump (e.g., a linear pump) coupled between the first portion the said second portion, the hydraulic pump driving a hydraulic fluid therein when the first portion moves with respect to the second portion due to wave energy. A fluid rectifier is provided in the AWECS and is in fluid communication with at least one hydraulic or pneumatic pump that generates a unidirectional hydraulic or pneumatic fluid flow. A rotary vane pump is coupled to the fluid rectifier.

The rotary vane pump uses the unidirectional flow to generate a rotational motion via a drive member. A rotating electrical generator (e.g., a DC generator) is coupled to that drive member, so that the drive member causes the rotating electrical generator to generate electricity when the drive member is rotating. A filter anchor is provided that includes a filter housing for filtering seawater prior to entry into a water desalination system for placement on a sea floor. The filter housing has an exterior and an interior chamber, at least one inlet for providing the seawater to the interior chamber, and at least one outlet for providing filtered water to exit the interior chamber. A sand filter is disposed in the filter housing, separating the exterior from the interior chamber. The filter housing has at least one water conduction outlet conduit for allowing filtered water to exit the interior chamber to provide filtered water. The inlets for providing seawater may provide for a surface intake velocity of <0.5 ft/s to restrict incursion of fish larva and macro or microvertebrae. The filter anchor may be of a size to permit container transportable via truck transportation.

The interior chamber of the filter anchor may be substantially filled with clean, washed, coarse sand, from either a local beach or shoreline source or from sand obtained from a commercial sand source. The filter housing may have hatches between the exterior and the interior chamber which, when opened, provide for submersion of the filter housing via flooding of the interior chamber and controlled sinking of the filter anchor to the sea floor. The filter anchor, prior to use as a filter, may be floatable and towable to a deployment site in the sea. At least one submersible pump and submersible air snorkel may be included such that the filter anchor is refloatable when the hatches are in a closed position, wherein the interior chamber is substantially filled with air, wherein the submersible pump and air snorkel can be activated to float the filter anchor.

A wave energy conversion system is also provided that includes an articulated-barge system for converting wave energy into energy used to pump water to a desalinization system to generate potable water. At least one filter anchor is included. Each filter anchor includes a filter housing and a filter disposed therein for filtering seawater prior to entry into a water desalination system for placement on a sea floor. The filter housing has an exterior and an interior chamber, at least one inlet for providing the seawater to the interior chamber, and at least one outlet for providing filtered water to exit the interior chamber. The filter is disposed in the filter housing, separating the exterior from the interior chamber. The filter housing has at least one water conduction outlet conduit providing for filtered water to exit the interior chamber to provide filtered water to the desalinization system on the articulated barge. A mooring buoy is attached to each filter anchor by a mooring line. The desalinization system may include an RO membrane. The filter may be a sand filter. The filter anchor may include at least one feed line in the interior chamber to provide the filtered water to the water conduction outlet conduit. The filter housing may be constructed using a steel sheet. At least one inlet may be a manually controlled hatch or an automatically controlled hatch. A method of anchoring a wave energy conversion system and providing filtered water to a desalination system is also provided. The method also includes the steps of providing a mooring buoy for each filter anchor at the location, attaching each mooring buoy to one of the filter anchors by a mooring line, attaching each filter anchor to the articulated-barge system, and supplying filtered water to the articulated-barge system. The desalinization system may use a RO membrane.

Besides the patent by Murtha, McCormick, and Washington [129,133,134], a new company called Atmocean Inc. [135] hopes to bring to market an ocean-wave-powered process that supplies potable water—cutting costs and limiting pollution, as compared to traditional desalination systems that require power plants. Atmocean Inc., Santa Fe, N.M., has conducted 29 sea trials since its founding in 2006. In collaboration with Sandia National Laboratory and Reytek Corp., both of Albuquerque, its recent trials were in the Pacific Ocean, off Peru, in 2015. Following modifications, the company tested a one-eighth-scale model at Texas A&M in February.

Reytek is preparing a full-scale test unit for deployment in 2017 off Marystown, Newfoundland.

The deployed marine energy system could have minimal surface expression. In fact, some technologies are fixed bottom or anchor mounted below the surface, eliminating any surface expression. However, minimal surface expression implies that the device must be robust enough to withstand the marine environment. But unlike electricity production, low-cost storage in the form of water tanks can mitigate the challenges associated with resource intermittency, providing an opportunity to offset costs resulting from reliability constraints. Technology developers are designing systems that range from hybrid water and electric systems. The competition for marine desalination is diverse and site-specific. For large water utilities (e.g., San Diego Water Authority), other water sources will typically be considered before desalination technologies (i.e., surface water, groundwater, advanced water treatment for water reuse, water recycling, and water conservation portfolio options). In smaller, remote, or isolated locations where desalination is prominent, diesel-powered generators are typically used [136,137]. This is primarily driven by the reliability of diesel generation, and the perception that RO technologies must have an electricity input. Other renewables (e.g., wind, solar, geothermal) have been proposed and used in certain parts of the world for both membrane and thermal desalination technologies, although membrane technologies are the most common because they are the most energy efficient.

Marine energy has some specific advantages compared to other renewables or even diesel-powered systems. Given that marine energy technologies are inherently offshore, they will not be competing with land use as is the case with solar. In areas where social acceptance is a larger driver than water accessibility, fully or mostly submerged marine energy technologies will have less line-of-sight permitting and siting challenges than wind. Fully submerged technologies may even be designed at depths that can allow local fishing boats to travel through the water safely.

10.7 DESALINATION BY COGENERATION BY SMALL MODULAR NUCLEAR REACTORS

There are several reasons why use of nuclear heat from small modular nuclear reactors for desalination makes sense [1,138–146]: (a) the overall cost of fossil heat generation is being dominated by the cost of the fuel itself; (b) the current trends in fossil-fuel prices and supply uncertainties; (c) concerns about GHG emissions; and (d) the new generation of modular nuclear power plant (NPP) systems have highly enhanced safety levels and competitive economics. The coupling of desalination with nuclear systems is not technically difficult [1,138–146] but needs the following considerations: (a) avoiding radioactivity cross-contamination; (b) providing backup heating energy sources in case the nuclear system is not in operation (e.g., for refueling and maintenance); and (c) incorporating certain design features in case the thermal desalination option is used. At present, most desalination plants use fossil fuels and to a lesser degree clean energy. For some time now nuclear energy has been used for desalination processes; however, the enormous potential of nuclear energy is as yet not fully explored. It can generate much more freshwater than what is being produced, and also it is more affordable and does not release GHGs. Nuclear desalination is generally cost-competitive compared with the desalination using fossil fuels.

The IAEA [138,142,143] says that only nuclear reactors are capable of delivering copious quantities of energy which will be required for desalination projects of the future. Along with desalination of brackish or seawater, treatment of urban wastewater will be increasingly important and undertaken. The cost of water, however, varies significantly in different areas of the world. According to the United States Environmental Protection Agency, it is estimated that tap water costs an average of $2 per 1,000 gallons. However, the desalination project at Coquina Coast in Florida calculates the cost at $6.27–7.74 per 1,000 gallons. The difference in cost depends on the source of energy used in the process. With nuclear energy, it is possible to achieve an enormous economy of scale, which lowers the cost. A study carried out in Tunisia discovered that the costs of nuclear desalination were from one-third to less than half of those related to desalination via fossil fuels, depending on the desalination technology that is used [1,138–146]. Furthermore, desalination plants which mostly use fossil fuels contribute to increase in the levels of greenhouse effect gases. Several countries have implemented nuclear desalination, including India, Japan, and Kazakhstan. The latter operated a 750-MW thermal power plant for over 25 years. This plant not only generated desalinated water but also produced heat and electricity. Desalination of water with nuclear energy is an established, proven, and known technology with thousands of men hours.

As mentioned earlier, there are a number of processes that have been demonstrated for producing clean water from seawater; however, global experience is dominated by three primary processes: two distillation-based technologies [MED and MSF] and one membrane-based technology [RO] [1]. All seawater desalting processes—MSF, MED, and SWRO—consume significant amounts of energy and materials. In view of the rising fuel costs, the amount and cost of fuel consumed to desalinate seawater are some of the main factors determining the operational cost of water desalination. Similarly, the materials selected and the increased cost of materials for desalination have a significant impact on the capital cost. These rising costs, in turn, become a major factor in choosing the method and technology to be used [148]. The RO process employs a pressure-driven separation technique where water is forced under high pressure through a water-permeable membrane. No heating or phase change takes place. The main energy is the electricity required for the initial pressurization of the feedwater, 5–7 MPa for seawater, or 2–3 MPa for brackish water. The advantages are simple processing, low installation, and maintenance costs. The drawbacks are the necessary pretreatment of the feedwater, the short lifetime of the membranes, and the comparatively high content (1%–2%) of salt passing through the membrane [147,148].

Another key distinction in the three methods is the way that they couple with a power source. The RO plant has the most straightforward coupling since it can operate using only electricity, which is needed to run the high-pressure pumps. Therefore, it is not essential to colocate the desalination plant with the power plant so long as a grid connection is available. However, there may be an advantage for colocation of the power and RO desalination plant in terms of shared infrastructure and protection against grid disruption. Also, low-grade steam or warm wastewater from the power plant can be used to preheat the saline feedwater of the RO plant to improve its clean water production efficiency, although the quality of the distillate may be adversely impacted. Both MED and MSF plants require a thermal heat source, such as a steam line from the secondary side of the nuclear plant. This steam is typically extracted from a low-pressure turbine stage, which results in a commensurate decrease in the electrical output of the power plant and may have implications on the reliability and flexibility of operations for both the power plant and the desalination plant. Also, the use of a tertiary heat transport loop is typically required to ensure that no radio-nuclides such as tritium are carried over from the reactor's secondary loop to the distillation plant [1,138–146]. The choice of desalination method(s) is determined primarily by the characteristics of the source water and the water quality required by the end user. For example, RO technology typically has a lower capital cost but is less effective with feedwater that contains high level of organic materials that can foul the membranes or that have high salinity levels and can only produce potable water with-out further treatment. The two thermal distillation processes are much more tolerant of "dirty" or "salty" feedwater and produce high-purity water.

Progress in nuclear plants and desalination system is growing considerably, and numerous countries are getting involved in it. There is no restriction for choosing nuclear reactors for the desalination process, and any reactor has the potential cou-pled with desalination systems [79,149]. Nuclear energy can also be used for thermal desalting. The heat used in the nuclear reactor is used for this purpose. The cost of desalinizing seawater with nuclear energy is roughly equivalent to fossil energy. The first nuclear-powered MSF-RO unit was built in India. This unit is based on MSF technology developed in India and has a capacity of about 6,300 m^3/d [150]. Intending to design and develop hybrid systems including nuclear and desalination systems can increase the capacity of freshwater production. These interests have been increased significantly among the members of the IAEA for the last two decades.

There is numerous integration of nuclear systems with small-size desalination units [151–153,154,108]. An innovative combined system consisted of the nuclear heating reactor (NHR) with low-temperature MED + RO and NHR with low-temperature MED/ VC + MED has been presented [155]. Results from performance analysis indicated new aspects of the combination of the MED process with NHR. Investigated results from Misra and Kupitz [156] and Misra [157] have designated the role of nuclear desalination for the utilization of potable water needs in arid areas. The research also reported few nuclear reactors that can be best coupled to different desalination plants. In Egypt, an advanced hypothetical model for calculat-ing the effective parameters of the RO process is cogenerating with a nuclear reactor [158]. The cost of freshwater production by nuclear desalination was expected to be in the range of 0.4–1.8 USD/m^3 based on the reactor type and the desalination process [159].

Small nuclear reactors, known as Small Modular Reactors, are the most suitable kind for desalination, often with electricity cogeneration with low-pressure steam from the turbine and hot seawater feed from the final cooling system. A small reac-tor can produce between 80 and 100,000 m^3 a day. As seawater desalination tech-niques improve and more countries opt to build double-purpose electric plants (cogeneration), nuclear plants will need more advanced technologies plants, as well as more economical and efficient desalination systems. A mini review of small/medium nuclear reactors (SMRs) for potential desalination applications is recently published by Ahmed et al. [140]. This review suggests that SMRs are a promising alternative for powering large-scale desalination plants. The modern generations of these systems manifest cost-effectiveness and built-in safety features. The compatibility with geological and topological challenges is an added advantage. Moreover, funding opportunities and packages could be easily arranged for SMR. In U.S., modular NuScale reactor is preferred to be used with various desalination technologies. The NuScale small modular reactor design is especially well suited to support water desalination due to its high degree of modularity, enhanced safety and robustness, and flexible plant design. The NuScale plant can easily and effectively couple to a variety of desalination technologies and can be economically competitive for simultaneously producing clean electricity and clean water [1,142,143].

Worldwide, the accumulated operating experience of nuclear desalination has exceeded 200 reactor-years. All nuclear reactor types can provide the energy required by the various desalination processes. However, SMRs may offer, when available, the largest potential as coupling options to nuclear desalination systems in developing countries [1,142,143]. The development of innovative reactor concepts and fuel cycles with enhanced safety features as well as their attractive economics are expected to improve public acceptance and further the prospects of nuclear desalination. As mentioned above, nuclear heat-assisted modular desalination plants have been operated in Kazakhstan and Japan for many years. In Aktau, Kazakhstan, the liquid metal-cooled fast reactor BN-350 has been operating as an energy source for a multi-purpose energy complex since 1973, supplying regional industry and population with electricity, potable water, and heat. The complex consists of a nuclear reactor, a gas and/or oil fueled thermal power station, and MED and MSF desalination units. The seawater is taken from the Caspian Sea. The nuclear desalination capacity is about 80,000 m^3/d. A part of this capacity has now been decommissioned. In Japan, all of the NPPs are located at the seaside. Several NPPs of the electric power companies of Kansai, Shikoku, and Kyushu have seawater desalination systems using heat and/or electricity from the nuclear plant to produce feedwater for the steam generators and for on-site supply of potable water. MED, MSF, and RO desalination processes are used. The individual desalination capacities range from about 1,000–3,000 m^3/d. The experience gained so far with nuclear desalination is encouraging. There are nine nuclear units in Japan and one in Kazakhstan as mentioned above. In Japan, the desalination plants are constructed on-site at the NPPs with the aim of supplying the required make-up cooling water to these NPPs. Such desalination plants have clear-based desalination plant in the world [160]. The optimization of water desalination using nuclear reactors has been analyzed [161].

Small- and medium-sized nuclear reactors are suitable for desalination, often with cogeneration of electricity using low-pressure steam from the turbine and hot seawater feed from the final cooling system. The main opportunities for nuclear plants have been identified as 80–100,000 m^3/d and 200–500,000 m^3/d ranges. The U.S. Navy nuclear powered aircraft carriers reportedly desalinate 1,500 m^3/d each for use onboard. A 2006 IAEA report based on country case studies showed that costs would be in the range 50–94 c/m^3 for RO, 60–96 c/m^3 for MED, and \$1.18–1.48/m^3 for MSF processes, with marked economies of scale. These figures are consistent with later reports. Nuclear power was very competitive at 2006 gas and oil prices. A French study for Tunisia compared four nuclear power options with CC gas turbine and found that nuclear desalination costs were about half those of the gas plant for MED technology and about one-third less for RO. With all energy sources, desalination costs with RO were lower than MED costs. At the April 2010 Global Water Summit in Paris, the prospect of desalination plants being co-located with NPPs was supported by leading international water experts. As seawater desalination technologies are rapidly evolving and more countries are opting for dual-purpose-integrated power plants (i.e., cogeneration), the need for advanced technologies suitable for coupling to NPPs and leading to more efficient and economic nuclear desalination systems is obvious [1,142,143].

Large-scale deployment of nuclear desalination on a commercial basis will depend primarily on economic factors. Indicative costs are 70–90 c/m^3, much the same as fossil-fueled plants in the same areas. One strategy is to use power reactors which run at full capacity, but with all the electricity applied to meeting grid load when that is high and part of it to drive pumps for RO desalination when the grid demand is low. The BN-350 fast reactor at Aktau, in Kazakhstan, successfully supplied up to 135 MW of electric power while producing 80,000 m^3/d of potable water over 27 years, about 60% of its power being used for heat and desalination. The plant was designed as 1,000 MW but never operated at more than 750 MW, but it established the feasibility and reliability of such cogeneration plants. (In fact, oil/gas boilers were used in conjunction with it, and total desalination capacity through ten MED units was 120,000 m^3/d.) In Japan, some ten desalination facilities linked to pressurized water reactors operating for electricity production yield some 14,000 m^3/d of potable water, and over 100 reactor-years of experience have accrued. MSF was initially employed, but MED and RO have been found more efficient there. The water is used for the reactors' own cooling systems [1,142,143].

In 2002, a demonstration plant coupled to twin 170 MW nuclear power reactors (PHWR, pressurized heavy water reactor) was set up at the Madras Atomic Power Station, Kalpakkam, in southeast India. This hybrid Nuclear Desalination Demonstration Project comprises a RO unit with 1,800 m^3/d capacity and an MSF plant unit of 4,500 m^3/d costing about 25% more, plus a recently added barge-mounted RO unit. This is the largest nuclear desalination plant based on hybrid MSF-RO technology using low-pressure steam and seawater from a nuclear power station. In 2009, a 10,200 m^3/d MVC plant was set up at Kudankulam to supply freshwater for the new plant. It has four stages in each of the four streams. A low-temperature evaporation (LTE) nuclear desalination plant which uses waste heat from the nuclear research reactor at Trombay has operated since about 2004 to supply makeup water in the reactor. Pakistan commissioned a 4,800 m^3/d MED desalination plant in 2010, coupled to the Karachi NPP (KANUPP, a 125 MW PHWR) near Karachi, though in 2014 it was quoted as 1,600 m^3/d. South Korea has developed a small nuclear reactor (SMART) design for cogeneration of electricity and potable water. The 330 MW SMART reactor (an integral PWR) has a long design life and needs refueling only every 3 years. The main concept has the SMART reactor coupled to four

MED units, each with TVC (MED-TVC) and producing 40,000 m³/d in total, with 90 MW. Argentina has designed an integral 100 MW PWR (CAREM) suitable for cogeneration or desalination alone, and a prototype is being built next to Atucha. A larger version is envisaged, which may be built in Saudi Arabia. *NHR-200:* China's INET has developed this, based on a 5 MW pilot plant [1,142,143].

For seawater desalination and electricity generation, Russian design organizations have developed and can supply to customers floating nuclear power/desalination stations based on the small modular KLT-40 type reactor plant derived from Russian icebreakers. ATETs-80 is a twin-reactor cogeneration unit using KLT-40 and may be floating or land-based, producing 85 MW plus 120,000 m³/d of potable water. The small ABV-6 reactor is 38 MW thermal, and a pair mounted on a 97 m barge is known as Volnolom floating NPP, producing 12 MW plus 40,000 m³/d of potable water by RO. The industrial enterprises of Russia are working on the development of a floating nuclear cogeneration plan for the country's northern regions. This can serve as a prototype for the floating nuclear power station (FNPS) desalination complex. The electric energy generated by the FNPS is partially transmitted to the ship for seawater desalination and its excess is used for supply to coastal users. Advantages of floating power/desalination complexes are [138,153]: (a) convenient maintenance by a floating base at a mooring site and decommissioning by tugging to the Supplier's country; (b)commercial production and long-term confirmation of service life characteristics of the KLT-40-type reactor plants and desalination units; (c) possibility of installation in different coastal regions of the world; (d) high fabrication quality at a shipyard and "turnkey" delivery to the customer in a short period of time; and (e) for any of the options, the supply of electrical energy to users within the plant is provided by the shipboard electric power station.

Excess electricity can be used either for the production of additional desalinated water using RO or for sale. The cost of desalinated water can be materially reduced by compensating part of the production costs out of the profits from sale of electric energy. Construction of desalination plants using distillation plants with film-type horizontal evaporators seems to be the most practical at this time. The arrangement, which separates power generation and desalinated water production, has certain advantages over an arrangement in which they are combined on one floating structure as illustrated in Figure 10.8. This arrangement simplifies a solution to the problem of preserving a high efficiency of desalinated water production from the complex when the reactors are shut down by supplying the desalination plant with electric energy from the external grid. The ship for RO desalination is a non-self-propelled structure housing systems and equipment providing for the supply of seawater, its pretreatment, desalination, supply of desalinated water to users, and cleaning of the desalination units.

At present, new technologies are being developed for seawater desalination using RO. For example, in the Canadian CANDESAL desalination program, the use of RO technology is accompanied by preheating of the seawater in the turbine condenser. Preheating allows considerable reduction in the specific power consumption for desalination and in the cost of desalinated water. In this connection, it seems beneficial to develop a joint Canada–Russian project for floating nuclear power/ desalination complex, with the FNPS based on the KLT-40 shipboard reactor plant

FIGURE 10.8 Model of the Project 20870 (back) with a desalination unit (front)—Russian floating nuclear desalination [1,162].

and the new application of RO seawater desalination. Floating nuclear reactors have high potential due to the possibility of installation in different coastal regions of the world, high fabrication quality at a shipyard, and "turnkey" delivery to the customer in a short period of time. The supply of electrical energy to users within the plant is provided by the shipboard electric power station [1,77,138–153,155–161].

10.8 FUTURE PROSPECTS OF DESALINATION BY HYBRID RENEWABLE ENERGY SOURCES AND COGENERATION PROCESSES

It is important to note that RO and MED techniques are the leading methods in the desalination market, and extensive studies are carried out to increase efficiency and reduce investment costs for these two processes. Additionally, most desalination plants in the world use solar and wind energy (in the renewable sector). Developments of energy networks across countries would be certainly costly, and by supplying demanded energy for remote areas, the problem is multiplied. By using renewables such as solar, wind, geothermal, or other forms of local energy resources, dealing with this problem becomes solvable. It is important to mention that renewable sources have variable behaviors in each region, whereas in some areas, access to them is problematic. Therefore, utilizing a system that has been driven by an individual renewable source has a high risk in terms of cost and performance. A true combination of two or more energy systems with particular resources called a hybrid system is an authenticated option that would be suitable to face this problem.

As most of the energy generation processes entail significant amounts of water, and water requires energy for treatment and desalination, subsequently these two resources are inseparably connected. Researches trend represents that the desalination systems are improving and their progress is endless. Reducing fossil fuel reserves and raising concerns about environmental and economic aspects have increased the focus on hybrid RE. Investigating the researches and studies on desalination processes and RE systems, the future prospects for these systems include the following [1,2]:

1. Future trends consider the use of hybrid renewables in addition to solar-powered FO and dew vaporation. Upcoming outlooks of solar desalination contain shifting to other less explored desalination methods. Dew vaporation is an innovative method in solar-powered desalination. In dew vaporation process, saturated steam is used as a carrier gas to vaporize water from the saline feed as distillate. Focusing on this process and utilizing it in hybrid renewables would increase the efficiency in desalination systems [2,53].

2. Increasing the attention to FD systems can provide a new opportunity for increasing the efficiency of the desalination process. The main advantage of indirect freeze desalination is the production of refrigerant-free water, which determines the future usage of product water [163].

3. Using ocean thermocline energy for the desalination system is an innovative strategy for producing freshwater. In this method, by exploiting low-temperature differential between surface hot water and deep-sea cold water, freshwater would be produced [164].

4. At present, desalination processes are very expensive with low energy efficiency, which suggesting innovative methods that will reduce the costs would make them more affordable and economical.

5. RO is a practical and conventional method in this field. RO is the widely used method for producing freshwater and 69% of desalination plants use this system.

6. Considered strategies in the fields of construction and development are expected to reach 20,000 desalination plants until 2020. In this case, increasing the attention on the renewables (especially hybrid plants) has economic and environmental benefits.

7. In terms of energy and economic, the productivity of RE systems in coupling with RO depends on numbers of conditions, e.g., renewables sources, availability of components, and economic restrictions.

8. Batteries and energy storage devices are one of the important parameters for designing hybrid systems. These facilities have a direct impact on the economic aspects of the systems.

9. Solar energy is a vital option for hybrid systems, but wind energy has the best performance. Combination of solar and wind is most widely used because they complement each other.

10. The geothermal has a very small contribution in the desalination process. However, due to the constant temperature at a specified depth, the energy provided from this source is more stable than other renewable sources. Hybrid solar-geothermal can significantly improve the energy capacity for desalination.

11. While solar-thermal and solar-PV can individually help desalination, a hybrid solar thermal-solar PV makes the most sense.

12. The design and deployment of hybrid systems are tailored to the area's potentials, economic conditions, and technical capabilities.

13. Regardless of the techno-economic conditions of hybrid renewables systems, these types of systems are crucial solutions for providing power and freshwater for remote areas.

14. Desalination is an energy-intensive process because of the energy required to separate salts and other dissolved solids from water. In operation, the actual pressure required is approximately two times the osmotic pressure; for seawater, this translates to about 800–1,000 pounds per square inch. The energy required to run pumps that can achieve these high pressures account for approximately 25%–40% of the overall cost of water [136].

15. Wave- or tidal-powered desalination could be used to directly pressurize seawater without generating electricity for a reverse-osmosis system, eliminating one of the largest cost drivers for the production of desalinated water.

16. There are two primary market segments for desalination: water utilities and isolated or small-scale distributed systems. Large-scale desalination systems require tens of megawatts to run and provide tens of million gallons of desalinated water per day. Small-scale systems vary in size from tens to hundreds of kilowatts and provide hundreds to thousands of gallons of water per day.

17. Marine energy resources are inherently located near potential desalination water supplies and high population concentrations along the coast, therefore areas that have unreliable grid connections or water infrastructure may receive dual benefits from marine energy systems. In the long term, marine energy could provide low-cost, emission-free, drought-resistant drinking water to larger municipalities.

18. The National Renewable Energy Laboratory's (NREL's) simulation results suggest a direct pressurization application could be more cost-competitive when producing water than a wave-energy system producing electricity given current cost estimates [117]. This finding clearly signals a near-term market opportunity for wave energy, requiring smaller cost reductions than grid-power applications.

19. The connections between small modular nuclear reactors and desalination are very important. Since modular reactors are capable of generating waste heat at high temperatures, its use for desalination by cogeneration is very viable.

20. Floating nuclear reactors generating power for ship and heat and power for desalination have been very successful both in Russia and China.

10.9 HYBRID ENERGY SYSTEMS FOR WASTEWATER TREATMENT

As mentioned throughout this book, in recent years, the use of RES within the energy sector has been increasing, currently accounting for 13.7% of the global energy supply. At present, however, about 80% of all primary energy in the world is still derived from fossil fuels, namely, 31.7% from oil, 28.1% from coal, and 21.6%

from natural gas [3]. Although great amounts of energy can be obtained from fossil fuels, their use has a high environmental impact. Furthermore, climate change linked to fossil fuel combustion presents an important long-term impact on the availability and quality of water worldwide [5]. Just like for desalination, several options are also available for hybrid RE systems for wastewater treatment. These hybrid systems help decarbonize wastewater industry.

The study by Del Moral and Petrakopoulou [165] presented the simulation and analysis of a 20 MW hybrid solar/biomass power plant combined with an advanced wastewater facility for a. most demanding areas of the Iberian Peninsula, the Spanish region of Andalusia. This plant aimed to provide the area with potable water and electricity. The plant used molten salts as the working fluid and included two thermal storage tanks of a total capacity of 600 MWh for operational autonomy of up to 4 hours. The biomass (olive pomace) combustor was used as an auxiliary system, in the event of limited solar radiation or inadequate thermal energy in the storage tanks. The water treatment plant consisted of a direct potable reuse system with the objective of treating urban sewage and producing clean water for the selected region. The specific treatment train consisted of physical, biological, and chemical procedures including membrane bioreactors, ozonization, biological and granular activated carbon, RO, UV radiation, and chlorination.

The study found that the biomass combustor had the highest exergy destruction among all plant components due to the chemical reaction taking place there. Some heat exchangers, such as the feedwater preheaters, also displayed relatively low exergetic efficiencies due to the mixing of streams with great temperature differences. A sensitivity analysis was conducted to determine the feasibility of the cogeneration of electricity and water in the area. With a capacity factor of 85% and an annual operation of 7,446 hours, the hybrid solar/biomass power plant generated 148.92 GWh. Exergetic analyses were realized for two extreme cases: exclusive use of the solar block and exclusive use of the biomass system. The full-load global exergetic efficiency of the power plant was found to be 15% when the solar part is used and 34% when the biomass support system was required. The net capital investment required for the construction and implementation of the plant was found to be 211,526,000 euro. This cost included both the hybrid power plant with thermal storage and the required advanced treatment technologies used to supply the region with potable water. The levelized cost of electricity of the combined plant was found to be 0.25 EUR/kWh, a value somewhat higher than existing solar tower power plants. Lastly, a sensitivity analysis of the water reclamation plant was conducted in order to evaluate the feasibility and viability of the project. The required selling price of the generated potable water was found to be 14.61 EUR/m. Accounting for present conditions, this cost is considered relatively high. However, foreseen future water limitations are expected to drive water prices up, especially in arid regions with intense lack of water resources. In essence, the combination of RE plants with water generation processes will provide a valuable and profitable alternative for local communities to create environmentally safe facilities with continuous and stable power supply and the additional sustainable management of fresh water resources. Bustamante [166] also evaluated a solar-biomass hybrid energy generation system for self-sustainable wastewater treatment.

Soni et al. [167] examined wind/solar power hybrid system for household wastewater treatment. In this study, an adaptable, affordable, and sustainable wastewater treatment system powered by wind/solar energy was proposed based on proven theory and technology. A household in India was singled out to illustrate the workings of the proposed system, where the wastewater was recirculated through a hybrid of water purifiers powered by solar/wind energy. The system demonstrated in this study was specifically designed for small-scale applications, i.e., for a single household. The solar still was divided into four stages. Partial vacuum is created inside the still so as to obtain boiling point temperatures of 70°C, 67°C, 62°C, and 50°C in the four stages. Dhanbad, India 23.79°N, 86.43°E, with an average solar intensity of 850 W/m² for 6 hours a day, was used for this study. A lumped parameter mathematical model was developed for this study. With an aperture area of 2.5 m², the total amount of water distilled was found to be 43.3 kg/d. The system proposed is more efficient than existing systems as it was able to achieve efficiencies as high as 53%. The effect of wind speed on distillate output yield was also examined.

The wind-solar hybrid system presented by Soni et al. [167] was a self-sustaining system for pumping domestic wastewater from ground level using wind energy and making it reusable with the aid of solar energy. The system was designed in such a way so as to be installed in households without disturbing the existing plumbing systems. In most Indian households, the household water, excluding the sewage, collects at a common point so as to be disposed of to drains. The system presented in this study did not incorporate sewage waste, due to sanitary and environmental issues. This wastewater, excluding sewage, needed to be pumped from ground level to roof level where it was distilled in order to obtain pure water. Thus, there were two major tasks to be performed by the system: first, lifting water from ground level and, second, making it consumable by purifying it. Water was lifted from ground level with the help of a wind pump. If the wind did not blow for a couple of days, the storage tank may get empty. To counter this, a crank was attached to the wind driven rotor. When the wind did not blow, a hand-driven wheel achieved the rotary motion of the rotor.

The distillation system presented in the study was a combination of flat plate collector, tubular heat exchanger, and an evaporative condenser unit. The distiller consisted of multiple reservoirs created by a stacked array of distillation trays that acted as condensers for the tray below. Each stage consisted of an extruded cylindrical opening which was used to create the required partial vacuum. A valve was provided in each stage in order to regulate the pressure. There were multiple stages inside the distiller. Perfect sealing was maintained between the stages to prevent any vapor loss through the contact surfaces between the stages. Sunlight captured by the glass cover was concentrated on a black surface, heating the water in the topmost chamber and, thus, evaporating it. This vapor was then condensed to form droplets. Wastewater was fed into each stage from the tank. Each stage consisted of an evaporator and a condenser surface.

There was a pressure gradient inside the chamber. The pressure decreased in each stage in order to obtain a lower boiling point in upper chambers. Based on economic and productivity analysis, the study concluded that optimum number of stages to be four. These temperatures were used as the solar still worked best in this range [167–171]. A pump was used to create a partial vacuum inside the distillation chamber. To achieve high temperatures, heat was supplied to the wastewater in the lowest reservoir via a heat exchanger through which water was circulated from the solar collector, passed through the heat exchanger tubes, and then returned back to the solar collector again. Cold water got heated and moved to the surface so that it can evaporate faster. Vapor generated in the lower stage condenses on the bottom surface of the intermediate stage, thus giving its heat to the water in the intermediate stage. Vapor from the intermediate stage condenses on the upper stage, transferring its latent heat of condensation to the water in the upper stage. Water in the top-most reservoir, which was painted black to maximize radiation capture, was also heated directly by solar radiation. In the intermediate stages, heat transfer apart from radiation and convection occurred by evaporation and condensation, thus utilizing the latent heat of condensation and improving the system's efficiency. Radiation and convection constitute minor energy transfer between the stages and, hence, were ignored.

Reddy et al. [170] showed that the gap between stages should be 10 cm. Evacuated solar stills have already been demonstrated in the past but no one had used a pressure gradient along the stages to get a variable evaporating point of water in the stages. The suggested concept can improve the efficiency of the existing systems to a great extent. The only problem associated with the working of the still was that the plates need to be cleaned daily. To prevent algae and scaling on inner black surfaces, a still would be required to be dried completely once a week. Bleaching or chlorination can also be used to prevent algae formation. Ahmed et al. [53] experimentally investigated the multistage evacuated solar still. The productivity of this new system was found to be about threefold greater than the maximum productivity of the basin type solar still. While designing the above system, some keys points were taken into consideration: (a) simple, appropriate technology was adopted, as an overly complex system would be challenging due to its maintenance problems; (b) the system was flexible as per the demand of the people in a particular area, i.e., its cost varied as per the quantity of water required; (c) the system operated in a sustainable manner. This means being funded, owned, and operated by the individuals using the water supply; (d) the system was independent of any external power source. The basic principle used was reducing the air pressure in order to reduce the boiling point of water.

The water purification system presented above utilized two modes of RE, solar, and wind. Wind energy was used to drive a vacuum pump which reduced the air pressure inside the system. The number of stages was optimized to be four, as any further increase in the number of stages was not justifiable from an economic point of view. The vapor pressures maintained in the successive stages of the still were 31, 27, 20, and 18 kPa. Constricting nozzles were used to connect the household drain with the recirculating loop and still. Solar energy was used to heat the water lying in the chambers of the still. Solar energy was transferred to the system from the bottom chamber via a heat exchanger and from the top by direct heating of the water. The fresh water production capacity of the investigated solar-collector four-stage solar still, when operated for 6 hours a day at a constant flux of 850 W/m², was found to be 17.4 kg/m²/d at $V_w = 1$ m/s, which was greater than conventional multistage solar stills [168–171]. The annual cost of the system was approximately Rs. 7450, with the per unit water cost in the range of 0.5–1.2 Rs/kg for the wind speed range of 1–5 m/s.

Water evaporated in four different stages, each separated by a distance of 10 cm. In the absence of wind, a hand-driven wheel can be used to drive the reciprocating pump to propel the water from ground to roof level. The suggested multistage solar desalination system can meet the fresh water needs of rural and urban communities by distilling 25–45 kg/d, considering wind speed is in the range of 1–5 m/s.

Gandiglio et al. [172] examined enhancement of energy efficiency of wastewater treatment plants (WWTPs) through codigestion and FC systems. The study provided an overview of technological measures to increase the self-sufficiency of WWTPs, in particular, for the largely diffused activated sludge (AS)-based WWTP. The operation of WWTPs entails a huge amount of electricity. Thermal energy is also required for preheating the sludge and sometimes exsiccation of the digested sludge. On the other hand, the entering organic matter contained in the wastewater is a source of energy. Organic matter is recovered as sludge, which is digested in large stirred tanks (anaerobic digester) to produce biogas. The onsite availability of biogas represents a great opportunity to cover a significant share of WWTP electricity and thermal demands. Especially, biogas can be efficiently converted into electrical energy (and heat) via high-temperature FC generators. The final part of this study reported a case study based on the use of sewage biogas into a solid oxide FC. However, the efficient biogas conversion in CHP devices was not sufficient. Self-sufficiency required a combination of efficient biogas conversion, the maximization the yield of biogas from the organic substrate, and the minimization of the thermal duty connected to the preheating of the sludge feeding the anaerobic digester (generally achieved with prethickeners). Finally, the codigestion of the organic fraction of municipal solid waste into digesters treating sludge from WWTPs represents an additional opportunity for increasing the biogas production of existing WWTPs, thus helping the transition toward self-sufficient plants.

Maktabifard et al. [173] examined methods for achieving energy neutrality in WWTPs through energy savings and enhancing RE production. WWTPs consume high amounts of energy which is mostly purchased from the grid. During the past years, many ongoing measures have taken place to analyze the possible solutions for both reducing the energy consumption and increasing the RE production in the plants. This review contained all possible aspects which may assist to move toward energy neutrality in WWTPs. The sources of energy in wastewater were introduced and different indicators to express the energy consumption were discussed with examples of the operating WWTPs worldwide. Furthermore, the pathways for energy consumption reductions were reviewed including the operational strategies and the novel technological upgrades of the wastewater treatment processes. Then, the methods of recovering the potential energy hidden in wastewater were described along with application of renewable energies in WWTPs. The available assessment methods, which may help in analyzing and comparing WWTPs in terms of energy and GHG emissions, were introduced. Eventually, successful case studies on energy self-sufficiency of WWTPs were listed and the innovative projects in this area were presented.

Finally, Tee et al. [174] outlined various hybrid wastewater treatment processes that can be effective pollutant removal and simultaneously generate bioenergy. The study classified hybrid wastewater systems which typically include physical–biological hybrid, physical–chemical hybrid, chemical–biological hybrid, and physical–chemical–biological hybrid system. The study showed that based on the literature, hybrid systems have demonstrated some potential advantages compared to stand-alone systems such as more stable and sustainable in the voltage generated, better overall treatment efficiency, and energy savings all leading to decarbonization of wastewater industry. The study showed that hybrid treatment system can be a great choice of treatment options in wastewater in terms of effectiveness. Nevertheless, the overall cost of the hybrid system has to be taken into consideration in terms of capital costs, the operating costs, and maintenance costs [175]. Most costs are very site specific, and for a full-scale system, these costs strongly depend on the flow rate of the effluent, the configuration of the reactor, the nature (concentration) of the effluent as well as the pursued extent of treatment. Hybrid system with a combination of physical treatment such as membrane may pose a challenge because of the high operating cost in terms of energy consumption. If the hybrid system is not designed in a way to have positive energy gained in the overall system, membrane-based hybrid system may not be worth to invest. Besides, membrane technologies may require high maintenance cost depending on fouling frequency and the particular application of the membrane-based hybrid system. Frequent membrane replacement can be very costly.

The physical–chemical precipitation is simple to implement, reliable, and efficient but presents several disadvantages such as increased operating costs due to the consumption of chemical reagents and corrosiveness of some of the coagulants which may lead to other problems [176]. Besides, excess sludge production through coagulation, occultation, and precipitation may lead to problem in terms of disposal unless the sludge produced is able to be recycled for other purposes. Some of the biological process combinations such as AS process may require large amount of oxygen supply for the biological process. Such hybrid system may require high operating cost which makes the overall hybrid system not worth to be invested. There must be a balance made between the energy consumed with the energy gain from the hybrid system in order to make the overall system worthwhile.

10.10   FUTURE ROLE OF MFC FOR WASTEWATER TREATMENT

Some of the prior studies on hybrid schemes of FC systems by Abdullah et al. had demonstrated the feasibility and superiority of hybrid systems compared to stand-alone systems for various applications other than effluent or waste water treatment [177,178]. Some of the notable advantages are (a) more stable and sustainable voltage generated, (b) better overall treatment efficiency, and (c) energy saving potential [179]. MFCs are capable of recovering the potential energy present in wastewater and converting it directly into electricity [180]. Using MFCs may help offset wastewater treatment operating costs and make advanced wastewater treatment more affordable for both developing and industrialized nations [181].

Conversion of wastes into bioelectricity by using microbial electrochemical technologies is predicted to be the future trend of the hybrid system. MFC technologies represent the newest approach for generating electricity-bioelectricity generation from biomass using bacteria. Currently, most of the MFCs are done in lab-scale and people are moving toward the pilot-scale MFCs where lots of efforts have to be put in to make it happen. It is predicted that MFC hybrid system would replace the AS or trickling filter (TF) system [174]. The MFC is a chemical–biological treatment process, and thus, such hybrid treatment is able to remove organic contaminants in the same manner as accomplished by the AS aeration tank or the TF. The adoption of MFC into hybrid treatment system has several advantages which are listed below:

1. Bioenergy generation together with pollutant removal, where the current generated is dependent on the wastewater strength and the coulombic efficiency.
2. Elimination of aeration unit. For air-cathode MFC system, no aeration is required. Aeration system in AS is very costly where such system can consume 50% of the electricity used at a treatment plant.
3. Reduction of solids production. As compared to the aerobic system such as TF and AS, bacterial biomass production by MFC is much lower due to the anaerobic condition of MFC.
4. Minimize odor problem in the treatment system. MFC is a closed system where odor will not be a major problem in the overall treatment system.

There are four possible treatment process flows that can be envisaged in the near future. First, it is expected that the MFC process could be integrated into the process flow of a conventional system replacing the AS or TF systems. In this case, the MFC would be used in a manner similar to that of a TF in a TF/solids contact arrangement. Second, MFC can be a pretreatment unit for a membrane bioreactor (MBR) process. Tee et al. [174] predicts that MFC-adsorption hybrid system can be a great hybrid treatment system in the near future due to the great ability of the adsorption column in removing various types of pollutants in wastewater. MFC-adsorption hybrid system can be presented in the form of either ex-situ or in-situ system configurations.

On the other hand, energy production from salinity gradient can be a popular trend in the near future. Energy can be produced from a reverse electrodialysis (RED) due to the salinity gradient [174]. It is reported that the electrical potential of 0.1–0.2 V per pair of membrane can be produced from seawater and freshwater (or treated wastewater) through pairs of ion-exchange membranes in a RED. Globally, up to 980 GW of power could be generated from salinity gradient energy where freshwater flows into the sea [174]. A RED stack can be placed between the anode and cathode chambers of an MFC or microbial electrolysis cell, creating a hybrid technology called a microbial reverse electrodialysis cell (MRC).

Generally, hybrid systems are more stable and sustainable in terms of voltage generation and treatment efficiency as compared to stand-alone system. Bioenergy generated can help to offset the treatment operating costs of the overall system. In terms of energy balance, bioenergy generated from the hybrid system must be at least equal or greater than the energy used to operate the overall system. Conversion of wastes into bioelectricity by using microbial electrochemical technologies is predicted to be the future trend of the hybrid system. As mentioned above, MFC could be replacing the AS or TF systems or even have it as a pretreatment process for MBR. Energy production from salinity gradient such as RED can be another popular trend in the near future. Yet, integration of MFC with adsorption column can be an interesting option. Tee et al. [174] examines the details of all of these options.

REFERENCES

1. Shah, Y.T. (2020). *Modular Systems for Energy Usage Management.* CRC Press, New York.
2. Esmaeilion, F. (2020). Hybrid renewable energy systems for desalination. *Applied Water Science* 10: 84, Doi: 10.1007/s13201-020-1168-5.
3. Kucera, J. (2019) *Desalination: Water from Water.* Wiley, New York.
4. Shatat, M., Worall, M. and Riffat, S. (2013). Opportunities for solar water desalination worldwide. *Sustainable Cities and Society* 9: 67–80.
5. Shatat, M., Worall, M. and Riffat, S. (2013). Economic study for an affordable small scale solar water desalination system in remote and semi-arid region. *Renewable and Sustainable Energy Reviews* 25: 543–551.
6. Subramani, A. and Jacangelo, J.G. (2015). Emerging desalination technologies for water treatment: A critical review. *Water Research* 75: 164–187.
7. Guthrie, P. (2010). *Global Water Security – An Engineering Perspective.* The Royal Academy of Engineering, London.
8. Hanjra, M.A. and Qureshi, M.E. 2010. Global water crisis and future food security in an era of climate change. *Food Policy* 35(5): 365–377.
9. Jury, W.A. and Vaux, H.J. (2007). The emerging global water crisis: Managing scarcity and conflict between water users. *Advances in Agronomy* 95: 1–76.
10. Yang, H. and Abbaspour, K.C. (2007). Analysis of wastewater reuse potential in Beijing. *Desalination* 212(1): 238–250.
11. Tchobanoglous, G., Burton, F.L. and Stensel, H.D. (2003). *Wastewater Engineering Treatment and Reuse,* 4th ed. Metcalf and Eddy, McGraw-Hill, New York.
12. Munia, H., Guillaume, J.H.A., Mirumachi, N., Porkka, M., Wada, Y. and Kummu, M. (2016). Water stress in global transboundary river basins: Significance of upstream water use on downstream stress. *Environmental Research Letters* 11(1): 14002.
13. Falkenmark, M., Berntell, A., Jägerskog, A., Lundqvist, J., Matz, M. and Tropp, H. (2007). *On the Verge of a New Water Scarcity: A Call for Good Governance and Human Ingenuity.* Stockholm International Water Institute (SIWI), Stockholm.
14. Economic and Social Commission for Western Asia. (2001). *Energy Options for Water Desalination in Selected ESCWA Member Countries.* United Nations, New York.
15. CORDIS Database. (2006). CORDIS - EU research projects under FP6 (2002-2006). EU Open data portal published by European parliament, commission and council of the European Union.
16. Desalination. (2019, April). *A Report by Office of Energy Efficiency and Renewable Energy.* Department of Energy, Washington, DC.
17. Global Water Desalination Market. (2018, August 28). 2018–2025 current trends. A website report www.reuters.com/brandfeatures/venture-capital/article?id=48863.
18. Seawater Desalination Costs, by Water Reuse Association, White Paper by Water Reuse Association. (2011, September). Revised January 2012, Seawater Desalination Costs – WateReuse. A website report https://watereuse.org/wp-content/uploads/.../WateReuse_ Desal_Cost_White_Paper.pdf (2012).
19. Walton, M. (2019, January 21). *Commentary: Desalinated Water Affects the Energy Equation in the Middle East, WEO Energy Analyst.* IEA Paris, France. www.iea.org/.../ desalinated-water-affects-the-energy-equation-in-the-middle-ea.
20. Shahzad, M.W., Burhan, M., Ybyraiymkul, D. and Ng, K.C. (2019). Desalination processes' efficiency and future roadmap. *Entropy* 21: 84, Doi: 10.3390/e21010084, www. mdpi.com/journal/entropy.
21. Krishna, H.J. (2004). *Introduction to Desalination Technologies: Texas Water Development.* www.twdb.texas.gov/publications/reports/numbered_reports/doc/.../ C1.pdf, a website report

22. Desalination. (2019). Wikipedia, The free encyclopedia, last visited 23 July 2019.

23. Desalination Technology: An Overview | ScienceDirect Topics. (2016). A website report www. sciencedirect.com/topics/earth-and-planetary.../desalination-technology.

24. Ghalavand, Y., Hatamipour, M.S. and Rahimi, A. (2014). A review on energy consumption of desalination processes. *Desalination and Water Treatment*. Doi: 10.1080/19443994.2014.892837.

25. Lozier, J.C. (2011). Desalination technology overview. A website report by *Water Resources Research Center Conference*, Yuma, AZ. https://wrrc.arizona.edu/sites/wrrc.arizona.edu/files/programs/conf2011/.../Lozier.pdf.

26. Chaudhry, S. (2013). An overview of industrial desalination technologies ASME Industrial Demineralization (Desalination). *Best Practices and Future Directions Workshop*, January 28–29, Washington, DC. https://community.asme.org/.../An-Overview-of-Industrial-Desalination-Technologies, a website report.

27. Al-Karaghouli, A.A. and Kazmerski, L.L. (2011). *Renewable Energy Opportunities in Water Desalination*. National Renewable Energy Laboratory, Golden, CO.

28. Shahzad, M., Ybyraiymkul, D., Burhan, M. and Ng, K. (2018, September) Renewable energy driven desalination hybrids for sustainability. Open access Intech paper. Doi: 10.57772/Intechopen.77019.

29. Hamed, O.A. (2005). Overview of hybrid desalination systems: Current status and future prospects. *Desalination* 186(1–3): 207–214; *Presented at the International Conference on Water Resources and Arid Environment*, 5–8 December 2004, Riyadh, Doi: 10.1016/j.desal.2005.03.095.

30. Fath, H., Sadik, A. and Mezher, T. (2013). Present and future trend in the production and energy consumption of desalinated water in GCC countries. *International Journal of Thermal & Environmental Engineering* 5(2): 155–165.

31. Shahzad, M.W., Burhan, M., Li, A. and Ng, K.C. (2017). Energy-water-environment nexus underpinning future desalination sustainability. *Desalination* 413: 52–64.

32. Shahzad, M.W., Burhan, M., Son, H.S., Seung Jin, O. and Ng, K.C. (2018). Desalination processes evaluation of common platform: A universal performance ratio (UPR) method. *Applied Thermal Engineering* 134: 62–67.

33. Padrón, I., Avila, D., Marichal, G. and Rodríguez, J. (2019). Assessment of hybrid renewable energy systems to supplied energy to autonomous desalination systems in two islands of the canary archipelago. *Renewable and Sustainable Energy Reviews* 101: 221–230, Doi: 10.1016/j.rser.2018.11.009.

34. Griffiths, S. (2017). Renewable energy policy trends and recommendations for GCC countries. *Energy Transitions* 1 (3): 1–15.

35. Renewable Energy Market Analysis. (2016). The GCC Region, The International Renewable Energy Agency (IRENA) report. ISBN: 978-92-95111-81-3.

36. Ali, E., Orfi, J., Najib, A. and Saleh, J. (2018). Enhancement of brackish water desalination using hybrid membrane distillation and reverse osmosis systems. *PLoS One* 13(10): e0205012, Doi: 10.1371/journal.pone.0205012.

37. Khan, M.A.M., Rehman, S. and Al-Sulaiman, F.A. (2018). A hybrid renewable energy system as a potential energy source for water desalination using reverse osmosis: A review. *Renewable and Sustainable Energy Reviews*, Elsevier, 97(C): 456–477.

38. Al-Hayek, I. and Badran, O.O. (2004). The effect of using different designs of solar stills on water distillation. *Desalination* 169: 121–127.

39. Economic and Social Commission for Western Asia. (2009). *ESCWA Water Development Report 3. Role of Desalination in Addressing Water Scarcity*. United Nations, New York.

40. Sodha, M., Kumar, A., Tiwari, G. and Tyagi, R. (1981). Simple multiple wick solar still: Analysis and performance. *Solar Energy* 26: 127–131.

41. Fernandez, J.L. and Chargoy, N. (1990). Multi-stage, indirectly heated solar still. *Solar Energy* 44(4): 215–223.
42. Al-Hinai, H., Al-Nassri, M.S. and Jubran, B.A. (2002). Effect of climatic, design and operational parameters on the yield of a simple solar still. *Energy Conversion and Management* 43: 1639–1650.
43. Mathioulakis, E., Belessiotis, V. and Delyannis, E. (2007). Desalination by using alternative energy: Review and state-of-the-art. *Desalination* 203: 346–365.
44. Qiblawey, H.M. and Banat, F. (2008). Solar thermal desalination technologies. *Desalination* 220: 633–644.
45. Price, H. (2003). *Assessment of Parabolic Trough and Power Tower Solar Technology Cost and Performance Forecasts.* Sargent & Lundy LLC Consulting Group, National Renewable Energy Laboratory, Golden, CO. www.nrel.gov/solar/parabolic_trough. html. NREL/SR-550-34440 under contract no. DE-AC36-99-GO10337, NREL, Golden, CO.
46. Trieb, F. and Müller-Steinhagen, H. (2007). Concentrating solar power for seawater desalination in the Middle East and North Africa. Submitted to Desalination.
47. Al-Shammiri, M. and Safar, M. (1999). Multi-effect distillation plants: State of the art. *Desalination* 126: 45–59.
48. Reif, J.H. and Alhalabi, W. (2015) Solar-thermal powered desalination: Its significant challenges and potential. *Renewable and Sustainable Energy Reviews* 48: 152–165.
49. Quteishat, K. and Abu-Arabi, M. (2006). *Promotion of Solar Desalination in the MENA Region.* Middle East Desalination Centre, Muscat, Oman. www.menarec.com/ docs/Abu-Arabi.pdf. Accessed 28 March 2006, a website report.
50. Collares Pereira, M., Carvalho, M.J. and Correia de Oliveira, J. (2003). A new low concentration CPC type collector with convection controlled by a honeycomb TIM Material: A compromise with stagnation temperature control and survival of cheap fabrication materials. *Proceedings of the ISES Solar World Congress 2003*, 14–19 June, Göteborg.
51. El Saliby, I., Okour, Y., Shon, H.K., Kandasamy, J. and Kim, S. (2009 October). Desalination plants in Australia, review and facts, *Desalination*, 247(1–3): 1–14.
52. Zarza, E. and Blanco, M. (1996). Advanced M.E.D. solar desalination plant: Seven years of experience at the Plataforma Solar de Almería. *Proceedings of the Mediterranean Conference on Renewable Energy Sources for Water Production*, Santorini, 182.
53. Ahmed, F.E., Hashaikeh, R. and Hilal, N. (2019). Solar powered desalination–Technology, energy and future outlook. *Desalination* 453: 54–76.
54. Yılmaz, I.H. and Söylemez, M.S. (2012). Design and computer simulation on multi-effect evaporation seawater desalination system using hybrid renewable energy sources in Turkey. *Desalination* 291: 23–40.
55. Thomson, M., Gwillim, J., Rowbottom, A., Draisey, I. and Miranda, M. (2001). Battery less photovoltaic reverse osmosis desalination system. Technical Report, S/P2/00305/ REP, ETSU, DTI, UK.
56. Fiorenza, G., Sharma, V. and Braccio, G. (2003). Technoeconomic evaluation of a solar powered water desalination plant. *Energy Conversion and Management* 44: 2217–2240.
57. Finken, P. and Korupp, K. (1991). Water desalination plants powered by wind generators and photovoltaic systems. *Proceedings of the New Technologies for the Use of Renewable Energy Sources in Water Desalination Conference, Session II,* Athens, 65–98.
58. Adiga, M.R., Adhikary, S.K., Narayanan, P.K., Harkare, W.P., Gomkale, S.D. and Govindan, K.P. (1987). Performance analysis of photovoltaic electrodialysis desalination plant at Tanote in the Thar desert. *Desalination* 67: 59–66.
59. Kuroda, O., Takahashi, S., Kubota, S., et al. (1987). An electrodialysis seawater desalination system powered by photovoltaic cells. *Desalination* 67: 161–169.

60. Šály, V., Ružinský, M. and Baratka, S. (2006) Photovoltaics in Slovakia—status and conditions for development within integrating Europe. *Renewable Energy* 31(6): 865–875.
61. Pietruszko, S.M. and Gradzki, M. (2003). Performance of grid connected small PV system in Poland. *Applied Energy* 74(74): 177–184.
62. Kumar, S. and Tiwari, G.N. (2009). Estimation of internal heat transfer coefficients of a hybrid (PV/T active solar still. *Solar Energy* 83: 1656–1667.
63. Kumar, S. and Tiwari, G.N. (2009). Life cycle cost analysis of single slope hybrid (PV/T) active solar still. *Applied Energy* 86: 1995–2004.
64. Gaur, M.K. and Tiwari, G.N. (2010). Optimization of number of collectors for integrated PV/T hybrid active solar still. *Applied Energy* 87: 1763–1772.
65. Mittelman, G., Kribus, A., Mouchtar, O. and Dayan, A. (2009). Water desalination with concentrating photovoltaic/thermal (CPVT) systems. *Solar Energy* 83: 1322–1334, Doi: 10.1016/j.solener.2009.04.003.
66. Kalogirou, S.A. (2005) Seawater desalination using renewable energy sources. *Progress in Energy and Combustion Science* 31(3): 242–281.
67. Goosen, M., Mahmoudi, H. and Ghaffour, N. (2010) Water desalination using geothermal energy. *Energies* 3(8): 1423–1442.
68. Paulsen, K. and Hensel, F. (2005). Introduction of a new energy recovery system—optimized for the combination with renewable energy. *Desalination* 184(1–3): 211–215.
69. Ghaffour, N., Lattemann, S., Missimer, T., Ng, K.C., Sinha, S. and Amy, G. (2014) Renewable energy-driven innovative energy-efficient desalination technologies. *Applied Energy* 136: 1155–1165.
70. Segurado, R., Costa. M., Duić, N. and Carvalho, M.G. (2015). Integrated analysis of energy and water supply in islands. Case study of S. Vicente, Cape Verde. *Energy* 92: 639–648.
71. Park, G.L., Schäfer, A.I. and Richards, B.S. (2009). Potential of wind-powered renewable energy membrane systems for Ghana. *Desalination* 248(1–3): 169–176.
72. Garcia-Rodriguez, L., Romero-Ternero, V. and Gomez-Camacho, C. (2001). Economic analysis of wind-powered desalination. *Desalination* 137(1–3): 259–265.
73. Peñate, B., Castellano, F., Bello, A. and García-Rodríguez, L. (2011). Assessment of a stand-alone gradual capacity reverse osmosis desalination plant to adapt to wind power availability: A case study. *Energy* 36(7): 4372–4384.
74. Gökçek, M. and Gökçek, Ö.B. (2016). Technical and economic evaluation of freshwater production from a wind-powered small-scale seawater reverse osmosis system (WP-SWRO). *Desalination* 381: 47–57.
75. Ismail, T.M., Azab, A.K., Elkady, M.A. and Elnasr, M.M.A. (2016). Theoretical investigation of the performance of integrated seawater desalination plant utilizing renewable energy. *Energy Conversion and Management* 126: 811–825.
76. Soshinskaya, M., Crijns-Graus, W.H.J., van der Meer, J. and Guerrero, J.M. (2014). Application of a microgrid with renewables for a water treatment plant. *Applied Energy* 134: 20–34.
77. Mentis, D., Karalis, G., Zervos, A. et al. (2016). Desalination using renewable energy sources on the arid islands of South Aegean Sea. *Energy* 94: 262–272.
78. Moh'd A.A.-N., Kiwan, S.M. and Talafha, S. (2016). Hybrid solar-wind water distillation system. *Desalination* 395: 33–40.
79. Khan, S.U.-D., Khan, S.U.-D., Haider, S. et al. (2017). Development and techno-economic analysis of small modular nuclear reactor and desalination system across Middle East and North Africa region. *Desalination* 406: 51–59.
80. Smaoui, M., Abdelkafi, A. and Krichen, L. (2015) Optimal sizing of standalone photovoltaic/wind/hydrogen hybrid system supplying a desalination unit. *Solar Energy* 120: 263–276.

81. Henderson, C.R., Manwell, J.F. and McGowan, J.G. (2009). A wind/diesel hybrid system with desalination for Star Island, NH: Feasibility study results. *Desalination* 237(1–3): 318–329.
82. Nagaraj, R., Thirugnanamurthy, D., Rajput, M.M. and Panigrahi, B.K. (2016). Techno-economic analysis of hybrid power system sizing applied to small desalination plants for sustainable operation. *International Journal of Sustainable Built Environment* 5(2): 269–276.
83. Ackermann, T. and Soder, L. (2002). An overview of wind energy-status 2002. *Renewable and Sustainable Energy Reviews* 6: 67–128.
84. Carta, J.A., Gonzalez, J. and Subiela, V. (2004). The SDAWES project: An ambitious R&D prototype for wind powered desalination. *Desalination* 161: 33–48.
85. Maurel, A. (1991). Desalination by RO using RE (solar and wind): Cadarache Center Experience. *Proceedings of the New Technologies for the Use of RE Sources in Water Desalination*, 26–28 September, Greece, 17–26.
86. Peral, A., Contreras, G.A. and Navarro, T. (1991). IDM—project: Results of one year's operation. *Proceedings of the New Technologies for the Use of RE Sources in Water Desalination*, 26–28 September, Greece, 56–80.
87. De La Nuez Pestana, I., Francisco Javier, G., Celso Argudo, E. and Antonio Gomez, G. (2004). Optimization of RO desalination systems powered by RE: Part I. Wind energy. *Desalination* 160: 293299.
88. Miranda, M. and Infield, D. (2002). A wind powered seawater RO system without batteries. *Desalination* 153: 9–16.
89. Rahal, Z. and Infield, D.G. (1997, October). Wind powered stand alone desalination. *Proceedings of the European Wind Energy Conference*, Dublin. Ireland. Available at: http://info.lboro.ac.uk/departments/el/research/crest/publictn.html. ISBN 0953392201, 9780953392209
90. Rahal, Z. and Infield, D.G. (1997, August). Computer modelling of a large scale stand alone wind-powered desalination plant. *Proceedings of the British Wind Energy Conference*, Stirling. Available at: http://info.lboro.ac.uk/departments/el/research/crest/publictn.html.
91. ENERCON. (2006). Desalination units. www.enercon.de.
92. Vujcic, R. and Krneta, M. (2000). Wind-driven seawater desalination plant for agricultural development on the islands of the County of Split and Dalmatia. *Renewable Energy* 19: 173–183.
93. Stahl, M. (1991). Small wind powered RO seawater desalination plant design, erection and operation experience. *Seminar on New Technologies for the Use of Renewable Energies in Water Desalination*, 26–28 September, Commission of the European Communities, DG XVII for Energy, CRES (Centre for Renewable Energy Sources), Athens.
94. Kostopoulos, C. (1996). *Proceedings of the Mediterranean Conference on Renewable Energy Sources for Water Production*. European Commission, EURORED Network, 10–12 June, CRES, EDS, Santorini, 20–25.
95. Infield, D. (1997). Performance analysis of a small wind powered reverse osmosis plant. *Solar Energy* 61(6): 415–421.
96. Manwell, J.F. and McGowan, J.G. (1994). Recent renewable energy driven desalination system research and development in North America. *Desalination* 94(3): 229–241.
97. Valdez Salas, B. and Schorr Wiener, M. (2012). *Desalination, Trends and Technologies*. Taylor & Francis, Abingdon.
98. Barbier, E. (2002). Geothermal energy technology and current status: An overview. *Renewable and Sustainable Energy Reviews* 6: 3–65.
99. Barbier, E. (1997). Nature and technology of geothermal energy. *Renewable and Sustainable Energy Reviews* 1(1–2): 1–69.

100. Kalogirou, S.A. (2005). Seawater desalination using renewable energy sources. *Progress in Energy and Combustion Science* 31(3): 242–281.

101. Gude, V.G., Nirmalakhandan, N. and Deng, S. (2010) Renewable and sustainable approaches for desalination. *Renewable and Sustainable Energy Reviews* 14(9): 2641–2654.

102. Awerbuch, L., Lindemuth, T.E., May, S.C. and Rogers, A.N. (1976). Geothermal energy recovery process. *Desalination* 19: 325–336.

103. Boegli, W.J., Suemoto, S.H. and Trompeter, K.M. (1977). Geothermal desalting at the East Mesa test site. (Experimental results of vertical tube evaporator, MSF and high-temperature ED. Data about fouling, heat transfer coefficients, scaling.) *Desalination* 22: 77–90.

104. Ophir, A. (1982). Desalination plant using low grade geothermal heat. *Desalination* 40: 125–132.

105. Karytsas, K., Alexandrou, V. and Boukis, I. (2002). The Kimolos geothermal desalination project. *Proceeding of International Workshop on Possibilities of Geothermal Energy Development in the Aegean Islands Region*, Milos Island, Greece, 206–219.

106. Bourouni, K., Martin, R. and Tadrist, L. (1997). Experimental investigation of evaporation performances of a desalination prototype using the aero-evapo condensation process. *Desalination* 114: 111–128.

107. Bouchekima, B. (2003). Renewable energy for desalination: A solar desalination plant for domestic water needs in arid areas of South Algeria. *Desalination* 153: 65–69.

108. Belessiotis, V. and Delyannis, E. (2010). Renewable energy resources, in Encyclopedia of Life Support Systems (EOLSS). *Desalination: Desalination with Renewable Energies*. Available at: www.desware.net/DeswareLogin/LoginForm.Aspx. ISBN: 978-1-84826-429-8 (eBook). ISBN: 978-1-84826-879-1 (Print Volume).

109. Bourouni, K. and Chaibi, M.T. (2005) Application of geothermal energy for brackish water desalination in the south of Tunisia. *Ground Water* 2185(290): 225.

110. Sarbatly, R. and Chiam, C.-K. (2013). Evaluation of geothermal energy in desalination by vacuum membrane distillation. *Applied Energy* 112: 737–746.

111. Tomaszewska, B. and Bodzek, M. (2013). Desalination of geothermal waters using a hybrid UF-RO process. Part I: Boron removal in pilotscale tests. *Desalination* 319: 99–106.

112. Turchi, C., Akar, S., Cath, T., Vanneste, J., Gustafson, E. and Akerley, J. (2017). *Desalination of Impaired Water Using Geothermal Energy*. National Renewable Energy Lab, Golden, CO.

113. Goosen, M., Mahmoudi, H. and Ghaffour, N. (2010). Water desalination using geothermal energy. *Energies* 3(8): 1423–1442.

114. Manenti, F., Masi, M., Santucci, G. and Manenti, G. (2013). Parametric simulation and economic assessment of a heat integrated geothermal desalination plant. *Desalination* 317: 193–205.

115. Calise, F., Cipollina. A., d'Accadia, M.D. and Piacentino, A. (2014). A novel renewable polygeneration system for a small Mediterranean volcanic island for the combined production of energy and water: Dynamic simulation and economic assessment. *Applied Energy* 135: 675–693.

116. Missimer, T.M., Kim, Y.-D., Rachman, R. and Ng, K.C. (2013). Sustainable renewable energy seawater desalination using combined-cycle solar and geothermal heat sources. *Desalination and Water Treatment* 51(4–6): 1161–1170.

117. Yu, Y.-H. and Jenne, D. Analysis of a wave-powered, reverse-osmosis system and its economic availability in the United States, *National Renewable Energy Laboratory*, Conference Paper, NREL/CP-5000-67973 August 2017 Contract No. DEAC36-08GO28308

118. Sharmila, N., Jalihal, P., Swamy, A.K. and Ravindran, M. (2004). Wave powered desalination system. *Energy* 29(11): 1659–1672.

119. Burn, S., Hoang, M., Zarzo, D., et al. (2015). Desalination techniques—a review of the opportunities for desalination in agriculture. *Desalination* 364: 2–16.
120. Corsini, A., Tortora, E. and Cima, E. (2015). Preliminary assessment of wave energy use in an off-grid minor island desalination plant. *Energy Procedia* 82: 789–796.
121. Davies, P.A. (2005). Wave-powered desalination: Resource assessment and review of technology. *Desalination* 186(1–3): 97–109.
122. Falcão, A.F.O. (2014). *Modelling of Wave Energy Conversion.* Universidade Técnica de Lisboa, Instituto Superior Técnico, Lisbon.
123. Foteinis, S. and Tsoutsos, T. (2017). Strategies to improve sustainability and offset the initial high capital expenditure of wave energy converters (WECs). *Renewable and Sustainable Energy Reviews* 70: 775–785.
124. Viola, A, Franzitta, V., Trapanese, M., Curto, D. and Viola, D. (2016) Nexus water & energy: A case study of wave energy converters (WECs) to desalination applications in Sicily. *International Journal of Heat and Technology* 34(2): S379–S386.
125. Alkaisi, A., Mossad, R. and Sharifian-Barforoush, A. (2017). A review of the water desalination systems integrated with renewable energy. *Energy Procedia* 110: 268–274.
126. Franzitta, V., Curto, D., Milone, D. and Viola, A. (2016). The desalination process driven by wave energy: A challenge for the future. *Energies* 9(12): 1032.
127. Salter, S.H., Cruz, J.M.B.P., Lucas, J.A.A. and Pascal, R.C.R. (2010). *Wave powered desalination. Macro-engineering Seawater in Unique Environments.* Springer, Berlin, 657–674.
128. Dashtpour, R. and Al-Zubaidy, S.N. (2012). Energy efficient reverse osmosis desalination process. *International Journal of Environmental Science and Development* 3(4): 339.
129. Murtha, R., McCormick, M.E. and Washington, M.K. (2016). Modular sand filtration-anchor system and wave energy water desalination system and methods of using potable water produced by wave energy desalination. United States Patent Application 20160236950, Application Number: 15/023791.
130. Franzitta, V., Curto, D., Milone, D. and Viola, A. (2016). The desalination process driven by wave energy, a challenge for the future. *Energies* 9: 1032, Doi: 10.3390/en9121032, www.mdpi.com/journal/energies.
131. NikWB, W., Olanrewaju, S., Rosliza, R., Prawoto, Y. and Muzathik, A. (2011). Wave energy resource assessment and review of the technologies. *International Journal of Energy and Environment* 2: 1101–1112.
132. Lovo, R. (2001, May). Initial evaluation of the subfloor water intake structure system (SWISS) vs. conventional multimedia pretreatment techniques. Assistance Agreement No. 98-FC-81-0044. Desalination Research and Development Program Report No. 66, U.S. Department of Interior.
133. McCormick, M. (1981). *Ocean Wave Energy Conversion.* Wiley-Interscience, New York. Reprinted by Dover Publication, Long Island, New York in 2007.
134. Murtha, R., McCormick, M.E. and Washington, M.K. (2014). Modular sand filtration: Anchor system and wave energy water desalination system incorporating the same. Assignee Murtech Inc. Publication of US20140158624A1, 2014-07-15; Publication of US8778176B2 2019-07-25.
135. Armistead, T. (2017, July 12). Wave energy desalination is small, modular and cheap. A website report by Engineering News Record. www.enr.com/.../42354-wave-energy-desalination-is-small-modular-and-cheap.
136. Lantz, E., Olis, D. and Warren, A. (2011). *U.S. Virgin Islands Energy Road Map: Analysis* (Technical Report). NREL/TP-7A20-52360. National Renewable Energy Laboratory, Golden, CO. https://www.nrel.gov/docs/fy11osti/52360.pdf.
137. Lantz, E., Hand, M. and Wiser, R. (2012). *The Past and Future Cost of Wind Energy: Preprint.* NREL/CP-6A20-54526. National Renewable Energy Laboratory, Golden, CO. https://www.nrel.gov/docs/fy12osti/54526.pdf. August 2012 under Contract No. DE-AC36-08GO28308, NREL, Golden, CO.

138. IAEA. (2007). Economics of nuclear desalination: New developments and site specific studies, Vienna, IAEA-TECDOC-1561. ISBN 978-92-0-105607-8, ISSN 1011-4289, www-pub.iaea.org/MTCD/Publications/PDF/te_1561_web.pdf.

139. Nuclear desalination: World Nuclear Association. (2019). A website report www.world-nuclear.org/information.../non...nuclear.../nuclear-desalination.asp.

140. Ahmed, S.A., Hani, H.A., Al Bazedi, G.A., El-Sayed, M.M.H. and Abulnour, A.M.G. (2014). Small/medium nuclear reactors for potential desalination applications: Mini review. *Korean Journal of Chemical Engineering* 31(6): 924–929. Doi: 10.1007/s11814-014-0079-2.

141. Jones, E., Qadir, M., van Vliet, M.T.H., Smakhtin, V. and Kang, S. (2019). Review, the state of desalination and brine production: A global outlook. *Science of the Total Environment* 657: 1343–1356, Doi: 10.1016/j.scitotenv.2018.12.076.

142. IAEA. (1997). Nuclear desalination of sea water. *Symposium Proceedings*, Vienna.

143. International Atomic Energy Agency, (1998). Nuclear Heat Applications: Design Aspects and Operating Experience, IAEA-TECDOC-1056, IAEA, Vienna.

144. Konishi, T. and Misra, B.M. (2001). Freshwater from the Seas. *IAEA Bulletin* 43. Iaea.org.

145. Russian Floating Nuclear Power Station. (2019). Wikipedia, The free encyclopedia, last visited 11 August 2019.

146. KLT-40 Reactor. (2019). Wikipedia, The free encyclopedia, last visited 4 December 2017.

147. Spadaro, J., Langlois, L. and Hamilton, B. (2000). Greenhouse gas emissions of electricity generation chains: Assessing the difference. *IAEA Bulletin*, Vienna 42(2): 19–24.

148. Tewari, P.K. and Rao, I.S. (2002). LTE desalination utilizing waste heat from a nuclear research reactor. *Desalination* 150: 45–49.

149. Belkaid, A., Amzert, S.A., Bouaichaoui, Y. and Chibane, H. (2012). Economic study of nuclear seawater desalination for mostaganem site. *Procedia Engineering* 33: 134–145.

150. Dittmar, M. (2012). Nuclear energy: Status and future limitations. *Energy* 37(1): 35–40.

151. Wu, S. and Zhang, Z. (2003). An approach to improve the economy of desalination plants with a nuclear heating reactor by coupling with hybrid technologies. *Desalination* 155(2): 179–185.

152. Wu, S. and Zheng, W. (2002). Coupling of nuclear heating reactor with desalination process. *Desalination* 142(2): 187–193.

153. Wu, S., Dong, D., Zhang, D. and Wang, X. (2000). Seawater desalination plant using nuclear heating reactor coupled with MED process. *Nuclear Science and Techniques* 11(1): 6–12.

154. Khan, S.U.-D., Khan, S.U.-D., Danish, S.N., Orfi, J., Rana, U.A. and Haider, S. (2018) *Nuclear Energy Powered Seawater Desalination. Renewable Energy Powered Desalination Handbook*. Elsevier, New York, 225–264.

155. Kim, H.S. and No, H.C. (2012). Thermal coupling of HTGRs and MED desalination plants, and its performance and cost analysis for nuclear desalination. *Desalination* 303: 17–22.

156. Misra, B.M. and Kupitz, J. (2004). The role of nuclear desalination in meeting the potable water needs in water scarce areas in the next decades. *Desalination* 166: 1–9.

157. Misra, B.M. (2003). Advances in nuclear desalination. *International Journal of Nuclear Desalination* 1(1): 19–29.

158. Kavvadias, K.C. and Khamis, I. (2010). The IAEA DEEP desalination economic model: A critical review. *Desalination* 257(1–3): 150–157.

159. Al-Othman, A., Darwish, N.N., Qasim, M., Tawalbeh, M., Darwish, N.A. and Hilal, N. (2019). Nuclear desalination: A state-of-the-art review. *Desalination* 457: 39–61.

160. Tewari, P.K. and Khamis, I. (2007). Non-electrical applications of nuclear power. *Proceedings of International Conference*, 16–19 April, Oarai.

161. Lacomte, M. and Bandelier, P. (2002). Optimization of water desalination by high temperature reactor using electricity/heat cogeneration. *Proceedings of the International Conference on Nuclear Desalination: Options and Challenges*, 16–18 October, Marrakech.
162. Russian Floating Nuclear Power Station. (2019). Wikipedia, The free encyclopedia, last visited 11 August 2019.
163. Kalista, B., Shin, H., Cho, J. and Jang, A. (2018). Current development and future prospect review of freeze desalination. *Desalination* 447: 167–181.
164. Ng, K.C. and Shahzad, M.W. (2018). Sustainable desalination using ocean thermocline energy. *Renewable and Sustainable Energy Reviews* 82: 240–246.
165. del Moral, A. and Petrakopoulou, F. (2019). Evaluation of the coupling of a hybrid power plant with a water generation system, *Applied Science* 9: 4989, Doi: 10.3390/app9234989, www.mdpi.com/journal/applsci, an MDPI publication.
166. Bustamante, M.J. (2016). *Developing a Solar–Bio Hybrid Energy Generation System for Self-Sustainable Wastewater Treatment*, Ph.D. Thesis, Michigan State University, East Lansing, MI.
167. Soni, A., Stagner, J. and Ting, D. (2017). Adaptable wind/solar powered hybrid system for household wastewater treatment. *Sustainable Energy Technologies and Assessments*, Doi: 10.1016/j.seta.2017.02.015.
168. Clark, J. (1990). The steady state performance of a solar still. *Solar Energy* 44: 43–49.
169. Adhikari, R. and Kumar, A. (1990). Estimation of mass transfer rates in solar stills. *International Journal of Energy Research* 14: 737–744.
170. Reddy, K., Ravi Kumar, K., O'Donovan, T. and Malick, T. (2012). Performance evaluation of evacuated multi-stage solar water desalination system. *Desalination* 288: 80–92.
171. Eames, J., Maidment, G. and Lalzad, A. (2007). A theoretical and experimental investigation of small scale solar power barometric desalination system. *Applied Thermal Engineering* 27: 1951–1959.
172. Gandiglio, M., Lanzini, A., Soto, A., Leone, P. and Santarelli, M. (2017, October 30). Enhancing the energy efficiency of wastewater treatment plants through co-digestion and fuel cell systems. Review Article Frontiers in Environmental Science, Doi: 10.3389/fenvs.2017.00070.
173. Maktabifard, M., Zaborowska, E. and Makinia, J. (2018). Achieving energy neutrality in wastewater treatment plants through energy savings and enhancing renewable energy production. *Reviews in Environmental Science and Biotechnology* 17: 655–689, Doi: 10.1007/s11157-018-9478-x.
174. Tee, P.F., Abdullah, M.O., Tan, I.A.W., Abdul Rashid, N.K., Mohamed Amin, M.A. Nolasco-Hipolito, C. and Bujang, K. Review on hybrid energy systems for wastewater treatment and bio-energy production. *Renewable and Sustainable Energy Reviews* 54: 235–246.
175. Hai, F.I., Yamamoto, K. and Fukushi, K. (2007). Hybrid treatment systems for dye wastewater. *Critical Reviews in Environmental Science and Technology* 37(4): 315–377.
176. Balamane-Zizi, O. and Ait-Amara H. (2010). Study of the simultaneous elimination of phosphates and heavy metals contained In dairy wastewater by a physical– chemical and biological mixed process; consequences on the biodegradability. *Energy Procedia* 18: 1341–1360.
177. Abdullah, M.O. and Gan, Y.K. (2006). Feasibility study of a mini fuel cell to detect interference from a cellular phone. *Journal of Power Sources* 155: 311–318.
178. Abdullah, M.O., Yung, V.C., Anyi, M., Othman, A.K., Ab. Hamid, K.B. and Tarawe, J. (2010). Review and comparison study of hybrid diesel/solar/hydro/fuel cell energy schemes for a rural ICT telecenter. *Energy* 35: 639–646.
179. Abdullah, M.O., Tay, K.M. and Gan, Y.K. (2011). A multi-purpose mini hybrid fuel cell-solar portable device for rural application: Laboratory testing (paper accepted for publication in *International Journal of Research and Review in Applied Sciences*, Special Edition entitled "Energy sustainability for global techno-economic uplift").

180. Manickam, S.S., Karra, U., Huang, L., Bui, N.N., Li, B. and McCutcheon, J.R. (2013). Activated carbon nano!ber anodes for microbial fuel cells. *Carbon* 53: 19–28.
181. Liu, H., Ramnarayanan, R. and Logan, B.E. (2004). Production of electricity during wastewater treatment using a single chamber microbial fuel cell. *Environmental Science & Technology* 38(7): 2281–2285.

11 Hybrid Energy Systems for Hydrogen Production

11.1 INTRODUCTION

Hydrogen is the simplest and most abundant element on earth. Hydrogen combines readily with other chemical elements, and it is always found as part of another substance, such as water, hydrocarbon, or alcohol. Hydrogen is also found in natural biomass, which includes plants and animals. For this reason, it is considered as an *energy carrier* and not as an energy source.

The hydrogen industry is well established and has decades of experience in industry sectors using hydrogen as a feedstock. The hydrogen feedstock market has a total estimated value of USD 115 billion and is expected to grow significantly in the coming years, reaching USD 155 billion by 2022. In 2015 total global hydrogen demand was estimated to be 8 exajoules (EJ) [1]. The largest share of hydrogen demand is from the chemicals sector for the production of ammonia and in refining for hydrocracking and desulfurization of fuels. Other industry sectors also use hydrogen, such as producers of iron and steel, glass, electronics, specialty chemicals, and bulk chemicals, but their combined share of total global demand is small. In future, the demand of hydrogen in energy industry will rise significantly with success in fuel cells and their use in transportation sector.

Some believe that in the distance future, electricity and hydrogen are the answers for carbon-free energy. Both can be used without carbon emission. My previous book [2] examined a strategy to produce electricity with low or no use of fossil fuels through hybrid power systems. Hydrogen can also be produced using diverse, domestic resources, including nuclear, natural gas and coal, biomass, and other renewable sources. The latter includes solar, wind, hydroelectric, or geothermal energy. This diversity of domestic energy sources makes hydrogen a promising energy carrier and important for energy security. It is important that hydrogen be produced using a variety of resources and process technologies or pathways that do not include fossil fuels. The production of hydrogen can be achieved via various process technologies, including thermal (natural or bio gas reforming, renewable liquid and biooil processing, biomass, and coal gasification), electrolytic (water splitting using a variety of energy resources), and photolytic (splitting of water using sunlight through biological and electrochemical materials).

The annual production of hydrogen is estimated to be about 55 million tons with its consumption increasing by approximately 6% per year [3]. Hydrogen can be produced in many ways from a broad spectrum of initial raw materials. Nowadays, hydrogen is mainly produced by the steam reforming of natural gas, a process which leads to massive emissions of greenhouse gases [4,5]. Close to 50% of the global demand for hydrogen is currently generated via steam reforming of natural gas, about 30% from oil/naphtha reforming from refinery/chemical industrial off-gases, 18% from coal gasification, 3.9% from water electrolysis, and 0.1% from other sources [6]. Electrolytic and plasma processes demonstrate a high efficiency for hydrogen production, but unfortunately, they are considered energy-intensive and expensive processes [7]. Thermal dissociation of water is so far limited to laboratory level exploration.

Since hydrogen has a bright future in energy industry because of fuel cell and its wide variety of applications including the ones in transportation sector, it is important to decarbonize hydrogen production. This problem can be addressed by the utilization of alternative renewable raw materials and energy resources in place of fossil fuels. The use of hybrid nuclear energy process has also potential. There are two issues at stake: raw materials and the resources for the required energy for its production. As mentioned above, currently both raw materials and required energy are provided by fossil fuels resulting in excessive emission of carbon dioxide. The decarbonization of hydrogen production will require two strategies: (a) replace fossil fuel by biomass and water as raw materials and (b) provide required energy by either renewable sources or nuclear energy. Both of these strategies are examined in this chapter. The chapter will also show that concepts of hybrid energy and hybrid processes are essential in implementing these strategies.

The technical maturity of various options for hydrogen productions is shown in Table 11.1. This table clearly shows that currently steam reforming (also partial oxidation), gasification/pyrolysis, and electrolysis are the major commercially viable choices. We will, therefore, focus our discussion to these technologies. In line with strategy (a) mentioned above, we will also focus on the use of biomass and water asraw materials [3].

Thus, two best options to reduce carbon footprint for hydrogen production are (a) to use biomass as raw material accompanied by hybrid renewable energy system(s)

TABLE 11.1
Technology Maturity for Hydrogen Production [3]

Technology	Feedstock	Efficiency	Maturity
Steam reforming	Hydrocarbons	70%–85%	Commercial
Partial oxidation	Hydrocarbons	60%–75%	Commercial
Autothermal reforming	Hydrocarbons	60–75%	Near term
Plasma reforming	Hydrocarbons	9%–85%*	Long term
Biomass gasification	Biomass	35%–50%	Commercial
Aqueous phase reforming	Carbohydrates	35%–55%	Med. term
Electrolysis	H_2O + electricity	50%–70%	Commercial
Photolysis	H_2O + sunlight	0.5%*	Long term
Thermochemical water splitting	H_2O + heat	NA	Long term

* Hydrogen purification is not included

or nuclear reactor to provide required energy for various thermochemical processes like gasification, pyrolysis, and reforming of bio-syngas etc. Biomass can also be hybridized with other raw materials like coal, wastewater or other devices like fuel cell, electrolysis; and (b) use hybrid processes for electrolysis where required heat and electricity are supplied by either renewable energy or nuclear heat [8,9]. These two options are discussed in details in this chapter.

11.2   ROLE OF BIOMASS FOR HYDROGEN PRODUCTION

Because biomass is our only renewable source of hydrocarbons, conversion of a small portion of the planet's huge biomass resource to fuels is an important option for our transportation needs. Hydrogen can be produced from this renewable feedstock. In the near term, biomass is anticipated to become the most likely renewable organic substitute to petroleum. Biomass is available from a wide range of sources, such as animal wastes, municipal solid wastes, crop residues, short rotation woody crops, agricultural wastes, sawdust, aquatic plants, short rotation herbaceous species (e.g., switch grass), waste paper, corn, and many others [10,11].

A recent U.S. National Research Council (NRC) report (Transitions to Alternative Transportation Technologies: A Focus on Hydrogen, July 2008) asserts that central-ized production of hydrogen from biomass gasification is the renewable pathway that has the highest likelihood of commercial viability in the 2015–2035 timeframe. Biomass-to-hydrogen is complex, not only because of the technical details of the conversion processes themselves but also because of the many process types that could be employed. The conversion type with the most potential for large-scale cen-tralized production, as pointed out in the NRC report, is gasification, which in itself is but one of several technologies available within the larger category called ther-mochemical conversion. Besides gasification, pyrolysis and reforming of bio-syngas produced from gasification are also viable paths for hydrogen production.

Gasification—whether steam, air/oxygen, catalytic, or indirect—involves subject-ing the biomass to elevated temperatures and pressures in order to reduce the organic materials to hydrogen and carbon monoxide/dioxide gases (along with varying quan-tities of undesirable solid and gaseous byproducts). From there, the hydrogen can be separated out by membrane, chemical, or catalytic steps. Techno-economic analyses indicate that gasification biorefineries may have to be large in order to be economi-cally feasible. This will require significant capital investment, large access of raw materials, and corresponding storage and transportation infrastructure.

Biochemical conversion of biomass to hydrogen also presents several possible pathways. Ethanol produced from lignocellulosic materials could be further reformed to hydrogen, as could other biofuels or intermediate products of various biochemi-cal routes. Certain regional implications, feedstock types, or end-use requirements might make this a viable, if not a widespread, option. More interesting perhaps is dark fermentation, a process that uses anaerobic microorganisms to produce hydro-gen directly, much in the way that bacteria or yeast can produce ethanol via fermenta-tion. Such organisms might be enhanced to better perform the hydrogen production task. They typically need to start with glucose, so the cellulosic ethanol pretreatment and hydrolysis techniques that are being developed now to break down cellulose into glucose would also be required for the dark fermentation pathway. As mentioned before, aqueous phase reforming is another approach for hydrogen production from biomass. This process needs to be commercialized. Biochemical transformation process generally tends to be slow and at smaller scale.

Gasification technology commonly used with biomass and coal as fuel feedstock is very mature and commercially used in many processes. It is a variation of pyrolysis and, therefore, is based upon partial oxidation of the feedstock material into a mixture of hydrogen, methane, higher hydrocarbons, carbon monoxide, carbon dioxide, and nitrogen, known as "producer gas" [10]. The gasification process typically suffers from low thermal efficiency since moisture contained in the biomass must also be vaporized. It can be performed with or without a catalyst and in a fixed-bed or fluidized-bed reactor, with the latter reactor having typically better performance [11]. Addition of steam and/or oxygen in the gasification process results in the production of "syngas" with a H_2/CO ratio of 2/1, the latter used as feedstock to a Fischer-Tropsch reactor to make higher hydrocarbons (synthetic gasoline and diesel) or to a WGS reactor for hydrogen production [11]. Superheated steam (ca. 900°C) has been used to reform dry biomass to achieve high hydrogen yields. However, gasification process provides significant amounts of "tars" (a complex mixture of higher aromatic hydrocarbons) in the product gas even operated in the 800°C–1,000°C range. A secondary reactor, which utilizes calcined dolomite and/or nickel catalyst, is used to catalytically clean and upgrade the product gas [11]. Ideally, oxygen should be used in these gasification plants; however, oxygen separation unit is cost prohibitive for small-scale plants. This limits the gasifiers to the use of air resulting in significant dilution of the product as well as the production of NO_x. Low-cost, efficient oxygen separators are needed for this technology. For hydrogen production, a WGS process can be employed to increase the hydrogen concentration followed by a separation process to produce pure hydrogen [12]. Typically, gasification reactors are built on a large scale and require massive amounts of material to be continuously fed. They can achieve efficiencies in the order of 35%–50% based on the lower heating value [7]. One of the problems of this technology is that a tremendous amount of resources must be used to gather the large amounts of biomass to the central processing plant. Currently, the high logistics costs of gasification plants and the removal of "tars" to acceptable levels for pure hydrogen production limit the commercialization of biomass-based hydrogen production. Future development of smaller efficient distributed gasification plants may be required for this technology for cost-effective hydrogen production.

Another currently promising method of hydrogen production is pyrolysis or copyrolysis. Raw organic material is heated and gasified at a pressure of 0.1–0.5 MPa in the 500°C–900°C range [13–16]. The process takes place in the absence of oxygen and air, and therefore, the formation of dioxins can be almost ruled out. Since no water or air is present, no carbon oxides (e.g., CO or CO_2) are formed, eliminating the need for secondary reactors (WGS, PrOx, etc.). Consequently, this process offers significant emissions reduction. However, if air or water is present (the materials have not been dried), significant CO_x emissions will be produced. Among the advantages of this process are fuel flexibility, relative simplicity and compactness, clean carbon byproduct, and reduction in CO_x emissions [13–17].

Based on the temperature range, pyrolysis processes are divided into low (up to 500°C), medium (500°C–800°C), and high temperatures (over 800°C). Fast pyrolysis is one of the latest processes for the transformation of organic material into products with higher energy content. This is followed by catalytic steam reforming of the liquid (or its vapors) to hydrogen. The products of fast pyrolysis appear in the entire phases formed (solid, liquid, and gaseous). An advantage of this approach is that the biooil, as an intermediate product, has a higher energy density than the biomass feedstock and can more easily be transported. One of the challenges with this approach is the potential for fouling by the carbon formed, but proponents claim that this can be minimized by appropriate design. Since it has the potential for lower CO and CO_2 emissions, and it can be operated in such a way as to recover a significant amount of solid carbon, which is easily sequestered [14,17], pyrolysis may play a significant role in the future. This technique may prove to be applicable to smaller, distributed biorefineries, whereas the gasification process described above may cater to the large, centralized installations.

The application of the copyrolysis of a mixture of coal with organic wastes has recently received an interest in industrially advanced countries, as it should limit and lighten the burden of wastes in waste disposal (waste and pure plastics, rubber, cellulose, paper, textiles, and wood) [18,19]. Pyrolysis and copyrolysis are well-developed processes and could be used in commercial scale. Future prospects for methanol and hydrogen from biomass are also examined by Hamelinck and Faaij [20].

11.3 BIOMASS-BASED HYBRID SYSTEMS

As mentioned above, thermochemical processing (gasification and pyrolysis) of biomass has significant commercial viability both at large and small scales. Recent assessments show more than 400 million tons of biomass available per year in the United States [21]. This could be converted to roughly 30 million tons of hydrogen by thermochemical processing. Some estimates predict that, with relatively minor changes to land management and agricultural practices, as much as 1 billion tons of biomass could be available in the future [22]. In addition to great availability, thermochemical biomass plants provide many opportunities for system integration [23].

Dean et al. [24] at NREL evaluated the possibility of utilizing biomass' renewable and dispatchable characteristics in combination with other raw materials and energy technologies to improve the efficiency, reliability, or cost of producing electricity and hydrogen from renewable energy sources. This project addressed the definition and evaluation of opportunities for combined production of hydrogen and electric power by combining biomass conversion with other hydrogen-production technologies, including wind, solar, coal, and nuclear. Since biomass is hard to store and transport over a long distance, in order to build sustainable hybrid processes, an analysis was made to identify the areas of the United States that have greater than 2,000 t/d (TPD) of biomass available within a 50-mile radius. Multiple types of biomass were considered, including crop residues, forest residues, and primary and secondary mill residues. The dark area in Figure 11.1 illustrate these areas. With this as background, following hybrid options were considered by Dean et al. [24] and in my previous book [23].

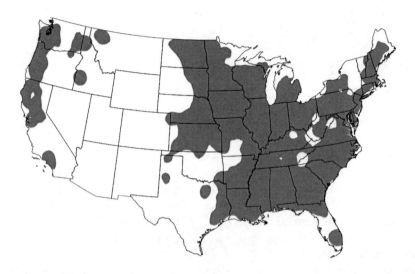

FIGURE 11.1   Biomass resources availability [24,26]. Dark area represents biomass resource availability.

11.3.1   COAL-BIOMASS

This hybrid combination was discussed earlier in my previous book [20]. This co-fuel presents many positive features. Due to the large existing coal infrastructure in the United States, early thinking of coal and biomass involves cofiring, cogasification, or copyrolyzing of biomass with coal. Biomass can be cofired in existing coal combustors but only in marginally small percentages due to several issues related to feed preparation [25]. There are numerous ways coal-biomass can be cogasified. One way to address feed problems is to separately prepare coal and biomass feed and then mix them and cogasify in a suitable gasifier. Feed preparation of biomass may involve processes such as water removal process, torrefaction, particle sizing, grinding, etc., all of which can get expensive. Finding the right operating conditions for the joint gasification of coal and biomass may also be problematic because biomass is more reactive than coal and will gasify under somewhat milder conditions than coal. Other way to address feed problems is to gasify biomass and then cofeed pulverized coal and biomass-produced syngas into existing combustors [27]. Another option is to torrefy (or thermally pretreat) the biomass, which produces a char that can be cofed with the coal slurry. Still another way is to gasify coal and biomass separately and then mix the syngas from the two gasifiers. This allows independent control of operating conditions for the two gasifiers. A good overview of the practical issues of duel-feed systems is provided in "Biomass Cofiring: Economics, Policy and Opportunities" [28]. In addition to Hughes article, the DOE white paper "Biomass Cofiring: A Renewable Alternative for Utilities" [29] provides information about existing plants operating on both fuels. This subject is also treated in great details in my previous book [20].

Cogasification of coal and biomass has been a focus of recent research and significant literature on the subject is available [23,30,31]. These dual-feed systems help to reduce the greenhouse gas emissions of the existing coal infrastructure while maintaining economies of scale and avoiding the difficulties of finding large, reliable quantities of biomass for power generation. As pointed out in my previous book, 70/30 coal/biomass mixture makes the system carbon neutral in its entire lifecycle. Since biomass is hard to store and transport, coal-biomass mixture allows the large-scale operation without excessive need of biomass. Biomass is often replaced by municipal waste as shown in my previous book [23]. When significant amounts of biomass are cofed, problems can result from increased fouling of downstream processes and high alkali content in the product ash. In the Netherlands, the Buggenum coal gasification plant has reported cofeed percentages of up to 30% with only minor changes in plant power and waste output [27]. Several other cogasification plants are described in Chapter 4 and in my previous book. Cormos [32] proposed coal gasification with biomass cofeed for production of fuel and power [32]. Once the syngas is produced from cogasification, it can be steam reformed to hydrogen as currently done commercially.

Another synthesis possibility is thermal integration of biomass gasification or biooil reforming facilities with existing coal-fired power plants. Biomass gasification, whether directly or indirectly heated, requires a steam source that could come directly from a coal power plant. The major challenges to this type of integration are the added capital cost, the low steam temperatures relative to gasification requirements, and the mismatch in scale between biomass availability and steam production. Biomass availability could be addressed by gasification of biooil produced from multiple off-site pyrolysis units. These and other options are also discussed in great detail my previous book [23].

It should be noted that syngas produced from biomass or coal-biomass mixture gasification can also be used for power generation. Biomass gasification for power generation is a more efficient route to power production than direct combustion of biomass. The concept of integrated gasification combined-cycle (IGCC) used for coal can also be applied to biomass or coal-biomass mixture. Extensive research has been conducted on using biomass and the syngas produced by a biomass gasifier to create power using either a gas turbine alone, a steam turbine alone, or an integrated combined-cycle approach. As pointed out in Chapter 5 and in my previous book [23], IGCC is a more efficient way to generate power. At least two major studies have been released by NREL directly addressing the technology, economics, and life-cycle implications of this type of hybrid power-generation technology [33]. In addition to these assessments, many biomass-to-liquid fuel studies assume that unconverted syngas is burned in a gas turbine for power generation [34]. Waste heat from IGCC plant can also be converted to power (by thermoelectricity) to improve further its thermal efficiency.

From a greenhouse-gas emissions standpoint, biomass-based power plants produce significantly fewer emissions than do coal or natural gas systems. Even when carbon sequestration is used on fossil-fuel plants, a biomass IGCC plant produces fewer atmospheric greenhouse gas emissions [33]. The major challenge for biomass-based IGCC plants is the economies of scale limitations due to biomass availability. Biomass IGCC plants typically are in the 10–60-MW range, as compared to 500-MW coal gasification plants [27]. One possible option is to use cogasification of coal and biomass to improve the scale of the operation. This option is also discussed in my previous book [23].

11.3.2 Wastewater Treatment—Biomass

Every year approximately 5.6 million dry tons of solid waste (or sludge) are produced in the United States [35]. This number is increasing every year. A significant amount of this sludge is either land-filled or incinerated. Gasification could provide an alternative use for this readily available source of biomass.

The high water content of sludge is a significant challenge faced by traditional biomass gasification systems. Two options exist to overcome the water challenge. Municipal waste can be preprocessed and dried to levels acceptable to the reaction chamber before gasification, or the gasifier can be run at much lower efficiencies and the water can be vaporized in the reaction chamber itself. Using a directly heated gasifier and wet biomass would result in a significant percentage of the input carbon being burned to heat water rather than to produce syngas [24].

One promising alternative to traditional gasification is plasma gasification. This is typically carried out in electrically heated arc furnaces running at temperatures well above 950°C. It has been successfully used to produce high-quality syngas and power from sewage sludge in the United States, Canada, Malaysia, and Japan. The most well known of these plants is located in Japan. It produces approximately 4 MW of grid electricity by processing 138 tons of sewage sludge per day. A good overview of the state of the technology and references to existing plants can be found in "Plasma Gasification of Sewage Sludge: Process Development and Energy Optimization," by Mountouris et al. [36]. Another promising alternative is to use microwave for drying and gasification. The use of microwave for treating industrial waste has been successfully examined [9,23,37]. This approach poses some intriguing opportunities. Third approach is to mix coal and waste. As pointed out in my previous book, this has been successfully commercialized. So far, however, only small concentration of waste has been investigated. The concept has also been tested for some industrial organic and plastic wastes [24].

Though less glamorous, biogas digesters are another option for turning sludge into useful gas. This 100-year-old technology has been—and is—used by households in China, India, and other countries to produce natural gas for combustion in lanterns and stoves. Because the technology is simple and implementation is low cost, this approach is being adopted by many wastewater treatment facilities in the United States. Unlike thermochemical transformation, fermentation technology tends to be slow and sometimes restricted in size [24].

11.3.3 Concentrated Solar—Biomass

Using solar energy to provide the heat for thermochemical biomass processing would reduce the environmental impact of both gasification and pyrolysis facilities. As pointed out by Dean et al. [24], there are two main areas of research in this hybridization option—direct thermal transfer and indirect thermal transfer.

In direct thermal transfer, solar concentrators are focused into the reaction chamber of a pyrolyzer or gasifier. To date, several bench-scale systems have been designed and tested with disappointing results [38]. These systems suffer from several technical problems including the amount of solar concentration needed to reach plausible reaction temperatures, solar intermittency, the need for a clear window into the reaction chamber, scalability concerns, and the severe solar diffusion caused by particle movement within the reactor. In addition to technical challenges, capital costs for building a plant are expected to be significant. Although technical hurdles remain, valuable research is ongoing. A good summary of the state of technology is provided by Steinfeld in "Solar Thermochemical Production of Hydrogen—A Review" [39]. Research on this topic also is ongoing at the University of Colorado. Attempts are also made to use solar energy for steam reforming process which is highly endo-thermic. Currently, about 10%–15% of fossil fuel is used to provide process heat for this process. Solar energy-driven European reforming process called SOLREF (solar reforming or solar energy assisted reforming) [40] has shown some promise. In this process, the use of concentrated solar power for supplying high-temperature process heat to the steam-reforming of natural gas has the potential of avoiding up to 35% of the CO_2 emissions derived from the conventional fossil-fuel-based method. The cost of hydrogen is estimated at 0.05 EUR/kWh_{LHV-H2}. The project aims at developing the technology to a precommercial phase. Eidgenössische Technische Hochschule (ETH) is responsible for the thermodynamic analysis and reactor/process dynamic modeling. The project is, however, currently focus on natural gas. In future, use of biogas for reforming needs to be undertaken. The SOLREF technology has effi-ciency benefits compared to other technologies using renewable fuel for hydrogen production, since it allows the production of hydrogen and storage of the produced CO_2 in the same location using only solar energy. This also allows the process to have carbon footprint much smaller compared to its competitors based on fossil fuel resources.

Indirect thermal transfer relies on heating the outer walls of a reaction chamber or heating an intermediate used for thermal storage. These systems can provide a more consistent heat source at the expense of reduced absolute temperature. Because of temperature limitations, indirect thermal transfer systems are more likely to be used for biomass pyrolysis than for gasification. One interesting option is the use of con-centrated solar energy to heat molten salts, which then can be used as a pyrolyzing medium. Preliminary research suggests that, using this approach, a pyrolysis reactor could be run at steady state on solar energy alone [41].

Whether indirect or direct thermal transfer is used, both technologies require significant solar radiation and concentration. Rough concentration ratios for trough, tower, and dish concentrators are 100, 1,000, and 3,000 suns, respectively [42]. A parabolic (near) and tower (background) solar concentrators are illustrated in Figure 11.2 [24]. All three technologies are feasible for power generation using a heated fluid in a traditional thermal cycle. Trough concentration systems typically run at temperatures ranging from 300°C to 4000°C. The temperatures needed for pyrolysis and gasification (500°C and 860°C, respectively) favor tower or dish con-centration systems. Tower concentrator systems are the most likely candidate for use with a stationary chemical reactor. The design of a continuous flow reactor which can harness very high solar radiation to be captured by reacting fluid is a challenge.

The United States has significant solar resources at its disposal, but they gen-erally are concentrated in the deserts of the southwest where biomass availabil-ity is low. Plant location is further restricted because tower concentration systems require not only high solar radiation but also large areas of flat land for construction. Transportation costs are one of the major obstacles to using biomass as an energy source; therefore, the lack of local resources is problematic. As reported by Dean et al. [24], Figure 11.3 shows the areas with biomass resources greater than 2,000 TPD within 50 miles (dark area). Light grey areas denote solar resources in the south-western United States of 6 kWh/m²/d or greater direct normal radiation. The solar

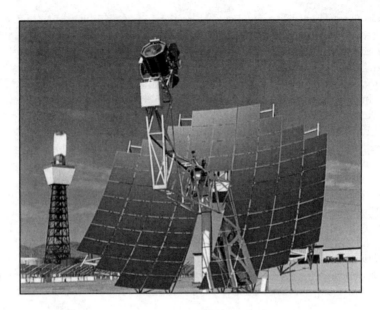

FIGURE 11.2   Parabolic (near) and tower (background) solar concentractions [24].

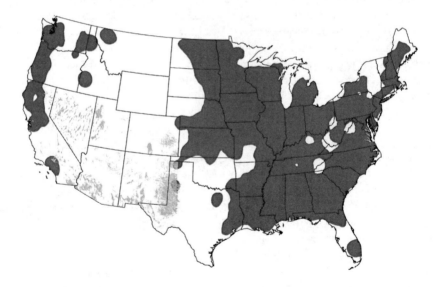

FIGURE 11.3   Solar tower plant locations (grey area) versus biomass resources (dark area) [24].

resources are further constrained to flat sites (areas with less than 1% land slope) excluding environmentally protected lands, urban areas, and water features. In short, the orange areas show sites that might be capable of supporting a solar tower concentrator plant. One solution to avoid location problem is to use syngas from gasification of coal-biomass mixture.

Use of photovoltaic (PV) electricity has several advantages over direct use of the radiation. The most significant advantage is the possibility of bringing the electricity to the biomass resources rather than having to ship biomass large distances. In addition, there is significant potential for distributed PV installation throughout the United States.

Dean et al. [24] points out that the availability of low-cost solar generated electricity could be a challenge due to many factors. Peak solar radiation generally coincides with peak electricity demand, making the cost of the renewable electricity too great for cost-effective biorefinery usage. In addition, capital costs remain high for solar installations, driving up the baseline cost of solar-generated electricity. Photo-electrochemical water splitting is a future possibility for hydrogen production via direct water splitting. This technology currently is not considered viable for hybridization because of its high cost and low efficiency, even at bench scale [43]. In the future, it could provide a way to produce oxygen and hydrogen for biomass gasification without the significant electricity requirements of both cryogenic air separation units and electrolyzers. New developments in solar energy technology may make this option viable in some parts of US. Thermochemical dissociation of water by solar energy is another possibility. This technology is not location-dependent. The technology is, however, currently at the development stage.

11.3.4 NUCLEAR—BIOMASS

Thermal integration and colocation of biomass processing with nuclear energy is a promising hybridization option. The presence of near carbon-neutral power and steam from the reactor could significantly increase the efficiency of a biomass plant. Charles Forsberg makes a strong case for this concept in his paper "Meeting U.S. Liquid Transport Fuel Needs with a Nuclear Hydrogen Biomass System" [44]. Since the publication of his paper, significant progress has been made in the linking of nuclear heat to various industrial applications. In particular, since the development of small modular nuclear reactors (particularly various forms of generation IV high-temperature gas-cooled HTGR reactors as described in detail in my previous books [2,9,23,37,45,46]), nuclear heat can provide high enough temperature for coal or biomass gasification, reforming, and pyrolysis.

Nuclear energy currently provides 20% of the electricity in the United States [47]. According to the Nuclear Energy Institute, there are 104 nuclear reactors in the United States, and another 30 plants currently are seeking federal license approval. All of the existing plants provide a reliable source of electricity and could provide low-pressure, low-temperature steam to a biomass-processing facility. Currently, this steam is a waste stream that must be condensed after the last turbine cycle for U.S. plants. Most gasification plants would need to upgrade the steam quality before it entered the reactor, but ethanol plants could use the low-quality steam directly [32].

During last century, the centralized power generated from large-scale nuclear power plants was inefficient and this forced measures to use waste heat for industrial or commercial purposes. The development of generation IV nuclear reactors also allowed nuclear waste heat to reach high temperatures. Unfortunately, since heat cannot be transported at long distance, waste heat needed to be used in local industries which was not always possible. For economic reasons, to benefit from hybridization of nuclear heat and biomass, the nuclear plant would need to either be near large biomass resources or have access to low-cost barge transportation. An examination of locations of nuclear facility and biomass availability in U.S. by Deans et al. [24]

indicated some mismatch. Even with significant biomass resource availability, a scale mismatch between biomass availability and steam production remains. As mentioned earlier, this mismatch can be partially handled by raising scale of biomass gasification using coal-biomass mixture or by gasifying bio-oil produced from multiple off-site pyrolysis facilities (the "hub-and-spoke" concept). One way to take advantage of the scale mismatch might be to pull a slipstream of super-critical steam from the power plant steam cycle for use in gasification. Significant plant modifications would be required for this type of integration.

Nuclear power plants rely on nonrenewable uranium resources to create heat and subsequently power. For any biomass-nuclear hybridization to be reasonable, sufficient domestic uranium resources must be available in the long term. According to Nuclear Energy Association estimates, fuel availability is not a concern for several centuries [48]. Reserves could last significantly longer with improvements in mining technology and reactor design, and increased fuel-rod recycling.

Recent development of small modular reactors can be the answer to resolve the mismatch between nuclear heat and biomass availability. These reactors are flexible, modular, and movable. NuScale and several VHTR (very high temperature gas-cooled reactor) and SMR (small modular reactor) can generate heat at the levels required for biomass gasification and pyrolysis as well for reforming of bio-syngas, and they (among other reactors) have solved this issue since these reactors are flexible and movable and can be located near industry. SMRs are also more energy efficient and require less capital to install. In a later section, the use of nuclear heat for high-temperature electrolysis (HTE) is described. Major challenges to such a system include the resistance of many people in the United States to the building of new nuclear plants, and U.S. security concerns involved with additional onsite processing. It might be possible to address both issues with biomass. Placing the biomass plant outside of the secure perimeter and piping the steam over the fence could negate security concerns. Creating an additional source of farm income in rural areas could go a long way toward overcoming local resistance to nuclear reactors. Extensive developments of SMR can certainly facilitate nuclear-biomass hybridization. This subject is also discussed in Chapters 6 and 8 and in my several previous books [2,9,23,37,45,46].

11.3.5 FUEL CELL-BIOMASS

Dean et al. [24] suggested that coupling a fuel cell directly with the syngas output of a biomass gasification plant is a highly efficient way to produce electricity from biomass. With no moving parts and freedom from the Carnot limit, fuel cells can achieve much greater efficiencies than conventional turbines. In addition to high efficiencies, fuel cells run on a variety of fuels and typically have low maintenance requirements.

Molten carbonate (MCFC) and solid oxide (SOFC) fuel cells are the most likely candidates for combination with gasification because of their relatively low fuel-quality demands, high operating temperatures, and tolerance of carbon monoxide [49]. Owing to the high operating temperatures of these fuel cells (600°C–1,000°C), it is typically most economical to use them to produce combined heat and power.

Several studies have examined the possibility of combining gasification with high temperature fuel cells. Total plant electrical efficiencies of approximately 40% have been reported in literature [25].

The major obstacles to both biomass SOFC and MCFC systems are cost, syngas cleaning, and durability. Both SOFC and MCFC systems are extremely sensitive to sulfur and some of the corrosive tars produced by gasification. The cost of syngas increases with increasing purity requirements, thus cleaning the syngas for fuel cell use could be a significant burden. Dean et al. [24] points out that high-temperature fuel cells are commercially available but cost still is a major barrier to large-scale deployment, especially when combined with the high costs of gasification equipment. New developments in fuel cell can certainly make this option a viable one.

11.3.6 Electrolysis—Biomass

Directly heated gasification systems require a source of pure oxygen if they are to be used for fuel production. Currently, plants that use oxygen produce it with cryogenic air separation units (ASUs) [50]. Electrolysis could provide an alternative to air separation units with the added benefit of producing a pure hydrogen stream. The research carried out by Deans et al. [24] indicates that this hybridization option could be promising from both a technical and economic perspective [51]. The feasibility of producing oxygen and hydrogen with electrolysis, however, is heavily dependent on both the price of electricity and the value of the end products to the plant.

At standard temperature and pressure, an ideal electrolyzer would use 39 kWh of electricity to produce 1 kg of hydrogen. The actual state of technology limits system efficiencies to between 56% and 73% meaning that approximately 53–70 kWh of electricity is needed for every kilogram of hydrogen produced [52]. Replacing a single ASU unit for oxygen production requires multiple electrolyzers. Dean et al. [24] pointed out that the largest commercial electrolyzer is produced by StatoilHydro (formerly NorskHydro); it has a maximum flow rate of 43.6 kg/h of hydrogen (174.4 kg/h of oxygen) [53]. A 2,000 TPD biomass gasifier would require a large bank of these electrolyzers running at full capacity. According to NREL's most recent H2A (hydrogen analysis) forecourt (refueling station) electrolysis analysis, one 174.4-kg/h electrolyzer installed with hardware costs approximately $2.5 million. Therefore, the electrolyzer bank for a 2,000 TPD gasification plant would cost significantly more than a comparable ASU unit. One possibility for addressing these high capital costs is to use enriched air for gasification rather than pure oxygen.

Along with the high capital cost of electrolysis, there are other concerns with this hybridization. Water usage is a key concern with electrolyzer systems and would be especially pronounced when combined with the generally high water requirements of biomass processes. Also, pressurized gasification plants use nitrogen from the ASU unit for pressurizing the biomass feed system. Another source of inert pressurization would be needed. New developments in HTE may also change the picture. HTE can be supported by nuclear heat or waste heat from industrial processes. This topic is further discussed in subsequent sections.

11.3.7 WIND—BIOMASS

Biomass gasification and pyrolysis plants typically require external power for operation when the plants are optimized for fuel production. Because of this requirement, many of the biomass gasification and pyrolysis literature reviewed mentioned that the use of renewable sources of power would further add to the environmental benefits of thermochemical biomass processing. Although many papers mentioned using electricity produced by renewables, few examined how to directly couple intermittent wind power with thermochemical processing.

Wind turbines (WTs) have quickly become a widely accepted, commercial source of renewable energy in the United States. Over the last 29 years, U.S. utilities have vastly improved their knowledge and ability to manage intermittent electricity sources. Significant issues remain, however, if large-scale wind power is pursued in the United States. As pointed out by Dean et al. [24], these issues are addressed in detail in the Department of Energy report "20% Wind Energy by 2030" [54].

The two most significant issues with wind power are its location and its intermittency. Figure 11.4 shows the nation's wind-resource distribution. The vast majority of land-based wind resources are found in the rural areas of the middle United States. To successfully utilize these resources, power must be transported long distances to demand centers. Additionally, the intermittency of wind means that installing too much capacity will create grid instability unless suitable grid leveling or backup storage options are available.

Transportation of wind-generated power can be accomplished via the electrical grid or by converting the electricity to a transportable fuel. Using the national electric grid to transport the power would require significant updates to the national infrastructure. Additional high-voltage transmission lines would be needed to connect wind resources with urban areas [54]. Another option is to convert intermittent electricity into a fuel. Several studies recently have been conducted on using electrolyzers to create hydrogen from wind-generated electricity [55].

Dean et al. [24] point out that intermittency of wind electricity can cause challenges for the power grid if proper leveling options are not available. One option commonly used today is to use natural gas turbines to maintain system reliability. Turbines are readily available and can be brought online and offline very rapidly. Another option is to use batteries and electrolyzers to store power during peak winds for use during low-wind or no-wind conditions [56]. Third option is to use solar-wind-storage hybrid option.

Denholm of NREL (57) has pointed out challenges related to wind-biomass hybrid option. The proposed system would use compressed-air energy storage to store off-peak electricity generated by wind. This energy then would be used as needed by a properly designed biomass gasification plant [57]. It appears that significant portions of national wind resources are in areas that also have biomass availability. Because of this, the shift from viewing wind electricity as an external source of electricity to trying to find direct synthesis between the two technologies appears to be a promising area of research and development. NREL developed geographical locations of class 4 or better wind resources and biomass resources of more than 2000 TPD of certain type within 50 mile radius. Their study indicated that there are small pockets of the northwest and northeast

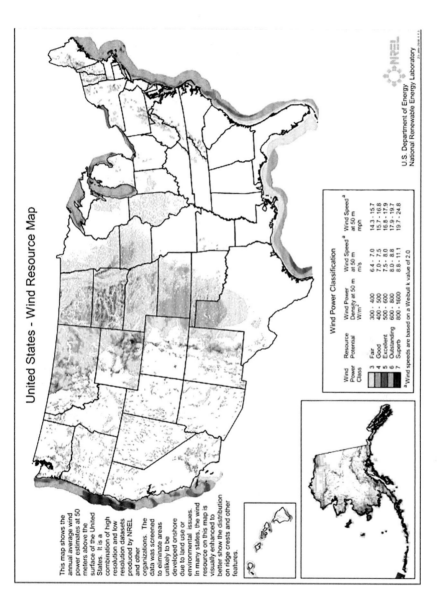

FIGURE 11.4 Wind resource availability map [24,54].

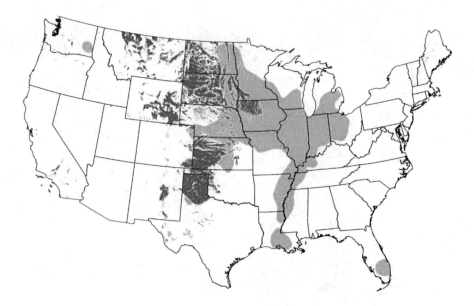

FIGURE 11.5 Wind resources (dark area) versus agricultural biomass resources (grey area) [24].

United States where both class 4 or greater wind and sufficient woody biomass exist for a colocated, combined system. NREL study also compared the agricultural (crop) residue biomass resources versus available wind. Crop residues considered included corn, wheat, soybeans, cotton, sorghum, barley, oats, rice, rye, canola, beans, peas, peanuts, potatoes, safflower, sunflower, sugarcane, and flaxseed. It is important to note that estimates of residue were adjusted down to allow for soil-erosion control, animal feed, bedding, and other existing farm uses [21]. As shown in Figure 11.5, there is significantly more overlap of agricultural biomass with wind than woody biomass and wind.

11.3.8 INDUSTRIAL WASTE HEAT-BIOMASS HYBRIDIZATION

It could be advantageous to look to industries that currently have high heat processes, biomass waste streams, or large steam requirements for synthesis possibilities. One example is the replacing of industrial gas in limekilns with synthesis gas from the gasification of hog fuel [58]. More study in this area is needed. Hybrid processes involving waste heat from industry and biomass are examined in Chapter 9. Once again the major challenge is finding the locations where high level of industrial waste heat and sizable biomass resources coexists.

11.4 HYBRID BIOMASS SYSTEMS RECOMMENDED BY NREL

Based on their analysis, Dean et al. [24] gave strong support to direct wind and wind-electrolyzer combinations with biomass gasification due to several factors. In addition to novelty, the use of renewable wind resources to power a renewable biomass process lays the foundation for truly renewable fuel production. Finally, rec-ommended cases have the potential to increase syngas and fuel yields from a given amount of biomass. Thus, two recommended options by NREL were as follows:

1. Direct grid leveling of intermittent wind power with an indirectly heated biomass gasification plant. The plant will produce both electricity and fuel.
2. Using an electrolyzer in place of an air separation unit for a directly heated biomass gasifier for coproduction of fuel and power.

These options were favored because wind availability significantly overlaps biomass resource availability, making the use of locally produced wind electricity for gasifi-cation feasible. Additionally, gasification plants provide multiple opportunities for electricity use. Although wind power is a promising and largely commercial renew-able source of energy, its penetration of the grid poses some unique challenges. These challenges include management of intermittency with peaking units and, in the extreme case, finding use for electricity produced by wind when there is no demand. Managing intermittency will drive utilities to invest in additional peaking units and will increase the need for interruptible customers and dispatchable loads. Finally, wind in many parts of the country is a stranded resource because of a lack of grid access. Finding direct synthesis between the two technologies could allow a hybrid system to manage local intermittency or capture stranded resources. We examine these two hybrid systems in more detail below. The discussion follows the report by Dean et al. [21].

11.5   INDIRECT GASIFIER HYBRID SYSTEM

This concept investigates two possible changes to a biomass-to-hydrogen plant based on indirect gasification architecture. The first modification is to allow switching between fuel production and electricity production based on grid demand. This is accomplished by routing some or all of the synthesis gas from the gasifier to a gas turbine instead to the fuel-production reactors. In addition to power production, modifications that enable use of additional cheap or surplus electricity by the gasifier are investigated. A sche-matic of indirect gasifier hybrid system is illustrated in Figure 11.6.

Indirectly heated gasification is a two-stage fluidized-bed process where the heat needed for reaction is produced by burning char in a separate chamber to heat sand. The hot sand then is circulated through the reaction chamber to drive reaction kinet-ics. The layout is shown in Figure 11.7.

Electricity supplied to the gasifier during periods of low demand (lesser purchase price) will be used to heat the gasifier reaction chamber. As the temperature of the gasifier is increased, the proportions of syngas, char, and tar produced by the gasifier from a given amount of biomass change via a known relationship for a given system. Adding heat energy will create additional syngas, which will increase plant efficiency.

The ideal plant would continuously adjust both feed use and fuel production to optimize the plant economics. Electricity would be produced instead of hydrogen only when electricity was the more profitable product and vice versa. Similarly, electricity would be used for heating (or be sunk) only when electricity costs were

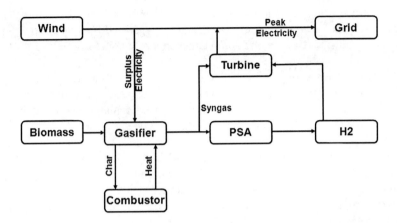

FIGURE 11.6   Indirect hybrid block diagram [24].

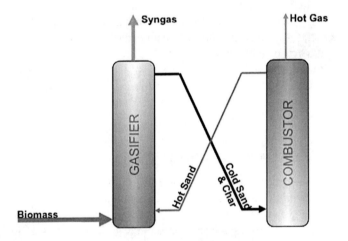

FIGURE 11.7   Indirect gasifier diagram: hybrid gasifier-combustor process [24].

low enough that the additional efficiency provided by the heat offsets the cost of that electricity. The feed and product selection decision is summarized in Figure 11.8.

Analysis of this concept was separated into the peaking and sinking modifications. The two modifications were analyzed individually to highlight the effect of each on plant economics.

11.5.1   PEAKING MODIFICATIONS

Previous NREL studies examined the possibility of using a 2,000 TPD woody biomass plant for dedicated power production and for dedicated hydrogen production (33,59). These studies assumed steady-state operation of the biomass plant. The analysis by Dean et al. [24] differed from previous studies by alternating between hydrogen production and electricity production based on market demand. This effectively combined the existing NREL Biomass-to-Hydrogen and biomass integrated

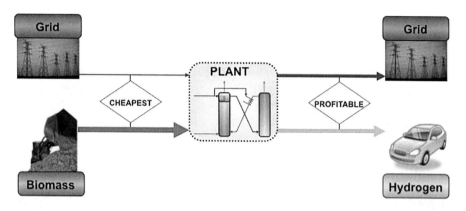

FIGURE 11.8   Feed and product selection for the concept [24].

gasification combined cycle studies into one hybrid system. This combination referred as peaking modification is described in detail by Dean et al. [24].

The proposed modification could increase the economic promise of biomass utilization. The syngas produced by the gasifier can be used to produce hydrogen fuel or it can be used in a gas turbine to provide peaking electricity, depending on which option will maximize profit. The synthesis gas composition available for use in all calculations was based on the biomass-to-hydrogen indirectly heated gasifier study previously completed by NREL [59]. More details on the plant design based on this concept are given by NREL report [24].

11.6   DIRECT GASIFIER HYBRID SYSTEM

The direct gasifier hybrid system concept illustrated in Figure 11.9 is based on directly heated gasifier architecture. Directly heated gasifiers typically have a single combustion/reaction chamber and burn a small portion of the biomass feed to create heat. A source of pure oxygen is required for combustion if the syngas is to be used

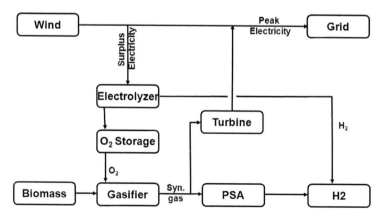

FIGURE 11.9   Direct hybrid system block diagram [24].

for fuel production. Electrolysis could provide an alternative to an air separation unit, with the added benefit of producing an additional pure hydrogen stream.

To date, some research has been conducted on the feasibility of combining electrolysis with gasification [51]. This research concluded that the economic feasibility of this combination was greatly dependent on the price of available electricity. The proposed hybrid system directly addresses electricity price dependence by running the electrolysis system intermittently. Electricity available during periods of low demand (low purchase price) is used by electrolyzers to produce oxygen and hydrogen for use by the gasifier or stored for later use. During periods of peak electricity demand, the stored oxygen is used to create syngas rather than for running the electrolyzers.

Most directly heated gasifiers have a single combustion/reaction chamber and burn a small portion of the biomass feed to create heat. They typically are run at high pressure to improve overall plant efficiencies and reduce tar production. A source of pure oxygen is required for combustion if the syngas is to be used for fuel production so that nitrogen dilution does not affect the downstream processing. Currently, plants that use oxygen produce it with cryogenic air separation units. The biomass-to-hydrogen plant design for a direct gasifier is similar to that of the indirectly heated gasifier described earlier. The major differences between the two plants involve the addition of the air separation equipment and biomass feeding/prep equipment. An inert gas stream is needed to pressurize lock-hoppers for feeding biomass into the reactor because the gasifier is run at high pressure (approximately 24 bar). The ASUs create a stream of pure nitrogen that typically is compressed and used for this purpose. Otherwise, the same clean-up processes usually can be used.

Electrolysis could provide an alternative to air separation units with the added benefit of producing a pure hydrogen stream. Key changes to the plant include replacing the entire ASU with an electrolyzer bank and replacing the LO-CAT (a trademark flexible hydrogen sulfide removal process)/ZnO sulfur removal steps with a two-stage Selexol plant. The sulfur removal change was driven by the need for an inert gas for feed pressurization. Selexol is a well-proven process that uses a dimethyl ether-based solvent to remove both sulfur and CO_2 from the gas stream. The plant envisioned by Dean et al. [24] is shown in Figure 11.10.

11.7 HYBRID ENERGY SYSTEMS FOR HYDROGEN PRODUCTION BY ELECTROLYSIS

A promising green method for the production of hydrogen in the future could be water electrolysis. Currently, approximately only 4% of hydrogen worldwide is produced by this process [3]. The electrolysis of water or its breaking into hydrogen and oxygen is a well-known method which began to be used commercially in 1890 [60,61]. There are two types of electrolysis: low temperature which includes alkaline electrolyzers and proton exchange membrane electrolyzers and both are commercially available today [3,62,63]. High-temperature electrolyzers SOEC are also currently used because it can use different types of raw materials such as natural gas, syngas, etc. These electrolyzers are constantly improved because of their importance in flexible fuel cell development [64–69] and hydrogen storage devices [64]. Here, we briefly examine hybrid energy systems using renewable sources and nuclear energy.

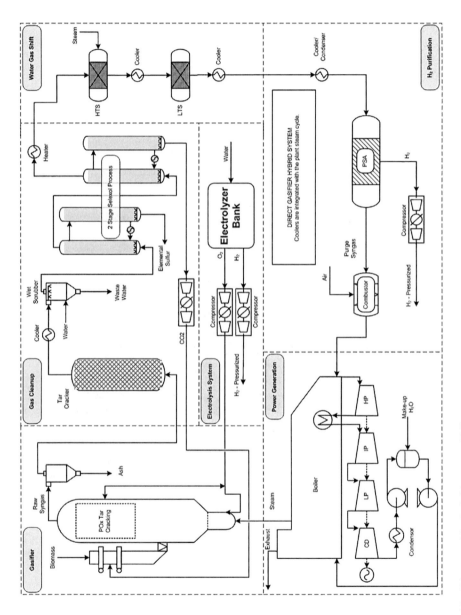

FIGURE 11.10 Direct gasifier hybrid concept [24].

11.7.1 Hybrid Renewable (Wind, Solar) Electrolysis

The worldwide electricity production potential from renewables and nuclear is staggering. If addressed and utilized aggressively, there is sufficient resource to support not only large inputs to the electrical grids across the planet but also significant hydrogen production. As an example, by itself the available wind power resource in the USA is estimated to be more than 2,800 GW (today, total US electricity generation capacity is roughly 1,100 GW), enough to produce over 150 billion kg/y of hydrogen, which exceeds the US gasoline quantity consumed annually in terms of energy equivalency.

Both renewable (like solar and wind) and nuclear energy can be connected to electrolysis in two ways. As mentioned above, in the first method, these sources can be directly injected into the grid and subsequently electricity is withdrawn from the grid for electrolysis as needed. There are advantages and disadvantages to this approach. As pointed out in my previous book [2], insertions of intermittent sources like solar and wind can be challenging for grid stability and power quality. Often this requires an insertion of additional storage device, particularly if only one source (solar and wind) is used. Battery has been the most logical choice. As pointed out earlier, this method can be cost-effective and easy to implement.

The second method is the direct insertion of wind or solar PV energy into low-temperature electrolysis. For wind, this was described earlier. Several renewables-to-hydrogen electrolysis test projects are underway in the USA and worldwide. At the U.S. National Renewable Energy Laboratory (NREL) in Colorado, a partnership between NREL and the local utility, Xcel Energy, has resulted in a pilot-scale project using wind and PV. The hydrogen is stored, then used to fuel NREL's Mercedes Benz F-Cell FCV, or converted into electricity for injection back onto the grid during times of peak electrical loads. In the 1920s and 1930s, MW-scale alkaline electrolyzers were built next to hydroelectric facilities in several locations around the world. So, we know how to do renewable hydrogen through electrolysis, have done it in the past, and now need to overcome the relatively modest technical and economic barriers to renewable hydrogen electrolysis for future transportation needs.

Sun et al. [68] investigated solar-driven alkaline water electrolysis with multifunctional catalysts. They indicated that the main challenge arises from the serious partial loading issue when intermittent and unstable renewable energy is coupled to water electrolyzers. An energy storage device can mitigate this incompatibility between water electrolyzer and renewable energy sources. The study demonstrates an alkaline water electrolysis (AWE) device driven by solar PV through a full cell of lithium-ion battery (LIB) as an energy reservoir (PV–LIB–AWE). Stable power output from LIB drives the water electrolyzer for steady hydrogen production and thus overcomes the partial loading issue of AWE. Moreover, a multifunctional hierarchical material, porous nickel oxide decorated nitrogen-doped carbon (NC) support, with excellent electrochemical performances for LIBs, oxygen evolution reaction (OER), and hydrogen evolution reaction (HER) for the PV–LIB–AWE system was developed. Density functional theory calculations showed that the strong interaction between metal oxide and NC tailors the electronic structure and then optimizes activation energy of OER process. PV–LIB–AWE-integrated system demonstrated here offers an alternative approach to drive water electrolysis with intermittent renewable energy for a truly sustainable energy future.

Jia et al. [69] examined solar water splitting by PV-electrolysis with a solar-to-hydrogen (STH) efficiency over 30%. Hydrogen production via electrochemical water splitting is a promising approach for storing solar energy. For this technology to be economically competitive, it is critical to develop water splitting systems with high STH efficiencies. Jia et al. [69] report a PV-electrolysis system with the highest STH efficiency for any water splitting technology to date, to the best of our knowledge. The system consists of two polymer electrolyte membrane electrolyzers in series with one InGaP/GaAs/GaInNAsSb triple-junction solar cell, which produces a large-enough voltage to drive both electrolyzers with no additional energy input. The solar concentration is adjusted such that the maximum power point of the PV is well matched to the operating capacity of the electrolyzers to optimize the system efficiency. The system achieves a 48-h average STH efficiency of 30%. These results demonstrate the potential of PV-electrolysis systems for cost-effective solar energy storage.

The use of PV/wind hybrid system for electrolysis has been theoretically examined in the literature. Khalilnejad et al. [70] designed and simulated a wind and PV hybrid electrolyzer system, which maximizes the hydrogen production for a diurnal operation of the system. The operation of the system was optimized using imperialist competitive algorithm. The objective of this optimization was to combine the PV array and WT in a way that, for minimized average excess power generation, maximum hydrogen would be produced. Actual meteorological data of Miami were used for simulations. A framework of the advanced alkaline electrolyzer with the detailed electrochemical model was used. This optimal system comprised a PV module with a power of 7.9 kW and a WT module with a power of 11 kW. The rate of hydrogen production was 0.0192 mol/s; an average Faraday efficiency of 86.9%. The electrolyzer worked with 53.7% of its nominal power. The availability of the wind for longer periods of time reflected the greater contribution of WT in comparison with PV toward the overall throughput of the system.

Rezaei et al. [71] study evaluated wind and solar energy potentials in prone areas of Iran by the Weibull distribution function and the Angstrom-Prescott equation for hydrogen production. To this end, the meteorological data of solar radiation and wind speed recorded at 10 m height in the time interval of 3 h in a 5-year period were used. The findings indicated that Manjil and Zahedan with yearly wind and solar energy densities of 6004 (kWh/m^2) and 2247 (kWh/m^2), respectively, have the greatest amount of energy among the other cities. After examining three different types of commercial WTs and PV systems, it becomes clear that by utilizing one set of Gamesa G47 turbine, 91 kg/d of hydrogen, which provides energy for 91 car/week, can be produced in Manjil and will save about 1347 L of gasoline in the week. Besides, by installing one thousand sets of X21–345 PV systems in Zahedan, 20 kg/d of hydrogen, enough for 20 cars per week, can be generated and 296 L of gasoline can be saved. Finally, the RETScreen software is used to calculate the annual CO_2 emission reduction after replacing gasoline with the produced hydrogen.

A blend of nuclear and wind energy ("Nuwind") to provide emission-free and economically feasible hydrogen production has been suggested [72,73]. The nuclear plant is assumed to deliver electricity in periods of low wind electricity production. Recent development small modular reactors like NuScale reactor can significantly facilitate low-temperature electrolysis. The intermittency of renewable energy sources coupled with varying demand for electricity requires a kind of energy mediator in the form of storage, conditioning, and regeneration. The "Hydrolyzer" currently being developed by H2Green, Inc., can effectively mediate energy production and consumption. It uses electricity to charge materials, generates kinetic or potential energy, separates and stores the H_2 and O_2 from water, and generates heat. The intermittent peaks and valleys of electricity supply and demand can be smoothened with H_2 as a storage medium [72,73].

11.7.2 COMMERCIAL-SCALE HYBRID NUCLEAR HEAT-BASED HTE

There is an increasing level of interest in the development of large-scale nonfossil hydrogen production technologies [74,75]. In terms of the transportation sector, this interest is driven by the near-term demand for hydrogen for refining of increasingly low-quality petroleum resources, the expected intermediate-term demand for carbon-neutral synthetic fuels, and the potential long-term demand for hydrogen as an environmentally benign direct transportation fuel [76–78]. Additional important nontransportation markets for large-scale hydrogen production include ammonia production and (potentially) carbon-free steel production [79]. As mentioned earlier, there is a need for nonfossil option to meet future hydrogen demand since methane reforming and other fossil-fuel conversion processes emit large quantities of greenhouse gases to the environment [80]. Nonfossil carbon-free options for hydrogen production include conventional water electrolysis coupled to either renewable (e.g., wind, solar) energy sources or nuclear energy. The renewable-hydrogen option may be viable as a supplementary source, but would be very expensive as a large-scale stand-alone option [81,82]. Conventional electrolysis coupled to nuclear baseload power can approach economical viability when combined with off-peak power, but the capital cost is high [83]. To achieve higher overall hydrogen production efficiencies, high-temperature thermochemical [84] or electrolytic [85] processes can be used. The required high-temperature process heat can be based on concentrated solar energy [86] or on nuclear energy from advanced high-temperature reactors [87].

High-temperature nuclear reactors have the potential for substantially increasing the efficiency of hydrogen production from water, with no consumption of fossil fuels, no production of greenhouse gases, and no other forms of air pollution. Advanced nuclear hydrogen production can be accomplished via HTE or thermochemical processes, using high-temperature nuclear process heat [88].

In order to achieve their best efficiencies, these processes require high-temperature operation (~850°C) and are therefore tied to the development of advanced high-temperature nuclear reactors. A conceptual depiction of a high-temperature gas-cooled reactor coupled to a HTE system is shown in Figure 11.11. In this scheme, the primary helium coolant serves as the working fluid to drive a gas-turbine power cycle, which provides the electrical energy required for the HTE process. In addition, some of the hot helium is used to deliver high-temperature nuclear process heat directly to the endothermic HTE process. High-temperature electrolytic water-splitting

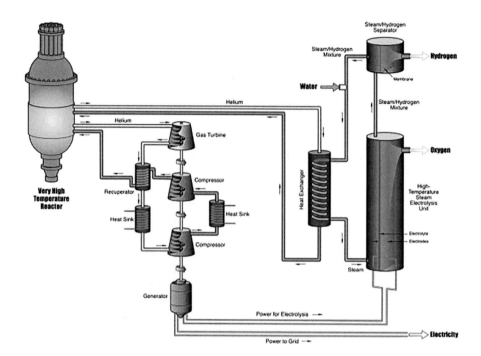

FIGURE 11.11 Concept for HTE system coupled to an advanced nuclear reactor [75].

supported by nuclear process heat and electricity has the potential to produce hydrogen with overall thermal-to-hydrogen efficiencies of 50% or higher, based on high heating value. This efficiency is similar to that of the thermochemical processes [89,90], but without the severe corrosive conditions of the thermochemical processes and without the fossil fuel consumption and greenhouse gas emissions associated with hydrocarbon processes. Furthermore, based on a detailed life-cycle analysis (LCA) [91], nuclear HTE is far superior to the conventional steam reforming process for hydrogen production with respect to global warming and acidification potential. LCA of nuclear power generation have reached similar conclusions, according to a series of papers [e.g., 92] produced by the EC-sponsored ExternE project. Another recent study at Stanford [93] recognized the potential contribution of nuclear energy in the context of global warming and air pollution external costs. However, this study assigned a certain elevated risk of nuclear war associated with the development of nuclear energy, which resulted in a low overall ranking for nuclear.

From 2003 to 2009, development and demonstration of advanced nuclear hydrogen technologies were supported by the U.S. Department of Energy under the Nuclear Hydrogen Initiative [94]. During 2009, this program sponsored a technology down-selection activity by which an independent review team recommended HTE as the most appropriate advanced nuclear hydrogen production technology for near-term deployment [95]. The INL HTE program also includes an investigation of the feasibility of direct syngas production by simultaneous electrolytic reduction of steam and carbon dioxide (coelectrolysis) at high temperature using solid-oxide cells.

Syngas, a mixture of hydrogen and carbon monoxide, can be used for the production of synthetic liquid fuels via Fischer-Tropsch or other synthesis processes. This concept, coupled with nuclear energy, provides a possible path to reduced greenhouse gas emissions and increased energy independence, without the major infrastructure shift that would be required for a purely hydrogen-based transportation system [96–99]. Furthermore, if the carbon dioxide feedstock is obtained from biomass, the entire concept would be climate-neutral.

As an alternative to centralized large-scale systems with direct coupling to high-temperature reactors, distributed hydrogen production could be accomplished using modular HTE units powered from grid electricity and an alternate high-temperature heat source such as concentrated solar energy [100] or a biomass gasifier [101]. This approach could be quite economical if off-peak electricity is used [102]. Furthermore, recent developments of small modular nuclear reactors, nuclear energy from these types of reactors is an ideal source of energy for HTE process. HTE makes use of solid oxide electrolysis cells to electrochemically split steam into hydrogen and oxygen at about 850°C. The overall thermal-to-hydrogen efficiency of HTE can approach 50%, based on the lower heating value of the produced hydrogen. The feasibility of this technology has been demonstrated up to the 15 kW scale.

More discussion on issues related to hybrid nuclear reactor-electrolysis process is given in IAEA report [103] and other literature [104,105]. The principles described here also apply to the use of nuclear heat for biomass or coal-biomass mixture gasification or pyrolysis. More discussions on generation IV high temperature reactors are also given in IAEA report [103].

11.8 ECONOMIC ASPECTS OF GREEN HYDROGEN PRODUCTION

At present, the most widely used and cheapest method for hydrogen production is the *steam reforming of methane (natural gas)*. This method includes about half of the world hydrogen production, and hydrogen price is about 7 USD/GJ. A comparable price for hydrogen is provided by partial oxidation of hydrocarbons. However, greenhouse gases generated by thermochemical processes must be captured and stored, and thus, an increase in the hydrogen price by 25%–30% must be considered [106].

The further used thermochemical processes include gasification and pyrolysis of biomass. The price of hydrogen thus obtained is about three times greater than the price of hydrogen obtained by the SR process. Therefore, these processes are generally not considered as cost-competitive of steam reforming. The price of hydrogen from gasification of biomass ranges from 10 to 14 USD/GJ and that from pyrolysis 8.9–15.5 USD/GJ. It depends on the equipment, availability, and cost of feedstock [3,4].

Electrolysis of water is one of the simplest technologies for producing hydrogen without byproducts. Electrolytic processes can be classified as highly effective. On the other hand, the input electricity cost is relatively high and plays a key role in the price of hydrogen obtained. HTE using nuclear heat from small modular nuclear reactors shows some promises.

By the year 2030, the dominant methods for hydrogen production will be *steam reforming of natural gas* and *catalyzed biomass gasification*. In a relatively small extent both coal gasification and electrolysis will be used. The use of solar energy in a given context is questionable but also possible. Probably, the role of solar energy will increase by 2050 [3,4]. The use of nuclear heat from small modular reactors has potential.

Ruth et al. [107] analyzed N-R HESs using two different hydrogen production technologies. They chose hydrogen as the industrial product in this analysis for four reasons: (a) stakeholders expressed interest in hydrogen as an industrial product from N-R HESs; (b) hydrogen produced via electrolysis of very low-priced electricity is a key aspect of the H2@Scale Big Idea suggested by DOE; (c) hydrogen is a potential energy carrier that both the transportation and industrial sectors can use while reducing emissions of air pollutants such as SO_x, NO_x, and particulates; and (d) hydrogen has been proposed as a means to provide long-duration, seasonal energy storage for electricity and other uses. The first hydrogen production technology they analyzed was HTE and the second was low-temperature electrolysis (LTE). LTE required only electricity to convert water to hydrogen. HTE required less electricity because it uses both electricity and heat to provide the energy necessary to electrolyze water. This analysis was built upon two case studies of N-R HES applications for synthetic gasoline production in West Texas and desalination plant in Arizona [108] described earlier in Chapter 6. The economic analysis of both LTE and HTE using nuclear energy indicated that if natural gas prices are higher than the projection of $6.98/MMBtu, the N-R HES is potentially profitable. Additional analysis is required to determine the level of profitability at various natural gas price projections. Two reports by Ruth et al. [107,108] for case studies in Texas and Arizona show significant viability of N-R HES options. As shown in Chapter 6, analysis was, however, carried out with various hypotheses.

REFERENCES

1. Hydrogen Council. (2017). Hydrogen scaling up—A sustainable pathway for the global energy transition. http://hydrogencouncil.com/wp-content/uploads/2017/11/Hydrogen-Scaling-up_Hydrogen-Council_2017.compressed.pdf. Contact secretariat@hydrogencouncil.com. www.hydrogencouncil.com, a website report.
2. Shah, Y.T. (2021). *Hybrid Power-Generation, Storage and Grids*. CRC Press, New York.
3. Kalamaras, C. and Estathiou, A. (2013). Hydrogen production technologies: Current state and future developments, editor Al-Assaf, Y., Conference Paper | Open Access Volume 2013 |Article ID 690627 |, Doi: 10.1155/2013/690627.
4. Balat, M. and Balat, M. (2009). Political, economic and environmental impacts of biomass-based hydrogen. *International Journal of Hydrogen Energy* 34(9): 3589–3603, View at: Publisher Site | Google Scholar.
5. Konieczny, A., Mondal, K., Wiltowski, T. and Dydo, P. (2008). Catalyst development for thermocatalytic decomposition of methane to hydrogen. *International Journal of Hydrogen Energy* 33(1): 264–272, View at: Publisher Site | Google Scholar.
6. Muradov, N.Z. and Veziroğlu, T.N. (2005). From hydrocarbon to hydrogen-carbon to hydrogen economy. *International Journal of Hydrogen Energy* 30(3): 225–237, View at: Google Scholar.
7. Holladay, J.D. Hu, J. King, D.L. and Wang, Y. (2009). An overview of hydrogen production technologies. *Catalysis Today* 139(4): 244–260, View at: Publisher Site | Google Scholar.
8. Shah, Y.T. (2014). *Water for Energy and Fuel Production*. CRC Press, New York.
9. Shah, Y.T. (2018). *Thermal Energy-Sources, Recovery and Applications*. CRC Press, New York.

10. Demirbas, M.F. (2006). Hydrogen from various biomass species via pyrolysis and steam gasification processes. *Energy Sources A* 28(3): 245–252, View at: Publisher Site | Google Scholar.

11. Asadullah, M., Ito, S.I., Kunimori, K., Yamada, M. and Tomishige, K. (2002). Energy efficient production of hydrogen and syngas from biomass: development of low-temperature catalytic process for cellulose gasification. *Environmental Science and Technology* 36(20): 4476–4481, View at: Publisher Site | Google Scholar.

12. Weber, G., Fu, Q. and Wu, H. (2006). Energy efficiency of an integrated process based on gasification for hydrogen production from biomass. *Developments in Chemical Engineering and Mineral Processing* 14(1–2): 33–49, View at: Google Scholar.

13. Ni, M., Leung, D.Y.C., Leung, M.K.H. and Sumathy, K. (2006). An overview of hydrogen production from biomass. *Fuel Processing Technology* 87(5): 461–472, View at: Publisher Site | Google Scholar.

14. Muradov, N. (2003). Emission-free fuel reformers for mobile and portable fuel cell applications. *Journal of Power Sources* 118(1–2): 320–324, View at: Publisher Site | Google Scholar

15. Demirbaş, A. and Arin, G. (2004). Hydrogen from biomass via pyrolysis: Relationships between yield of hydrogen and temperature. *Energy Sources* 26(11): 1061–1069, View at: Publisher Site | Google Scholar

16. Demirbaş, A. (2005). Recovery of chemicals and gasoline-range fuels from plastic wastes via pyrolysis. *Energy Sources* 27(14): 1313–1319, View at: Publisher Site | Google Scholar.

17. Zhagfarov, F.G., Grigor'Eva, N.A., and Lapidus, A.L. (2005). New catalysts of hydrocarbon pyrolysis. *Chemistry and Technology of Fuels and Oils* 41(2): 141–145, View at: Publisher Site | Google Scholar.

18. Sakurovs, R. (2003). Interactions between coking coals and plastics during co-pyrolysis. *Fuel* 82(15–17): 1911–1916, View at: Publisher Site | Google Scholar

19. Oriňák, A., Halás, L., Amar, I., Andersson, J.T. and Ádámová, M. (2006). Co-pyrolysis of polymethyl methacrylate with brown coal and effect on monomer production. *Fuel* 85(1): 12–18, View at: Google Scholar.

20. Hamelinck, C.N. and Faaij, A.P.C. (2002). Future prospects for production of methanol and hydrogen from biomass. *Journal of Power Sources* 111(1): 1–22.

21. Milbrandt, A. (2005). *A Geographic Perspective on the Current Biomass Resource Availability in the United States.* TP-560–39181. Golden, CO: National Renewable Energy Laboratory.

22. Perlack, R.D. (2005). *Biomass as Feedstock for a Bioenergy and Bioproducts Industry: The Technical Feasibility of a Billion-Ton Annual Supply.* Oak Ridge National Laboratory, Oak Ridge, TN.

23. Shah, Y.T. (2016). *Energy and Fuel Systems Integration*, CRC Press, New York.

24. Dean, J., Braun, R., Munoz, D., Penev M. and Kinchin, C. Analysis of hybrid hydrogen systems. *Technical Report* NREL/TP-560-46934 January 2010, NREL, Golden, CO.

25. Wang, L., Weller, C.L., Jones, D.D. and Hanna, M.A. (2008). Contemporary issues in thermal gasification of biomass and its application to electricity and fuel production. *Biomass and Bioenergy* (32): 573–581.

26. NREL GIS data is available online at http://www.nrel.gov/gis/solar.html. Accessed 8 November 2009.

27. Electric Power Research Institute. (2006). Gasification Technology Status—December 2006. ID 1012224.

28. Hughes, E. (2000). Biomass Cofiring: Economics, policy and opportunities. *Biomass and Bioenergy* 19: 457–465.

29. U.S. Department of Energy. (2000, June). Biomass Cofiring: A renewable alternative for utilities. DOE. GO-102000-1055, a report by NREL, Golden CO.

30. McLendon, T.R., Lui, A.P., Pineault, R.L., Beer, S.K. and Richardson, S.W. (2004). High-pressure co-gasification of coal and biomass in a fluidized bed. *Biomass and Bioenergy* 26: 377–388.

31. Valero, A. and Uson, S. (2006). Oxy-co-gasification of coal and biomass in an integrated gasification combined cycle (IGCC) power plant. *Energy* 31: 1643–1655.

32. Cormos, C. (2009). Assessment of hydrogen and electricity co-production schemes based on gasification process with carbon capture and storage. *International Journal of Hydrogen Energy* 34: 6065–6077.

33. Craig, K. and Mann, M. (1996). *Cost and Performance Analysis of Biomass-Based Integrated Gasification Combined-Cycle (BIGCC) Power Systems.* TP-430–21657. National Renewable Energy Laboratory, Golden, CO.

34. Larson, E., Jin, H. and Celik, F. (2005). *Gasification-Based Fuels and Electricity Production from Biomass, without and with Carbon Capture and Storage.* PEI, Princeton University, Princeton, NJ.

35. Bagchi, B., Rawlston, J., Counce, R.M., Holmes, J.M. and Bienkowski, P.R. (2006). Green production of hydrogen from excess biosolids originating from municipal waste water treatment. *Separation Science and Technology* 41–11: 2613–2628.

36. Mountouris, A., Voutsas, E. and Tassios, D. (2008). Plasma gasification of sewage sludge: Process development and energy optimization. *Energy Conversion and Management* 49: 2264–2271.

37. Shah, Y.T. (2017). *Chemical Energy from Natural and Synthetic Gas.* CRC Press, New York.

38. Lede, J. (1999). Solar thermochemical conversion of biomass. *Solar Energy* 65: 3–13.

39. Steinfeld, A. (2005). Solar thermochemical production of hydrogen—A review. *Solar Energy* 78: 603–615.

40. Solar Steam Reforming of Methane Rich Gas for Synthesis Gas Production (SOLREF), Final Report Summary—SOLREF (Solar Steam Reforming of Methane Rich Gas for Synthesis Gas Production) CORDISEU research results. (2009). Coordinated by DEUTSCHES ZENTRUM FUER LUFT-UND RAUMFAHRT E. V. Germany.

41. Adinberg, R., Epstein, M. and Karni, J. (2004). Solar gasification of biomass: A molten salt pyrolysis study. *Transactions of the ASME* 26: 850–857.

42. Masters, G. (2004). *Renewable and Efficient Electric Power Systems.* Wiley-Interscience, Hoboken, NJ.

43. Turner, J., Sverdrup, G., Mann, M.K., Maness, P.C., Kroposki, B., Ghirardi, M., Evans, R.J. and Blake, D. (2008). Renewable hydrogen production. *International Journal of Energy Research* 32: 379–407.

44. Forsberg, C. (2007, November). Meeting U.S. liquid transport fuel needs with a nuclear hydrogen biomass system. *Presented at the American Institute of Chemical Engineers Annual Meeting,* 2007 November, Salt Lake City.

45. Shah, Y.T. (2019). *Modular Systems for Energy and Fuel Recovery and Conversion.* CRC Press, New York.

46. Shah, Y.T. (2020). *Modular Systems for Energy Fuel Energy Usage Management.* CRC Press, New York.

47. Nuclear Energy Institute. General statistical information. http://www.nei.org/resourcesandstats/nuclear_statistics/us nuclear power plants/. Accessed 9 November 2009, a website report.

48. Price, R. and Blaise, J. (2002). Nuclear fuel resources: Enough to last? *NEA News,* 20.2.

49. Seitarides, T., Athanasiou, C. and Zabaniotou, A. (2008). Modular biomass gasification-based solid oxide fuel cells (SOFC) for sustainable development. *Renewable and Sustainable Energy Reviews* 12: 1251–1276.

50. Ciferno, J. and Marano, J. (2002). *Benchmarking Biomass Gasification Technologies for Fuels, Chemicals and Hydrogen Production.* U.S. Department of Energy, National Energy Technology Laboratory, Pittsburgh, PA.

51. Gassner, M. and Marechal, F. (2008). Thermo-economic optimization of the integration of electrolysis in synthetic natural gas production from wood. *Energy* 33: 189–198.

52. Kroposki, B., Levene, J., Harrison, K., Sen, P.K. and Novachek, F., (2006). *Electrolysis: Information and Opportunities for Electric Power Utilities.* NREL/TP-581-40605. National Renewable Energy Laboratory, Golden, CO.

53. StatoilHydro. Hydrogen technologies. http://www.electrolysers.com/. Accessed 8 November 2009.

54. U.S. Department of Energy, Energy Efficiency and Renewable Energy Laboratory (2008). *20% Wind Energy by 2030—Increasing Wind Energy's Contribution to U.S. Electricity Supply.* GO-102008-2567, a report by Department of Energy, Washington DC.

55. Levene, J., Kroposki, B. and Sverdrup, G. (2006). *Wind Energy and Production of Hydrogen and Electricity—Opportunities for Renewable* – Preprint, a report by OSTI. GOV, Department of Energy, Washington, DC.

56. Fingersh, L.J. (2004). *Optimization of Utility-Scale Wind-Hydrogen-Battery Systems.* CP-500-36117. National Renewable Energy Laboratory, Golden, CO.

57. Denholm, P. (2006). Improving the technical, environmental and social performance of wind energy systems using biomass–based energy storage. *Renewable Energy* 31: 1355–1370.

58. Gribik, A.M., Mizia, R.E., Gatley, H. and Phillips, B. (2007). *Economic and Technical Assessment of Wood Biomass Fuel Gasification for Industrial Gas Production.* EXT-07-13292. Idaho National Laboratory, INL, Idaho Falls, Idaho.

59. Spath, P., Aden, A., Eggeman, T., Ringer, M., Wallace, B. and Jechura, J. (2005). *Biomass to Hydrogen Production Detailed Design and Economics Utilizing the Battelle Columbus Laboratory Indirectly Heated Gasifier.* TP-510-37408. National Renewable Energy Laboratory, Golden, CO.

60. FCH JU (Fuel Cells and Hydrogen Joint Undertaking). (2017). Program review days report. www.fch.europa.eu/page/programme-posters-and-presentations-0, a website report.

61. FCH JU. (2014). Development of water electrolysis in the European Union. www.fch,Europa.eu/node/783, a website report.

62. NREL (National Renewable Energy Laboratory). (2016). Economic assessment of hydrogen technologies participating in California electricity markets. Authors: Joshua Eichman, Aaron Townsend and Marc Melaina, Technical Report NREL/TP-5400-65856, February 2016, Under Contract No. DE-AC36-08GO28308, NREL, Golden CO.

63. NREL. (2016). California power-to-gas and power-to-hydrogen near-term business case evaluation. Authors: Josh Eichman and Francisco Flores-Espino. A technical report Technical Report NREL/TP-5400-67384, December 2016, Under Contract No. DE-AC36-08GO28308, NREL, Golden CO.

64. FCH JU. (2017). Study on early business cases for H2 in energy storage and more broadly power to H2 applications. study by HINICIO and Tractebel, www.fch.europa.eu/sites/default/files/P2H_Full_Study_ FCHJU.pdf.

65. Hydrogen from renewable power-technology outlook for the energy transition, an IRENA report, Abu Dhabi (2016, September). www.irena.org ISBN 978-92-9260-077-8.

66. O'Brien, J.E., McKellar, M.G. and Herring, J. S. Performance predictions for commercial-scale high-temperature electrolysis plants coupled to three advanced reactor types. *2008 International Congress on Advances in Nuclear Power Plants*, 8–12 June, 2008, Anaheim, CA.

67. High Temperature Electrolysis. (2020). Wikipedia, The free encyclopedia, last visited 3 July, 2020.

68. Zixu, S., Wang, G., Koh, S. et al., (2020). Solar-driven alkaline water electrolysis with multifunctional catalysts. *Advanced Functional Materials*. 30, Doi: 10.1002/adfm.202002138.

69. Jia, J., Seitz, L., Benck, J., Huo, Y., Chen, Y., Ng, J.W.D., Bilir, T., Harris, J.S. and Jaramillo, T.F. (2016). Solar water splitting by photovoltaic-electrolysis with a solar-to-hydrogen efficiency over 30%. *Nature Communications* 7: 13237, Doi: 10.1038/ncomms13237.

70. Khalilnejad, A., Sundararajan, A. and Sarwat, A.I. (2018). Optimal design of hybrid wind/photovoltaic electrolyzer for maximum hydrogen production using imperialist competitive algorithm. *Journal of Modern Power Systems and Clean Energy* 6: 40–49. https://doi.org/10.1007/s40565-017-0293-0

71. Rezaei, M., Mostafaeipour, A., Qolipour, M. and Momeni, M. (2019). Energy supply for water electrolysis systems using wind and solar energy to produce hydrogen: A case study of Iran. *Frontiers in Energy* 13, Doi: 10.1007/s11708-019-0635-x.

72. Miller, A.I. and Duffey, R.B. (2005). Co-generation of hydrogen from nuclear and wind: The effect on costs of realistic variations in wind capacity and power prices. *Proceedings of the 13th International Conference on Nuclear Engineering ICONE-13*, Beijing.

73. Naterer, G.F., Fowler, M., Cotton, J. and Gabriel, K. (2008). Synergistic roles of off-peak electrolysis and thermochemical production of hydrogen from nuclear energy in Canada. *International Journal of Hydrogen Energy* 33: 6849–6857.

74. O'Brien, J.E., McKellar, M.G. and Herring, J.S. Performance predictions for commercial-scale high-temperature electrolysis plants coupled to three advanced reactor types. *2008 International Congress on Advances in Nuclear Power Plants*, 8–12 June 2008, Anaheim, CA.

75. O'Brien, J.E. (2010, August). *Large-Scale Hydrogen Production from Nuclear Energy Using High Temperature Electrolysis*. IHTC 14, INL/CON-10-18016 PREPRINT, INL, Idaho Falls, Idaho

76. Forsberg, C.W. (2005, December). The hydrogen economy is coming. The question is where? *Chemical Engineering Progress*: 20–22.

77. Lewis, D. (2008). Hydrogen and its relationship with nuclear energy. *Progress in Nuclear Energy* 50: 394–401.

78. Kruger, P. (2009). Nuclear production of hydrogen as an appropriate technology. *Nuclear Technology* 166: 11–17.

79. Forsberg, C.W. (2007). Future hydrogen markets for large-scale hydrogen production systems. *International Journal of Hydrogen Energy* 32: 431–439.

80. Duffey, R.B. (2009). Nuclear production of hydrogen: When worlds collide. *International Journal of Energy Research* 33: 126–134.

81. Granovskii, M., Dincer, I. and Rosen, M.A. (2007). Greenhouse gas emissions reduction by use of wind and solar energies for hydrogen and electricity production: economic factors *International Journal of Hydrogen Energy* 32: 927–931.

82. Rand, D.A.J. and Dell, R.M. (2008). *Hydrogen Energy: Challenges and Prospects*. Royal Society of Chemistry, London.

83. Floch, P.-H., Gabriel, S., Mansilla, C. and Werkoff, F. (2007). On the production of hydrogen via alkaline electrolysis during off-peak periods. *International Journal of Hydrogen Energy* 32: 4641–4647.

84. Schultz, K.R., Brown, L.C., Besenbruch, G.E. and Hamilton, C.J. (2003, February). Large-scale production of hydrogen by nuclear energy for the hydrogen economy. Report GA-A24265, 22 p.

85. O'Brien, J.E., Stoots, C.M., Herring, J.S. and Hartvigsen, J.J. (2007, May). Performance of planar high-temperature electrolysis stacks for hydrogen production from nuclear energy. *Nuclear Technology* 158: 118–131.
86. Steinfeld, A. (2005, May). Solar thermochemical production of hydrogen. *Solar Energy* 78(5): 603–615.
87. Southworth, F., Macdonald, P.E., Harrell, D.J., Park, C.V., Shaber, E.L., Holbrook, M.R., and Petti, D.A. (2003). The next generation nuclear plant (NGNP) project. *Proceedings, Global 2003*, 276–287, New York.
88. Elder, R. and Allen, R. (2009). Nuclear heat for hydrogen production: Coupling a very high/high temperature reactor to a hydrogen production plant. *Progress in Nuclear Energy* 51: 500–525.
89. Yildiz, B. and Kazimi, M.S. (2006). Efficiency of hydrogen production systems using alternative nuclear energy technologies. *International Journal of Hydrogen Energy* 31: 77–92.
90. O'Brien, J.E., McKellar, M.G. and Herring, J.S. Performance predictions for commercial-scale high-temperature electrolysis plants coupled to three advanced reactor types. *2008 International Congress on Advances in Nuclear Power Plants*, 8–12 June 2008, Anaheim, CA.
91. Utgikar, V. and Thiesen, T. (2006). Life cycle assessment of high temperature electrolysis for hydrogen production via nuclear energy. *International Journal of Hydrogen Energy* 31: 939–944.
92. Friedrich, R., Rabl, A. and Spadaro, J.V. (2001, December). Quantifyingthe costs of air pollution: The ExternE project of the EC. *Pollution Atmospherique*: 77–104.
93. Jacobsen, M.Z. (2009). Review of solutions to global warming, air pollution, and energy security. *Energy and Environmental Science* 2: 148–173.
94. Schultz, K., Sink, Pickard, P., Herring, J.S., O'Brien, J.E., Buckingham, R., Summers, W. and Michele Lewis, M. Status of the US nuclear hydrogen initiative. *Proceedings of ICAPP 2007*, Paper 7530, 13–18 May 2007, Nice, France, *The Nuclear Renaissance at Work*, V. 5, Societe Francaise d'Energie Nucleaire – ICAPP 2007, 2932–2940.
95. Varrin, R.D., Reifsneider, K., Scott, D.S., Irving, P. and Rolfson, G. (2009, August). NGNP hydrogen technology down-selection; results of the independent review team evaluation. Dominion Engineering report# R-6917-00-01.
96. O'Brien, J.E., McKellar, M.G., Stoots, C.M., Herring, J.S. and Hawkes, G.L. (2009, May). Parametric study of large-scale production of syngas via high temperature electrolysis. *International Journal of Hydrogen Energy* 34: 4216–4226.
97. Stoots, C.M. and O'Brien, J.E. (2009, May). Results of recent high-temperature co-electrolysis studies at the Idaho national laboratory. *International Journal of Hydrogen Energy* 34(9): 4208–4215.
98. Jensen, S.H., Larsen, P.H. and Mogensen, M. (2007). Hydrogen and synthetic fuel production from renewable energy sources. *International Journal of Hydrogen Energy* 32: 3253–3257.
99. Mogensen, M., Jensen, S. H., Hauch, A., Chorkendorff, lb. and Jacobsen, T. (2008). Reversible solid oxide cells. *Ceramic Engineering and Science Proceedings*, V 28, n 4, Advances in Solid Oxide Fuel Cells III—A Collection of Papers Presented at the 31st International Conference on Advanced Ceramics and Composites, 91–101.
100. Hawkes, G.L. and McKellar, M.G. Liquid fuel production from biomass via high temperature steam electrolysis. *2009 AIChE Annual Meeting*, 8–13 November 2009, Nashville, TN.
101. Arashi, H., Naito, H. and Miura, I. (1991). Hydrogen production from high-temperature steam electrolysis using solar energy. *International Journal of Hydrogen Energy* 16(9): 603–608.

102. Forsberg, C.W. (2009). Economics of meeting peak electricity demand using hydrogen and oxygen from base-load nuclear or off-peak electricity. *Nuclear Technology* 166: 18–26.
103. Hydrogen production using nuclear energy. A report by IAEA Nuclear Energy series No. NP-T4.2. (2013). IAEA, Vienna.
104. Hino, R., Haga, K., Aita, H. and Sekita, K. (2004). R&D on hydrogen production by high temperature electrolysis of steam. *Nuclear Engineering and Design* 233: 363–375.
105. Hauch, A., Ebbesen, S.D., Jensen, S.H. and Mogensen, M. (2008). Highly efficient high temperature electrolysis. *Journal of Materials Chemistry* 18: 2331–2340.
106. Balat H. and Kirtay, E. (2010). Hydrogen from biomass—present scenario and future prospects. *International Journal of Hydrogen Energy* 35(14): 7416–7426.
107. Ruth, M., Cutler, D., Flores-Espino, F. and Stark, G. (2017). The economic potential of nuclear-renewable hybrid energy systems producing hydrogen. Technical Report NREL/TP-6A50-66764 April 2017 Contract No. DE-AC36–08GO28308, NREL, Golden CO.
108. Ruth, M., Cutler, D., Flores-Espino, F., Stark, G., Jenkin, T., Simpkins, T. and Macknick, J. (2016). The economic potential of two nuclear-renewable hybrid energy systems. Technical Report NREL/TP-6A50-66073 August 2016 Contract No. DE-AC36-08GO28308, NREL, Golden CO.

Index

AAEC 216
ABB 388
absorption
    chillers 37
    refrigerators 12
ABV-6 473
AC 388, 453
AC and DC microgrids 146
AccD 305, 306
AC-DC-AC double conversion 397, 399
AC/DC full bridge inverter system 150
AC grid 422
AC systems 147
AD 444, 446
ADNOC 350
advanced materials 279
advantages of coal-solar hybridization 188
AEA 138
AERODESA 461
AEROGEDESA 461, 462
AES, all electric ship 142
AGC 168
AGM batteries 131
air vehicles 154
the Amal oil field 342
AmbD 305, 306
Amezon's Echo 26
AMI, advanced metering infrastructure 26
ancillary services 238
antifouling technologies 279
antilock braking 168
applications 36
APU 168
AQ 359, 360, 364, 366
Aquarius eco ship 133
Aquarius MAS 132
Aquarius MRE solution 133
AQUASOL 452
aqueous phase reforming 494
AREVA 189
Arizona desalination plant 253
ASHRAE 19
AST-500 98
ASU 213, 214, 216, 505, 512
ATES 114
ATET-80 473
atom 383
automobiles 129
autothermal reforming 494

aviation
    biofuel 350
    industry 158
AWE 514
AWECS 466
AWS 380

backup power 390
balance of system 31
balancing of power grid 84
barge-wave and below water energy conversion 465
baseload laborer 391
baseload power 390
basin still 449
BAT 315
battery cycle life 417
battery lifetime 395
BCF 342
the Beatrice oil field 340
benefits 389
BFC 407, 408
BGL 204
BIG SOLAR GRAZ 91
bikes 129
biofuels 135
biogas 363
biomass 184, 295, 296, 306, 307, 495, 497, 498, 499, 500, 503, 504, 505, 506, 508
biomass-coal integration 209
biomass gasification 494
BIPV 46
BIPVT 44
BiPVT/a 43
BiPVT/w 47
BIST 36
BN-350 471
BNSF 168, 173
boats 129
brown field 335
building
    efficiency 60
    industry 19
    integrated window systems 36, 47
buses 129
BWR 286
BWRO 453

Caihong 156
CANDESAL 473

capacitive storage 419
CAPEX 335
carbon 296
    capture 194
    free 71
    neutral cycle 210
    tax impact 210
carbon dioxide 195
CAREM 473
the Caolinga oil field 342
CASC 156
case studies 57, 253, 257, 333, 339
CBGTL 205, 206
CBTL 205
CCGT 187, 188, 327, 445
CCHP 12, 358
CCS 184, 192, 193, 194, 214
CCTV 173
centralized 2
CERTS 146
CETO 465
CFD analysis 44
challenges to renewable integration 328
chemical industry 300
chemical interconnections for N-Res 239
chemicals 206, 216, 227, 292
clean energy alternatives 308
CLFRs 192
closing perspectives 312, 342, 352, 357
CMEA 94, 96
CMOS 416
CMOS devices based on harvesters and systems 415
CNNC 98, 99
CNOOC 340, 350
CNPC 350, 352
CNT 414
coal
    based hybrid power plant 184
    and biomass 184, 498
    to chemicals 206
    industry 183
    -natural gas 184
    -solar hybrid for power and fuels 187
    subsidy 331
CO_2 capture and conversion 210
Co-combustion 184
cofired plants 194
cofiring 184
cogasification 196, 204
cogeneration 12, 15, 16, 93, 285, 347, 468, 474
coke 295
cold ironing 143
collaborative market redesign 354
combined cycle 195
combined heat and power 3, 12, 15, 71, 73, 75, 79,
        81, 91, 92, 93, 97, 99, 110, 274, 283,
        295, 315, 480

commercial aspects 50, 494
commercial scale operations 39, 50, 452, 516
composite PNG 413
compressor electrification 344
compressors 345
compute-intensive 382
computing industries 379
concentrated solar 500
concentrated solar-biomass 500
concentrating mirror still 449
concentration solar thermal desalination 451
control 253
control sensitivity 395
conversion 195
COP 35, 49, 53, 293, 294
COP21 1, 446, 447
COS 203
costs 327
coupling mode 227
CPBT 47, 455
CPU 382, 394
CPV 453
c-PVT 455
CREST 461, 462
CRL energy 214, 215
CS mode 143
CSP 188, 191, 232, 241, 347, 451, 453
CSP/MED 451
CSP/RO 451
CV 203

data centers 381
data mining 382
DC/AC 454
DC/AC inverters 147
DC/AC power converters 147
DC/DC boost converters 150
DC/DC converters 399
DC-DC converters 422
DC electricity 359, 453
DC powered system 147
DC section 145
decarbonizing district heating 88
DECC 73
deck space 335
deep electrification 324
DEIM converters 465
DE industry growth 114
demand response 28, 400
depletion 325
DER, DE 28, 60, 71, 111, 113
desalination 441, 468
desalination process alternatives 442
design optimization 250
DG systems 30
DH 84, 88, 89, 92, 93, 95, 96, 97, 99, 102
    by biomass 78

by hybrid industrial waste heat 101
with CHP 79
with hybrid wind energy 93
with small modular nuclear reactors 93
DHAPP 98
DHC 84
DHR-400 98
DHW 52
diesel
-electric ship 131
fuel 195
train 169
direct gasifier hybrid system 511
direct grid leveling 509
direct solar thermal desalination 449
disadvantages of coal-solar hybridization 190
distillation 300
district heating and cooling (DHC) 71, 72, 76
diversity of hybrid energy systems 224
DNV GL 140, 334
DOE 32, 115–120, 169, 203, 248, 250, 271, 519
domestic wastewater 440
DOT 169
downstream integration 345
DP, dynamic programming 164
DPM 419
DPP 465
DRAM 382
drivers 76
drying 299
DSSC 405, 407, 408, 415
DVFS 394
DWT 137, 154
dynamic modeling 250
dynamic pricing 28

E&A 152
EC 517
ECA, emission control area 143
ECN 197
economic(s)
competitiveness 210
potential 305
of renewable hybrid system 61, 518
ED 442, 443, 454, 460, 461, 462, 463
EDLC 175, 414
EDP group 10, 334
EES 150
EFACEC 10
EFASOLAR3 334
efficiency 195, 443
E-HAPI 159
EIA 271
ELCOGAS IGCC plant 201
electric aircraft 162
electrical grid 30, 60
electrically driven 32

electricity interconnections 251
electricity interconnections for N-Res 236
electricity subsidy 331
electric locks 168
electric railway 170, 174
electrification 307
electrification of drilling 336
electrochemical process 349
electrolysis 212, 214, 495, 512
electrolysis-biomass 505
electronic industries 379
electrostatic energy 404
EMP 131, 132, 138
ENABLEH2 162
ENERCON 462
ENERGIX program 140
energy
balance challenge 392
central 113
conversion 465
efficiency 276, 384, 395
harvesters 335, 403
harvesters integrated with energy storage and/
or end users 411
intensive sectors 307
keeper 391
management 27, 29, 152, 153, 162
sail 132
sources 474
supply options 22
supply priorities 22
tax code 57
tax credit 57
energy storage (ESS) 137, 140, 174, 175
environmental concerns 326
EOR, enhanced oil recovery 323, 338, 339
EPA 308
EPBT 45, 455
EPRI 190
ESD 394, 396, 397, 398, 420
ETC 303
ETH 501
EU 76, 100
EUR 90, 303, 457, 461, 477, 501
EURAC 102
Europe 81
EV, electric vehicles 7
evaluation 333
evaporation 300
execution time 395

facebook 384
falling renewable energy costs 327
faradaic storage 419
FAWN 384
FCS 135
FC system 480

FDV 204
federal Energy regulatory commission 354
FEP 405
ferries 129
flame stability 365
flash video 382
FLEXYNETS 102
fluctuating renewable electricity 84
fluidized bed 295, 509
flywheel system 242
FNPS 473
food
    and Beverage 292
    processing 299
forest products 292
formaldehyde 239
forms of hybrid energy in data centers 400
the fort liard oil and ngas field 341
fossil fuels 3, 323, 476
FPC 303
FPSO 335
FPV systems 334
FT 204, 214, 239
fuel cell
    based ships 153
    -biomass 504
    technology 195, 459
fuels 187
fuel switching 22
fugitive methane emission 324
future electrical grid 60
future prospects 474
fuzzy control logic 417
FVB 80

Gazprom 351
GCC 443, 445, 446, 447
GDP 1, 141, 271
GE 7, 140
GEA 298
geothermal energy 299, 341
geothermal heat 300
geothermal heat pump 50
geothermal wells 463
GFR 286
GHG emissions 19, 52, 61, 129, 192, 208, 224,
        259, 271, 356, 358, 442, 468
GHGM 366
GHGRP 308
GHP 86, 87
GHX 86, 87, 88
glass 292
glazing 47
global assessment 96
global demand 439
global trade 137
global warming 441

Google 379, 398
Google's Home 26
Government policies 330
GPS 168
green field 335
green hydrogen production 518
green-works framework for data centers 390, 391,
        392, 393
grid
    connection 21, 31
    management 28
    structure 2
GTHTR 445
Gulf of Mexico 325
GWP 85

harvester
    -sensor integrations 410
    -storage integrations 411
HC 403, 407
    for harvesting biomechanical and biochemical
        energy 407
    for harvesting solar and mechanical energy 407
    for harvesting solar and thermal energy 408
H_2/CO 496
H/C ratio 208, 215
HD 443, 449, 450
HD300 167
heat
    integration 400
    pipe 35, 49
    recovery 344
heat pump
    integration 48
    for process heat 113, 293
HELIOS 156
HELIOS plant 91
HER 514
HES 274, 312
HESS 175, 417
HET, hybrid electric train 172, 173
HFC 175
HHV 363
Hier-homo architecture 398, 399
Hier-hybrid architecture 398, 399
high quality oil reserves 325
Hi-Z technology 289
HMP 131, 138
HOMER 146
the hour of power 139
HP-PVT 49
HRES-RO 457
HSB 83, 84
HTE 241, 505, 516, 517, 519
HTGR 100, 248
HTGTR 100
HTR 286, 445

Hunt aviation 155
HVAC 28, 111
HY 152, 153
hybrid aircraft 154
hybrid approach 381
hybrid biomass 508
hybrid biomass-based DH 81
hybrid DE in US 115
hybrid electric aircraft 162
hybrid electric building design 59
hybrid electric railway 170, 174
hybrid energy
    for buildings 59
    defined 4
    (wind, solar)-electrolysis 514
    module(s) 60, 61
    via microscale waste heat applications 408
    storage for low-power embedded systems
        applications 416
hybrid ferries 131
hybridized oil and gas with renewable energy 350
hybrid microgrids for ships 145
hybrid modular geothermal heat pump 86
hybrid nanogenerators 411
hybrid nuclear heat based HTE 516
hybrid power 5, 9, 198
hybrid power module 152
hybrid processes 384, 443
hybrid PV/solar thermal concept 42
hybrid PV-wind 457
hybrid renewable energy for data centers 384, 391
hybrid ships 137
hybrid solar-biomass DH 82
hybrid solar energy for desalination 448
hybrid solar-thermal energy 88
hybrid solar thermal-solar PV 455
hybrid solar-wind 8
hybrid storage devices for data centers 396
hybrid trains and railways 167
hybrid wave energy 448, 464
hybrid wind energy 93
hybrid wind energy for desalination 462
hydrocarbon 129
hydrogen 216, 348, 493
hydrogen aircraft 158
hydro-generator 8
hydrogen fuel cell 134, 135
hydrogen interconnections for N-Res 241
hydrogen production 211, 257, 348, 493, 495, 512
hydropower 6
hyperscale data 386
HY4 powered 155

IAEA 247, 442, 471, 518
ICAO 158, 160
ICE 129
ICP 418

IEA 100, 194, 301
IGCC 198, 200, 201, 202, 203, 499
IMO 138
impact assessment 250
implementation challenges 388
IMPS 151
indirect gasifier hybrid system 509
indirect solar thermal desalination 449
indirect storage of heat 113
industrial 195
    applications 246
    electricity use 304
    heat pumps 279
    processes 295
    waste heat 101
    waste heat-biomass 508
INET 99
information interconnections for N-Res 243
INL 226, 250
INL HTE program 517
innovations 69
innovative condensing heat exchangers 279
innovative use of backup power 402
inorganic PNG 413
integrated EMS 162
integrated heat recovery technologies 279
integrating Faradaic and Capacitive Storage
        Mechanisms 419
intermittent wind power 509
inverters 422
IPCC 138, 194
IRENA 75, 301, 306
iron and Steel 292
issues 59
ITC, investment tax credit 330, 461
ITO 405

Jack pump 337
JOULE program 461

Kalina cycle 287
KANUPP 472
Kern River oil field 340
kinetic and solar energy 405
KLT-40 473
knock phenomenon 365
KPC 351
KSA 457

landfills 363
large ships 141
Lawrence Berkley National Laboratory 379
LBT 446
LCA 517
LCHP 350
LCOE 290
LEDs 405

levelized cost 327
LH2 161, 162
LIB 416, 417, 514
licensing 253
life cycle 6, 143, 356
Li-ion batteries 396, 411
liquid synthetic fuels 204
liquid-to-liquid heat exchangers 279
LMFR 286
LNG 137, 140, 154
load boosting 390
load broker 391
load following 390
load shedding 390
LO-CAT 512
Louisiana Bayou oil field 340
low and zero emission hybrid generation 357
low carbon renewable fuel storage and
        transportation 361
low powered embedded system applications 416
low-rank coal 247
LRV 174
LTDH 112
LTE 472, 519

magnetic and kinetic energy 403
manned solar aircraft 157
manufacturing 292
manufacturing industry 4, 236, 245, 271, 276,
        305, 307, 312, 313
marine energy resources 47
marine power systems 147
MATLAB/SIMULINK 10, 152, 334
MBR 482
MCFC 136, 504
the Mckittrick oil field 342
MD 443, 449, 450
mechanical interconnections for N-Res 242
MECS 279, 304
MED 442, 445, 446, 448, 449, 450, 451, 452, 453,
        457, 463, 469, 470, 471, 472
MEDAD 446
MED+AD 444
Medaka Eco ferry 131
MED+RO 470
MED-TVC 445, 473
MEMS 404, 411
meritime industry 176
meritime microgrids 149
methanol 239
MFC 481, 482
Microsoft 398
microthermal electric power 410
midstream integration 343
midway sunset oil field 339
millennials 60
MIRDC 173

MIT 466
MNSs 403, 411
modeling and simulation 250
modern grid 25
modular units 34
molten salt 236
motor driven systems 280
MPG 59
MPPT 416
MRC 482
MRE 138
MSF 443, 445, 448, 449, 450, 452, 463, 469,
        470, 471
MSF/MED hybrid 444, 446
MSF+RO 445, 470
MSR 286
MS solar hybrid 131
MSW 204
MTCO 308
MTG 204
MTOW 160, 161
multi-crystal 38
multifunctional energy façade 36
multilevel converters 421
multiple-tray tilted still 449
multi-source energy harvesters 403
multi-source hybrid concept 87
MVA 146
MVC 442, 445, 457
MWP 466

nanogenerators 411
national association of manufacturers 271
natural gas 78
    additions 186
    subsidy 331
    -renewable sources 353
naval vessel electrification 140
NC 514
NEMSs 411
NER 325
NEST 26
net demand 223
Netflix 380
NE train 167
New Zealand 214
NG 407
NGCC 253, 358, 363
NGNP 248
NHR 99, 470, 473
NLOC 351
NOC 350
novel hybrid processes 211
NOx 186, 187, 203, 354, 366
NPP 468, 471, 472, 473
NREL 20, 21, 35, 212, 250, 476, 505, 506, 508
NREL hybrid concept 212

N-R HES 228, 229, 230, 247, 519
nuclear
 -biomass 503
 -coal integration 207
 heat 96, 285
 industry 223
 -renewable energy systems 226, 229
NuWind 515
NZEB 21, 22, 23, 24

Oak Ridge national lab 78
OCIUS technology 136
OECD countries 1, 272, 297, 298
OER 514
off-grid 21
offloading 335
offshore units 334
OGCI 352
O&G industries 323, 325
oil 6
 and gas
  production 335
  transportation 343
 refining 345
 spills 326
 subsidy 331
onshore wind 327
operational considerations 329
OPRODES-JORCT98-0274 461
optimization models 107
options 186
ORC 79, 80

parallel hybrid 139
partial oxidation 494, 496
pathfinder 156
PCs 382
PDO 342, 351
peaking modification 510
PEM electrolyzer 365
PENG, PNG 411, 412, 413
PEPG 293
performance 335
perspectives 452
PET 298
petroleum refining 292
P2G 354, 364, 367
phase change 442
photolysis 494
photovoltaic hybrid systems 339
photovoltaics (PV) 6, 22, 41, 448, 453, 455, 458,
  459, 460, 476, 502, 503, 514, 515
PHWR 47
piezoelectric devices 287, 293
pilot and Commercial scale operations 452
pipeline 346
Pipistrel Taurus 160

PKE 212
PLA 298, 405
plasma reforming 494
plastic 298
plug and play 26
plug in hybrid electric vehicle 139
PM, particulate matter 132
PMs 386, 419
POD 34
POLK IGCC plant 203
polymer PNG 413
portable electronics 402
potential 304
power 187
 delivery architecture 397
 electronics for renewable energy systems 420
 generation 252
 grid 84
 management technologies 388
 module 152
primary recovery 336
Prinsesse Benedikte 131
priority areas 307
process 30
 heating systems 276
 intensification 280
production 346
pros and cons 57
PSA 452
PTO/PTI 144
PTT 298
PUE 384, 385
pulverized coal 327
PVDF 405, 406, 407, 412
PV-LIB-AWE 514
PV/RO 453, 454, 457
PVT/heat pump 48
PVT
 integrated heat pipe 49
 system 42, 43
 trigeneration 49
PVTRAIN 168
PVT/w 455
PV-wind energy system 57, 58, 59, 515
PV/wind/storage 54
PWR 286
PZT cantilever 404, 412

Qinetiq Zephyr 156

railways 167, 170
RBMK 94
RE 447, 480
reactor design 250
RED 482
reduction 308
regenerative braking 174

regional aspects 307
regulatory frameworks 238
renewable energy
    building 30
    credits 22
    integration 305
    technologies 304
renewable fuel injection in the grid 362
renewable sources 134
RES 447
residential 39, 61
RF 406
RO 441, 442, 443, 446, 451, 455, 456, 460, 461,
        462, 463, 464, 468, 469, 472, 475
RO+AD 444
the Rocky mountain oil field 341
ROI 145
role of renewable energy in desalination 447
role of storage in hybrid generation 360

safety 253
SAHP 36, 48
SASOL 211
Saudi Aramco 352
SB-HPWH 36
SCPU 414
SCR 358
SCWR 286
SDAWES 461
SDH 88
SE4ALL 274
secondary recovery 337
security 253
self powered hybrid micro/nano systems 410
serial hybrid 139
sewer heat recovery 87
SFC 141
SFOC 142
shedding and Boosting 394, 396
sheffield 103, 104
shipping industry 134
SHIP project 135
ships 140
single crystal 38
SIPH 308
SLAs 381, 382, 396
small and medium size enterprises (SMEs) 175,
        307, 417
small hybrid fossil-renewable heating and cooling
        grids 76
small modular nuclear reactors 93, 94, 99, 100,
        223, 228, 247, 312, 468, 471, 503
SMART 472
SMES *see* small and medium size enterprises
        (SMEs)
SNG 216, 239
SOEC 512

SOFC 359, 504, 505
solar
    boosted heat pump 35
    cooling 303
    electric hybrid 158
    electric PV 37
    powered boats 130
    powered train systems 168
Solar Impulse 2 157
solar, kinetic and radio frequency energy 406
solar-PV based applications 135, 453
solar PV canopy 170
solar PVT with geothermal heat pump 50
solar sailor 130
Solar ship Inc. 155
solar thermal 34, 42, 207
    recharge 87
    technologies 36
solar-wind-battery hybrid system 9
SOLREF 501
Sox 186
SPF 293
steam 276, 348
steam reforming 494
sterilizing 299
STH 515
storage 34, 252
successful N-Res applications 250
Suizhong 36-1 oil field 340
sun21 130
supercapacitors 396
SWCC 441, 445
SWISS 465
SWRO 441, 444, 445, 446, 469
SWRO+AD 446
SWRO/MSF hybrid 444
SWRO PV/RO 453
system level interconnections for N-Res 243
system reliability 329

TBT 446
T-C cycle 241
TDS 454
TE, thermoelectric 292, 293, 386, 408, 409
technology capabilities 388
TEG technology 290, 408, 409
temporal framework 354
TENG 411, 412, 413, 415
TEOR, thermally enhanced oil recovery 15
tertiary recovery 338
TES, thermal energy storage 109, 110, 111,
        112, 113
TEU 135, 137
textile industry 299
theoretical 57
thermal energy 301
thermal hydraulics 251

thermal interconnections for N-RES 230
thermally driven 32
thermal processes 442
thermal systems 342
thermionic devices 292
thermochemical water splitting 494
thermodynamic cycle 286
thermoelectric generation 287
thermoelectric power 287
thermophotovoltaic devices 287, 291
thin film 38
tightly coupled N-RHES systems for power and
heat 230
tilted-wick solar still 449
tools 250
traditional grid 25
traditional steam cycle 287
trains 167, 168
TRANSYS 43, 52
trigeneration 13, 49
TRL 162
Trucks 129
TU-155, TU-156 161
TUBITAK 205
turbo expanders 344
TVA 195
TVC 442, 445
TXU energy 26

UAE 74
UK 106, 286
UN 138
UNIDO 184
unit operations 276
unmanned aerial vehicle 156
UPS 395, 396, 402
UPS backup time 395
USD 90, 145, 314, 315, 493, 518
USGCRP 323
USSR 94, 98
utility scale 39
utility scale solar 327
utility service interruptions 60
the Utsira Nord oil field 341

vapor recompression 300
VAR 150
variability of generation 328
VC 442, 445, 460, 461
VCD 445

VCSs 44
VDU 347
vehicle industry 129
VHTR 286, 503
Vili 169
VRR 349
VTOL 160, 161
VVER 95

washing 299
waste heat 195
  to power 286
  recovery and conversion 280
  recovery systems 290
wastewater treatment 476, 481, 499
watch battery 409
water industry 439
wearable devices of ESS 415
WEB/PHP 382
web services 382
well pad and platform 332
west Texas synthetic gasoline 253
WGC 203
WGS 215, 216, 496
Willem Alexander IGCC plant 202
wind and thermal energy 405
wind-biomass 506
wind energy 135, 193
Wind2H2 212
windows media 382
window systems 46
wind power systems 340
wind resource classification 55
wind/RO 457, 461
wind-solar 478
wind/solar-coal integration 209
wind turbine (WT) 506, 515
Wobbe index 365
WWL 135
WWTP 480

Xcel energy 212
Xeon 383

Yachts 129

ZAK 212
zero emission hybrid generation 357
zero energy buildings 20
ZT 290